FM 5-34

ENGINEER FIELD DATA

HEADQUARTERS, DEPARTMENT OF THE ARMY

DISTRIBUTION RESTRICTION: Distribution is authorized to US Government agencies only to protect technical or operational information from automatic dissemination under the International Exchange Program or by other means. This protection applies to publications required solely for official use and to those containing valuable technical or operational information. This determination was made 1 April 1999. Other requests for this document will be referred to Commandant, US Army Engineer School, ATTN: ATSE-TD-D, Fort Leonard Wood, MO 65473-8929.

DESTRUCTION NOTICE: Destroy by any method that will prevent disclosure of contents or reconstruction of the document.

C3, FM 5-34

Change 3

Headquarters
Department of the Army
Washington, DC, 10 April 2003

ENGINEER FIELD DATA

1. Change FM 5-34, 30 August 1999, as follows:

Remove Old Pages	Insert New Pages
4-1 and 4-2	4-1 and 4-2
5-11 and 5-12	5-11 and 5-12
8-53 and 8-54	8-53 and 8-54

2. A bar (|) marks new or changed material.

3. File this transmittal sheet in front of the publication.

DISTRIBUTION RESTRICTION: Distribution is authorized to US Government agencies only to protect technical or operational information from automatic dissemination under the International Exchange Program or by other means. This protection applies to publications required solely for official use and to those containing valuable technical or operational information. This determination was made 1 April 1999. Other requests for this document will be referred to Commandant, US Army Engineer School, ATTN: ATSE-DOT-DD, 320 MANSCEN Loop, Suite 336, Fort Leonard Wood, MO 65473-8929.

DESTRUCTION NOTICE: Destroy by any method that will prevent disclosure of contents or reconstruction of the document.

By Order of the Secretary of the Army:

ERIC K. SHINSEKI
General, United States Army
Chief of Staff

Official:

Joel B. Hudson
JOEL B. HUDSON
Administrative Assistant to the
Secretary of the Army
0307111

DISTRIBUTION:

Active Army, USAR, and ARNG: To be distributed in accordance with the initial distribution number 110026, requirements for FM 5-34.

This publication is available on the General Dennis J. Reimer Training And Doctrine Digital Library at www.adtdl.army.mil

Field Manual
No. 5-34

*FM 5-34
Headquarters
Department of the Army
Washington, DC, 30 August 1999

Engineer Field Data

Table of Contents

	Page
LIST OF FIGURES AND TABLES	x
Figures	x
Tables	xviii
PREFACE	xxiii
CHAPTER 1. Combat Operations	1-1
Troop-Leading Procedures (TLPs)	1-1
Combat Orders	1-1
Warning Order	1-1
Operation Order (OPORD)	1-1
Fragmentary Order (FRAGO)	1-1
Movement Order	1-12
March Rates	1-13
Bivouac and Assembly Aeas	1-13
Mounted/Dismounted Operations	1-15
Movement Techniques	1-15
Local/Job Site Security	1-17
Patrolling	1-17
Reconnaissance Patrol	1-17
Combat Patrol	1-17
Ambush	1-18
Raid	1-20

DISTRIBUTION RESTRICTION: Distribution is authorized to US Government agencies only to protect technical or operational information from automatic dissemination under the International Exchange Program or by other means. This protection applies to publications required solely for official use and to those containing valuable technical or operational information. This determination was made 1 April 1999. Other requests for this document will be referred to Commandant, US Army Engineer School, ATTN: ATSE-TD-D, Fort Leonard Wood, MO 65473-8929.

DESTRUCTION NOTICE: Destroy by any method that will prevent disclosure of contents or reconstruction of the document.

---

*This publication supersedes FM 5-34, 14 September 1987.

FM 5-34

	Page
Fire-Support Procedures and Characteristics	1-20
Call-for-Fire Elements	1-20
Observer Identification	1-20
Warning Order	1-20
Target Location	1-21
Target Description	1-21
Method of Engagement (Optional)	1-21
Method of Fire and Control (Optional)	1-22
Three Transmissions in a Call for Fire	1-22
Message to Observer	1-22
Adjustments	1-22
Spotting	1-22
Lateral (Right/Left)	1-23
Range Correction (Up/Down)	1-24
Range Deviation	1-24
Quick Smoke	1-25
Fire-Support Equipment	1-28
Nuclear, Biological, Chemical	1-30
Chemical Agents	1-30
NBC Reports	1-33
Alarms, Signals, and Warnings	1-39
MOPP Levels	1-39
NBC Markers	1-40
Unmasking Procedures	1-40
With Detector Kit	1-40
Without Detector Kit	1-41
Decontamination	1-41
Equipment	1-41
Personnel	1-42
Medical Procedures	1-44
General First-Aid Procedures	1-44
Cardiopulmonary Resuscitation (CPR) Procedures	1-47
Medical Evacuation (MEDEVAC)	1-48
Field-Sanitation Facilities	1-52
Water Disinfection and Quantity Requirements	1-53
Calcium Hypochlorite	1-53
Iodine Tablets	1-54
Boiling	1-54
Daily Water Requirements	1-54
Communications	1-54
Antenna Locations	1-55
Expedient Antennas	1-55
Communications Equipment	1-57
Authentication	1-58
Standard Radio-Transmission Format	1-60
Single-Channel, Ground-to-Air Radio System (SINCGARS)	1-60

FM 5-34

	Page
Loading Frequencies—Manual (MAN), CUE, and 1-6	1-61
Clearing Frequencies	1-62
Loading Frequency Hop Data (Local Fill)	1-63
Loading Communications-Security (COMSEC) Keys (Local Fill)	1-63
Cold-Start Net Opening	1-64
CUE Frequency	1-65
Late Net Entry	1-66
Passive	1-66
CUE/ERF	1-66
Operator's Troubleshooting Checklist	1-66
Visual Signals	1-68
Rehearsals	1-71
Rehearsal Types	1-71
Principles	1-71
Techniques	1-71
Participant Levels	1-71

CHAPTER 2. Threat **2-1**
- Stability Operations and Support Operations Threat 2-1
  - Terrorism 2-1
  - Harassment 2-2
  - Protective Measures 2-2
  - Mid- to High-Intensity Threat 2-3
- Threat Organization 2-5
  - Military Districts 2-5
  - Motorized Infantry Division 2-5
  - Infantry and Mechanized Infantry Division 2-6
  - Motorized Infantry and Infantry Brigade 2-6
  - Mechanized Infantry and Tank Brigade 2-6
  - Armor and Mechanized – Based Threat 2-8
    - Maneuver Divisions 2-8
    - Maneuver Brigades 2-8
- Major Threat Equipment 2-10
- Threat Offensive Operations 2-11
- Crossing Capabilities and Characteristics 2-11
- Threat's Offensive River Crossing 2-15

Chapter 3. Reconnaissance **3-1**
- Route Classification 3-1
  - Critical Features 3-1
  - Slopes and Radius Computation 3-2
  - Obstruction (OB) 3-3
  - Report and Overlay 3-3
- Road Reconnaissance 3-5
  - Classification 3-5
  - Recording 3-5
- Bridge Reconnaissance 3-6
  - Hasty 3-6

	Page
Deliberate	3-6
Bridge Reconnaissance Report	3-18
Tunnel Reconnaissance	3-18
Water-Crossing Reconnaissance	3-18
Ford Reconnaissance	3-24
Engineer Reconnaissance	3-24

Chapter 4. Mobility ... 4-1
Mine Detection ... 4-1
Minefield Indicators ... 4-1
Detection and Removal ... 4-2
    Visual Inspections ... 4-2
    Probing ... 4-2
    Electronic Mine Detector ... 4-2
    Manual Clearing ... 4-2
Obstacle-Breaching Theory ... 4-6
Obstacle-Reduction Techniques ... 4-6
Standard Lane Marking ... 4-9
    Initial Lane Marking ... 4-10
    Intermediate Lane Marking ... 4-10
    Full Lane Marking ... 4-12
North Atlantic Treaty Organizaiton (NATO) Standard Marking ... 4-17
Combat Roads and Trails ... 4-18
Expedient Surfaces Over Mud ... 4-19
    Chespaling Mats ... 4-19
    Corduroy ... 4-19
    Tread Roads ... 4-21
    Other Surfaces ... 4-21
Expedient Surfaces Over Sand ... 4-24
    Wire Mesh ... 4-24
    Sand Grid ... 4-24
Forward Aviation ... 4-25
    Army Aircraft and Helicopter Characteristics ... 4-25
    Construction of Forward Landing Zone or Airstrip ... 4-28
    Maintenance and Repair ... 4-28

Chapter 5. Defensive Operations and Obstacle Integration
Framework ... 5-1
Planning ... 5-1
    Procedures ... 5-1
    Maneuver TF Responsibilities ... 5-2
        TF Commander ... 5-2
        Company Commander ... 5-2
        TF Operations and Training Officer (US Army) (S3) ... 5-2
    Specific Engineer Coordinations ... 5-3
        With a TF Commander ... 5-3
        With a Maneuver Company Commander ... 5-3
        With a TF S2 ... 5-3
        With a TF S3 ... 5-3

	Page
With a TF Supply Officer (US Army) (S4)	5-4
With a Fire-Support Officer (FSO)/ADA	5-4
Obstacles	5-4
Obstacle Classification	5-4
Obstacle Command and Control	5-4
Reports	5-11
Report of Intention to Lay	5-12
Report of Initiation	5-12
Report of Progress	5-12
Report of Completion	5-13
Report of Transfer	5-13
Report of Change	5-13

Chapter 6. Constructed and Preconstructed Obstacles 6-1

Wire Obstacles	6-1
Barbed-Wire Obstacles	6-3
Triple-Standard Concertina	6-4
Four-Strand Cattle Fence	6-7
Other Wire Obstacles	6-7
Antivehicular Obstacles	6-10
AT Ditches and Road Craters	6-10
Log Cribs	6-10
Log Hurdles	6-12
Log/Steel Post Obstacle	6-12
Tetrahedrons, Hedgehogs, and Other Barriers	6-12

Chapter 7. Landmine and Special-Purpose Munition Obstacles ... 7-1

Conventional Minefields	7-1
Row Mining	7-1
Standard-Pattern Minefields	7-7
Hasty Protective Row Minefields	7-8
Scatterable Minefields	7-9
Modular Pack Mine System (MOPMS)	7-9
Volcano	7-11
ADAM/RAAM	7-12
Gator	7-13
Special-Purpose Munitions	7-14
M86 Pursuit Deterrent Munition (PDM)	7-14
M18A1 Claymore	7-14
Selectable Lightweight Attack Munition (SLAM)	7-15
M93 Hornet	7-15
Raptor Intelligent Combat Outpost	7-19
Recording	7-20
Minefield Markings	7-26
Marking sets	7-26
Marking procedures	7-26
US Mines and Fuses	7-28
Foreign Mines	7-37

FM 5-34

	Page
Chapter 8. Survivability	**8-1**
Weapons Fighting Positions	8-1
Individual Fighting Positions	8-1
Siting to Engage the Enemy	8-2
Preparing by Stages	8-2
Crew-Served-Weapons Fighting Positions	8-9
Range Card	8-14
Details	8-14
Sectors of Fire	8-14
Target Reference Points (TRPs)/Reference Points (RPs)	8-15
Dead Space	8-15
Maximum Engagement Line (MEL)	8-15
Weapon Reference Point (WRP)	8-15
Preparation Procedures	8-16
Vehicle Positions	8-23
Hasty Fighting Positions	8-37
Modified Fighting Positions	8-37
Deliberate Fighting Positions	8-40
Protective Fighting Positions	8-42
Artillery and Parapet	8-42
Deep-Cut	8-43
Trenches, Revetments, Bunkers, and Shelters	8-45
Trenches	8-45
Revetments	8-45
Retaining Wall	8-45
Facing Revetments	8-47
Bunkers	8-48
Shelters	8-48
Camouflage	8-48
Position Development Stages	8-48
Lightweight Camouflage Screen	8-53
Estimation	8-53
Emplacement	8-54
Checkpoint Construction	8-55
Tower Construction	8-58
Chapter 9. Demolitions and Modernized Demolition Initiators (MDI)	**9-1**
Safety Considerations	9-1
Misfires	9-2
Nonelectric-Misfire Clearing Procedures	9-2
Electric-Misfire Clearing Procedures	9-4
Explosive Characteristics	9-4
Waterproofing	9-4
Priming	9-5

FM 5-34

	Page
Firing Systems	9-7
Charge Calculations	9-8
Steel-Cutting Charges	9-8
Timber-Cutting Charges	9-11
Abatis	9-13
Breaching Charges	9-14
Counterforce Charges	9-18
Boulder-Blasting Charges	9-19
Cratering Charges	9-20
Breaching Procedures	9-22
Bridge Demolitions	9-25
Abutment and Intermediate-Support Demolitions	9-39
Demolition Reconnaissance	9-41
Equipment/Ammunition Destruction	9-46
Ammunition	9-46
Guns	9-46
Armored Fighting Vehicles (AFVs)	9-46
Wheeled Vehicles	9-47
Expedient Demolitions	9-47
Cratering Charge	9-47
Shaped Charge	9-48
Platter Charge	9-49
Grapeshot Charge	9-49
Ammonium Nitrate Satchel Charge	9-50
Bangalore Torpedo	9-51
Detonating-Cord Wick (Borehole Method)	9-51
Time Fuse	9-51
Gregory Knot (Branch-Line Connection)	9-51
MDI Firing Systems	9-53
Stand-Alone System	9-55
Combination Firing System	9-57
Splicing the Shock Tube	9-58
Safety Procedures	9-58
MDI Misfire Clearing Procedures	9-59
Chapter 10. Bridging	**10-1**
River-Crossing Operations	10-1
Bridging/Rafting	10-2
Boats	10-2
Improved Float Bridge (Ribbon)	10-2
Long-Term Anchorage Systems	10-8
Approach Guys	10-8
Upstream Anchorage	10-8
Downstream Anchorage	10-8
Installation	10-8
Overhead-Cable Design Sequence	10-10

FM 5-34

	Page
Cable Data	10-10
Tower data	10-14
Deadman Data	10-15
Medium Girder Bridge (MGB)	10-20
MGB Design—SS, 4 Through 12 Bays	10-22
MGB Design—DS, 2E + 1 Through 12 Bays	10-28
MGB Design—DS, 2E + 13 Through 22 Bays (w/o LRS)	10-33
MGB Design—DS, 2E + 13 Through 22 Bays (w/LRS)	10-38
Bailey Bridge, Type M-2	10-43
Truss	10-43
Site Reconnaissance	10-44
Bridge Design	10-46
Engineer Multirole Bridge Company	10-65

Chapter 11. Roads and Airfields 11-1

- Soils and Geology 11-1
  - Characteristics 11-1
  - Moisture Content 11-4
  - Stabilization 11-4
- Engineering Properties of Rocks 11-5
- Drainage 11-8
  - Runoff Estimate 11-8
  - Culverts 11-9
    - Design 11-9
    - Length 11-11
    - Installation 11-11
  - Open-Ditch Design 11-13
  - Expedient Airfield Surfaces 11-15
    - Minimum Operating Strip (MOS) 11-15
    - Work Priority 11-15
  - Membrane and Mat Repair 11-17
    - Membranes 11-17
    - Mats 11-17
      - M8A1 11-17
      - AM2 11-17
      - M19 11-17
  - Other Repairs 11-21
- Road Design 11-23
  - Elements of a Horizontal Curve 11-23
  - Degree of Curvature 11-24
    - Arc Definition 11-24
    - Chord Definition 11-24
  - Equations for the Simple, Horizontal-Curve Design 11-25
    - Radius of Curvature 11-25
    - Tangent Distance 11-25
    - External Distance 11-26

	Page
Long Chord (C)	11-26
Middle Ordinate	11-26
Length of Curve (L)	11-26
Designing HorizontalCurves	11-27
Horizontal-Curve Design Examples	11-28

Chapter 12. Rigging 12-1
Rope 12-1
Knots and Attachments 12-3
Rope Bridges 12-14
    One-Rope Bridge 12-14
    Two-Rope Bridge 12-15
Chains and Hooks 12-16
Slings 12-18
Picket Holdfasts 12-20

Chapter 13. Environmental-Risk Management 13-1
Purpose 13-1
Legal and Regulatory Responsibilities 13-1
Risk-Management Principles 13-2
Environmental Benefits of Risk Management 13-2
The Risk-Management Process 13-3
Summary 13-11

Chapter 14. Miscellaneous Field Data 14-1
Weight and Gravity 14-1
Construction Material 14-3
    Electrical Wire 14-3
    Lumber Data 14-5
Trigonometric Functions and Geometric Figures 14-8
US Equipment and Weapons Characteristics 14-12
    Vehicle Dimensions and Classifications 14-12
    Expedient Vehicle Classification 14-14
        Wheeled 14-14
        Tracked 14-14
        Nonstandard Combinations 14-15
        Other-Than-Rated Load 14-15
    US Weapons 14-16
Operational Symbols 14-19
Conversion Factors 14-30
Levels of Risk Management 14-31

Glossary Glossary-1

References References-1

Index Index-1

FM 5-34

Page

List of Figures and Tables

Figures

Figure 1-1.	WO format	1-2
Figure 1-2.	Sample of a company OPORD	1-4
Figure 1-3.	Halt formation	1-12
Figure 1-4.	Sectors of fire	1-12
Figure 1-5.	Traveling dismounted elements	1-15
Figure 1-6.	Movement formations	1-16
Figure 1-7.	Traveling and traveling overwatch	1-16
Figure 1-8.	Bounding overwatch	1-16
Figure 1-9.	Typical organization and employment-point (linear) ambush	1-18
Figure 1-10.	Typical organization and employment-point) (vehicular) ambush	1-19
Figure 1-11.	Multi-claymore-mine mechanical ambush	1-19
Figure 1-12.	Typical organization for a raid patrol	1-20
Figure 1-13.	Examples of observer identification and WO	1-21
Figure 1-14.	Sample missions	1-23
Figure 1-15.	Adjusting field artillery fires	1-24
Figure 1-16.	Adjusting points for quick smoke	1-26
Figure 1-17.	Hasty method for estimating angle	1-27
Figure 1-18.	NBC markers	1-40
Figure 1-19.	CPR in basic life support	1-47
Figure 1-20.	Field latrines	1-52
Figure 1-21.	Hand-washing device, using No. 10 can	1-53
Figure 1-22.	Shower unit, using metal drums	1-53
Figure 1-23.	Jungle-expedient antenna (FM)	1-55
Figure 1-24.	Long-wire antenna (FM)	1-56
Figure 1-25.	Expedient, suspended, vertical antennas (FM)	1-56
Figure 1-26.	Improvised, center-fed, half-wave antenna (AM)	1-57
Figure 1-27.	Authentication procedures	1-59
Figure 1-28.	RT front panel	1-62
Figure 1-29.	ECCM fill device connected to RT	1-64
Figure 1-30.	Visual signals	1-68
Figure 2-1.	Threat's minefield, using track-width mines	2-3
Figure 2-2.	Threat's minefield, using full-width mines	2-4
Figure 2-3.	Threat's antipersonnel (AP) minefield	2-4

FM 5-34

		Page
Figure 3-1.	Route-classification formula	3-2
Figure 3-2.	Radius-of-curvature calculation	3-2
Figure 3-3.	Slope computation (road gradient)	3-3
Figure 3-4.	Route-classification overlay	3-4
Figure 3-5.	Sample, road-reconnaissance report (front)	3-7
Figure 3-6.	Sample, road-reconnaissance report (back)	3-8
Figure 3-7.	Dimensions for concrete bridges	3-9
Figure 3-8.	Dimensions for a simple stringer bridge	3-10
Figure 3-9.	Dimensions for steel-truss bridges	3-11
Figure 3-10.	Dimensions for plate-girder bridges	3-12
Figure 3-11.	Dimensions for arch bridges	3-13
Figure 3-12.	Dimensions for suspension bridges	3-14
Figure 3-13.	Span types	3-15
Figure 3-14.	Sample, bridge-reconnaissance report (front)	3-18
Figure 3-15.	Sample, bridge-reconnaissance report (back)	3-19
Figure 3-16.	Tunnel sketch with required measurements	3-21
Figure 3-17.	River or stream measurements	3-22
Figure 3-18.	Measuring stream width with a compass	3-23
Figure 3-19.	Measuring stream velocity	3-23
Figure 3-20.	Sample, engineer-reconnaissance report (front)	3-25
Figure 3-21.	Sample, engineer-reconnaissance report (back)	3-26
Figure 3-22.	Overlay symbols	3-27
Figure 3-23.	Material, facility equipment, and service symbols	3-29
Figure 4-1.	Squad-size sweep team	4-2
Figure 4-2.	Platoon-size sweep team	4-3
Figure 4-3.	Sweep teams in echelon	4-4
Figure 4-4.	MICLIC skip zone	4-6
Figure 4-5.	Using a MICLIC (depth is less than 100 meters)	4-7
Figure 4-6.	Using a MICLIC (depth is uncertain or greater than 100 meters)	4-7
Figure 4-7.	Mine-plow width compared to tracked-vehicle widths	4-8
Figure 4-8.	Mine-roller width compared to tracked-vehicle widths	4-8
Figure 4-9.	Engineer-blade skim pattern	4-10
Figure 4-10.	Initial lane-marking pattern	4-11
Figure 4-11.	Intermediate lane-marking pattern	4-12
Figure 4-12.	Full lane-marking pattern	4-13
Figure 4-13.	Nonstandard marking devices	4-16
Figure 4-14.	NATO standard marker	4-17
Figure 4-15.	Combat roads and trails process	4-18
Figure 4-16.	Typical cross-section showing road nomenclature	4-18
Figure 4-17.	Chespaling-surface road construction	4-19
Figure 4-18.	Corduroy road surfaces	4-20
Figure 4-19.	Fascine corduroy	4-21
Figure 4-20.	Plank tread road	4-22
Figure 4-21.	Army track	4-22
Figure 4-22.	Component parts of a Sommerfield truck	4-23

FM 5-34

		Page
Figure 4-23.	Other expedient surfaces	4-23
Figure 4-24.	Chain-link wire-mesh road	4-24
Figure 4-25.	Sand grid	4-25
Figure 4-26.	Geometric layout of landing zones	4-30
Figure 4-27.	Panel layout of landing zones	4-30
Figure 4-28.	Inverted Y	4-31
Figure 4-29.	Standard flight and landing formations	4-31
Figure 5-1.	Obstacle classification	5-5
Figure 5-2.	Obstacle-control measure graphics	5-7
Figure 5-3.	Obstacle-effect graphics	5-8
Figure 5-4.	Example of enemy obstacle-tracking chart	5-9
Figure 6-1.	Schematic layout of barbed-wire obstacles (defense)	6-3
Figure 6-2.	Perimeter wire (defense)	6-4
Figure 6-3.	Double-apron fence	6-5
Figure 6-4.	Triple-standard concertina fence	6-6
Figure 6-5.	Installing concertina	6-6
Figure 6-6.	Joining concertina	6-7
Figure 6-7.	Four-strand cattle fence	6-7
Figure 6-8.	Tanglefoot	6-8
Figure 6-9.	Knife rest	6-8
Figure 6-10.	Trestle-apron fence	6-9
Figure 6-11.	Eleven-row antivehicular wire obstacle	6-9
Figure 6-12.	AT ditches	6-10
Figure 6-13.	Rectangular log-crib design	6-11
Figure 6-14.	Triangular log crib	6-11
Figure 6-15.	Log hurdles	6-13
Figure 6-16.	Post obstacles	6-13
Figure 6-17.	Steel hedgehog and tetrahedron	6-14
Figure 6-18.	Concrete tetrahedron and cubes	6-14
Figure 6-19.	Heavy equipment tires	6-15
Figure 6-20.	Jersey barrier	6-15
Figure 6-21.	Concrete-obstacle placement	6-15
Figure 7-1.	Standard disrupt and fix row minefields	7-5
Figure 7-2.	Standard turn row minefield	7-6
Figure 7-3.	Standard block row minefield	7-6
Figure 7-4.	Hasty protective row minefield record	7-8
Figure 7-5.	MOPMS dispenser emplacement and safety zone	7-10
Figure 7-6.	Standard MOPMS disrupt minefield	7-10
Figure 7-7.	Standard MOPMS fixed minefield	7-11
Figure 7-8.	Ground/air Volcano disrupt and fixed minefields	7-11
Figure 7-9.	Ground/air Volcano turn and block minefields	7-12
Figure 7-10.	M86 PDM	7-14
Figure 7-11.	M18A1	7-15

FM 5-34

		Page
Figure 7-12.	SLAM	7-16
Figure 7-13.	M93 Hornet	7-16
Figure 7-14.	Hornet reinforcing a conventional minefield	7-17
Figure 7-15.	Hornet reinforcing a Volcano minefield	7-17
Figure 7-16.	Hornet area-disruption obstacle	7-18
Figure 7-17.	Hornet gauntlet obstacle (one cluster)	7-18
Figure 7-18.	Hornet gauntlet obstacle (platoon)	7-19
Figure 7-19.	PIP Hornet	7-20
Figure 7-20.	Sample DA Form 1355 (front) (standard-pattern minefield)	7-21
Figure 7-21.	Sample DA Form 1355 (inside) (standard-pattern minefield)	7-22
Figure 7-22.	Sample DA Form 1355 (front side) for a Hornet minefield/ munition field	7-23
Figure 7-23.	Sample DA Form 1355 (back side) for a Hornet minefield/ munition field	7-24
Figure 7-24.	Standard marking signs	7-27
Figure 7-25.	Minefield marking fence	7-27
Figure 7-26.	AP mines (Korea only)	7-29
Figure 7-27.	AT mines	7-31
Figure 7-28.	Firing devices and trip flares	7-33
Figure 7-29.	AP SCATMINEs	7-36
Figure 7-30.	AT SCATMINE	7-37
Figure 7-31.	Foreign AT mines	7-38
Figure 7-32.	Foreign AP mines	7-43
Figure 8-1.	Stage 1, preparing a fighting position	8-3
Figure 8-2.	Stage 2, preparing a fighting position	8-4
Figure 8-3.	Stage 3, preparing a fighting position	8-5
Figure 8-4.	Stage 4, preparing a fighting position	8-5
Figure 8-5.	Hasty prone position	8-7
Figure 8-6.	Two-soldier fighting position	8-7
Figure 8-7.	Two-soldier fighting position	8-8
Figure 8-8.	Three-soldier *T*-position	8-8
Figure 8-9.	Planning the fighting position	8-9
Figure 8-10.	Traverse and elevation mechanism	8-10
Figure 8-11.	Digging the fighting position	8-11
Figure 8-12.	Digging grenade sumps	8-12
Figure 8-13.	Half of a position	8-12
Figure 8-14.	Two firing platforms with overhead cover	8-13
Figure 8-15.	Ammo bearer covering front	8-13
Figure 8-16.	Dug-in position for an MK19	8-14
Figure 8-17.	Placement of weapon symbol and left and right limits	8-16
Figure 8-18.	Circle value	8-17
Figure 8-19.	Terrain features for left and right limits	8-18
Figure 8-20.	Target reference points/reference points	8-19
Figure 8-21.	Dead space	8-20
Figure 8-22.	Maximum engagement lines	8-21

FM 5-34

		Page
Figure 8-23.	Weapon reference point	8-22
Figure 8-24.	Example of a completed range card	8-23
Figure 8-25.	Hasty fighting positions for combat vehicles	8-37
Figure 8-26.	Modified, two-tiered hiding position	8-38
Figure 8-27.	Modified, two-tiered artillery position	8-39
Figure 8-28.	Deliberate fighting positions for fighting vehicles	8-40
Figure 8-29.	105-mm parapet-position construction detail	8-42
Figure 8-30.	Deep-cut position	8-43
Figure 8-31.	Standard trench traces	8-46
Figure 8-32.	Sandbag revetment	8-46
Figure 8-33.	Retaining-wall anchoring method	8-47
Figure 8-34.	Brushwood hurdle	8-47
Figure 8-35.	Typical bunker	8-50
Figure 8-36.	Log fighting bunker with overhead cover	8-51
Figure 8-37.	Typical cut-and-cover shelter	8-51
Figure 8-38.	Air-transportable prefab shelter	8-52
Figure 8-39.	Hasty module determination chart	8-53
Figure 8-40.	Lightweight camouflage screens	8-54
Figure 8-41.	Placing net over vehicle	8-55
Figure 8-42.	Typical hasty checkpoint	8-56
Figure 8-43.	Typical one-way deliberate checkpoint	8-56
Figure 8-44.	Typical two-way deliberate checkpoint	8-57
Figure 8-45.	11- x 11-foot guard tower	8-58
Figure 8-46.	12- x 12-foot guard tower	8-60
Figure 9-1.	Priming with detonating cord	9-6
Figure 9-2.	Combination dual-firing system	9-7
Figure 9-3.	Calculation steps for explosives	9-8
Figure 9-4.	Steel-cutting charge emplacements	9-10
Figure 9-5.	Special steel-cutting charges	9-11
Figure 9-6.	Timber-cutting charges	9-12
Figure 9-7.	Stump-blasting charge placement	9-13
Figure 9-8.	Abatis	9-14
Figure 9-9.	Breaching-charge calculations	9-15
Figure 9-10.	Breaching radius	9-16
Figure 9-11.	Counterforce charge	9-18
Figure 9-12.	Boulder blasting	9-19
Figure 9-13.	Hasty crater	9-20
Figure 9-14.	Deliberate crater	9-20
Figure 9-15.	Relieved-face crater	9-21
Figure 9-16.	Backfilled log-wall breaching	9-22
Figure 9-17.	Log-crib breaching	9-22
Figure 9-18.	Placement of charges	9-23
Figure 9-19.	Explosive packs for destroying small concrete obstacles	9-24
Figure 9-20.	Methods of attack on simply supported bridges	9-28
Figure 9-21.	Methods of attack on continuous bridges	9-33
Figure 9-22.	Placement of 5-5-5-40 charge (triple-nickle forty)	9-39

	Page
Figure 12-24. Preparing a picket holdfast	12-22
Figure 13-1. Environmental hazard relationship to the risk-management process	13-2
Figure 13-2. Sample risk-management work sheet, all blocks filled in	13-4
Figure 13-3. Common environmental hazards	13-5
Figure 13-4. Hazard probability	13-6
Figure 13-5. Hazard severities	13-7
Figure 13-6. Risk-assessment matrix	13-9
Figure 13-7. Environmental-related controls	13-10
Figure 14-1. Trigonometric functions	14-8
Figure 14-2. Geometric figures and formulas	14-11
Figure 14-3. Single-vehicle expedient-class overload	14-15
Figure 14-4. Unit size and installation indicator	14-19
Figure 14-5. Unit identification symbols	14-20
Figure 14-6. Obstacle symbols	14-21
Figure 14-7. Weapon symbols	14-28
Figure 14-8. Risk management	14-31
Figure 14-9. Levels of decision matrix	14-31
Figure 14-10. Risk-assessment matrix	14-32
Figure 14-11. Steps in risk management	14-32
Figure 14-12. Risk-management work sheet	14-33

Tables

Page

Table 1-1.	Average march rates	1-13
Table 1-2.	Target bracketing	1-24
Table 1-3.	Artillery and mortar smoke	1-27
Table 1-4.	Artillery and mortar flares	1-27
Table 1-5.	Fire-support munitions	1-28
Table 1-6.	Fire-support system capabilities	1-29
Table 1-7.	Chemical agents' characteristics and defense	1-30
Table 1-8.	Line-item definitions	1-33
Table 1-9.	Types of NBC reports	1-35
Table 1-10.	Alarms and signals	1-39
Table 1-11.	MOPP levels	1-39
Table 1-12.	Natural decontaminants	1-42
Table 1-13.	First-aid, symptoms with treatment	1-44
Table 1-14.	First-aid, treatments	1-45
Table 1-15.	MEDEVAC report entries	1-48
Table 1-16.	MEDEVAC request form	1-49
Table 1-17.	Daily water requirements	1-54
Table 1-18.	Communication equipment, tactical radio sets AN/VRC-12 series	1-57
Table 1-19.	Communications equipment, auxillary	1-58
Table 1-20.	Communications equipment, wire	1-58
Table 1-21.	SINGCARS, general information	1-60
Table 1-22.	SINCGARS radio sets	1-61
Table 1-23.	Voice transmission maximum planning ranges	1-61
Table 1-24.	Data transmission maximum planning ranges	1-61
Table 2-1.	Normal parameters for threat's minefields	2-5
Table 2-2.	Threat organization, infantry based	2-6
Table 2-3.	Principal items of equipment for infantry-based threat	2-7
Table 2-4.	Threat organization, armor and mechanized based	2-8
Table 2-5.	Principal items of equipment for armor- and mechanized-based threat	2-9
Table 2-6.	Threat's defensive engineer equipment	2-10
Table 2-7.	Threat's defensive ditching and digging equipment	2-10
Table 2-8.	Light armored vehicles—wheeled capabilties and characteristics	2-12
Table 2-9.	Threat's bridging and rafting equipment	2-12
Table 2-10.	Threat's vehicle obstacle-crossing capabilities and characteristics	2-13
Table 2-11.	Threat's amphibious and ferry equipment	2-14

		Page
Table 2-12.	Threat's minefield-reduction equipment	2-14
Table 2-13.	Sample, enemy's obstacle report	2-15
Table 2-14.	Threat's river-crossing time line	2-15
Table 3-1.	Traffic-flow capability based on route width	3-1
Table 3-2.	Road-limiting characteristics and symbols	3-5
Table 3-3.	Road-surface materials and symbols	3-6
Table 3-4.	Dimensions required on the seven basic bridges	3-16
Table 3-5.	Engineer-reconnaissance checklist	3-20
Table 3-6.	Ford-site trafficability	3-24
Table 4-1.	Personnel and equipment requirements for a sweep team	4-3
Table 4-2.	Route-clearance team organization	4-5
Table 4-3.	Nonexplosive obstacle-breaching equipment	4-9
Table 4-4.	Lane-marking levels, unit responsibilities, and trigger events	4-14
Table 4-5.	Guidelines for lane-marking devices	4-15
Table 4-6.	Army helicopter characteristics	4-26
Table 4-7.	Combat-area airfield requirements	4-27
Table 4-8.	Dust-control requirements for heliports	4-28
Table 4-9.	Minimum geometric requirements for landing zones in close battle areas	4-29
Table 5-1.	SCATMINE emplacement authority	5-5
Table 5-2.	Obstacle-control measures	5-6
Table 5-3.	Obstacle numbers	5-10
Table 5-4.	Obstacle-type abbreviations	5-11
Table 5-5.	Report of intention to lay	5-12
Table 5-6.	Report of initiation	5-12
Table 5-7.	Report of progress	5-12
Table 5-8.	Report of completion of minefield	5-13
Table 6-1.	Wire and tape obstacle material	6-1
Table 6-2.	Requirements for 300-meter sections of various wire obstacles	6-2
Table 6-3.	Post requirements (post opposing/offset post)	6-12
Table 7-1.	Standard minefield characteristics	7-1
Table 7-2.	Class IV/V haul capacity	7-3
Table 7-3.	Platoon organization for row mining	7-7
Table 7-4.	SCATMINEs' sizes and safety zones	7-9
Table 7-5.	SCATMINEs' self-destruct times	7-9
Table 7-6.	Volcano minefield's characteristics	7-12
Table 7-7.	ADAM/RAAM minefield's density and size	7-13
Table 7-8.	ADAM/RAAM minefield's safety zones	7-13
Table 7-9.	Hornet minimum emplacement distances	7-19

		Page
Table 7-10.	Scatterable minefield's report and record	7-25
Table 7-11.	SCATMINE's warning report	7-26
Table 7-12.	Scatterable minefield's marking requirements	7-28
Table 7-13.	Characteristics of AP SCATMINEs	7-36
Table 7-14.	Characteristics of AT SCATMINEs	7-37
Table 8-1.	Material thickness for protection against direct and indirect fires	8-1
Table 8-2.	Characteristics of individual fighting positions	8-6
Table 8-3.	Characteristics of crew-served-weapons fighting positions	8-9
Table 8-4.	Dimensions of field artillery vehicle positions	8-24
Table 8-5.	Dozer team TDP calculations	8-25
Table 8-6.	Dozer team HDP calculations	8-28
Table 8-7.	ACE/ACE team TDP calculation	8-31
Table 8-8.	ACE/ACE team HDP calculations	8-34
Table 8-9.	Dimensions of field artillery vehicle positions	8-43
Table 8-10.	Dimensions of typical deep-cut positions	8-44
Table 8-11.	Recommended requirements for slope ratios in cuts and fills	8-45
Table 8-12.	Center-to-center spacing for wood-supporting soil cover to defeat various contact bursts	8-49
Table 8-13.	Expedient paints	8-52
Table 9-1.	Minimum safe distances for personnel in the open	9-2
Table 9-2.	Minimum safe distance from transmitter antennas	9-3
Table 9-3.	Military explosive characteristics	9-5
Table 9-4.	Steel-cutting formulas	9-9
Table 9-5.	C4 required to cut rectangular steel sections of given dimensions	9-9
Table 9-6.	Values of K for breaching charges	9-17
Table 9-7.	Thickness of breaching charge	9-18
Table 9-8.	Minimum E_R values for bottom attack (percent)	9-25
Table 9-9.	Minimum L_c values for top attack (midspan)	9-26
Table 9-10.	Minimum L_c values for arch and portal with pinned-footing bridge attacks	9-27
Table 9-11.	Gun-destruction charge sizes	9-46
Table 9-12.	MDI components	9-54
Table 10-1.	Assault-crossing equipment	10-1
Table 10-2.	BEBs	10-2
Table 10-3.	Ribbon-bridge allocations (L-series TOE)	10-2
Table 10-4.	Launch restrictions	10-3
Table 10-5.	Bridge classification	10-5
Table 10-6.	Boat requirements for anchoring a ribbon bridge	10-5
Table 10-7.	Ribbon-raft design	10-6
Table 10-8.	Planning factors for rafting operations, raft's centerline data	10-7

		Page
Table 10-9.	Unit rafting requirements	10-7
Table 10-10.	Design of upstream (primary) anchorage systems	10-8
Table 10-11.	Design of downstream (secondary) anchorage systems	10-9
Table 10-12.	Procedures for installing long-term anchorage systems	10-9
Table 10-13.	Data for overhead-design sequence	10-10
Table 10-14.	Size and number of master cables (C_D) for float bridges	10-12
Table 10-15.	Weight and breaking strengths for common cables (cable capacity)	10-13
Table 10-16.	Tower heights	10-14
Table 10-17.	Anchorage-cable capacities	10-16
Table 10-18.	Required HP (lb/sq ft)	10-16
Table 10-19.	O_2 ft factor	10-17
Table 10-20.	Flat bearing-plate dimensions	10-18
Table 10-21.	Flat bearing-plate dimensions	10-19
Table 10-22.	Dimensions for SS bridges, 4 through 8 bays	10-22
Table 10-23.	Dimensions for SS bridges, 9 through 12 bays	10-22
Table 10-24.	RB setup and packing	10-26
Table 10-25.	RB setup and packing (LNCG setting)	10-27
Table 10-26.	SS pallet loads	10-27
Table 10-27.	Manpower and time requirements	10-27
Table 10-28.	Dimensions for DS, 2E + 1 through 12 bays	10-28
Table 10-29.	Rule 1 for LNCG, 2E + 1 through 12 bays	10-30
Table 10-30.	Rule 2 for LNCG, 2E + 1 through 12 bays	10-31
Table 10-31.	Rule 3 for N and T, 2E + 1 through 12 bays	10-31
Table 10-32.	Rules 4A and 4B for N and T, 2E + 1 through 12 bays	10-32
Table 10-33.	DS pallet loads, 1 through 12 bays	10-32
Table 10-34.	Manpower and time requirements, 1 through 12 bays	10-33
Table 10-35.	DS dimensions, 2E + 13 through 22 bays w/o LRS	10-34
Table 10-36.	Rule 1 for LNCG, 2E + 13 through 12 bays w/o LRS	10-36
Table 10-37.	Rule 2, identifying N, 2E + 13 through 22 bays w/o LRS	10-36
Table 10-38.	Rule 3A and 3B for N and T, 2E + 13 through 22 bays w/o LRS	10-37
Table 10-39.	DS pallet loads, 13 through 22 bays w/o LRS	10-37
Table 10-40.	Manpower and time requirements, 13 through 22 bays w/o LRS	10-38
Table 10-41.	Dimensions for DS, 2E + 13 through 22 bays w/LRS	10-39
Table 10-42.	Minimum distances	10-40
Table 10-43.	Rule 1 for LNCG, 2E + 13 through 22 bays w/ LRS	10-41
Table 10-44.	Rule 2, identifying N, 2E + 13 through 22 bays w/ LRS	10-42
Table 10-45.	DS pallet loads, 2E + 13 through 22 bays with LRS	10-42
Table 10-46.	Manpower and time requirements, 2E + 13 through 22 bays w/ LRS	10-43
Table 10-47.	Truss/story configuration	10-43
Table 10-48.	Classification of Bailey bridge	10-51
Table 10-49.	Safe bearing capacity for various soils	10-53
Table 10-50.	Safe soil pressures	10-54

FM 5-34

	Page
Table 10-51. Roller clearance and grillage height	10-55
Table 10-52. Rocking-roller requirements	10-63
Table 10-53. Plain-roller requirements	10-63
Table 10-54. Jack requirements	10-63
Table 10-55. Organization of an assembly party	10-64
Table 10-56. Estimated assembly times	10-65
Table 11-1. Soil characteristics	11-1
Table 11-2. Recommended initial stabilizing agent (percent of weight)	11-4
Table 11-3. Soil conversion factors	11-5
Table 11-4. Rock characteristics	11-7
Table 11-5. Determining pipe diameter in relation to Qp	11-10
Table 11-6. Recommended gauges for nestable corrugated pipe	11-13
Table 11-7. Strut spacing using 4- by-4 inch timbers with compression caps	11-13
Table 11-8. Recommended requirements for slope ratios in cuts and fills— homogeneous soils	11-14
Table 11-9. Mat characteristics	11-15
Table 12-1. Properties of manila and sisal rope	12-1
Table 12-2. Breaking strength of 6 by 19 standard wire rope	12-2
Table 12-3. Wire-rope FS	12-2
Table 12-4. Knots	12-3
Table 12-5. Assembling wire-rope eye-loop connections	12-13
Table 12-6. Properties of chains (FS 6)	12-16
Table 12-7. Safe loads on hooks	12-17
Table 12-8. SWCs for manila-rope slings (standard, three-strand, splice in each end)	12-18
Table 12-9. SWCs for chain slings (new wrought-iron chains)	12-19
Table 12-10. SWCs for wire-rope slings (new IPS wire rope)	12-20
Table 12-11. Holding power of wooden picket holdfasts in loamy soil	12-21
Table 14-1. Specific weights and gravities	14-1
Table 14-2. Wire sizes for 110-volt single-phase circuits	14-3
Table 14-3. Wire sizes for 220-volt three-phase circuits	14-4
Table 14-4. Properties of southern pine	14-5
Table 14-5. Wood-screw diameters	14-6
Table 14-6. Nail and spike sizes	14-6
Table 14-7. Trigonometric functions	14-10
Table 14-8. Time-distance conversion	14-12
Table 14-9. Vehicle dimensions and classification	14-13
Table 14-10. Ranges of common weapons	14-16
Table 14-11. US tanks	14-17
Table 14-12. US antiarmor missiles	14-17
Table 14-13. US field artillery and air-defense weapons	14-18
Table 14-14. Conversion factors	14-30

FM 5-34

PREFACE

Field Manual (FM) 5-34 provides engineer soldiers at all levels with a source of reference for doctrine; technical data; and tactics, techniques, and procedures (TTP). It also provides a source of reference for information most commonly needed by engineers. Although this manual contains some information that cannot be found in other manuals, most of the information is taken from the manuals that engineers most commonly use.

FM 5-34 addresses combat operations, the threat engineer, reconnaissance operations, mobility operations, defensive operations, demolitions, bridging, roads and airfields, and rigging. The most pertinent information on these topics is included in this manual; however, for more detailed information, users of this manual should check the appropriate manuals in each subject area.

NOTE: United States (US) policy regarding the use and employment of antipersonnel land mines (APLs) outlined in this FM is subject to the convention on Certain Conventional Weapons and Executive Orders (EOs). Current US policy limits the use of non-self-destructing APLs to (1) defending the US and its allies from armed aggression across the Korean demilitarized zone and (2) training personnel engaged in demining and countermine operations. The use of the M18A1 claymore in the command-detonation mode is not restricted under international law or EO.

All references to US employment of non-self-destructing APLs (such as row mining) in this manual are intended to provide doctrine for use in Korea only. Detailed doctrine on APLs is also provided to ensure that US forces recognize how the enemy can employ these weapons.

As the US military seeks to end its reliance on APLs, commanders must consider the increased use of other systems such as the M18A1 claymore, nonlethal barriers (such as wire obstacles), sensors and surveillance platforms, and direct and indirect fires.

The proponent of this publication is HQ TRADOC. To submit changes for improving this publication use Department of the Army (DA) Form 2028 (Recommended Changes to Publications and Blank Forms) and forward to Commandant, United States Army Engineer School (USAES), ATTN: ATSE-TD-D, Fort Leonard Wood, Missouri 65473-6650.

The provisions of this publication are the subject of international standardization agreements (STANAGs) 2002 NBC (Edition 7), *Warning Signs*

FM 5-34

Page

for the Marking of Contaminated or Dangerous Land Areas, Complete Equipments Supplies and Stores; 2021 ENGR (Edition 5), *Computation of Bridge, Ferry, Raft, and Vehicle Classifications;* 2036 ENGR (Edition 5), *Land Mine Laying, Marking, Recording, and Reporting Procedures*; and 2047 NBC (Edition 6), *Emergency Alarms of Hazard or Attack (NBC and Air Attack Only)*.

Unless otherwise stated, masculine nouns and pronouns do not refer exclusively to men.

Chapter 1

Combat Operations

TROOP-LEADING PROCEDURES (TLPs)

The eight steps of troop leading are—

- Receive the mission.
- Issue a warning order (WO).
- Make a tentative plan that will accomplish the mission.
- Start the necessary movement.
- Reconnoiter.
- Complete the plan.
- Issue orders.
- Supervise and refine the plan.

COMBAT ORDERS

Combat orders are written or oral communications used to transmit information pertaining to combat operations.

Warning Order

A WO is a preliminary notice of an order or action that is to follow (see Figure 1-1, page 1-2). WOs help subordinate units and their staffs prepare for new missions.

Operation Order (OPORD)

An OPORD is a directive a commander issues to subordinate commanders to coordinate the execution of an operation. An OPORD always specifies an execution time and date. Figure 1-2, pages 1-4 through 1-11, shows the format for a company OPORD.

Fragmentary Order (FRAGO)

A FRAGO provides timely changes to existing orders to subordinate and supportive commanders while providing notification to higher and adjacent commands.

FM 5-34

Classification
(Change from oral orders, if any (optional))
A WARNING ORDER DOES NOT AUTHORIZE EXECUTION UNLESS SPECIFICALLY STATED

Copy___of___copies
Issuing headquarters
Place of issue
Date-time group of

signature

Message reference number

WARNING ORDER _____

References: Refer to higher headquarters OPLAN/OPORD, and identify map sheet for operation.

Optional

Time Zone Used Throughout the Order: (Optional)

Task Organization: (Optional) (See paragraph 1c.)

1. **SITUATION.**

   a. Enemy Forces. Include significant changes in enemy composition dispositions and courses of action. Information not available for inclusion in the initial WO can be included in subsequent WOs.

   b. Friendly Forces. (Optional) Only address if essential to the WO.

   (1) Higher commander's mission.

   (2) Higher commander's intent.

   c. Attachments and Detachments. Initial task organization, only address major unit changes.

2. **MISSION.** Issue headquarters' mission at the time of the WO. This is nothing more than higher headquarters' restated mission or commander's decisions during MDMP.

3. **EXECUTION.**

   Intent:

   a. Concept of Operations. Provide as much information as available; there may be none during the initial WO.

   b. Tasks to Maneuver Units. Any information on tasks to units for execution, movement to initiate, reconnaissance to initiate, or security to emplace.

   c. Tasks to Combat-Support Units. See paragraph 3b.

Figure 1-1. WO format

1-2 Combat Operations

d. Coordinating Instructions. Include any information available when the WO is issued. It may include the following:

- CCIR.
- Risk guidance.
- Deception guidance.
- Specific priorities, in order of completion.
- Time line.
- Guidance on orders and rehearsals.
- Orders group meeting (attendees, location, and time).
- Earliest movement time and degree of notice.

4. **SERVICE SUPPORT.** (Optional) Include any known logistics preparation for the operation.

   a. Special Equipment. Identify requirements and coordinate transfer to using unit.

   b. Transportation. Identify requirements and coordinate for pre-position of assets.

5. **COMMAND AND SIGNAL** (Optional)

   a. Command. State the chain of command if different for the unit SOP.

   b. Signal. Identify current SOI edition, and pre-position signal assets to support operation.

ACKNOWLEDGE: (Mandatory)

NAME (Commander's last name)

RANK (Commander's rank)

OFFICIAL: (Optional)

Classification

Figure 1-1. WO format (continued)

Combat Operations 1-3

FM 5-34

Classification
(Place the classification at the top and bottom of every page of the OPORD).

                                  Copy__of__copies
                                  Issuing headquarters
                                  Place of issue (coordinates)
                                  Date-time group of signature

OPERATION ORDER NUMBER ___ (code name, if used)

Reference(s): Map(s) or other references required.

Time Zone Used Throughout the Order:

Task Organization. By phase, accounts for all platoons and special equipment. Includes the command or support relationship.

Example:

                          Phases I - III
      Team Alpha   *Team Bravo*   *Company Control*
        1/A/45th Engr (OPCON)   *2/A/45th Engr (OPCON)*   *A&O/A/45th Engr*
(-)
        Volcano/A&O/A/45th Engr   *AVLM/A&O/A/45th Engr*

                              Phase IV
      1/A/45th Engr   *2/A/45th Engr*   *Company Control*
        AVLB/A&O/A/45th Engr   *Volcano/A&O/A/45th Engr*   *A&O/A/45th Engr*
(-)
        Volcano/A&O/A/45th Engr

1. SITUATION.

    a. Enemy Forces.

        (1) Terrain and weather. Include—

- Important terrain characteristics and their significance (OCOKA).
- Advantages and disadvantages to enemy/friendly maneuver and engineer operations.
- Light data and expected weather and their impact on a mission.

        (2) Enemy composition, disposition, and strength.

- Ensure that the focus is on the enemy that a supported unit expects to fight in a sector (or from a BP or strong point) or in a zone. Also identify adjacent enemy units—those that can reinforce an enemy's attack or defense.
- List the type of enemy unit; how it is equipped; and its designation, location, size, and strength.
- List current enemy activities that are pertinent.
- Distinguish known and templated locations of enemy forces/activities.

        (3) Capability. List the—

- Combat capability (range and orientation of direct/indirect fires, CATK forces, reserves, NBC, and ability to reposition).

Figure 1-2. Sample of a company OPORD

1-4 Combat Operations

- Mobility, countermobility, and survivability capability. This includes the amount, type, location, and expected employment of breaching equipment; the amount, type, location, and expected employment of tactical and protective obstacles; the amount, type, and expected use of scatterable mines; and the level of expected fortification for vehicles and infantry.

(4) **Intentions.** Include—

- The most probable and most dangerous enemy COA.
- How an enemy will probably react to a friendly attack or defense (especially the expected employment of mobility, countermobility, and survivability assets).
- The critical enemy events that platoon leaders should look for during a battle.

NOTE: When briefing an OPORD, use a sketch or sand table to explain the enemy's situation or use a map with overlay for very small groups.

   b. **Friendly Forces.**

   (1) **Higher.** Include the—

- TF mission, TF commander's intent, and TF scheme of maneuver/concept of the operation. This must be complete enough that the platoon leaders understand the fire (to include the indirect-fire plan) and maneuver plans of the supported unit.
- SOEO to support the TF's scheme of maneuver (same as in a TF OPORD and a TF engineer annex).

   (2) **Adjacent.**

- Include the maneuver missions/events/forces of adjacent units as they affect a supported unit and an engineer company's mission, to include specifics of adjacent engineer units, if appropriate.
- Identify the units at the flanks, to the front, and possibly to the rear.

   c. **Attachments and Detachments.**

- Do not include this subparagraph if the attached/detached units are clear in the task organization briefed at the beginning of an OPORD.
- Include the attachments and detachments to/from the engineer company's TOE for a mission and the effective time period.

Example:

Attachments: Maintenance contact team and medic team are attached to the company effective _____.

Detachments: 1/A/45th is OPCON to Tm Alpha during Phases I-III effective _____.
2/A/45th is OPCON to Tm Bravo during Phases I-III effective _____.

Figure 1-2. Sample of a company OPORD (continued)

NOTES:

1. When briefing an OPORD, use a sketch or sand table to explain the friendly situation, or use a map with an overlay for very small groups. This may be combined with the enemy-situation sketch.

2. When briefing an OPORD, use a sketch or sand table to explain the SOEO, or use a map with an overlay for very small groups. This may be combined with the friendly-situation sketch.

2. MISSION.

- A clear, concise statement of the who, what, where, when, and why of the engineer company's mission. The <u>who</u> is the engineer company. The engineer company commander decides <u>what</u>, when, where, and why based on his mission analysis. The essential tasks the engineer company commander identifies for the engineer company form the basis for a mission statement.

- An engineer company commander should be as specific as possible. Task organization, command or support relationships, or other factors may limit the specificity of a mission statement.

The following are examples of typical engineer company mission statements:

Offense: D/51st Engr Bn creates two lanes on Axis Red and at Obj Zulu and emplaces situational obstacles vic PL Green, 030500 DEC 199_ to support TF 5-21 attack and allow FPOL of follow-on forces.

Defense: D/51st Engr Bn constructs obstacles and prepares fighting positions to support the TF 2-51 defense in sector 030500 DEC 199_ to allow TF 2-51 to defeat an MRR attack.

3. EXECUTION.

Intent:

- Include a clear, concise statement of what the force must do to succeed with respect to the enemy and the terrain and to the desired end state.

- Provide a link between the mission and the concept of operation by stating the key tasks that, with the mission, are the basis for subordinates to exercise an initiative when unanticipated opportunities arise or when the original concept of operation no longer applies.

- Express intent in four or five sentences. This is mandatory for all orders.

Example:

The purpose of our operation is to overcome the effects of the enemy's tactical obstacles, by breaching or bypassing, to get the combat forces of TF 5-79 onto Obj Frank. The end state, from my perspective, will be two bypasses or breaching lanes cleared and marked for the TF's assault force, Tm Charlie. We will be consolidated forward of the enemy's obstacles, but to the rear of the objective. Be prepared to move forward to support the TF in establishing a hasty defense.

a. **Concept of Operations.** Ensure the concept of operations—

- Is a single paragraph. It may be divided into two or more subparagraphs.
- Is concise and understandable.
- Describes—
  — The employment of subordinate elements.

Figure 1-2. Sample of a company OPORD (continued)

1-6 Combat Operations

— The integration of other elements or systems within the operation.

— Any other aspects of the operation that a commander considers appropriate to clarify the concept and to ensure unity of effort.

NOTE: Depending on the operation, the following subparagraphs may be required within the concept of operations.

    (1) **Maneuver.**

    (2) **Fires.**

    (3) **Engineer.** Focus on how the forces under company control will accomplish their assigned tasks.

    (4) **Air defense.**

NOTE: A sketch or sand table should be used to explain the concept of operation when briefing the OPORD, or a map with an overlay should be used for very small groups.

    b. **Tasks to Subordinate Units.**

- List specific tasks to subunits retained under company control (platoons, the TOC, combat trains, company field trains, and others, as the commander determines).
- List subunits in the same order as in the task organization.
- Include O/O and B/P tasks, and list them in the subunit's paragraph in the order that they will likely be performed.
- Put missions/tasks common to two or more subunits in coordinating instructions.

Example:

    (1) *1st Plt*
        a) Construct directed-obstacle groups A1A and A1D.
        b) ...
        c) ...
    (2) *2d Plt...*
    (3) *A&O Plt*
        a) Construct fighting positions (see survivability matrix).
        b) ...

    c. **Coordinating Instructions.**

- List tasks, reporting requirements, and instructions for coordination that apply to two or more subunits within the company.
- Do not include SOP items unless they are required for emphasis or are a change from the normal SOP.
- Include, as a minimum, the—
  - References to obstacle-execution or survivability matrixes.
  - CCIR.
  - OEG.

Figure 1-2. Sample of a company OPORD (continued)

FM 5-34

- MOPP status (level and effective time period) and any changes in MOPP level.
- Air-defense warning and weapons-control status.
- Directed coordination between subunits or with adjacent units.
- Sleep plan.
- Priorities of work.
- Lane-marking system.
- Obstacle restrictions, belts, or zones that affect a TF.
- Rehearsals.
- ROE.
- Environmental considerations.
- Instructions about consolidation or reorganization.

NOTES:

1. The sum of all subunit tasks and coordinating instructions balances with the specified and implied tasks that the commander identified during the planning process.

2. The OPORD should refer to the appropriate obstacle or the other execution matrixes, survivability matrixes, time lines, and so forth instead of listing the same information in paragraph 3.b. or 3.c. These items are annexes to the OPORD.

Example:

    (1) Coordinating Instructions.

        (a) Details for directed-obstacle groups are in the directed-obstacle matrix.

        (b) ...

    (2) ...

4. SERVICE SUPPORT.

    a. Support Concept.

NOTE: Include items only if different from the SOP. Much of the information in paragraph 4 can easily be included in SOPs. SOPs must be understood and rehearsed.

- Include the concept for providing subunits with CSS before, during, and immediately after an operation.
- Designate primary and back-up channels for logistical support for each platoon. (For example, through the company's organic CSS assets, through the supported unit's CSS system, or through a combination of company and supported unit.)
- Ensure that the support concept is consistent with the company's task organization for the mission and command or support relationships.

Figure 1-2. Sample of a company OPORD (continued)

1-8 Combat Operations

FM 5-34

- State what method of company resupply/LOGPAC will be used (service-station or tailgate) and give the location of resupply points and times, when appropriate.
- Use the supported unit's CSS graphics to help integrate the company's CSS plan into the supported unit's plan.
- Give the location, movement, and subsequent locations of critical CSS nodes before, during, and after a battle. These include the—
  - Engineer company trains.
  - Engineer battalion trains.
  - TF combat and field trains.
  - TF main and jump aid stations, patient-collection points, and AXPs.
  - TF and engineer UMCP.
  - TF and engineer CCPs and EPW collection points.
  - TF logistics release points.
  - Class IV/V supply points.
  - Decontamination sites.
  - Location of parent engineer CSS assets pushed forward.
  - Collocation of engineer and supported unit CSS assets/nodes.
  - Hazardous material/waste collection points.

NOTE: When briefing the OPORD, do not brief the CSS node locations if providing a CSS overlay or hard copy that would give the same information. Tell the platoon leaders that they have the information on an overlay or a hard copy.

b. **Materiel and Services.**

- Outline the platoon allocations of command-regulated materials.
- State what services are available to the platoons through the company and the supported unit.
- Include the special allowances/plans made for sustaining the special engineer equipment or forces (for example, fuel tanker dedicated to fueling dozers/ACEs located at the Class IV/V supply point).

(1) **Supply.** List the—

- Basic loads that the unit maintains.
- Method of obtaining supplies if different from the support concept.

(a) **Class I.**

- Ration cycle.
- Basic load that the platoons (days of supply) and the company trains or field trains maintain.

(b) **Class III.**

- Top-off times and locations.
- Location of emergency Class III at the company and the TF.

(c) **Classes IV and V.**

- Platoon allocation/basic-load small arms.
- Platoon allocation/basic-load demolitions.

Figure 1-2. Sample of a company OPORD (continued)

Combat Operations 1-9

- Platoon allocation/basic-load mines/Class IV supplies.
- Class IV/V stockages at Class IV/V supply point (on-hand and allocation from higher) and the planned platoon allocations by obstacle group.
- Type of mine resupply to be used.
- Location, type, and amount of emergency Class V at the company and the TF.
- Volcano/MICLIC/MOPMS reload plan.

(d) **Other classes of supply.** As necessary.

(2) **Transportation.** Include—

- TF and engineer company haul assets allocated to the platoons and their priority by subunit.
- Primary, alternate, and dirty MSRs.
- Designated routes from the Class IV/V supply points to the obstacle groups.

(3) **Maintenance.**

- Include the maintenance/recovery support from the engineer company, the parent engineer battalion, or the supported maneuver unit.
- State the maintenance priorities by vehicle, unit, or a combination of both.
- Include the authority for controlled substitution.

c. **Medical Evacuation and Hospitalization.** Include the—

- Wounded-in-action evacuation plan (primary and alternate)—through the supported unit or through the engineer company.
- Routine sick call location and time.
- Class VIII resupply location, time, and allocation.

d. **Personnel Support.** Include the—

- Method of handling EPWs—through the supported unit or the engineer company.
- Mail.
- Religious services.
- Graves registration.

e. **Civil-Military.** Identify engineer supplies, services, or equipment provided by the HN.

Figure 1-2. Sample of a company OPORD (continued)

FM 5-34

5. **COMMAND AND SIGNAL.**

   a. **Command.** Include the—

   - Key leader locations during each phase of the battle (company and TF levels).
   - C^2 node locations during each phase of the battle (company and TF levels).
   - Succession of command that supports the continuity of command during battle.

Example:

   (1) Command
   (a) I will be with 2d Plt during Phases I and II. During Phase III, I will be vic CP 43. During Phase IV, I will be vic CP 46. The TF commander...
   (b) The company CP will be with the TF main CP. Initial location is...
   (c) The succession of command is A&O Plt leader; 2d Plt leader...
   (2) ....

   b. **Signal.** Include the—

   - Communications/signal peculiarities for an operation (specific code words).
   - Visual/audio signals critical to the battle or for use in emergencies.
   - SOI index and times when radio listening silence in is effect.
   - Method for communications and priority. FM nets that the commander wants the subunits on to simplify C^2.
   - Reports that the engineer company commander wants from the subunits.

Acknowledge:

Commander's signature
Commander's rank

OFFICIAL:
(Authentication)

ANNEXES: Possible annexes include—
  • OPORD-execution matrix
  • Directed-obstacle-execution matrix
  • Situational-obstacle-execution matrix
  • Reserve-obstacle-execution matrix
  • Company time line
  • Survivability-execution matrix)
  • Overlays (TF maneuver, fire-support, SITEMP, engineer company operations graphics, scheme-of-obstacle overlay, and CSS)
  • Environmental considerations

Distribution:

CLASSIFICATION

Figure 1-2. Sample of a company OPORD (continued)

Combat Operations 1-11

FM 5-34

MOVEMENT ORDER

A movement order or briefing should include, as a minimum, the following:

- Enemy and friendly situation.
- Destination.
- Start, critical, release, and rally points.
- Rate of march and catch up speed (see Figure 1-3 for halt formations).
- Support (indirect, direct, and medical) and communications.
- Actions on contact.
- Order of march.
- Route/alternate route.
- Distance between vehicles (50 meters, daytime; 25 meters, nighttime).
- Departure time.
- Location of commander.
- Lead vehicle (security/reconnaissance, see Figure 1-4).

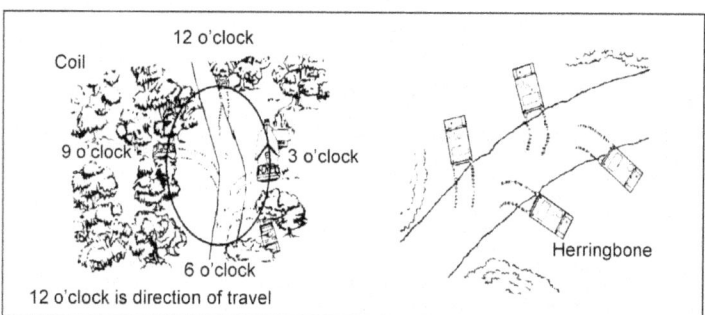

Figure 1-3. Halt formation

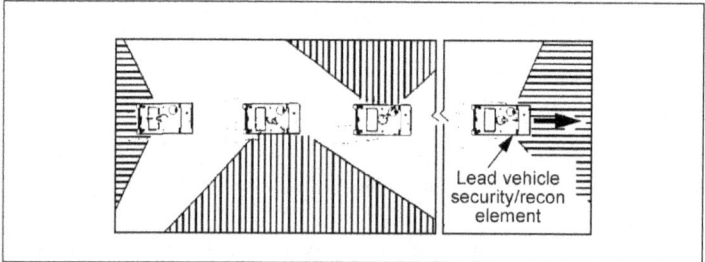

Figure 1-4. Sectors of fire

1-12 Combat Operations

MARCH RATES

Table 1-1. Average march rates

Unit	Average Rates (kmph)				Days March (kmph)
	On Roads		Cross-Country		
	Day	Night	Day	Night	
Foot troops	4	3.2	2.4	1.6	20-32
Trucks, general	40	40 w/lights 16 w/bo	12	8	280
Tracked vehicles	24	24 w/lights 16 w/bo	16	8	240
Truck-drawn artillery	40	40 w/lights 16 w/bo	16	8	280
Tractor-drawn artillery	32	32 w/lights 16 w/bo	16	8	240

NOTE: This table is for general planning and comparison purposes. All the rates are variable according to the movement conditions as determined by reconnaissance. The average rates include periodic rest halts.

BIVOUAC AND ASSEMBLY AREAS

An area must be organized to provide a continuous 360° perimeter security. When any element leaves the perimeter, either shrink the perimeter or redistribute the perimeter responsibilities. Crew-served weapons are the basis for a unit's defense. Individual weapons provide security for the crew-served weapons and must have overlapping sectors of fire.

Site selection characteristics are—

- Concealment.
- Cover from direct and indirect fire.
- Defendable terrain.
- Drainage and a surface that will support vehicles.
- Exits and entrances and adequate internal roads or trails.
- Space for dispersing vehicles, personnel, and equipment.
- Suitable landing site nearby for supporting helicopters.

NOTE: Do not include environmentally sensitive areas, such as designated wetlands, archeological areas, or spawning and breeding areas for endangered species of flora and fauna, in a training site.

FM 5-34

Quartering party responsibilities are to—

- Reconnoiter the area.
- Check the area for nuclear, biological, chemical (NBC) hazards.
- Check the area for obstacles and mines and then mark or remove them.
- Check for and cordon off environmentally sensitive areas.
- Designate maintenance areas that are not in or along natural groundwater runoff routes.
- Choose routes and locations that prevent vehicles from crossing streams/rivers or bottom land at other-than-designated crossing points.
- Mark platoon and squad sectors.
- Select a command post (CP) location.
- Select a company trains location.
- Provide guides for the incoming unit(s) to accomplish immediate occupation.

Recommended priority of work is to—

- Post local security (listening post [LP]/observation post [OP]).
- Position crew-served weapons (vehicle-mounted weapons, antitank (AT) weapons, and machine guns) and chemical alarms.
- Assign individual fighting positions.
- Clear fields of fire, prepare range cards, and camouflage vehicles.
- Prepare hasty fighting positions.
- Install/change to land-line communication.
- Emplace obstacles and mines.
- Construct primary fighting positions.
- Prepare alternate and supplementary fighting positions.
- Stockpile ammunition, food, and water.

Recommended actions at the bivouac and assembly area are to—

- Reorganize.
- Check weapons.
- Maintain vehicles and supplies.
- Distribute supplies.
- Rest and attend to personal hygiene.

1-14 Combat Operations

FM 5-34

- Consume rations.
- Rehearse.
- Perform precombat checks/inspections.
- Check communications.

MOUNTED/DISMOUNTED OPERATIONS

MOVEMENT TECHNIQUES

Regardless of the means of transportation, units could eventually move on foot to accomplish their mission. Since a unit is vulnerable while moving on foot, it must use proper movement techniques and constant security to avoid unplanned enemy contact. A dismounted squad moves with one fire team following the other fire teams in the wedge formation (see Figure 1-5). Larger elements move in a column, wedge, V-line or echelon left or right (see Figure 1-6, page 1-16). The enemy's situation determines which of the following three techniques will be used: when enemy contact is not likely, TRAVELING (see Figure 1-7, page 1-16); when enemy contact is possible, TRAVELING OVERWATCH (see Figure 1-7); when enemy contact is expected, BOUNDING OVERWATCH (see Figure 1-8, page 1-16). Leaders, except fire-team leaders, move within a formation where they can best control the situation and do their job.

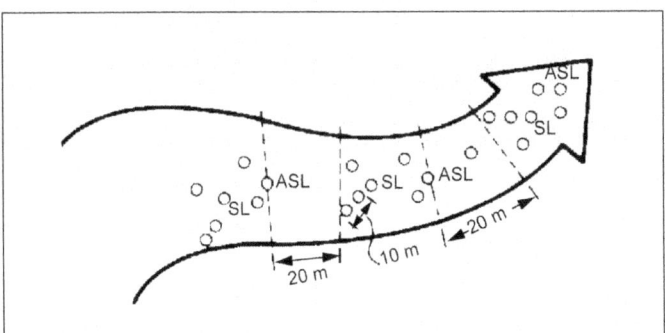

Figure 1-5. Traveling dismounted elements

Combat Operations 1-15

FM 5-34

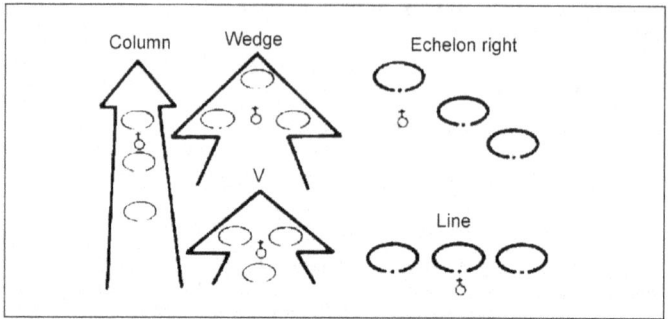

Figure 1-6. Movement formations

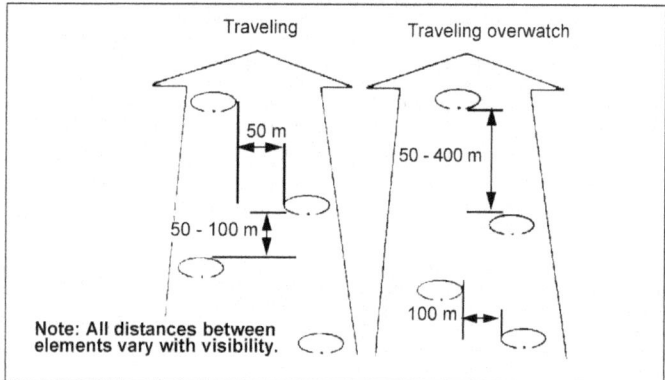

Figure 1-7. Traveling and traveling overwatch

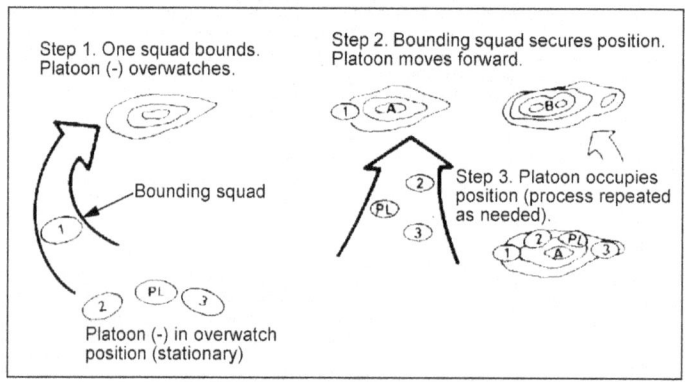

Figure 1-8. Bounding overwatch

1-16 Combat Operations

FM 5-34

LOCAL/JOB SITE SECURITY

Before moving to a job site or a new position, leaders should inform everyone of warning signals, code words, and pyrotechnics. Upon arrival at the new location, the unit should—

- Occupy an overwatching position.
- Dispatch a reconnaissance/mine sweeping/NBC team and establish a hasty perimeter.
- Establish the escape routes and identify the avenues of approach (AAs), LP/OPs, and crew-served weapons positions.
- Place the LP/OP and NBC alarms.
- Position the crew-served, AT, and automatic weapons and prepare the range cards.
- Divide the job site/position into defensive sectors and assign sectors of responsibility.
- Maintain communications with the parent unit.

PATROLLING

The two types of patrols are reconnaissance (zone, area, or route) and combat (ambush, raid, or tracking). The five key principles of a successful patrol are security, surprise, coordinated fire, violence, and control. To prepare for a patrol—

- Issue a WO.
- Conduct the required coordination.
- Issue an OPORD.
- Inspect and rehearse.

RECONNAISSANCE PATROL

Reconnaissance patrols provide the commander with timely, accurate information of the enemy and the terrain he controls. The information should be collected following the size, activity, location, unit, time, and equipment (SALUTE) report format. The gathered information must be shared with all patrol members. For more information, see FM 5-170.

COMBAT PATROL

There are many missions a combat patrol can perform. This section discusses an ambush and a raid, as the techniques for these patrols apply in general to other combat patrols.

Combat Operations 1-17

Ambush

An ambush is a surprise attack from a concealed position on a moving or temporarily halted target. See Figures 1-9 through 1-11. Key planning points for a successful ambush are—

- Covering the entire kill zone by fire.
- Using existing or reinforcing obstacles (claymores and other mines) to keep the enemy in the kill zone.
- Protecting the assault and support elements with claymores, other mines, or explosives.
- Using security elements or teams to isolate the kill zone.
- Assaulting into the kill zone to search dead or wounded, assembling prisoners, and collecting equipment.
- Timing the actions of all elements to preclude loss of surprise.
- Using only one squad to conduct the entire ambush and rotating squads over time from the objective release point (ORP) (this technique is useful when the ambush must be manned for a long time).

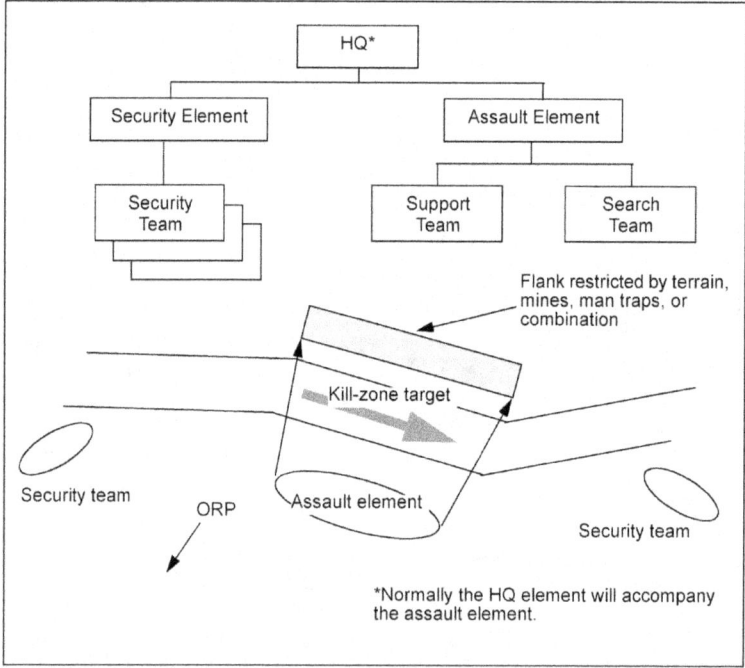

Figure 1-9. Typical organization and employment-point (linear) ambush

1-18 Combat Operations

FM 5-34

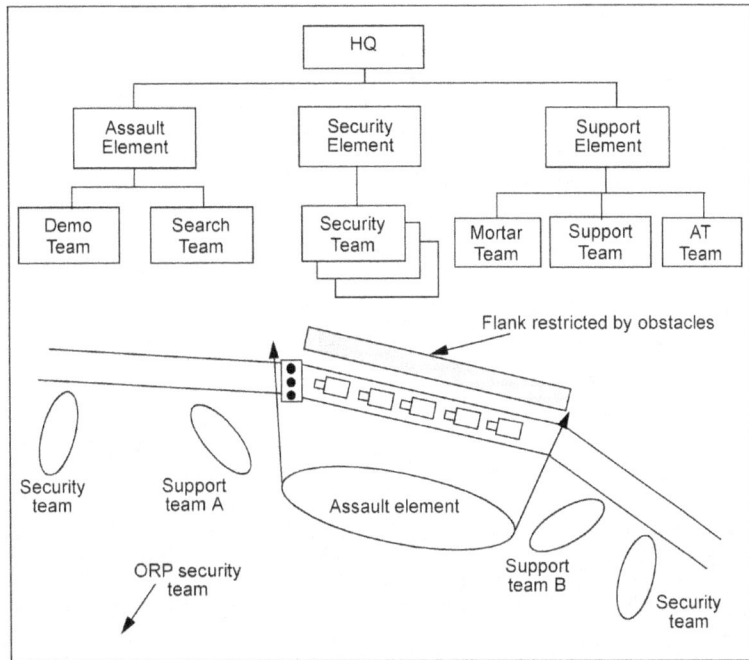

Figure 1-10. Typical organization and employment-point (vehicular) ambush

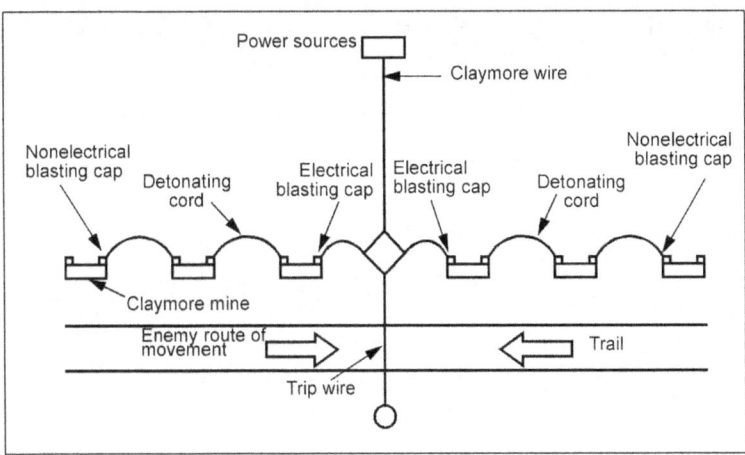

Figure 1-11. Multi-claymore-mine mechanical ambush

Combat Operations 1-19

FM 5-34

Raid

A raid (see Figure 1-12) is an attack on a position or an installation followed by a preplanned withdrawal. (Squads do not execute raids.) The sequence of actions for a raid is similar to those of an ambush. The assault element may have to conduct a breach of an obstacle. It may have additional tasks to perform on the objective; for example, demolition of fixed facilities.

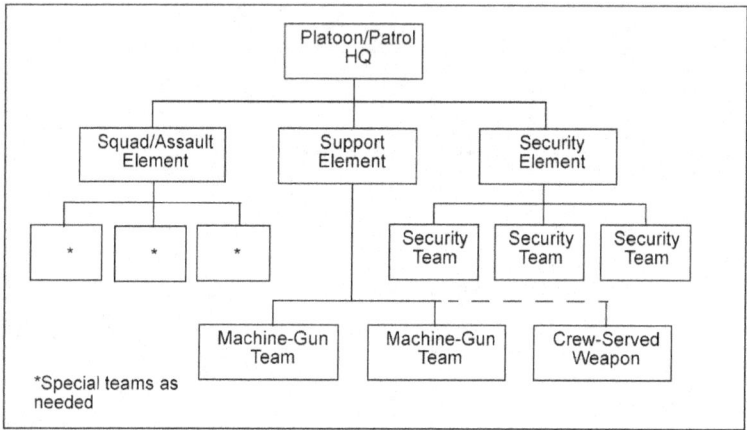

Figure 1-12. Typical organization for a raid patrol

FIRE-SUPPORT PROCEDURES AND CHARACTERISTICS

(For more information, see FM 6-30.)

CALL-FOR-FIRE ELEMENTS

Observer Identification

Call sign.

Warning Order

- Type of mission (adjust fire, fire for effect, suppression, immediate suppression, and immediate smoke).
- Size of element to fire for effect (battery, battalion).
- Method of target location (polar plot, laser polar plot, shift from a known point, grid).
- Figure 1-13 shows examples of an observer identification and a WO.

1-20 Combat Operations

Adjust Fire Mission
Grid method: **A57**, this is **A71**, adjust fire, over.

Fire-For-Effect Mission
Polar-plot method: **A57**, this is **A71**, fire for effects (battalion call sign is **B6S13**), polar, over.
Shift-from-a-known-point method: **A57**, this is **A71**, fire for effect, shift known point 3, over.

Suppression Mission
F28, this is **F72**, suppress **AA7749**, over

Immediate Suppression Mission
F28, this is **F72**, immediate suppression, grid **NK453215**, over.

Figure 1-13. Examples of observer identification and WO

Target Location

- Grid (six-digit grid).
- Shift from a known point (lateral shift "left or right," range shift "add or drop," vertical shift "up or down").
- Polar plot (direction and distance from the observer). **NOTE: The fire-direction center (FDC) must know the observer's location.**

Target Description

- What it is (troops, equipment, supply dump, trucks).
- What it is doing (digging in, in an assembly area).
- How many elements it has (squad, platoon, three trucks, six tanks).
- What the degree of protection is (in open, in foxholes, in bunkers with overhead cover).
- What the size and shape are, if significant.

Method of Engagement (Optional)

- Type of adjustment (precision or area).
- Danger close.
- Mark.
- Trajectory.
- Ammunition (projectile, high explosive [HE], illumination, improved capability missile [ICM], smoke; fuze; volume of fire).
- Distribution.

FM 5-34

Method of Fire and Control (Optional)

At my command, cannot observe, time on target, continuous illumination, cease loading, check firing, continuous fire, repeat.

Three Transmissions in a Call for Fire

- Observer identification and WO (see Figure 1-13, page 1-21).
- Target location.
- Description of target, method of engagement, and method of fire and control.

Message to Observer

After the FDC receives the call for fire, it determines how the target will be attacked. That decision is announced to the observer in the form of a message to observer (MTO) (see Figure 1-14). The MTO consists of the following items:

- Unit(s) to fire.
- Changes to the call for fire.
- Number of rounds.
- Target number.
- Authentication.

When nonsecure communications (except unique fire support such as suppressive fires posture) are used, challenge and reply authentication is considered a normal element of initial requests for indirect fire. The FDC challenges the observer after the last read back of the fire request.

ADJUSTMENTS

The adjustments that may be required to obtain a round on target are spotting, lateral, and range.

Spotting

Spotting refers to where a round lands in relation to a target, such as short or long and how many mils right or left of a target. Spotting examples are *short 40 right* or *long 50 left*.

FM 5-34

EXAMPLES

Fire Mission (Grid)
Initial Fire Request

Observer	FDC
Z57, this is Z71, adjust fire, over.	This is Z57, adjust fire, out.
Grid NK180513, over.	Grid NK 180513, out.
Infantry platoon in the open, ICM in effect, over.	
I authenticate Charlie, over.	Infantry platoon in the open, ICM in effect, authenticate Papa Bravo, over.
Message to Observer	
	Z, 2 rounds, target AF1027, over.
Z, 2 rounds, target AF1027, out direction 1680, over.	Direction 1680, out.

NOTE: Direction is sent before or with the first subsequent correction.

Fire Mission (Shift)
Intial Fire Request

Observer	FDC
H66, this is H44, adjust fire, shift AA7733, over.	This is H66, adjust fire, shift AA7733, over.
Direction 5210, left 380, add 400, down 35, over.	Direction 5210, left 380, add 400, down 35, out.
Combat OP in open, ICM in effect, over.	Combat OP in open, ICM in effect, authenticate Lima Foxtrot, over.
I authenticate Papa, out.	
Message to Observer	
	H, 1 round, target AA7742, over.
H, 1 round, target AA7742, out.	

Fire Mission (Polar)
Initial Fire Request

Observer	FDC
Z56, this is Z31, fire for effect, polar, over.	This is Z56, fire for effect, polar, out.
Direction 4520, distance 2300, down 35, over.	Direction 4520, distance 2300, down 35, out.
Infantry company in open, ICM, over.	Infantry company in open, ICM, authenticate Tango Foxtrot, over.
I authenticate Echo, out.	
Message to observer	
	Y, VT, 3 rounds, target AF2036, over.
Y, VT, 3 rounds, target AF2036, out.	

Figure 1-14. Sample missions

Lateral (Right/Left)

Adjustment for the lateral shift is from impact to observer target (OT) line in meters. Corrections of 20 meters or less will be ignored until firing for effect.

$$W = Rm$$

where—

W = lateral shift correction, in meters

$R = OT\ factor = \dfrac{target\ range\ (to\ nearest\ 1,000\ meters)}{1,000}$

m = distance between burst and target, in mils

Combat Operations 1-23

FM 5-34

NOTE: If target range is less than 1,000 meters, round to nearest 100 meters.

Range Correction (Up/Down)

Mechanical time fuze only. Initial range shift correction is used to bracket target (see Table 1-2).

Table 1-2. Target bracketing

Distance to Target	Change
Less than 1,000	+/- 100 meters
1,000 to 1,999	+/- 200 meters
2,000 or greater	+/- 400 meters

Range Deviation

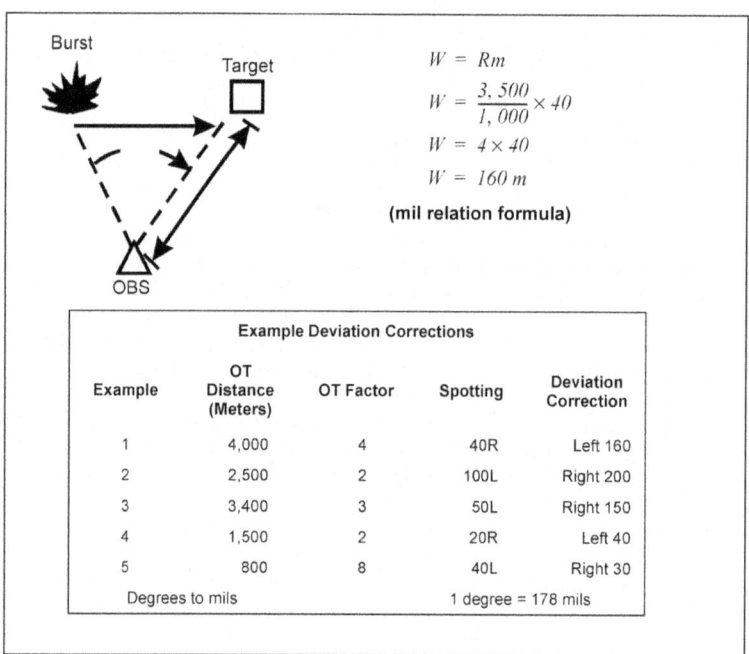

$$W = Rm$$
$$W = \frac{3,500}{1,000} \times 40$$
$$W = 4 \times 40$$
$$W = 160\ m$$

(mil relation formula)

		Example Deviation Corrections		
Example	OT Distance (Meters)	OT Factor	Spotting	Deviation Correction
1	4,000	4	40R	Left 160
2	2,500	2	100L	Right 200
3	3,400	3	50L	Right 150
4	1,500	2	20R	Left 40
5	800	8	40L	Right 30
Degrees to mils			1 degree = 178 mils	

Figure 1-15. Adjusting field artillery fires

FM 5-34

QUICK SMOKE

When using quick smoke, consider the wind speed and direction, the required smoke duration, and other friendly units in the area (see Figure 1-16).

EXAMPLE:

QUICK SMOKE

"M6J41, this is B5T36, adjust fire/fire for effect, over."

"Grid BS612327, direction 1600, over."

"Enemy OP, HC smoke in effect, over."

FM 5-34

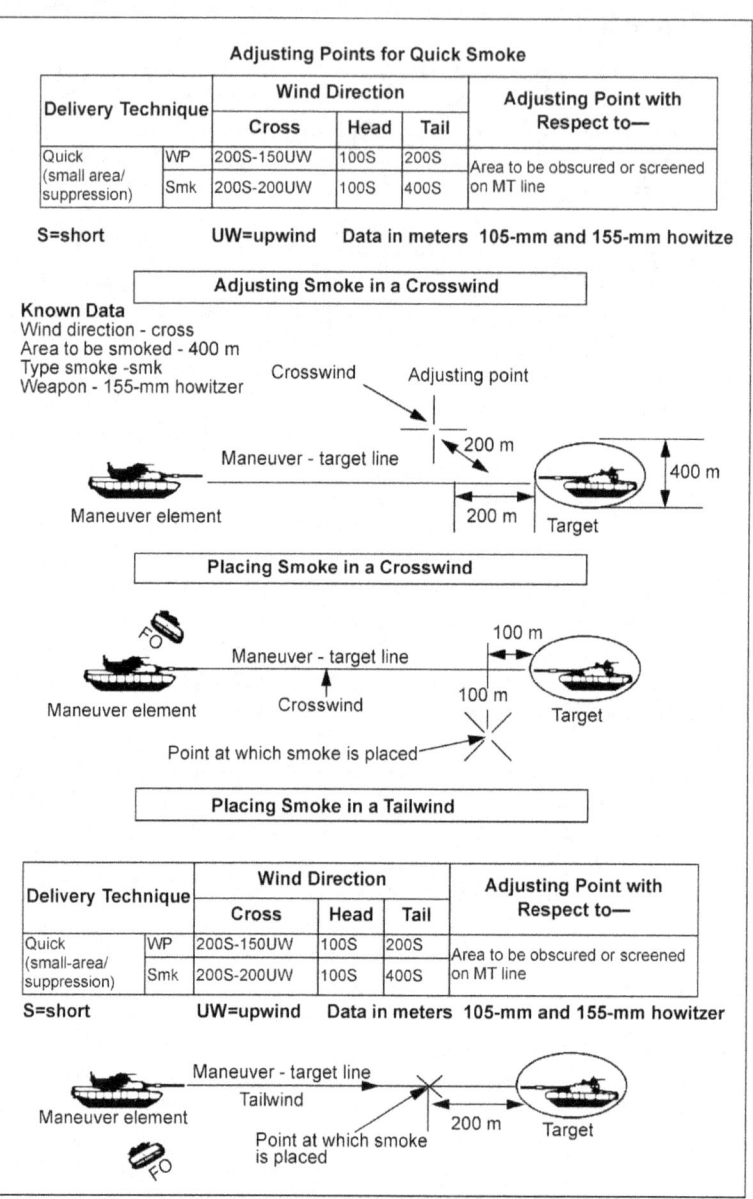

Figure 1-16. Adjusting points for quick smoke

1-26 Combat Operations

FM 5-34

Table 1-3. Artillery and mortar smoke

Delivery System	Type of Round	Time Needed to Build Effective Smoke (min)	Average Burning Time (min)	Average Obscuration Length, Wind Direction (m per round)		
				Cross	Quartering	Head/Tail
155 mm	WP	0.5	1-1.5	100	75	50
	HC	1-1.5	4	350	250	75
105 mm	WP	0.5	1-1.5	75	60	50
	HC	1-1.5	3	250	175	50
107 mm	WP	0.5	1	200	80	40
81 mm	WP	0.5	1	100	60	40

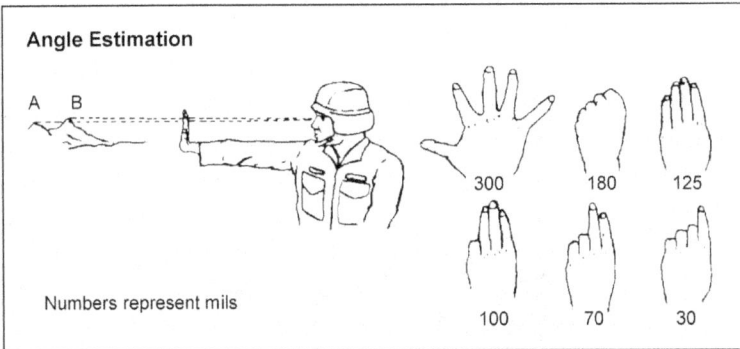

Figure 1-17. Hasty method for estimating angle

Table 1-4. Artillery and mortar flares

Type Weapon/Round	Range (m)	Illum Time (sec)	Continuous Illum (rounds per min)	Diameter of Illum Area (m)	Candle-power
81 mm/M301A3	3,300	75	2	1,100	500,000
107 mm/M355A2	5,500	90	2	1,500	850,000
105 mm/M314	8,500	60	2	1,000	600,000
155 mm/M118	11,600	60	2	1,000	500,000
155 mm/M485	14,000	120+	1	2,000	1,000,000

Combat Operations 1-27

FM 5-34

FIRE-SUPPORT EQUIPMENT

Table 1-5. Fire-support munitions

Ammunition		Fuzes	
Type	Typical Targets	Type	Typical Targets
HE	Personnel, light armor, crew weapons	Impact (quick)	Surface targets
HEAT/HEP-T (105 only)	Light armor, light skin vehicles	Delay	Cratering, heavily wooded
DPICM	All targets	Mechanical time	Dug-in, defilade positions
APERS (105 only)	Personnel	Proximity (VT)	Dug-in, defilade positions
WP	Vehicles, fuel/ammo stores (also used as quick smoke)	Concrete piercing	Bunkers
Smoke	Screening		
Illum	Night/darkness		
Copperhead	Armor, point targets		
RAP	Long-range area targets		
Scatterable mines (ADAM/RAAM)	Mines, area denial (long and short duration)		
Nuclear			
Chemical			

1-28 Combat Operations

FM 5-34

Table 1-6. Fire-support system capabilities

	105-mm Howitzer, M119A1	105-mm Howitzer, M101A1	105-mm Howitzer, M102	144-mm Howitzer, M198	155-mm Howitzer, M109A2-5	155-mm Howitzer, M109A6 Paladin	203-mm Howitzer, M110A2	MLRS M70 ATACMS
Max range w/ RAP (m)	20,100	15,100	15,300	30,100	23,500 (A2-4) 30,000 (A5)	30,000	30,000	32,000+ (rockets) 164,000 (ATACMS)
Max range w/o RAP (m)	14,000	11,000	11,400	24,000	18,000	22,000	22,900	NA
Max FPF width (m)	210 6 tubes	210 6 tubes	210 6 tubes	200 4 tubes	200 4 tubes	150 3 tubes	NA	NA
Max rate of fire (rd/min) (first 3 min)	6	10	10	4	4	4	1.5	NA
Sustained rate of fire (rd/min)	3	3	3	2	1	1	0.5	12 (rockets) 1 or 2 (10 sec (ATACMS)

Combat Operations 1-29

FM 5-34

NUCLEAR, BIOLOGICAL, CHEMICAL
CHEMICAL AGENTS

Table 1-7. Chemical agents' characteristics and defense

Type of Agent	Dissemi- nation Method	Means of Detection	Symptoms in Soldiers	Effects on Soldiers	Action Rate	Individual First Aid	Individual Decontami- nation	Protection Require- ment	US Agents Equivalent Symbol/ Name	Field Character- istics
Nerve	Aerosol or vapor	Chemical- agent detection kits and paper to detect liquids	Difficult breathing, drooling, nausea, vomiting, convulsions, and sometimes dim vision	Incapaci- tates; kills if high concen- tration is inhaled	Very rapid by inhala- tion; slow through skin	Give nerve agent antidote injection.	Nonpersis- tent—none needed	Protective mask and clothing	GA/Tabun GB/Sarin GD/Soman	Colorless
	Liquid droplet			Incapaci- tates; kills if conta- minated skin is not deconta- minated rapidly	Delayed through skin; more rapid through eyes	Artificial respira- tion may be neces- sary.	Persistent— Flush eyes with water. Decontami- nate skin using M258A1 kit or M291 skin decontami- taion kit (SDK).		VX Thickened G-agent	

1-30 Combat Operations

Table 1-7. Chemical agents' characteristics and defense (continued)

Type of Agent	Dissemination Method	Means of Detection	Symptoms in Soldiers	Effects on Soldiers	Action Rate	Individual First Aid	Individual Decontamination	Protection Requirement	US Agents Equivalent Symbol/Name	US Agents Equivalent Field Characteristics
Blister	Liquid droplet or vapors	Chemical-agent detector kits to detect vapors and aerosols; chemical-agent detector paper to detect liquids	Mustard, nitrogen mustard—no early symptoms. Lewisite, mustard lewisite—searing of eyes and stinging of skin. Phosgene oxime—irritation of eyes and nose.	Blisters, skin is destructive to respiratory tract; can cause temporary blindness. Some agents sting and form wheals on skin.	Blistering delayed hours to days, eye effects more rapid mustard lewisite and phosgene oxime very rapid	None	Flush eyes with water. Decontaminate skin with M258A1 kit or M291 SDK or wash with soap and water	Protective mask and clothing	Mustard (HD)	Pale yellow droplets
									Nitrogen-mustard (HN)	Dark droplets
									Lewisite (L), Mustard-lewisite (HL)	Dark, oily droplets
									Phosgene oxime (CX)	Colorless droplets

Combat Operations 1-31

FM 5-34

Table 1-7. Chemical agents' characteristics and defense (continued)

Type of Agent	Dissemination Method	Means of Detection	Symptoms in Soldiers	Effects on Soldiers	Action Rate	Individual		Protection Requirement	US Agents Equivalent	
						First Aid	Decontamination		Symbol/Name	Field Characteristics
Blood	Vapor (gas)	Chemical-agent detector kits to detect vapors and aerosols; chemical-agent detector paper to detect liquids	Convulsions and coma	Incapacitates; kills if high concentration is inhaled	Rapid	Mask, artificial respiration may be necessary	None	Protective mask	Hydrogen cyanide (AC) Cyanogen chloride (CK)	Colorless
Choking	Vapor (gas)		Coughing, choking, nausea, and headache	Damages and floods lungs	Immediate to 3 hours	For severe symptoms avoid movement and keep warm.	None	Protective mask	Phosgene (CG)	Colorless

1-32 Combat Operations

NBC REPORTS

Table 1-8. Line-item definitions

Line	Nuclear	Chemical & Biological	Remarks
A	Strike serial number	Strike serial number	Assigned by div NBC center
B	Position of observer	Position of observer	Use grid coordinates (UTM) or place.
C	Direction of attack from observer; include unit of measure	Direction of attack from observer	Nuc: DMN or MMN, DTN or MTN, DGN or MGN Chem: Direction measured clockwise from GN or MN (state which) in degrees or mils (state which)
D	DTG of detonation	DTG for start of attack	Nuc: Use Zulu time. Chem: State time zone used.
E	NA	DTG for end of attack	State time zone used.
F	Location of area attacked	Location of area attacked	Use grid coordinates (or place). State whether location is actual or estimated.
G	Suspected or observed event and means of delivery or kind of attack	Kind of attack	State whether attack was by artillery, mortars, rockets, missiles, bombs, or spray.
H	Type of burst	Type of agent/ type of burst: P or NP	Nuc: Specify air, surface, or subsurface. Chem: State if ground, air, or spray attack.
I	NA	Number of munitions or aircraft	If known
J	Flash-to-bang time	NA	Use seconds.
K	Crater present or absent and diameter	Description of terrain and vegetation	Nuc: Send in meters. Chem: Send in NBC 6.
L	Cloud width at H+5 min	NA	State if measure is deg or mils.
M	Stabilized cloud top or bottom angle or cloud top or bottom height at H+10 min.	NA	Nuc: State whether angle is cloud top or bottom. Chem: State if height is cloud top or bottom and if measured in m or ft.
N	Est yield	NA	Send as kt.
O	Reference date-time for estimated contour line when not H+1.	NA	Use when contours are plotted at H+1.
P	Radar purposes only	NA	
PA	NA	Predicted HA (coordinates)	If wind speed is 10 kmph or less, item is 010 (radius of the HA in km)
PAR	Coordinates of external contours of radioactive cloud	NA	Six-digit coordinates; letter R for radar set
PB	NA	Duration of hazard in attack area and HA	In days, hours, minutes, and so on.

Combat Operations 1-33

FM 5-34

Table 1-8. Line-item definitions (continued)

Line	Nuclear	Chemical & Biological	Remarks
PBR	Downwind direction of radioactive cloud and unit of measure	NA	DGM or MMN, DTN or MTN, DGN or MGN; letter R for radar set
Q	Location of reading	Location of sampling and type of sample	Nuc: UTM or place Chem: UTM or place; state if test was air or liquid.
R	Dose rate or actual value of decay exponent	NA	State dose rate in cGyph. See sample NBC 4 for terms associated with this line.
S	DTG of reading	DTG contamination detected	State time initial ID test sample or reading was taken.
T	H+1 DTG	DTG of latest contamination survey of the area	NBC 5 and 6 reports only
U	1,000-cGyph contour line	NA	Plot in red.
V	300-cGyph contour line	NA	Plot in green.
W	100-cGyph contour line	NA	Plot in blue.
X	20-cGyph contour line (30 cGyph contour line used by other NATO forces)	Area of actual contamination	Nuc: Plot in black. Chem: Plot in yellow.
Y	Direction of left and right radical lines	Downwind direction of hazard and wind speed	Nuc: Direction measured clockwise from GN to the left and then to the right radial lines (deg or mils, state which) 4 digits each. Chem: Direction, 4 digits (deg or mils) and wind speed, 3 digits (kmph)
Z	Effective wind speed. Downwind distance of Zone I Cloud radius (Include unit of measure for each category.)	NA	3 digits: Effective wind speed (kmph) 3 digits: Downwind distance of zone 1 (km or NM) 2 digits: Cloud radius (km or NM) If wind speed is less than 8 kmph, this line contains only 3-digit radius of zone 1 (km)
ZA	NA	Significant weather phenomena	Air stability (2 digits), temperature in centigrade (2 digits), humidity (1 digit), significant weather phenomena (1 digit), and cloud cover (1 digit).
ZB	Used to transmit correlation or transmission factors	Remarks	Include any additional information.

FM 5-34

Table 1-9. Types of NBC reports

Line	Nuclear	Chemical	Biological
	NBC 1 Report (Observer's Report)		
B	NB062634	LB200300	LB206300
C	90° Grid		
D	201405ZMAR96	201405ZMAR96	200410ZMAR96
E		201412ZMAR96	200414ZMAR96
F	LB206300Est	LB206300 Est	LB206300 Act
G	Aircraft	Bomblets	Aerial spray
H	Surface	Nerve, V, air burst	Unknown
J	60 sec		
L	15°		
M			

NOTE: Line items B, D, H, and either C or F should always be reported; other line items may be used if the information is known.

	NBC 2 Report (Evaluated Data)		
A	A024	B002	C001
D	201405ZMAR96	200945ZMAR96	201395ZMAR96
F	LB187486 Act	LB128456 Act	LB206300 Act
G	Aircraft	Bomblets	Unknown
H	Surface	Nerve, V, air burst	Unknown
N	50		
Y		0270° 015 kmph	
ZA		518640	

NOTES:
1. This report is normally based on two or more NBC 1 reports. It includes an attack location and, in the case of a nuclear detonation, an evaluated yield.
2. Refer to the chemical downwind message to determine cloud cover, significant weather phenomena, and air stability.

Combat Operations 1-35

FM 5-34

Table 1-9. Types of NBC reports (continued)

Line	Nuclear	Chemical	Biological
NBC 3 Report (Immediate Warning of Expected Contamination)			
A	A024	B002	C003
D	201405ZMAR96	201415ZMAR96	200530ZMAR96
F	WQ360540 Est	WQ350560 Act	WQ360540 to WQ368548
H		Nerve, V, air burst	Unknown
N	50		
PA		WQ555046	WQ555046
		WQ554050	WQ554050
		WQ635045	WQ635045
		WQ645044	WQ645044
PB		WQ060040	WQ060040
		In attack area, 2-4 days	
		In hazardous area, 1-2 days	
Y	027720312	0270°, 015 kmph	0270°, 015 kmph
Z	01902505		
ZA		518640	

NOTES:
1. If the effective windspeed is less than 8 kmph, line Z of the NBC 3 (nuclear) consists of three digits for the radius of zone 1.
2. If the windspeed is less than 10 kmph, line PA of the NBC 3 (chemical) is 010, which is the radius of the hazardous area.

NBC 4 Report (Reconnaissance, Monitoring, and Survey Results)

Line	Nuclear	Chemical	Biological
H		Nerve, V	Unk, susp, bio
Q	WQ354456	WQ354678, Liquid	WQ555046
R	35		
S	201535ZMAR96	170610ZMAR96	200630ZMAR96

NOTES:
1. Line items H, Q, R, and S may be repeated as often as necessary.
2. Radiation dose rates are measured in the open, with the instrument 1 meter above the ground.
3. In line R, descriptive words such as initial, peak increasing, decreasing, special, series, verification, or summary may be added.
4. If readings are taken inside a vehicle or shelter, also give the transmission factor.

1-36 Combat Operations

FM 5-34

Table 1-9. Types of NBC reports (continued)

NBC 5 Report (Areas of Actual Contamination)

Line	Nuclear	Chemical	Biological
A	A0012	B005	
D		200700ZMAR96	
H		Nerve, V, air burst	
S		201005ZMAR96	
T	201505ZMAR96	201110ZMAR96	
U			
V			
	ND651455		
	ND810510		
	ND821459		
	ND651455		
	ND604718		
W	ND991686		
	ND114420		
	ND595007		
X		ND206991	
		ND201576	
		ND200787	
		ND206991	

NOTE: This report is best sent as an overlay, if time and the tactical situation permits.

Combat Operations 1-37

FM 5-34

Table 1-9. Types of NBC reports (continued)

Line	Chemical or Biological
	NBC 6 Report (Detailed Information on Chemical or Biological Attacks)
A	B001
D	200945ZMAY96
E	200950ZMAY96
F	WQ450350, Act
G	Artillery
H	Nerve, V, air burst
I	20 rounds
K	Mostly small houses and barns, elevation 600 meters
M	Attack received as counterfire, enemy bypassed on right flank of attack area
Q	Liquid ground sample taken by detection team in attack area
S	201005ZMAY96
T	201110ZMAY96
X	As per overlay
Y	Downwind direction is 0090°, windspeed is 010 kmph.
ZB	This is the only chemical attack in our area.

NOTES:
1. This report is submitted only when requested.
2. This report is completed by battalion and higher NBC personnel. It is in narrative form, giving as much detailed information as possible for each line item.

1-38 Combat Operations

Alarms, Signals, and Warnings

Table 1-10. Alarms and signals

Type	Chemical/Biological	Nuclear
Vocal	Gas or Spray	Fallout
Sound	Succession of short signals • Metal to metal • Short horn blasts • Interrupted warbling siren	
Visual	Fists over shoulder or posted signs	
Audio/visual	M8A1, CAM, VDR2, M22 (CAA)	

MOPP Levels

Table 1-11. MOPP levels

MOPP Equipment	MOPP Ready	MOPP 0	MOPP 1	MOPP 2	MOPP 3	MOPP 4	Mask Only
Mask	Carried	Carried	Carried	Carried	Worn^1	Worn	Worn
Overgarment	Ready^3	Available^4	Worn^1	Worn^1	Worn^1	Worn	
Vinyl overboot	Ready^3	Available^4	Available^4	Worn	Worn	Worn	
Gloves	Ready^3	Available^4	Available^4	Available^4	Available^4	Worn	
Helmet protective cover	Ready^3	Available^4	Available^4	Worn	Worn	Worn	
CPU^2	Ready^3	Available^4	Worn^2	Worn^2	Worn^2	Worn^2	

^{1} In hot weather, coat or hood can be left open for ventilation.
^{2} The CPU is worn under the BDU (primarily applies to SOF, armor vehicle crewmen).
^{3} Must be available to the soldier within two hours; second set available in 6 hours.
^{4} Must be within arm's reach of soldier.

FM 5-34

NBC Markers

> This paragraph implements STANAGs 2002 and 2047

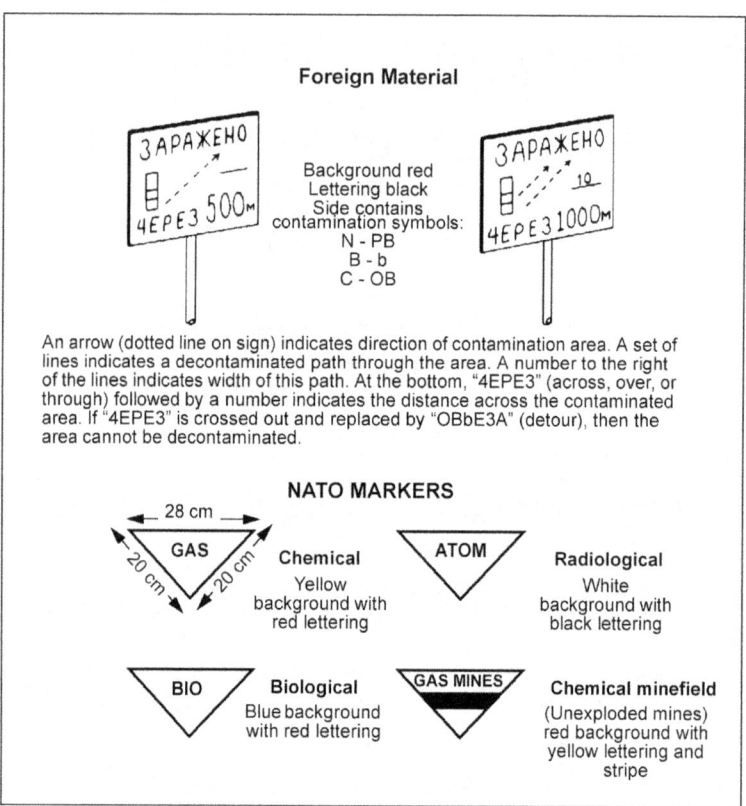

Figure 1-18. NBC markers

Unmasking Procedures

With Detector Kit

Use a chemical-agent detector kit (M256 series) to test for the presence or absence of chemical agents. After determining the absence of agents, use the following steps to check for chemical-agent symptoms:

1-40 Combat Operations

- Disarm the soldiers.
- Unmask one to two soldiers for 5 minutes and then remask them.
- Examine the soldiers in a shady area for chemical-agent symptoms for 10 minutes.
- Unmask the remainder of troops if no symptoms appear in the test soldiers. Continue to watch the soldiers for possible delayed symptoms.
- Have first-aid treatment immediately available, in case it is needed.

NOTE: **Bright light will cause the pupils to contract, which could be misinterpreted as a nerve-agent symptom.**

Without Detector Kit

Use the following steps for field-expedient unmasking:

- Use M8 paper to check for liquid contamination.
- Select one or two soldiers to take a deep breath, hold it, exhale, take a deep breath, and hold it.
- Breach the seal on their mask. Tell the soldiers to keep their eyes wide open for 15 seconds. Clear their masks and reestablish the seal.
- Wait for 10 minutes and watch for symptoms.
- Tell the selected soldiers to unmask for five minutes, if no symptoms appear. After 5 minutes, tell them to remask.
- Observe them for another 10 minutes for possible symptoms. If no symptoms develop, the rest of the soldiers can unmask. Keep watching for any chemical-agent symptoms.
- Have first-aid treatment immediately available, in case it is needed.

DECONTAMINATION

Equipment

Use issued items, whenever available, for expedient decontamination. Table 1-12 shows some natural decontaminants that are readily available and frequently occur in nature. While training on issue equipment (such as the DS-11 device) or concentrated bleaches (super tropical bleach [STB]), ensure that runoff does not lead directly to nearby streams or lakes. During combat operations, try to site or prepare deliberate decontamination areas to prevent runoff into drinking-water sources. Engineers typically assist chemical platoons which perform deliberate equipment decons by constructing sump pits or evaporation areas for containing contaminated runoff water.

Table 1-12. Natural decontaminants

Decontaminates	Use	Remarks	Cautions
Water	NBC	Flush contamination from surface with large amounts of water.	Effective in physically removing contamination; however, contamination is not neutralized.
Steam	NBC	Using steam and then scrubbing is more effective than using only steam.	Effective in physically removing contamination; however, contamination may not be neutralized.
Absorbents (earth, ashes, sawdust, rags, and similar materials)	Chemical	Use to physically remove gross contamination from surfaces.	The contamination is transferred from the surface to the absorbent. The absorbent becomes contaminated and must be disposed of accordingly. Sufficient contamination to produce casualties may well remain on surfaces.

Personnel

Decontaminate personnel using the buddy system and the following procedure:

Step 1. Remove and decontaminate the gear. Cover it with STB dry mix, and brush or rub it into the material. Shake off the excess. Set the gear aside on an uncontaminated surface.

Step 2. Decontaminate the hood. Use an M258A1 skin decontamination kit or M295 individual equipment decontamination kit (IEDK). Decontaminate any exposed areas of the protective mask. Use the instructions and directions in each IEDK for the decontamination steps and procedures. Lift the hood up off your buddy's shoulder by grasping the straps and pulling the hood over his head until the back of his head is exposed. Roll the hood tightly around the mask.

NOTE: Control contamination from spreading by putting all contaminated overgarments and towelettes in one pile.

Step 3. Remove the overgarment. Remove your buddy's jacket, placing it on the ground, black side up. Remove the trousers one leg at a time. Discard the trousers in a centralized pile to avoid spreading contamination.

Step 4. Remove the overboots and gloves. Cut the strips off your buddy's boots and pull them off. Have your buddy step on the jacket as you pull off the boots. Remove the gloves. Discard the boots and gloves into a centralized pile.

Step 5. Put on the overgarments. Open a package of new overgarments; do not touch them. Have your buddy dress while still standing on the old overgarments (Step 3).

Step 6. Put on the overboots and gloves. Open a package of new boots and gloves; do not touch them. Have your buddy put on new boots and gloves. He can step off the overgarments once his boots and gloves are on.

Step 7. Secure the hood. Decontaminate your gloves using an M258A1 skin decontamination kit. Unroll your buddy's hood and attach the straps. He should check all zippers and ties on the hood and overgarment to ensure that they are closed.

Step 8. Reverse roles. Repeat Steps 2 through 7 with your buddy helping you through the steps.

Step 9. Dig a large hole. Place all the contaminated clothing and discarded towelettes in the hole and cover them. Mark it as a contaminated area. You can burn all the contaminated clothing if you use a slow-burning fuel (kerosene or diesel fuel). DO NOT USE GASOLINE; it burns too quickly. Commanders must warn downwind units of possible downwind vapor hazards, if burning is accomplished.

Step 10. Secure the gear. Move to an assembly area, time and situation permitting. The unit can now perform unmasking procedures to get relief from the protective mask.

MEDICAL PROCEDURES

GENERAL FIRST-AID PROCEDURES

Table 1-13. First aid, symptoms with treatment

Problem	Symptom	First Aid
Common Wounds and Injuries		
Head wound	Possible scalp wound, headaches, recent unconsciousness, blood or fluid from ears or nose, slow breathing, vomiting, nausea, convulsions	Leave any brain tissue as is and cover with a sterile dressing. Secure the dressing, and make sure that the victim's head is higher than his body.
Jaw wound		Elevate the head slightly, clear the airway, control the bleeding, and protect the wound. Position the victim's head to allow for drainage from his mouth. **Do not give morphine.** Treat for shock, as needed.
Belly wound		Leave all organs as they are, and loosely place a sterile dressing over them. Do not give any food or liquid to the victim. Leave him on his back with his head turned to one side.
Chest wound (sucking)		Have the victim breathe out and hold his breath, if possible. Seal the wound, airtight, with plastic or foil. Cover the wound with a dry, sterile dressing. Secure it with bandages around the victim's body. Ensure that the wound is airtight and fully covered.
Burns and Heat Injuries		
Burns	1st degree—red skin 2nd degree—blistered skin 3d degree—destroyed tissue	Do not remove the clothes around the burn area. Do not apply grease or ointment. Cover the burn with a sterile dressing. Give the victim cool salt/soda water.
Heat cramps	Muscle cramps of the abdomen, legs, or arms	Move the victim to a shady area and loosen his clothing. Give him large amounts of cold salt water. (Dissolve two salt tablets or 1/4 teaspoon of table salt in a canteen of cool water.)
Heat exhaustion	Headache, excessive sweating, weakness, dizziness, nausea, and muscle cramps; pale, cool, and moist clammy skin	Lay the victim in a shady area and loosen his clothing. If he is conscious, ensure that he drinks three to five canteens of cool salt water in a 12-hour period. Prepare salt water as for heat cramps.
Heatstroke (sunstroke)	No sweating (hot, dry skin); collapse and unconsciousness may come suddenly or may be preceded by headache, dizziness, fast pulse, nausea, vomiting, and mental confusion.	Promptly immerse the victim in the coldest water possible. Add ice, if available. If you cannot immerse the victim, move him to a shady area, remove his clothing, and keep him wet by pouring water over his entire body. Fan his body continuously. Transport him to the nearest medical facility at once, keeping him cool during transport. If he becomes conscious, give him cool salt water, prepared as for heat cramps.

Table 1-13. First aid, symptoms with treatment (continued)

Wet- or Cold-Weather Injuries		
Problem	Symptom	First Aid
Frostbite	Skin is white, stiff, and numb.	Face—cover the frostbitten area with your warm hands until the victim feels pain in that area. Hands—place the frostbitten bare hands next to the skin in the victim's opposite armpits. Feet—seek a sheltered area. Place the victim's bare feet under the clothing or against the abdomen of another person. Deep frostbite—protect the part from additional injury and get the victim to a medical treatment facility immediately. Do not attempt to thaw deep frostbite. It is dangerous for a victim to walk on his frostbitten or thawed feet.
Immersion foot	Soles of the feet are wrinkled. Standing or walking is extremely painful.	Dry the victim's feet thoroughly, and get him to a medical treatment facility immediately. Try not to let him walk.
Trench foot	Numbness, possible tingling or aching sensation, cramping pain, and swelling	Treat as for immersion foot.
Snow blindness	Scratchy feeling in eyes	Cover the victim's eyes with a dark cloth. Transport him to a medical treatment facility at once.

Table 1-14. First aid, treatments

Basic Problems	
Problem	First Aid
Blocked airway	Extend the neck, turn the head to the side, and clear all refuse from the mouth. Open the airway; restore the breathing and heartbeat.
Bleeding	Apply direct pressure on the wound with a sterile dressing. Elevate the victim so that the wound is above the heart. Use a tourniquet as a last resort.
Wounds	Expose the wound, control the bleeding, apply a sterile dressing, and treat for shock. Do not clean the wound.
Fractures	Splint the break where and how it lies. Do not move the patient, if possible. Immobilize the joint above and below the fracture. Cover any exposed bones or open wounds.
Shock	Lay the patient on his back, elevate his feet, loosen his clothing, and keep him warm. Feed him hot liquids, if he is conscious. Turn his head to the side if he is unconscious.

Combat Operations 1-45

FM 5-34

Table 1-14. First aid, treatments (continued)

Problem	First Aid
Stings and Bites	
Black widow or brown recluse bite	Keep the victim quiet. Place an ice or a freeze pack, if available, around the area of the bite (helps stop the venom from spreading). Transport the victim to a medical treatment facility immediately.
Scorpion sting or tarantula bite	Apply an ice or a freeze pack, if available, for an ordinary scorpion sting or tarantula bite. (A baking-soda paste applied to the wound may relieve the pain.) If the sting or bite is on the face, neck, or genital organs or if the sting is from a South American scorpion, keep the victim as quiet as possible, and transport him to a medical treatment facility immediately.
Snake bite	Keep the victim quiet and reassure him. Place an ice or a freeze pack, if available, around the area of the bite. Immobilize the affected part in a position below the heart level, if possible. If the bite is on an arm or a leg, place a lightly constricting band (boot lace or strip of cloth) between the bite area and the heart, about 2 to 4 inches above the bite area. Tighten the band so as to stop blood flow near the skin but not so tight as to stop arterial flow or the pulse. Transport the victim to a medical treatment facility immediately. Kill the snake, if possible, without damaging the head; take it to the treatment facility.
Bee or wasp sting	Watch the victim. Treatment is usually not needed. Treat for shock if abnormal reactions occur.
Other Conditions	
Blisters	Do not open blisters unnecessarily; they are sterile until opened. If you must, be careful. Wash the area thoroughly with soap and water; apply an antiseptic to the skin. Sterilize a needle, in the open flame of a match, and puncture the blister at the edge. Use a sterile gauze pad and apply pressure along the margin of the blister to remove the fluid. Place a sterile dressing over the area. Do not attempt self-help for blisters that are in the center palm area.
Boils	Do not squeeze a boil; doing so may drive bacteria into the blood stream and cause internal abscesses or bone infection. This action is dangerous if the boil is around the nostrils, upper lip, or eyes, as the blood stream in these areas leads to the brain area. Relieve discomfort from small boils by applying warm compresses moistened in an Epsom salt solution (1 teaspoon salt to 1 pint of warm water) at 15-minute intervals. Do not apply compresses to facial boils unless you are under medical direction. If a boil breaks, wipe the pus away with a sterile pad that is moistened with rubbing alcohol. Work from healthy skin toward the boil and pus area. Apply a sterile dressing over the boil.
Unconsciousness	Apply lifesaving measures as appropriate. If the victim remains unconscious, place him on his abdomen or side with his head turned to one side to prevent choking on vomit, blood, or other fluid. If he has an abdominal wound, place him on his back with his head turned to one side. Transport him to a medical treatment facility immediately. Do not give an unconscious victim any fluids by mouth. If a victim has fainted, he should regain consciousness within a few minutes. If you have an ammonia inhalant capsule, break it and place it under his nose several times for a few seconds. If the victim is sitting up, gently lay him down, loosen his clothing, and apply a cool, wet cloth to his face. Ensure that he lies quietly. When a victim is in a sitting position and is about to faint, lower his head between his knees. Hold him so he doesn't fall.

1-46 Combat Operations

CARDIOPULMONARY RESUSCITATION (CPR) PROCEDURES

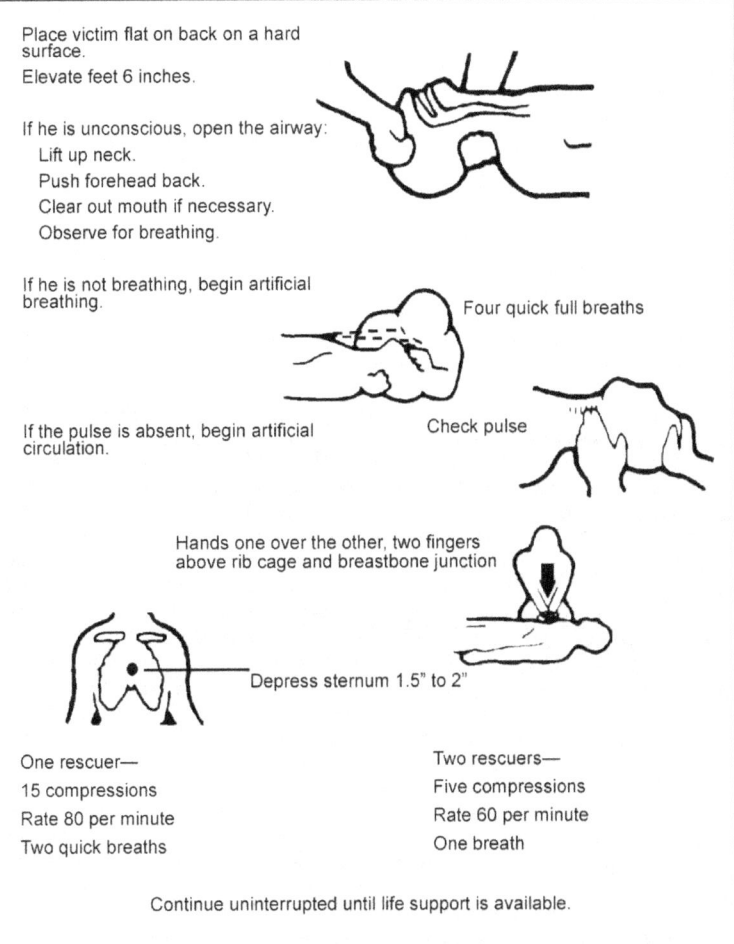

Figure 1-19. CPR in basic life support

MEDICAL EVACUATION (MEDEVAC)

Table 1-15 lists the precedence and types of MEDEVAC report entries. Table 1-16 lists line entries for the MEDEVAC request form.

Table 1-15. MEDEVAC report entries

Precedence	Type
Urgent—evacuation is required as soon as possible but NLT 2 hours to save life, limb, or eyesight.	**Peacetime**—actual patient/report may be transmitted in plain text.
Priority—evacuation is required within 4 hours or the patient's medical condition could deteriorate to **Urgent**.	**Wartime**—during wartime or training exercises, report must be transmitted, secured or encrypted.
Routine—evacuation is required within 24 hours.	
Tactical Immediate—a patient's medical condition is not **Urgent** or **Priority**, but evacuation is required ASAP so as not to endanger the unit's tactical mission.	

FM 5-34

Table 1-16. MEDEVAC request form

Line	Item	Explanation	Where/How Obtained	Provided By	Reason
1	Location of pickup site	Encrypt the grid coordinates of the pickup site. When using a numerical or digital cipher, use the same set line to encrypt the grid-zone letters and the coordinates. To avoid a misunderstanding, include a statement that grid-zone letters are included in the message (unless the unit SOP specifies its use at all times).	Map	Unit leader(s)	Required for the unit that is coordinating for the evacuation vehicle so that the vehicle is dispatched to location(s) of the casualty/patient
2	Radio frequency, call sign and suffix	Encrypt the frequency of the radio at the pickup site, not a relay frequency. Transmit the call sign in the clear (and suffix, if used) of the contact person at the pickup site.	SOI	RTO	Required so that the driver of the evacuation vehicle, while enroute, can contact the requesting unit for more information, such as change in the situation or directions
3	Number of patients by precedence	Includes only applicable information. The appropriate amount(s) and brevity numbers are encrypted: (#) - 1 - URGENT (#) - 2 - PRIORITY (#) - 3 - ROUTINE If you report two or more categories in the same request, insert "BREAK" between each category.	Evaluation of patient(s)	Medic or senior person present	Required by the unit that controls the evacuation vehicles to assist in prioritizing missions when more than one is received
4	Special equipment required	Encrypt the appropriate brevity number(s): 5 - None 6 - Hoist 7 - Stokes litter 8 - Forest/jungle penetrator	Evaluation of patient(s)	Medic or senior person present	Required so that the equipment can be placed on board the evacuation vehicle before the start of the mission (The semirigid litter is not part of a unit's TOE equipment and is not normally carried aboard the aircraft.)

Combat Operations 1-49

FM 5-34

Table 1-16. MEDEVAC request form (continued)

Line	Item	Explanation	Where/How Obtained	Provided By	Reason
5	Number of patients by type	Includes only applicable information. Appropriate amount(s) and brevity numbers are encrypted: (#) - 9 - Litter (#) - 0 - Ambulatory (sitting) If requesting MEDEVAC for both types, insert "BREAK" between each entry.	Evaluation of patient(s)	Medic or senior person present	Required so that the appropriate number of vehicles will be dispatched to the pickup site and configured to carry the patients requiring evacuation
6 Wartime	Security of pickup site	1 - No enemy troops in the area. 2 - Enemy troops possibly in the area, approach with caution. 3 - Enemy troops in the area, approach with caution. 4 - Enemy troops in the area, armed escort required	Evaluation of situation	Unit leader(s)	Required to assist the evacuation crew in determining if assistance is required to accomplish the mission; crews must be updated while enroute.
6 Peace-time	Number and type of wound, injury, or illness	Specific information regarding patient's wounds by type, such as gunshot and shrapnel; for serious bleeding patients, report blood type, if known.	Evaluation of patient(s)	Medic or senior person present	Required to assist evacuation personnel in determining treatment and special equipment needed
7	Method of marking pickup site	Appropriate brevity number(s) must be encrypted: 0 - Tree branches, pieces of wood, or stones place together 1 - Signal lamp or flashlight 2 - Vehicle lights 3 - Open flame 5 - Panels 6 - Pyrotechnic signal 7 - Smoke signal 8 - Signal person 9 - Strips of fabric or parachute	Situation and available materials	Unit leader(s)	Required to assist the evacuation crew in identifying the specific location of the pickup; note that the color of the panels and smoke should not be transmitted until the vehicle contacts the unit (just before its arrival). For security, the crew should identify the color, and the unit should verify it.

1-50 Combat Operations

Table 1-16. MEDEVAC request form (continued)

Line	Item	Explanation	Where/How Obtained	Provided By	Reason
8	Patient's nationality and status	The number of patients in each category need not be transmitted. The appropriate brevity number(s) is encrypted: 4 - US military 5 - US civilian 6 - Non-US military 7 - Non-US civilian 8 - EPW	Evaluation of patient(s)	Medic or senior person present	Required to assist in planning for destination the facilities and the need for guards; the unit requesting support should ensure that there is an English-speaking representative at the pickup site
9 Wartime	NBC contamination	Used only when applicable. The appropriate brevity number(s) is encrypted: 9 - Nuclear 0 - Biological 1 - Chemical	Evaluation of situation	Medic or senior person present	Required to assist in planning for the mission; which evacuation vehicle will accomplish the mission and when it will be used is determined.
9 Peacetime	Terrain description	Details of terrain features in and around the proposed landing site are included. If possible, the relationship of the site to prominent terrain features, such as a lake, mountain, or tower, should be described.	Area survey	Personnel at pickup site	Required to allow evacuation personnel to assess the route/AA into the area, of particular importance if hoist operation is required

Combat Operations 1-51

FIELD-SANITATION FACILITIES

Figure 1-20 shows field latrines. Keep all latrines at least 100 meters away from food operation, downhill, and at least 30 meters from groundwater sources. Keep latrines clean, and use residual insecticides to control insects. Once the latrine is full to 1 foot below the surface, or is to be abandoned, remove the box and spray 2 feet around the pit area. Fill the pit with successive 3-inch layers of compacted soil. Mound the pit with at least 1 foot of dirt and spray it with insecticide. Place a sign on top of the mound indicating the type, date closed, and unit. When high-water tables preclude the use of pit latrines, you may use burn-out latrines. Install half of a 55-gallon drum or barrel under each hole in the latrine box. Remove the drum daily, add fuel oil, and burn the contents to a dry ash. Add 1 inch of diesel fuel for insect control before replacing the drum in the latrine box. Construct the handwashing facilities and the shower unit (see Figures 1-21 and 1-22). For more information, see FM 21-10.

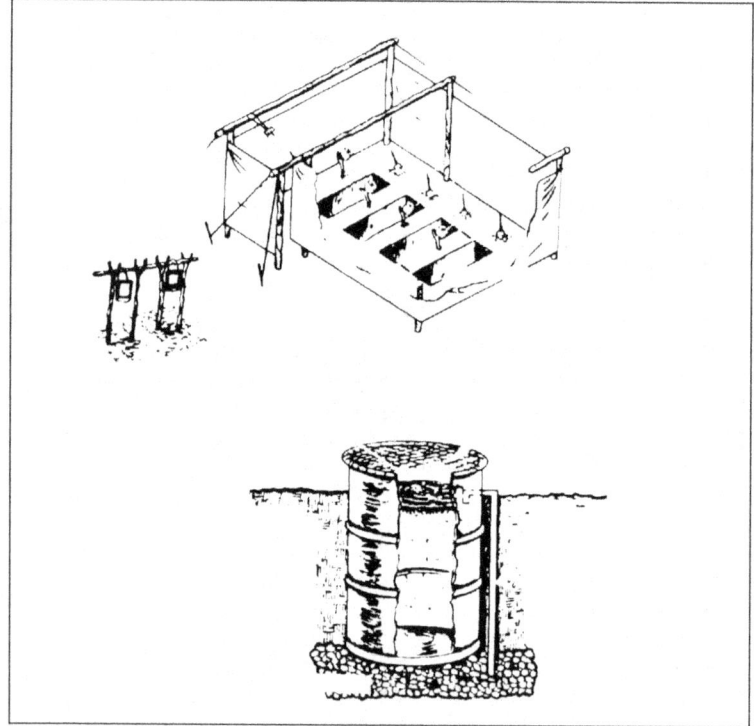

Figure 1-20. Field latrines

FM 5-34

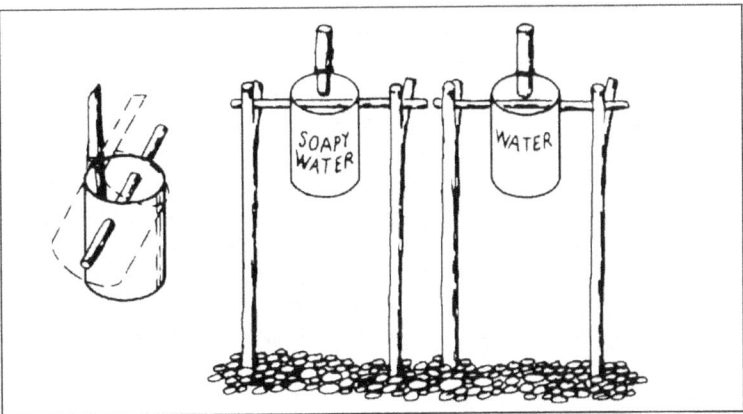

Figure 1-21. Hand-washing device, using No. 10 can

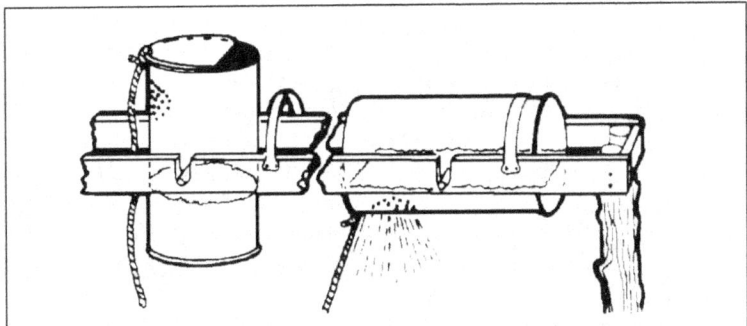

Figure 1-22. Shower unit, using metal drums

WATER DISINFECTION AND QUANTITY REQUIREMENTS

Calcium Hypochlorite

Use the following procedure to purify water in a 1-quart canteen with calcium hypochlorite ampules:

- Fill the canteen with the cleanest, clearest water available, leaving an air space of 1 inch or more below the neck of the canteen.

- Fill a canteen cup half full of water, and add the calcium hypochlorite from one ampule. Stir until dissolved.

- Fill the cap of a plastic canteen half full of the solution in the cup, and add it to the water in the canteen. Then place the cap on the canteen and shake it thoroughly.

Combat Operations 1-53

- Loosen the cap slightly, and invert the canteen, letting the treated water leak onto the threads around the neck of the canteen.
- Tighten the cap on the canteen, and wait at least 30 minutes before using the water for any purpose.

Iodine Tablets

Use one tablet per quart canteen for clear water and two tablets per quart canteen for cloudy water. Let the water stand for 5 minutes, shake well, allowing spill over to rinse canteen neck, and let stand another 20 minutes before using for any purpose.

Boiling

Bring the water to a rolling boil for 15 seconds.

Daily Water Requirements

Table 1-17. Daily water requirements

Element	Conditions of Use	Gallons per Day		Remarks
		Mild/Cold	Desert/Jungle	
Soldier	In combat— Minimum Normal March— Temporary camp Temporary camp Semipermanent camp Permanent camp	0.5 - 1 2 3 2 5 15 30-80 60-100	2 -3[1] 3 - 4[1] 6[2] 5[2]	Eating and drinking (3 days) When field rations are used Drinking plus cooking and personal hygiene Minimum for (all purposes does not include bathing) Waterborne sewage system and bathing
Vehicle	Level and rolling Mountainous	0.125 - 0.5 0.25 - 1		
Hospital	Drinking and cooking Water, waterborne sewage	10/bed 50/bed		Does not include bathing; includes medical personnel

[1]For unclimatized personnel or for all personnel when the dry-bulb reading exceeds 105° in the jungle
[2]Maximum consumption factor depends on the work performed, solar radiation, and other environmental stresses.

COMMUNICATIONS

Tactical communication responsibilities are—

- Senior to subordinates.
- Supporting to supported.

1-54 Combat Operations

- Reinforcing to reinforced.
- Lateral left to right if the standing operating procedures (SOP) or orders do not specify.

ANTENNA LOCATIONS

For maximum reception, place an antenna as high as possible and avoid any valleys. Locate an antenna away from built-up areas, metal obstructions, or electrical power lines.

EXPEDIENT ANTENNAS

Figure 1-23 and Figures 1-24 through 1-26, pages 1-56 and 1-57, show various antennas. To determine the antenna length (meters), use the following formula:

$$1/4 \, wave = {}^{234}/_{freq}; \quad 0.5 \, wave = {}^{468}/_{freq}; \quad full \, wave = {}^{936}/_{freq}$$

where—

freq = frequency, in megahertz

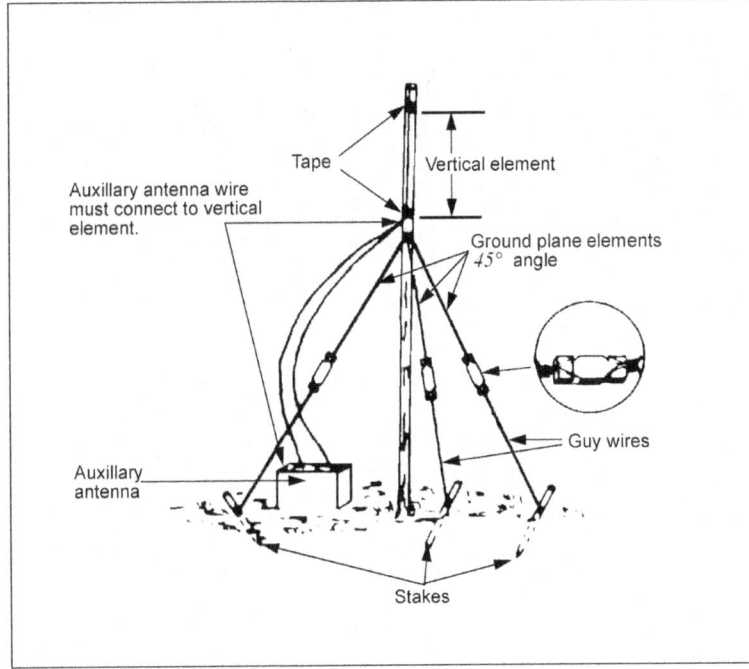

Figure 1-23. Jungle-expedient antenna (FM)

FM 5-34

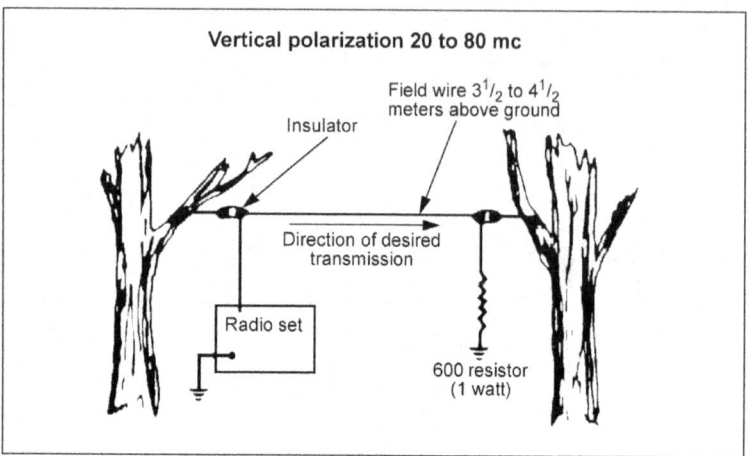

Figure 1-24. Long-wire antenna (FM)

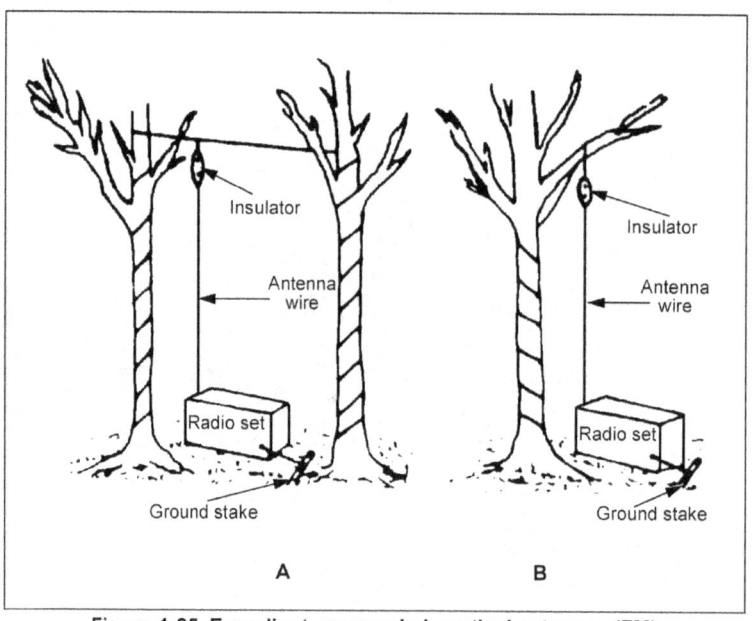

Figure 1-25. Expedient, suspended, vertical antennas (FM)

1-56 Combat Operations

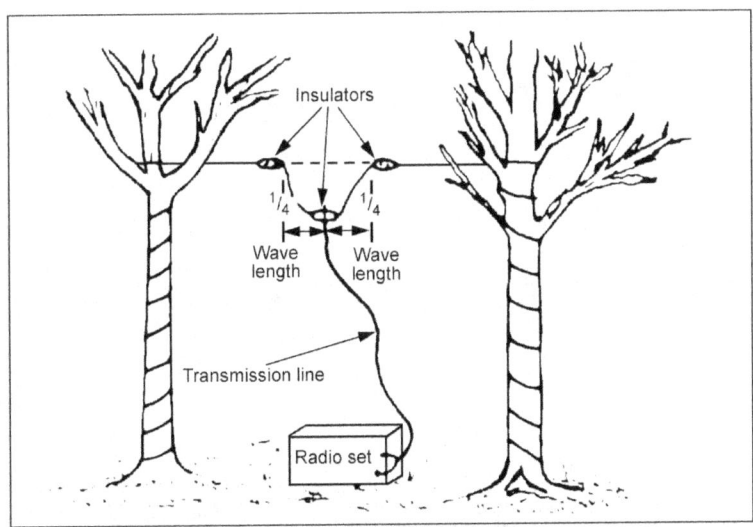

Figure 1-26. Improvised, center-fed, half-wave antenna (AM)

COMMUNICATIONS EQUIPMENT

Table 1-18 and Tables 1-19 and 1-20, page 1-58, list information on pieces of communication equipment.

Table 1-18. Communication equipment, tactical radio sets AN/VRC-12 series

Nomenclature	Frequency Range (MHz)	Range (km)
AN/PRC-77 series	30-75-95	8
AN/VRC-46	30-75-95	32
AN/VRC-47	30-75-96	32
AN/VRC-106	2.0-29.999	80
AN/VCR-142	2.0-29.999	80
AN/PRC-119A	30-87.975	10
AN/VRC-87A/C	30-87.975	10
AN/VRC-88A/C	30-87.975	10
AN/VRC-89A	30-87.975	10
AN/VRC-90A	30-87.975	40
AN/VRC-91A	30-87.975	10
AN/VRC-92A	30-87.975	40
Note: The AN/PRC-77 series includes the AN/VRC-64 (vehicular) and AN/GRC-160 (vehicular and man-pack).		

FM 5-34

Table 1-19. Communications equipment, auxillary

Nomenclature	Description	Range	Remarks
AN/GRA-39	Remoting set used with FM radio sets	Up to 3.2 km	Increases flexibility of radio sets; increases security. Radio and antenna can be exposed while operation is not.
RC-292 OE-254	General purpose, stationary ground, plane antenna	See remarks	Extends the range of tactical FM radio sets. Increases range of radio sets to about twice the stated planning range of the radio set. Radiating and ground plane elements must be of the proper length for a particular operating frequency.
AT-964	Long-wire, end-fed directional antenna	See remarks	Used with tactical FM radio sets. Good for reducing the enemy's ability to conduct interception and jamming. Can extend the planning range of radio sets by double or more, depending on the antenna used to receive/transmit at the distant site.

Table 1-20. Communications equipment, wire

Nomenclature	Description	Range	Remarks
TA-1/PT	Sound-powered telephone in handset form	16 km	Planning range depends on the condition of the wire (WD-1/TT). No batteries are required. Incoming signal is visual and adjustable audible. Telephone weighs 2.75 lb; case weighs 0.875 lb.
TA-312/PT	Tactical field telephone	35 km	Planning range depends on the condition of the wire (WD-1/TT). Batteries are required when operation is in LB position, as in local circuit to SB-22/PT. Incoming signal is adjustable audible. Has hands-free operation capability. Telephone weighs about 9.5 lb.
SB-22/PT	Lightweight, manual (monocord) switchboard, LB options		Switchboard has 12-circuit capability and may be expanded by stacking additional SB-22s. Each added SB-22 increases capability by 17 circuits, since only one operator's pack is necessary. Signaling may be audible and visual, or just visual.
SB-993-GT	Light, portable, emergency switchboard		Switchboard has 6-circuit capability for LB telephone lines, with an additional circuit plug for the operator's use. Incoming signal is visual only.

AUTHENTICATION

Authentication is mandatory in the following instances:

- Imitative deception is suspected.
- Initial enemy control and amplifying reports are made.
- Transmissions are made to order or end any radio silence.
- Plain messages are made to cancel other messages.

1-58 Combat Operations

FM 5-34

- A classified, uncoded message, such as changing frequencies and directing movements, is received.
- Initial radio contact is made, a net is opened and closed, or a transmission is made to a station that is under radio listening silence.
- A challenge is made
- A station's identity is doubted.

Figure 1-27 shows a chart to use in the authentication process.

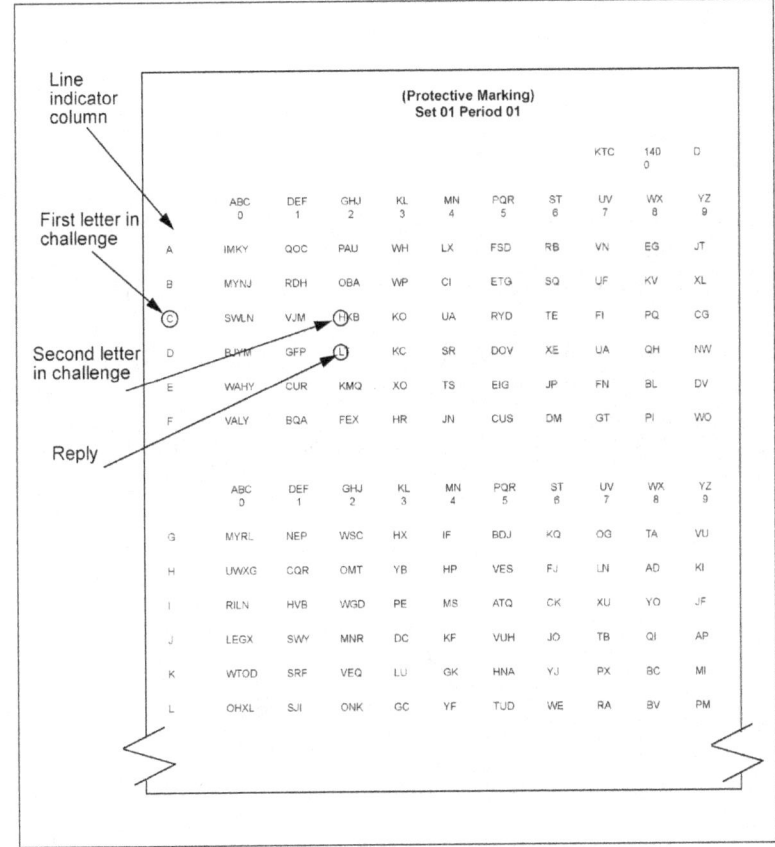

Figure 1-27. Authentication procedures

Combat Operations 1-59

When challenging, select two random letters, except Z, before transmitting. Make sure you know what the reply should be. Transmit a challenge, "AUTHENTICATE CHARLIE-HOTEL, OVER." The receiving station must reply, "I AUTHENTICATE LIMA, OVER." If authentication is incorrect or reply is not received promptly, transmit another challenge. If the next reply is incorrect or untimely, notify your supervisor, commander, or communication-electronics operation (CEO)

NOTE: When a challenge is from the last line, you must go to the first line for the reply.

STANDARD RADIO-TRANSMISSION FORMAT

CALL

MESSAGE - Indicates messages require recording

PRECEDENCE - Indicates priority of call

TIME - Followed by date-time group

FROM - Followed by call sign

TO - Followed by call sign of addressee

BREAK

TEXT - May consist of plain language, code, or cipher groups

BREAK

ENDING - Must include either OVER or OUT but never both in the same transmission.

Example: ZULU FOUR CHARLIE ONE SIX - THIS IS DELTA THREE X-RAY TWO NINE - MESSAGE - PRIORITY - TIME 181345Z - BREAK - FIGURES 6 STRINGERS NEEDED AT MY LOCATION ASAP - BREAK OVER.

SINGLE-CHANNEL, GROUND-TO-AIR RADIO SYSTEM (SINCGARS)

SINCGARS is a very high frequency (VHF) frequency-hopping tactical radio, available in manpack, vehicular, and airborne configurations. Figure 1-28, page 1-62, shows the radio transmitter's (RT's) front panel.

Table 1-21. SINGCARS, general information

Frequency	30 to 87.975 Mhz
Channels	2320
Presets	8 single channels 6 frequency-hopping channels/nets
Modes	2,400; 4,800; 9,600; and 16,000 bps; fixed and frequency-hopping; plain text; cipher-text encryption
Remote fill	Over-the-air rekeying

FM 5-34

Table 1-22. SINCGARS radio sets

AN/PRC-119	Manpack radio
AN/VRC-87	Short-range vehicular radio
AN/VRC-88	Short-range vehicular radio w/dismount
AN/VRC-89	Short-range/long-range vehicular radio
AN/VRC-90	Long-range vehicular radio
AN/VRC-91	Short-range/long-range vehicular radio w/dismount
AN/VRC-92	Long-range/long-range vehicular radio

Table 1-23. Voice transmission maximum planning ranges

Radio Type	RF Switch Position	Planning Ranges
Manpack/vehicular	LO (low)	200 m to 400 m
	M (medium)	400 m to 5 km
	HI (high)	5 km to 10 km
Vehicular only	PA (power amplifier)	10 km to 40 km

Table 1-24. Data transmission maximum planning ranges

Radio Type	Baud Rate Used	RF Switch Position	Planning Ranges
Manpack/vehicular (short range)	600 to 4800 bps 16,000 bps (16 kbps)	HI	3 km to 5 km
Vehicular (long range)	600 to 2400 bps 4800 bps 16,000 bps (16 kbps)	PA PA PA	5 km to 25 km 5 km to 25 km 3 km to 10 km

Loading Frequencies—Manual (MAN), CUE, and 1-6

- Obtain authorized operating frequency from signal operation instructions (SOI) or network control station (NCS).
- Set function (FCTN) to load (LD) (see Figure 1-28, page 1-62).
- Set MODE to single channel (SC).

Combat Operations 1-61

FM 5-34

- Set channel (CHAN) to MAN, CUE, or desired channel (1-6) where frequency is to be stored.
- Press frequency (FREQ); display will show "00000" or the RT's tuned-in frequency).
- Press clear (CLR) (display will show five lines).
- Enter the numbers of the new frequency (using keyboard buttons). (If you make a mistake while entering a frequency, press CLR to delete the last digit entered.)
- Press store (STO) (display will blink and show the frequency you just stored).
- Repeat steps 1 thru 8 for additional frequencies that you wish to load.
- Set FCTN to squelch (SQ) ON (or normal operating position).

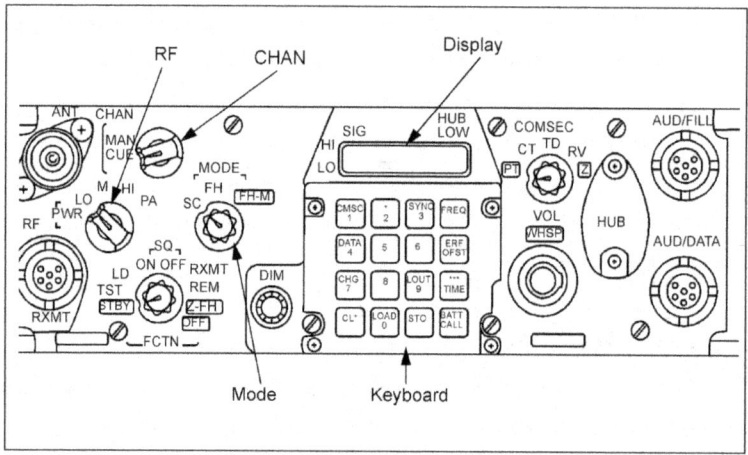

Figure 1-28. RT front panel

Clearing Frequencies

- Set MODE to SC.
- Set CHAN to MAN, CUE, or desired channel (1-6) where frequency is to be cleared.
- Press FREQ.
- Press CLR.
- Press LOAD; then press STO.
- Set FCTN to SQ ON (or normal operating position).

1-62 Combat Operations

FM 5-34

LOADING FREQUENCY HOP DATA (LOCAL FILL)

- Make sure that the electronic counter-countermeasure (ECCM) (see Figure 1-29, page 1-64) fill device is loaded.
- Set the ECCM fill-device function switch to OFF.
- Connect the ECCM fill device to the RT's connector audio (AUD)/FILL using a fill cable. (Always use a fill-device cable to connect a fill device to the RT. Equipment damage may result if you do not use a fill cable.)
- Set the RT's FCTN to LD.
- Set the RT's MODE to frequency hopping (FH).
- Set CHAN to where you load data (National Communications System [NCS] will direct you).
- Set the ECCM fill-device select switch to a position containing the desired data.
- Set ECCM fill-device function switch to ON.
- Press LOAD; the display will cycle, and you will hear a beep.
- Press STO; the display will blink and show STO followed by the first digit of the data.
- Change the ECCM fill-device select switch to a position containing the next data you want.
- Press LOAD; the display will cycle, and you will hear a beep.
- Press STO; then press the number button of the channel where you will store the data. The display will blink and show STO followed by the channel number in which the data was stored.
- Set the ECCM fill-device function switch to OFF.
- Disconnect the ECCM fill device.
- Set the RT's switches as needed for normal operation.

Loading Communications-Security (COMSEC) Keys (Local Fill)

- Turn off the fill device (KYK-13 or KYX-15); connect a fill cable to the RT's AUD/FILL connector.
- Set the RT's switches: FCTN to LD, COMSEC to cipher text (CT). The following applies to COMSEC alarms:
  —If you hear a COMSEC alarm (beeping alarm), key the handset twice for a minimum of $1/_2$ second each key. If a good COMSEC key is already in the radio, the alarm will clear to no alarm. If the radio does not have a good COMSEC key, the alarm will clear to a steady tone.

Combat Operations 1-63

FM 5-34

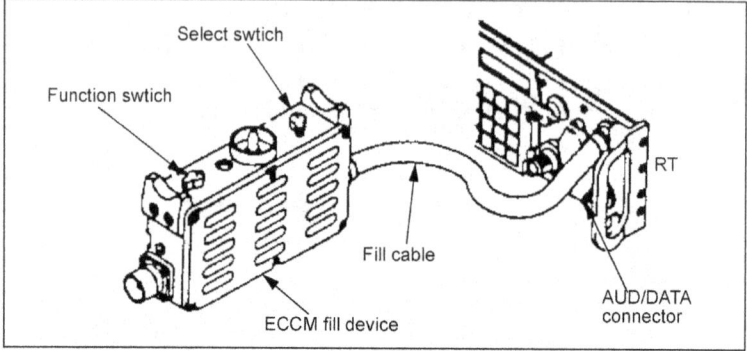

Figure 1-29. ECCM fill device connected to RT

—When you hear a steady tone at the handset, you can only load a COMSEC key. The steady tone will clear to no alarm if you successfully complete a fill procedure.

—If the COMSEC alarm will not clear, set FCTN to standby (STBY) and then to test (TST). If TST results in a *FAIL 5* display, there is a COMSEC failure. If *GOOD* is displayed, resume normal operation.

• Set the fill device controls (KYK-13, or KYX-15), set MODE to ON, and select the COMSEC key that you must load.

• Load the COMSEC key: Press the RT's LOAD; the display flashes LOAD then H TEK (or H KEK). Press STO and enter the channel number where you will store the COMSEC key. The display will blink, and you will hear two beeps.

NOTE: When changing the COMSEC key for during-operation updates, use the procedures as for a fill device.

Cold-Start Net Opening

After loading the necessary FH data into your RT, the NCS will send more data to your RT in preparation for a cold-start net opening. This is called an electronic remote fill (ERF). The NCS will direct you through the net opening.

• Load the following elements into your RT upon NCS opening alert:

—MAN channel frequency.

—CUE channel frequency, if designated by the commander.

—FH data from the fill device.

—COMSEC key from the fill device.

FM 5-34

- Set FCTN to SQ ON and then to LD.
- Set COMSEC to CT, if required.
- Set CHAN to MAN and MODE to FH; the display will show *COLD*.
- Stand by on MAN channel. The NCS will call you on MAN channel to verify communication and tell you where to store the ERF.
- Monitor the MAN channel and wait until the NCS comes back to you if you do not receive the ERF. The NCS will transmit the ERF to your RT. The display will show *HF234*, telling you that you received the ERF.
- Press STO to store the ERF.
- Enter the number 1 after the display readout asks you where you want to store the ERF. The display will change and you will hear a beep. Your own primary net is normally stored in CHAN 1. When entering other nets, use CHAN 2-6, as desired.
- Set CHAN to position 1, on command. The display will show *F234*.
- Stand by; the NCS will contact you to confirm communication on channel 1.
- Set, on command, the FCTN to SQ ON (or normal operating position).

CUE Frequency

Use the CUE frequency when you need to contact a FH radio net and are not an active member of that net. You can use CUE if you have missed your primary net's opening or if you need an ERF. CUE may also be used if you need a member of an alternate net or if you are operating a SC radio and wish to contact a FH net.

- Set COMSEC to plain text (PT).
- Set CHAN to CUE. Make sure it is loaded (if it is, do the loading procedures discussed in Loading Frequencies).
- Set the radio frequency (RF) to HI. If you are using a long-range RT, set the RF to power amplification (PA).
- Adjust the volume (VOL) as needed.
- Press the handset push-to-talk (PTT). Repeat this step if necessary; wait 15 seconds between tries. The NCS or designated member will contact you on CUE frequency.

Combat Operations 1-65

Late Net Entry

You can use two methods to join a net that is already operating, passive and CUE/ERF.

Passive

- Set CHAN to the channel that has the proper preset.
- Set other switches for normal FH operating positions.
- Press FREQ.
- Press synchronize (SYNC) (late entry)
- Monitor the channel for a least 3 minutes. (Do not press PTT.)
- Contact the net when you hear traffic. When you receive a net signal, late entry is canceled; L disappears from display when you press FREQ.
- Call the net when you hear traffic.
- Perform CUE /ERF if you do not make contact.

CUE/ERF

- Load the CUE frequency of the net to be entered. Set CHAN to CUE and FCTN to LD.
- Load the MAN frequency of the net to be entered. Set CHAN to MAN, leaving FCTN at LD.
- Set RF power to HI for manpack and PA for vehicular radio.
- Set COMSEC to PT.
- Set CHAN to CUE. Press PTT for 4 to 5 seconds.
- Immediately set COMSEC to CT and wait for a response.
- Repeat after 15 seconds until CUE call is answered. For each try, go to PT to send a CUE and CT to receive a reply.

    NOTE: A CUE call goes through only when the net is quiet. Because you do not know when the net is quiet, the solution is to repeat your CUE call until you get an answer.

- Wait for instructions from the NCS/Alt NCS regarding net entry and receiving an ERF when your CUE call is answered.
- You are ready to enter the net once you store the ERF.

Operator's Troubleshooting Checklist

If you have difficulty communicating, take the time to perform the following check before you decide that there is something wrong with your radio:

- Ensure that you have all the switches set properly.

- Ensure that all cable connections are tight.
- Ensure that the antenna is connected and positioned properly.
- Try to verify that you have line of sight (LOS) with other stations.
- Change your position to see if communications improve.
- Perform passive, late net entry if you have not heard net traffic in some time.
- Make sure that your radio has adequate power (especially manpack).
- Look and see if another net station is colocated in your area (called co-site interference).
- Determine if you are being jammed by the enemy. If so, take appropriate action.
- Do the following if your radio gives a strange, unexplained message that does not automatically clear: Set FCTN to STBY, then return to SQ ON. This action may clear your problem. If this does not, and the situation permits, set FCTN to Z-FH and wait for *GOOD*, then to OFF and wait 10 seconds, then back to Z-FH and again wait for *GOOD*. Now run a self-test. If *GOOD* results, reload and re-enter the net. If you still have a problem, contact unit maintenance.

FM 5-34

VISUAL SIGNALS

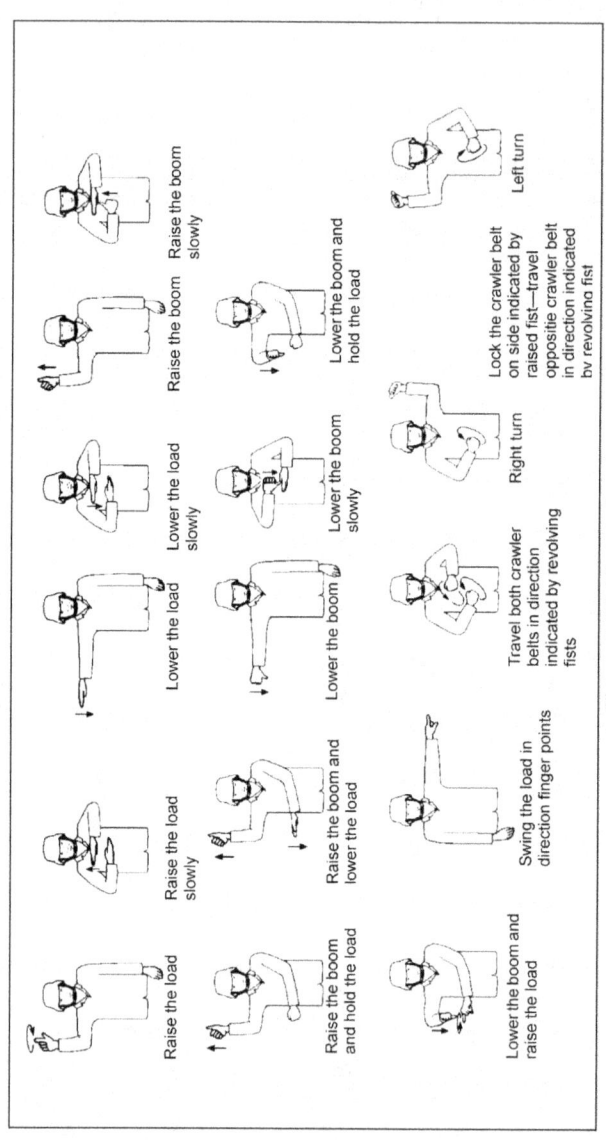

Figure 1-30. Visual signals

FM 5-34

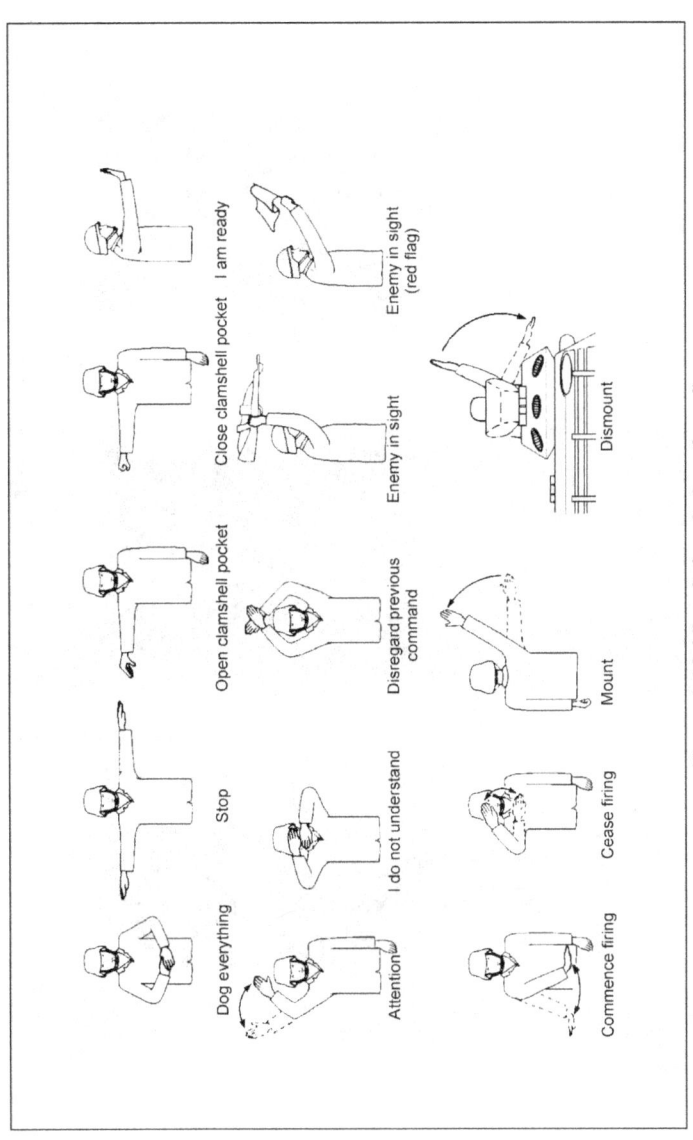

Figure 1-30. Visual signals (continued)

Combat Operations 1-69

FM 5-34

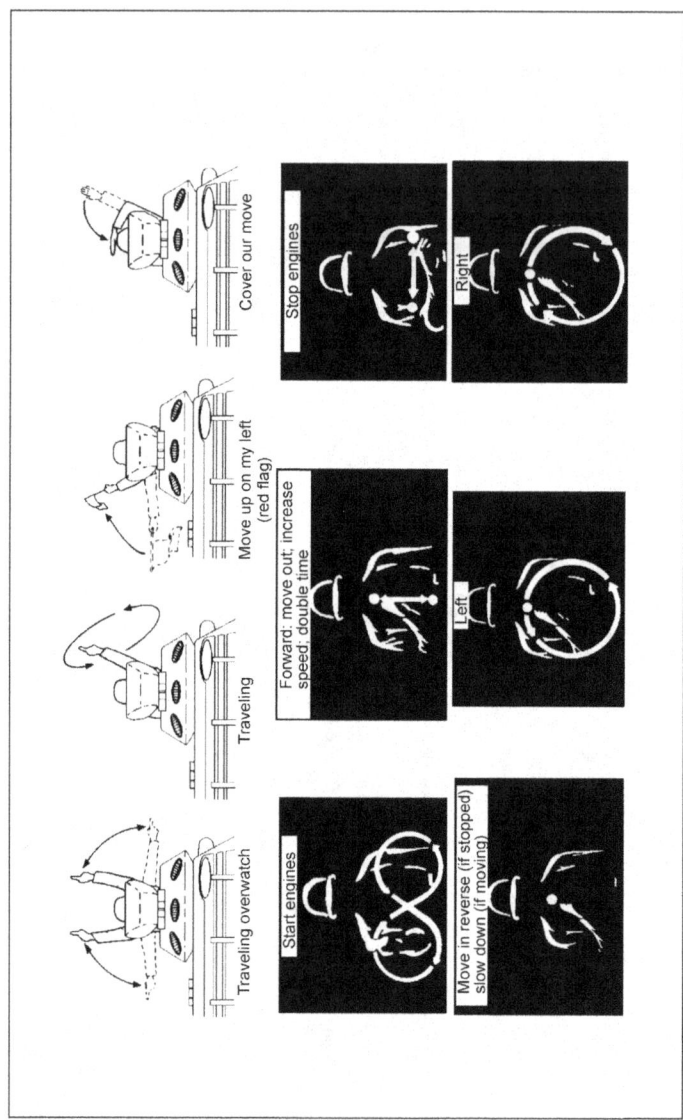

Figure 1-30. Visual signals (continued)

REHEARSALS

A rehearsal is the act or process of practicing an action to prepare for an actual performance of that action. Rehearsing key combat actions allows participants to—

- Become familiar with the operation.
- Gain a visual impression of the operation.
- Build a lasting mental picture of the sequence of key actions.
- Allow subordinate leaders to ascertain the tactical plan's feasibility, its common sense, and the adequacy of it C^2 measures.

REHEARSAL TYPES

The five type of rehearsals are confirmation brief, back brief, combined-arms, support, and battle drill or SOP.

PRINCIPLES

All rehearsals must adhere to the following principles:

- Support the scheme of maneuver and the commander's intent.
- Provide clear tasks/conditions/standards.
- Include multiechelon, combined-arms rehearsals.
- Include key participants.
- Enforce standards, conducted to standard.
- Provide feedback.
- Complement the preparation phase.
- Instill confidence in the plan and the leaders.

TECHNIQUES

The six categories of rehearsal techniques are the radio, map, sketch map, terrain model, reduced force, and full dress. Each technique follows the **crawl, walk, and run** training concept and increases in mission realism and a corresponding increase in rehearsal payoff. For more information on each technique, see FM 101-5, Appendix D.

PARTICIPANT LEVELS

The participant level details exactly who in the unit is required to attend the rehearsal. The enemy situation may have a significant impact on the participant level, as security must be maintained throughout.

Chapter 2

Threat

The threat that the Army faces has gone through a major change since the breakup of the Warsaw Pact and the former Soviet Union. The majority of the forces that the Army will face in mid- to high-intensity conflicts will continue to use the *Soviet* model for their operations. Regional wars will continue to challenge US vital interests. Stability operations and support operations will continue to require an increase in engineer support for psychological and tactical reasons. Many nations will increase qualitatively as technology becomes more affordable and available.

STABILITY OPERATIONS AND SUPPORT OPERATIONS THREAT

The stability and support area of operations (AO) routinely contains poorly developed road nets, installations, and airfields. These support structures must typically be constructed to accomplish the mission. Forces involved in stability operations and support operations will require engineer support for combat and sustainment engineering missions. To accomplish their mission, forces must understand and plan against four types of aggressors: criminals, protesters, terrorists, and subversives.

To accomplish their objectives, these aggressors employ a wide range of tactics, from harassment to terrorism. They may try to sever lines of communication (LOC) by various methods: mining roads, waterways, and railways; locating ambush sites next to LOC; or destroying bridges and tunnels with demolitions. Extended LOC cannot be fully secured; however, measures can be enforced to reduce the effect that the aggressor's activity has on them. All operations require US forces to maintain a high degree of force protection.

TERRORISM

Terrorism includes bombings, assassinations, kidnappings, threats, murder, mutilations, torture, and blackmail. If terrorism is used, it is usually to coerce or intimidate. Terrorism can also be used to discredit a government by provoking it to overreact to a situation. Such a reaction could alienate the people and show the government's inability to protect the local populace and its own installations. Terrorists usually believe that a successful operation against US forces involved in stability operations and support operations will provide greater legitimacy for themselves.

HARASSMENT

Harassment keeps forces on the defensive. If successful, it causes them to react to an aggressor's operations. As a result, a government cannot conduct offensive operations to stop an aggressor. Harassment also weakens a government's resources and disrupts its LOC. In stability operations and support operations, the aggressors seldom attempt to seize and defend objectives. During movements, they infiltrate. However, near a target area, small guerrilla elements mass to conduct operations. The most common techniques that the guerrillas use are the ambush, raid, and small-scale attacks. Their targets are security posts, small forces, facilities, and LOC.

PROTECTIVE MEASURES

The following is a list of measures to protect against aggressors:

- Eliminate potential hiding places near and within buildings.
- Ensure that there is an unobstructed view around all buildings.
- Place facilities within view of other occupied facilities.
- Place assets that are stored outside of the buildings within view of the occupied rooms of the buildings.
- Use signs or other indicators of where the assets are stored sparingly (includes exterior signs).
- Ensure that the buildings are at least 170 feet away from the installation's boundaries.
- Ensure that the lines of approach to the buildings are parallel.
- Minimize vehicle and personnel access points.
- Locate parking (to include public parking) areas as far from the buildings as practical but within view of the occupied rooms or buildings.
- Illuminate a building's exterior and the exterior areas where assets are located.
- Secure access to power and/or heat plants, gas mains, water supplies, and electrical services.
- Locate construction staging areas away from the asset areas.
- Locate buildings away from natural or man-made vantage points.
- Locate a building's critical assets within areas that do not have exterior walls, when possible.
- Minimize window areas.
- Cover windows that are next to doors so that the aggressors cannot unlock the doors through them.

2-2 Threat

FM 5-34

- Secure exposed exterior ladders and fire escapes.
- Design buildings so that the areas are not hidden from the view of control points or occupied spaces.
- Place assets in areas that are occupied 24 hours a day, when possible.
- Ensure that activities with large visitor populations take place away from protected assets, when possible.
- Locate protected assets in controlled areas where they are visible to more than one person.
- Place mail rooms on the perimeter of facilities.
- Provide emergency backup power generation for critical activities/buildings.

For information on protective obstacles, see Chapter 5.

MID- TO HIGH-INTENSITY THREAT

Figure 2-1 and Figures 2-2 and 2-3, page 2-4, show different threat minefields. Table 2-1, page 2-5, lists the parameters for threat's minefields.

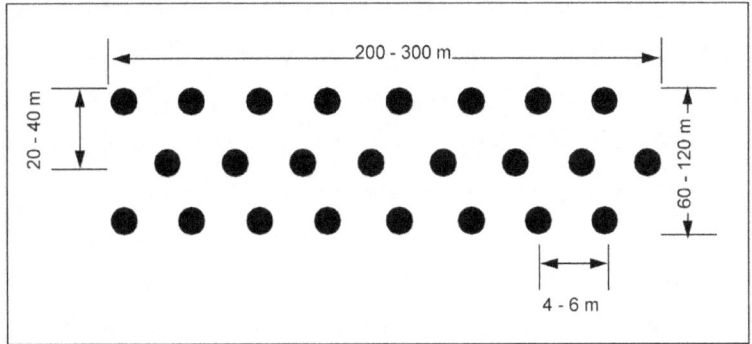

Figure 2-1. Threat's minefield, using track-width mines

Threat 2-3

FM 5-34

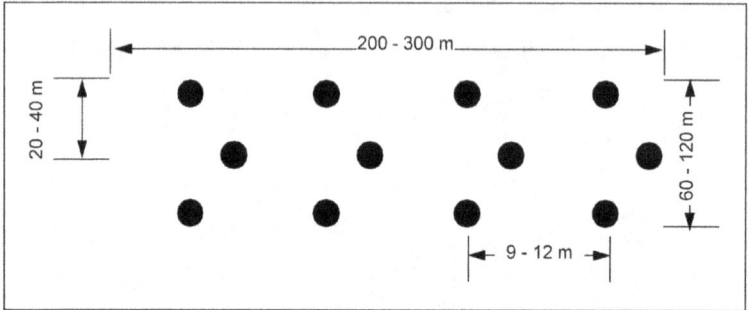

Figure 2-2. Threat's minefield, using full-width mines

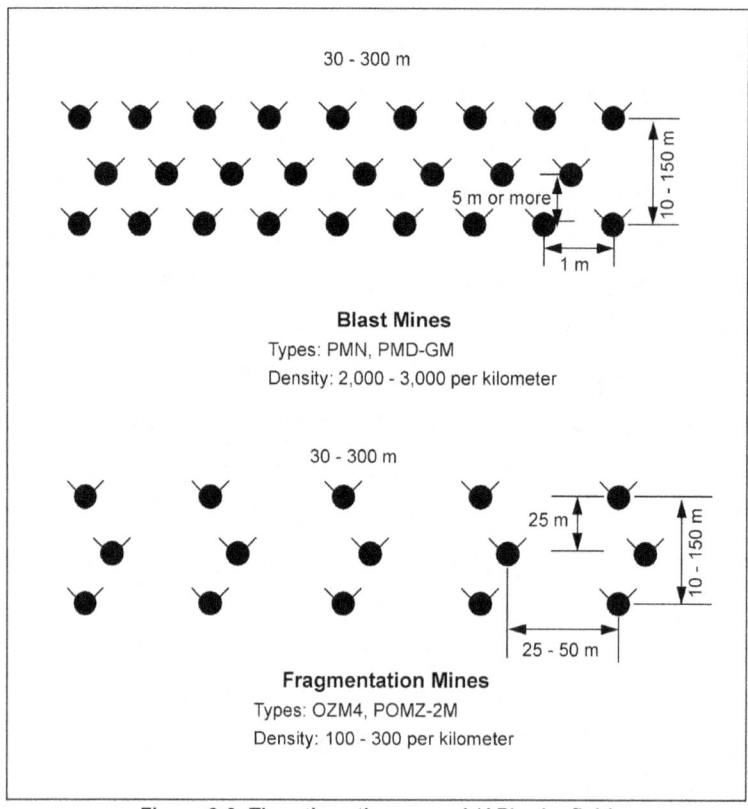

Figure 2-3. Threat's antipersonnel (AP) minefield

2-4 Threat

Table 2-1. Normal parameters for threat's minefields

AT Minefield	
Front (situation dependent)	200 to 300 meters
Depth	60 to 120 meters
Distance between rows	20 to 40 meters
Number of rows	3 to 4 rows
Distance between mines	4 to 6 meters for antitrack mines
	9 to 12 meters for antihull mines
Outlay, normal	550 to 750 antitrack mines/km
	300 to 400 antihull mines/km
Outlay, increased effect	1,000+ antitrack mines/km
	500+ antihull mines/km
Probability of destruction	0.57 for antitrack mines (750/km)
	0.85 for antihull mines (400/km)
AP Minefield	
Front	30 to 300 meters
Depth	10 to 150 meters
Distance between rows	5+ meters for blast mines
	25 to 50 meters for fragmentation mines
Number of rows	2 to 4 rows
Distance between mines	1 meter for blast mines
	50 meters to twice the lethal radius of fragmentation for fragmentation mines
Outlay, normal	2,000 to 3,000 for HE/blast mines (2,000/km)
	100 to 300 for fragmentation mines
Outlay, increased effect	2 to 3 times normal outlay
Probability of destruction	0.15 to 0.2 for HE/blast mines (2,000/km)
	0.1 to 0.15 for fragmentation mines (100/km)

THREAT ORGANIZATION

MILITARY DISTRICTS

Military districts and maneuver divisions are the highest-level tactical organizations in the ground forces (see Table 2-2, page 2-6). Military districts are not only geographical entities but also a level of command. Districts generally contain one or more separate brigades (infantry, motorized infantry, or mechanized infantry) and perhaps a tank brigade. An engineer battalion and possibly an engineer company per each separate brigade support each military district.

MOTORIZED INFANTRY DIVISION

The basic maneuver units in this division are three motorized infantry brigades (divisional). These divisional brigades differ structurally from their separate counterparts normally found within military districts. In addition to its motorized infantry brigades, this division may or may not have a separate tank battalion. Either an engineer company or engineer battalion supports a motorized infantry division.

Table 2-2. Threat organization, infantry based

Threat	Engineer Support Size
Military district	Battalion
Motorized infantry division	Company/battalion*
Infantry division	Company/battalion*
Mechanized infantry division	Battalion/company*
Motorized infantry brigade (separate)	Company
Infantry brigade (separate)	Company
Infantry brigade (divisional)	Company
Infantry brigade (militia)	None
Mechanized infantry brigade (separate)	Company
Mechanized infantry brigade (divisional)	Company
Tank brigade (94-Tank)	Company
Tank brigade (67-Tank)	Company
*Either a company- or battalion-sized force supports the threat organizations.	

INFANTRY AND MECHANIZED INFANTRY DIVISION

The majority of divisions in an infantry-based threat are either infantry or motorized infantry. Most infantry divisions have company-sized engineer assets; however, some better-equipped infantry divisions may have battalion-sized units of these types. Most mechanized infantry divisions have battalion-sized engineer assets but could be reduced to company-size assets.

MOTORIZED INFANTRY AND INFANTRY BRIGADE

The basic maneuver unit is the brigade, consisting of maneuver battalions and a wide array of combat support (CS) and combat service support (CSS) elements. A motorized infantry brigade is the most common in the infantry-based threat. Both motorized infantry and infantry brigades are supported by company-sized engineer assets.

MECHANIZED INFANTRY AND TANK BRIGADE

Even though motorized infantry brigades are the most common in an infantry based threat, some mechanized infantry and tank brigade threat is present. Since the infantry-based threat has no tank division, all tank brigades are actually separate.

The mechanized infantry brigade has two basic types: one equipped with armored personnel carriers (APCs) and one equipped with infantry fighting vehicles (IFVs). Each are supported with company-sized engineer assets.

An engineer company supports a 94-tank standard tank brigade, where a 64-tank standard has no dedicated engineer company (see Table 2-3).

Table 2-3. Principal items of equipment for infantry-based threat

Engineer Equipment	Engineer Battalion Mech Inf Div Mtzd Inf Div Inf Div Military Dist Army/CDF	Engineer Company Mtzd Inf Bde Inf Bde (Sep & Div)	Engineer Company Mech Inf Bde (Sep & Div) Mech Inf Div Military Dist	Engineer Company Tank Bde (94 Tank)
Minelayer, PMR/GMZ	3	3	3	3
Mineclearer, MTK/MTK-2	2			
Mine detector, DIM	3	1	1	1
Engineer recon vehicle, IRM	2			
Armored engineer tractor, IMR	2		1	1
Ditching machine, PZM/BTM/MDK	4	4	4	3
Route-clearing vehicle, BAT/PKT	8	1	1	1
Dozer blade, BTU				3
Mine clearing plow				27
Mine roller-plow				9
Bridge, tank-launched	4	4	1	3
Bridge, truck-launched	8		4	4
Tracked amphibian, K-61/PTS	12			
Trailer, amphibious, PKP	6			
Tracked ferry, GSP/PMM-2	6			
Bridge, PMP center	16			
Bridge, PMP ramp	2			
Assault boat	10			
Power boat	6			
Piledriver set, KMS	1			
Tractor	2			
Truck, sawmill	1			
Trailer, saw	1			
Grader	2			
Concrete mixer	1			
Truck, water purification	1			1

Threat 2-7

ARMOR- AND MECHANIZED–BASED THREAT

Maneuver Divisions

Ground forces have two basic types of maneuver divisions, the mechanized infantry division and the tank division (see Table 2-4). Both types are combined arms organizations. A mechanized infantry division has one tank brigade along with its three mechanized infantry brigades; whereas, a tank division has one mechanized infantry brigade along with its three tank brigades. CS and CSS units are basically the same for all mechanized infantry and tank divisions. Most motorized infantry divisions in an armor- and a mechanized-based threat have battalion-size engineer units. However, an engineer company may support some lesser-equipped motorized infantry divisions.

Table 2-4. Threat organization, armor and mechanized based

Threat	Engineer Support Size
Mechanized infantry division (IFV)	Battalion
Mechanized infantry division (APC)	Battalion
Tank division	Battalion
Motorized infantry division	Battalion/company*
Mechanized infantry brigade (IFV)(Div), MID	Company
Mechanized infantry brigade (IFV)(Div), TD	Company
Mechanized infantry brigade (IFV)(Sep)	Battalion
Mechanized infantry brigade (APC)(Div), MID	Company
Mechanized infantry brigade (APC)(Sep)	Battalion
Tank brigade (Div), MID	Company
Tank brigade (Div), TD	Company
Tank brigade (Sep)	Battalion
*Either a company- or battalion-sized force supports the threat organizations.	

Maneuver Brigades

Like the infantry-based threat, the basic maneuver unit is the brigade. Consisting of maneuver battalions and a wide array of CS and CSS elements. An armor- and a mechanized-based threat consists primarily of mechanized infantry units. The mechanized infantry brigade has two basic types: one equipped with IFVs and one equipped with APCs. There are also tank brigades.

An engineer company supports each mechanized infantry brigade (division), both IFV- and APC-equipped; whereas the separate mechanized infantry brigades is supported by an engineer battalion.

An engineer company supports each tank brigade (division), and like the mechanized infantry brigade, an engineer battalion supports the separate tank brigade (see Table 2-5).

Table 2-5. Principal items of equipment for armor- and mechanized-based threat

Engineer Equipment	Engineer Battalion Mech Inf Div Engr Bde Corps	Engineer Battalion Tank Div	Engineer Battalion Mech Inf Bde (Sep)	Engineer Company Tank Bde (Div)	Engineer Company Mech Inf Bde (Div)
Minelayer, PMR/GMZ	3	3	3	3	3
Minelayer, UMZ		3			
Mineclearer, MTK/MTK-2	2	2	2		
Mine detector, DIM	3	3	3	1	1
Engineer recon vehicle, IRM	2	2	2		
Armored engineer tractor, IMR	2	2	2	1	1
Ditching machine, BTM/MDK	4	4	2	1	1
Ditching machine, PZM/TMK				3	3
Route-clearing vehicle, BAT/PKT	8	8	2	1	1
Dozer blade, BTU					
Mine clearing plow					
Mine roller-plow			5	9	3
Bridge, tank-launched	4	4	4	3	1
Bridge, truck-launched, TMM	8	8		4	4
Tracked amphibian, K-61/PTS	12	12	12		
Trailer, amphibious, PKP	6	6	6		
Tracked ferry, GSP/PMM-2	6	6	6		
Bridge, PMP center	16	16			
Bridge, PMP ramp	2	2			
Assault boat	10	10	10		
Power boat	6	6			
Piledriver set, KMS	1	1			
Tractor	2	2	1		
Truck, sawmill	1	1			
Trailer, saw	1	1			
Grader	2	2			
Concrete mixer	1	1			
Truck, water purification	1	1	1		

MAJOR THREAT EQUIPMENT

Tables 2-6 and 2-7 list the common equipment used to prepare a threat's defensive position.

Table 2-6. Threat's defensive engineer equipment

Nomenclature	Type	Working Speed (kmph)	Distance Between Mines (m)	Depth of Mines (cm)
PMR-3/4	Single-chute trailer	Based on towing vehicle (5 kmph if burying mines)	4 to 5.5	6 to 12
GMZ (mine capacity is 208 mines)	Tracked minelayer	5 to 10	4 to 5.5	25
MI-4, MI-8 HIPC	Helicopter with chutes			Surface laid
UMZ	Tracked minelayer	10 to 40	Scattered	Surface laid

Table 2-7. Threat's defensive ditching and digging equipment

Ditching Equipment				
Nomenclature	Maximum Depth (m)	Digging Width (m)	Digging Capacity (cu m/hr)	Working Speed (m/hr)
MDK-2M	3.5	3.5	120 to 300	200 to 800
MDK-3	3.0	3.5	400 to 500	400 to 500
BTM-3	1.5	0.6	220 to 600	600
PZM-2	1.5	0.8 to 3.5	80 to 250	180
TMK-2	1.5	1.1	200 to 600	450
Digging Equipment				
Nomenclature		Digging Capacity (cu m/hr)		
BAT-M		200 to 250		
Self-entrenching blade on tanks		Hull-down position in 20 to 60 minutes		

FM 5-34

THREAT OFFENSIVE OPERATIONS

The threat emphasizes swift, efficient movement, or transfer of combat power from one point on the battlefield to another to reinforce success. This is accomplished by rapid column movement in march formation and successive deployment into prebattle and attack formations. A division is assigned either a march zone or march routes. As many as four routes are possible. A regiment is normally assigned one or two routes, and a battalion marches on one route. The distance between routes is about 3 kilometers to reduce vulnerability. The formation normally consists of reconnaissance, advance guard (or forward security element of a battalion), flank security elements, main force, and rear security element.

An attack formation normally is assumed about 1,000 meters from enemy positions. A division attacking with three regiments in the first echelon maintains an attack zone of 15 to 25 kilometers wide. A regiment attack front can vary from 3 to 8 kilometers; however, the most typical attack frontage of a regiment is 4 to 5 kilometers. A normal frontage for an attacking battalion is 1 to 2 kilometers, within a zone of 2 to 3 kilometers. A typical tank or motorized rifle company's attack frontage is 500 to 800 meters. Platoons normally attack on a frontage of 100 to 200 meters, with 50 to 100 meters between vehicles. In the attack formation, BTR's or BMP's normally follow between 100 to 400 meters behind the tanks. The speed during an attack is—

- 20 to 30 kilometers per hour on roads in column.
- 15 kilometers per hour cross-country in column.
- 200 meters per minute in attack formation.

CROSSING CAPABILITIES AND CHARACTERISTICS

Tables 2-8 and 2-9, page 2-12, and Tables 2-10 through 2-12, pages 2-13 and 2-14, list the characteristics and capabilities of the threat's vehicles and equipment. Table 2-13, page 2-15, is an example of an enemy's obstacle report.

FM 5-34

Table 2-8. Light armored vehicles—wheeled capabilties and characteristics

Characteristics	BRDM-2	BTR-60PB	BTR-70	BTR-80
Weight (metric ton)	7.0	10.2	11.5	11.0
Speed (kmph)	100	80	80	80 to 85
Water (kmph)	10	10	10	10
Trench crossing (m)	1.25 to 1.60	2.00	2.00	2.00
Vertical Step (m)	0.40	0.40	0.40	0.40
Gradability (deg)	30	30	30	30
Fording (m)	Amphibious	Amphibious	Amphibious	Amphibious
Main armament (mm)	14.5	14.5	14.5	14.5
Secondary armament (mm)	7.62	7.62	7.62	7.62

Table 2-9. Threat's bridging and rafting equipment

Nomenclature	Type	Load Carrying Capacity	Treadway Width (m)	Maximum Gap (m)	Assembly Time/ Meter (min)
PMP	Heavy pontoon	60/170[1]	6.5	Per set, 115	7
TMM	Truck mounted	60	3.8	Per span, 10.5	3.5
MTU-20	Tank mounted	50	3.3	18	5[2]

[1]Class 60 for bridge and up to class 70 for raft
[2]Emplacement time

FM 5-34

Table 2-10. Threat's vehicle obstacle-crossing capabilities and characteristics

	Light Armored Vehicles - Tracked					
	BTR-50	BMP-1	BMP-2	BMD-1	MT-LB	ACRV 1V12
Weight (mt)	14.2	13.5	14.3	7.5	9.7	11
Speed (kmph)	45	65	65	80	60	60
Water (kmph)	10	6	7	10	6	6
Trench crossing (m)	2.80	2.00	2.50	1.60	2.70	2.70
Vertical step (m)	1.10	0.80	0.77	0.80	0.70	0.70
Gradability (deg)	38	30	30	32	35	35
Fording (m)	Amphibious	Amphibious	Amphibious	Amphibious	Amphibious	Amphibious
Main armament (mm)	12.7	73	30	73	7.62	12.7
Effective range (50% probability of hit)	1,500	800 to 1,000	2,000 to 4,000	800 to 1,000	1,000	1,500
Secondary armament (mm)	NA	7.62	7.62	7.62	NA	NA
	Medium Tanks					
		T-54/55	T-62	T-64	T-72	T-80
Weight (mt)		36.0	37.5	39.0	41.0	42.0
Speed (kmph)		50	50	85	60	85
Water (kmph)		NA	NA	NA	NA	NA
Trench crossing (m)		2.70	2.80	2.70	2.70	2.70
Vertical step (m)		0.80	0.80	0.80	0.80	0.80
Gradability (deg)		30	30	30	30	30
Fording (m)		1.4 (5.5 w/ snorkel)	1.4 (5.5 w/ snorkel)	1.4 (5.5 w/ snorkel)	1.4 (5.5 w/ snorkel)	1.4 (5.5 w/ snorkel)
Main armament (mm)		100	115	125	125	125
Effective range (50% probability of hit)		1,500	1,600	2,100	2,100	2,400
Secondary armament (mm)		7.62	7.62	7.62	7.62	7.62

Threat 2-13

Table 2-11. Threat's amphibious and ferry equipment

Nomenclature	Type	Load Carrying Capacity (kg)	Personnel Load (soldiers)	Width (m)	Height (m)	Speed (kmph)	Allocation
K61	Amphibian track	5,000	50	3.2	2.1	36	12 per
PTS-M	Amphibian track	15,000	50	3.5	3.4	40	MRD/TD
PKP	Trailer	5,000		2.8	2.2		3 per MRD/TD
GSP	Ferry	50,000		21.5	3.2	7.7	6 per MRD/TD

Table 2-12. Threat's minefield-reduction equipment

Nomenclature	Type	Sweeping Speed (kmph)	Clearing Width (m)	Depth (cm)
UAZ69 DIM	Truck-mounted mine detector	10		25
KMT6M	Tank-mounted mine plow	6-12	2.2	10
KMT5M	Tank-mounted plow/roller combination	6-12	2.5	10
MTK-2	Armored line charge		2.5	
			3-m pressure mines, 8-m tilt rods, 180-m depth of minefield	

2-14 Threat

Table 2-13. Sample, enemy's obstacle report

ALPHA	Date and time of information collection
BRAVO	Location (grid to start and end points)
CHARLIE	Type of obstacle
DELTA	Enemy weapons or surveillance
ECHO	Depth of obstacle
FOXTROT	Estimated time required to reduce obstacle
GOLF	Estimated material and equipment required to reduce obstacle
HOTEL	Coordinates of obstacle bypass, if any
INDIA	Type of mines present, if any
JULIET	Buried and/or surface laid mines, if any
KILO	Antihandling devices, if any
LIMA	Depth of buried mines, if any
MIKE	Terrain restrictions on the use of the MICLIC or tank plow

THREAT'S OFFENSIVE RIVER CROSSING

The threat is well prepared to cross water obstacles. On the average, it anticipates that a formation on the offense will cross one water obstacle of average width (100 to 250 meters) and several narrower ones each day. It considers crossing water obstacles to be a complex combat mission but regards this as a normal part of a day's advance. Table 2-14 shows the time line for the threat's river crossing.

Table 2-14. Threat's river-crossing time line

Element	Crossing Time (hours)
Forward detachment (battalion)	1 to $1^1/_2$
First-echelon regiment	2 to 3
Division	5 to 6
NOTE: River width is from 100 to 250 meters.	

The threat uses two methods of assault crossing:

- Assault crossing from the line of march: A forward detachment reaches the water obstacle as quickly as possible, bypassing strong points and capturing existing bridges or river sections suitable for an assault crossing. It crosses the water, seizes a line on the opposite bank, and holds until the main force arrives.

- Prepared assault crossing: When the assault crossing from the line of march is not feasible, the threat uses the prepared assault crossing. The main force deploys at the water obstacle with subunits in direct contact with the opponent. The threat then makes more thorough preparations for the crossing. Success depends on covertness, so the crossing usually takes place at night.

> This Chapter implements STANAG 2021.

Chapter 3
Reconnaissance

ROUTE CLASSIFICATION

Engineers routinely assist maneuver units in the technical portion of a route reconnaissance. For more information on route classification, see FM 5-170. (On all reports, record all distances in metric dimensions.)

CRITICAL FEATURES

Consider the following features in route classification:

- Road width, slopes, and curves.
- Bridges, fords, tunnels, ferries, underpasses, swim sites, and other traffic-restricting features.
- Slide areas.
- Drainage.
- Natural and man-made features such as wooded, built-up, and possible dispersion areas.

Table 3-1 lists route widths; Figure 3-1, page 3-2, shows the formula for route classification.

Table 3-1. Traffic-flow capability based on route width

	Limited Access	Single Lane	Single Flow	Double Flow
Wheeled	At least 3.5 m	3.5 to 5.5 m	5.5 to 7.3 m	Over 7.3 m
Tracked and combination vehicles	At least 4.0 m	4.0 to 6.0 m	6.0 to 8.0 m	Over 8 m

SLOPES AND RADIUS COMPUTATION

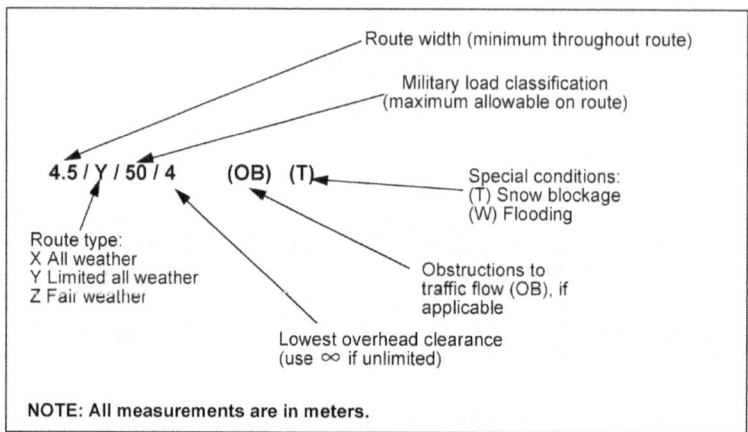

Figure 3-1. Route-classification formula

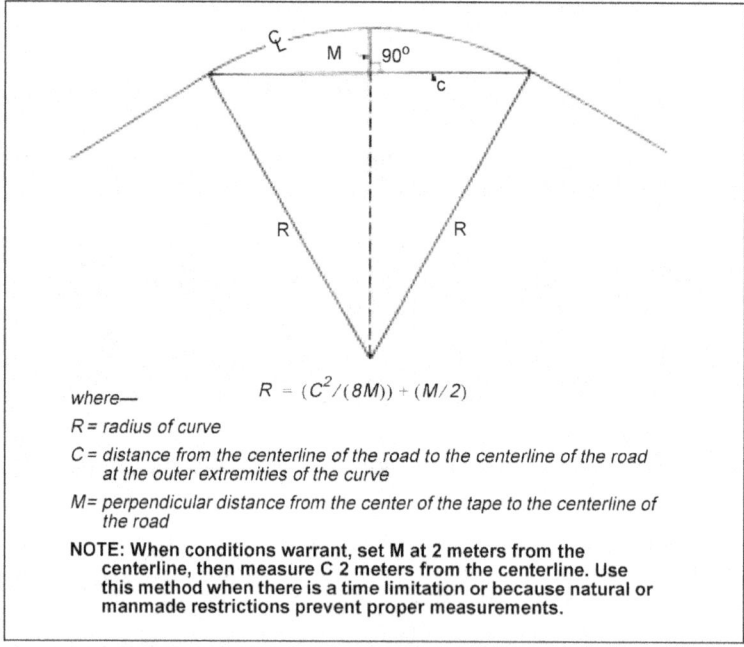

where—

$$R = (C^2/(8M)) + (M/2)$$

R = radius of curve

C = distance from the centerline of the road to the centerline of the road at the outer extremities of the curve

M = perpendicular distance from the center of the tape to the centerline of the road

NOTE: When conditions warrant, set M at 2 meters from the centerline, then measure C 2 meters from the centerline. Use this method when there is a time limitation or because natural or manmade restrictions prevent proper measurements.

Figure 3-2. Radius-of-curvature calculation

3-2 Reconnaissance

FM 5-34

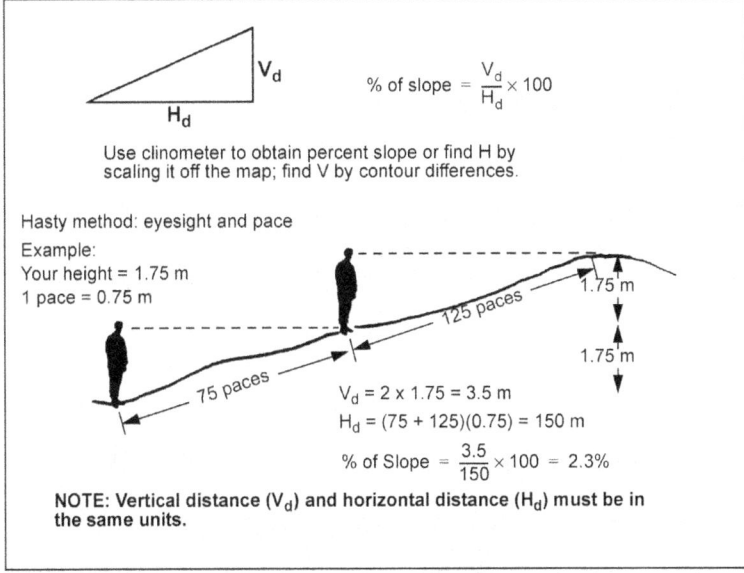

Figure 3-3. Slope computation (road gradient)

OBSTRUCTION (OB)

Obstructions are any factors which restrict type, amount, or speed of traffic flow. Whenever *(OB)* appears in the route formula, the overlay must show the exact nature of the OB. The most common obstructions are—

- Overhead clearance that is less than 4.3 meters.
- Width below minimum standard prescribed for the type of traffic flow in Table 3-1, page 3-1.
- Slopes of 7 percent or greater.
- Curves with a radius of 25 meters or less. Curves with a radius of 25.1 to 45 meters are not considered to be an obstruction; however, they must be recorded on the route overlay classification.
- Fords.
- Ferries.

REPORT AND OVERLAY

A route-classification report consists of an overlay; specific reconnaissance features (bridge, ford, or road); and any other supplementary overlays, reports, or sketches to support the report. As a minimum, the following information will be included on the route-classification overlay:

Reconnaissance 3-3

FM 5-34

- The route classification formula.
- The name, rank and social security number (SSN) of the person in charge of performing the classification.
- The unit conducting the classification.
- The date-time group (DTG) that the classification was conducted.
- The map's name, edition, and scale.
- Any remarks necessary to ensure complete understanding of the information on the overlay.

Figure 3-4 shows an example of a route-classification overlay.

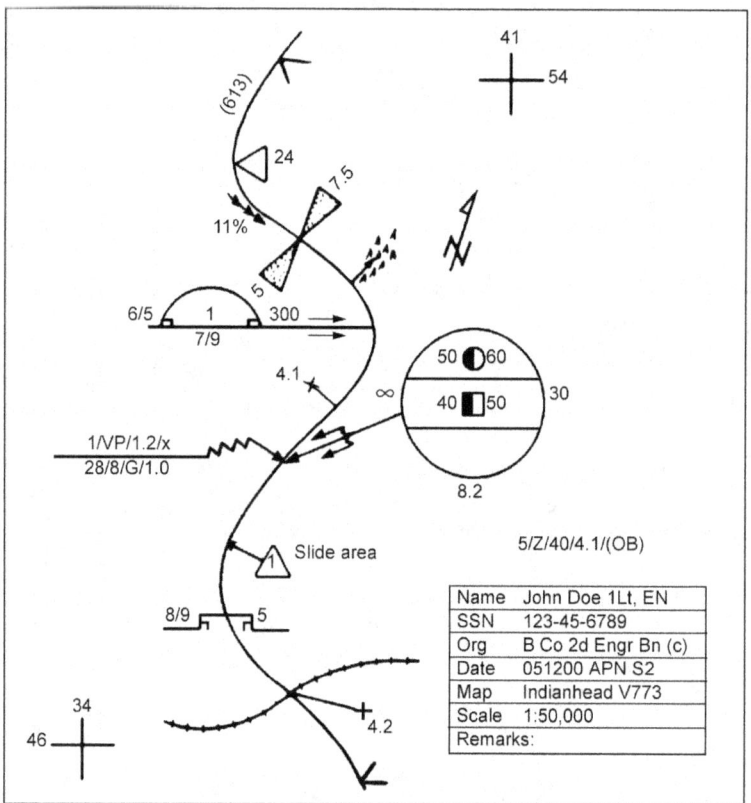

Figure 3-4. Route-classification overlay

3-4 Reconnaissance

ROAD RECONNAISSANCE

CLASSIFICATION

Road classification is expressed in a standardized sequence: (1) prefix, use A if there are no limiting characteristics and B if there are some limiting characteristics; (2) limiting characteristics, use the symbols in Table 3-2; (3) traveled way width/traveled way plus shoulder width; (4) road surface material, use the symbols in Table 3-3, page 3-6; (5) road length, obstructions, and special conditions, put each in parenthesis (see Figure 3-1, page 3-2).

Table 3-2. Road-limiting characteristics and symbols

Limiting Characteristics	Symbols
Curves (radius 25 meters or less)	c
Gradients (seven percent or greater)	g
Drainage (inadequate ditches, culverts)	d
Foundation (unstable)	f
Surface condition (bumpy, rutted, or pothole)	s
Camber or superelevation (excessive crown)	j
Unknown characteristics (used with other above symbols enclosed in parenthesis). Example: (c?) = unknown radius	
NOTE: All reports will be submitted in metric measurements.	

Example: *Bcgd(f?)s 3.2/4.8nb(4.3 kilometers)(OB)(T)* would indicate that the road has limits of sharp curves, steep grades, bad drainage, unknown foundation, and rough surface; has a traveled way width of 3.2 meters; has combined width and shoulders of 4.8 meters; has surface material that is bituminous surface treatment on natural earth, stabilized soil, sand-clay, or other selected material; is 4.3 kilometers long; contains obstructions; and is subject to snow blockage.

RECORDING

Record road-reconnaissance data on DA Form 1248 (see Figures 3-5 and 3-6, pages 3-7 and 3-8).

NOTE: Make note of the existing or potential environmental impacts due to heavy vehicular traffic.

Table 3-3. Road–surface materials and symbols

Surface Material	Symbol
Concrete	k
Bituminous or asphaltic concrete (bituminous plant mix)	kb
Bituminous surface treatment on natural earth, stabilized soil, sand-clay, or other select material	nb
Used when type of bituminous construction cannot be determined	b
Bituminous surface on paving brick, or stone	pb
Bitumen-penetrated macadam, water-bound macadam with superficial asphalt or tar cover.	rb
Pavement, brick, or stone	p
Water-bound macadam, crushed rock, or coral	r
Gravel	l
Natural earth, stabilized soil, sand-clay, shell, cinders, disintegrated granite, or other select material.	n
Other types not mentioned (indicate length when this symbol is used)	v

BRIDGE RECONNAISSANCE

Hasty

See Appendix B, FM 5-170, for hasty bridge classification.

Deliberate

To classify a bridge or prepare a bridge for demolition accurately, conduct a detailed reconnaissance. Use DA Form 1249, Figures 3-7 through 3-13 (pages 3-9 through 3-15), and Table 3-4 (page 3-16) to record the data. You can use Table 3-4 as a guide for developing a line-number report format for voice or digital transmission of bridge data. The information is used in conjunction with FM 5-446 for classification. The Sheffield Method for bridge destruction is discussed in FM 5-250.

ROAD RECONNAISSANCE REPORT
For use of this form, see FM 5-36, proponent agency is TRADOC.

DATE: 29 Aug 96

TO *(Headquarters ordering reconnaissance)*	FROM: *(Name, grade and unit of officer or NCO making reconnaissance)*
Cdr, ATTN: S-2, 21st Engr Bn	DOE, JOHN, 1LT, CoA, 522nd Eng

1. MAPS	a. COUNTRY	b. SCALE	c. SHEET NUMBER OF MAPS	2. DATE/TIME GROUP *(Of signature)*
	Ft. Leonard Wood Special	1:50,000	AMS V733 Sheet 5561 IV	291430Z Aug 96

SECTION I - GENERAL ROAD INFORMATION

3. ROAD GRID REFERENCE FROM	TO	4. ROAD MARKING *(Civilian or Military number of road)*	5. LENGTH OF ROAD *(Miles or kilometers, specify)*
UT 122864	UT 097999	MISSOURI ROUTE J	16 Km

6. WIDTH OF ROADWAY *(Feet or meters, specify)*	8. WEATHER DURING RECONNAISSANCE *(Include last rainfall, if known)*
6.7m to 9.3m	Fair – temp 79° Last rainfall – 15 Aug 96

7. RECONNAISSANCE DATE	TIME
29 Aug 96	0615

SECTION II - DETAILED ROAD INFORMATION *(When circumstances permit more detailed information will be shown in an overlay or on the mileage chart on the reverse side of this form. Standard symbols will be used.)*

	9. ALINEMENT *(Check one ONLY)*		10. DRAINAGE *(Check one ONLY)*
	(1) FLAT GRADIENTS AND EASY CURVES		(1) ADEQUATE DITCHES, CROWN/CAMBER WITH ADEQUATE CULVERTS IN GOOD CONDITION
	(2) STEEP GRADIENTS *(Excess of 7 in 100)*	✓	(2) INADEQUATE DITCHES, CROWN/CAMBER OR CULVERTS, ITS CULVERTS OR DITCHES ARE BLOCKED OR OTHERWISE IN POOR CONDITION
	(3) SHARP CURVES *(Radius less than 100 ft (30m))*		
✓	(4) STEEP GRADIENTS AND SHARP CURVES		

11. FOUNDATION *(Check one ONLY)*

✓	(1) STABILIZED COMPACT MATERIAL OF GOOD QUALITY		(2) UNSTABLE, LOOSE OR EASILY DISPLACED MATERIAL

12. SURFACE DESCRIPTION *(Complete items 12a and b)*

a. THE SURFACE *(Check one ONLY)*

✓	(1) FREE OF POTHOLES, BUMPS, OR RUTS LIKELY TO REDUCE CONVOY SPEED		(2) BUMPY, RUTTED OR POTHOLED TO AN EXTENT LIKELY TO REDUCE CONVOY SPEED

b. TYPE OF SURFACE *(Check one ONLY)*

	(1) CONCRETE		(6) WATERBOUND MACADAM
✓	(2) BITUMINOUS *(Specify type where known)*: Asphalt		(7) GRAVEL
			(8) LIGHTLY METALLED
	(3) BRICK *(Pave)*		(9) NATURAL OR STABILIZED SOIL, SAND CLAY, SHELL, CINDERS, DISINTEGRATED GRANITE, OR OTHER SELECTED MATERIAL
	(4) STONE *(Pave)*		(10) OTHER *(Describe)*:
	(5) CRUSHED ROCK OR CORAL		

SAMPLE

SECTION III - OBSTRUCTIONS *(List in the columns below particulars of the following obstructions which affect the traffic capacity of a road. If information of any factor cannot be ascertained, insert "NOT KNOWN")*
(a) Overhead obstructions, less than 14 feet or 4.25 meters, such as tunnels, bridges, overhead wires and overhanging buildings.
(b) Reductions in road widths which limit the traffic capacity, such as craters, narrow bridges, archways, and buildings.
(c) Excessive gradients (Above 7 in 100)
(d) Curves less than 100 feet (30 meters) in radius
(e) Fords

SERIAL NUMBER a	PARTICULARS b	GRID REFERENCE c	REMARKS d
	Steep Grade – 8%	UT 119872	200 m long
	Sharp Curve	UT 112877	Radius 21m
	Constriction	UT 112878	6.7m wide, 300 m long
	Constriction	UT 105896	7 m wide, 100 m long
	Built up area	UT 094856	7.3 m wide, 2000 long

DA FORM 1248, 1 JUL 60 — PREVIOUS EDITION IS OBSOLETE — USAPPC V1.00

Figure 3-5. Sample, DA Form 1248 (front)

FM 5-34

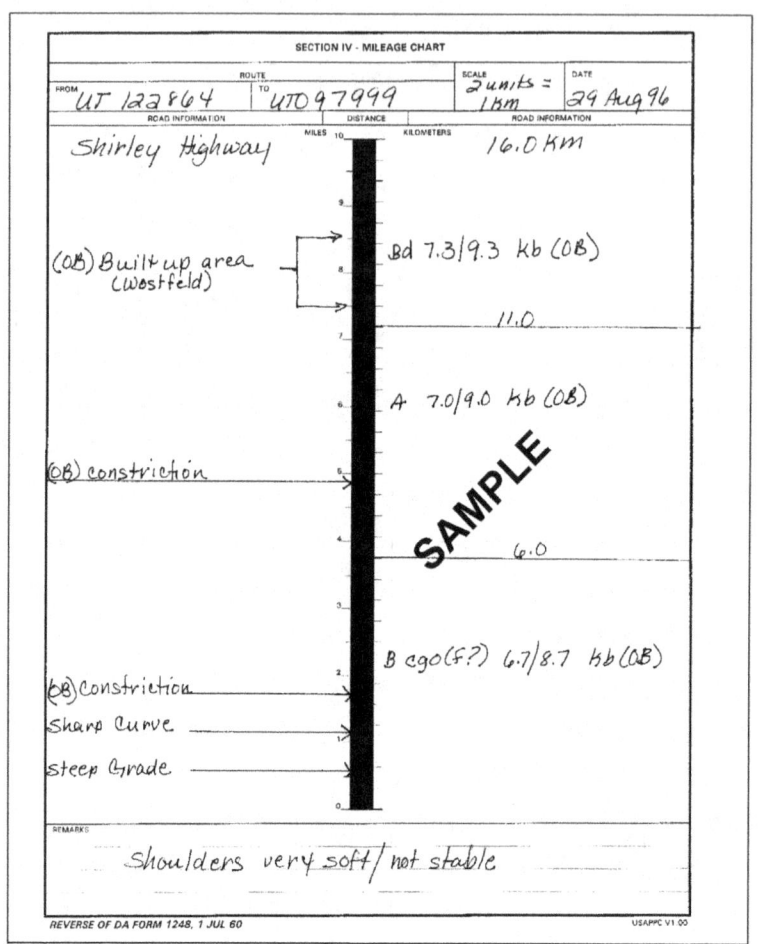

Figure 3-6. Sample, DA Form 1248 (back)

3-8 Reconnaissance

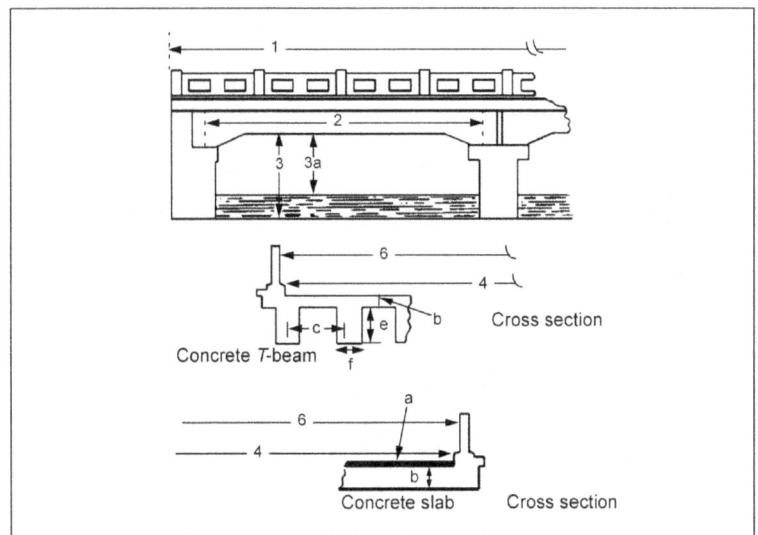

Figure 3-7. Dimensions for concrete bridges

FM 5-34

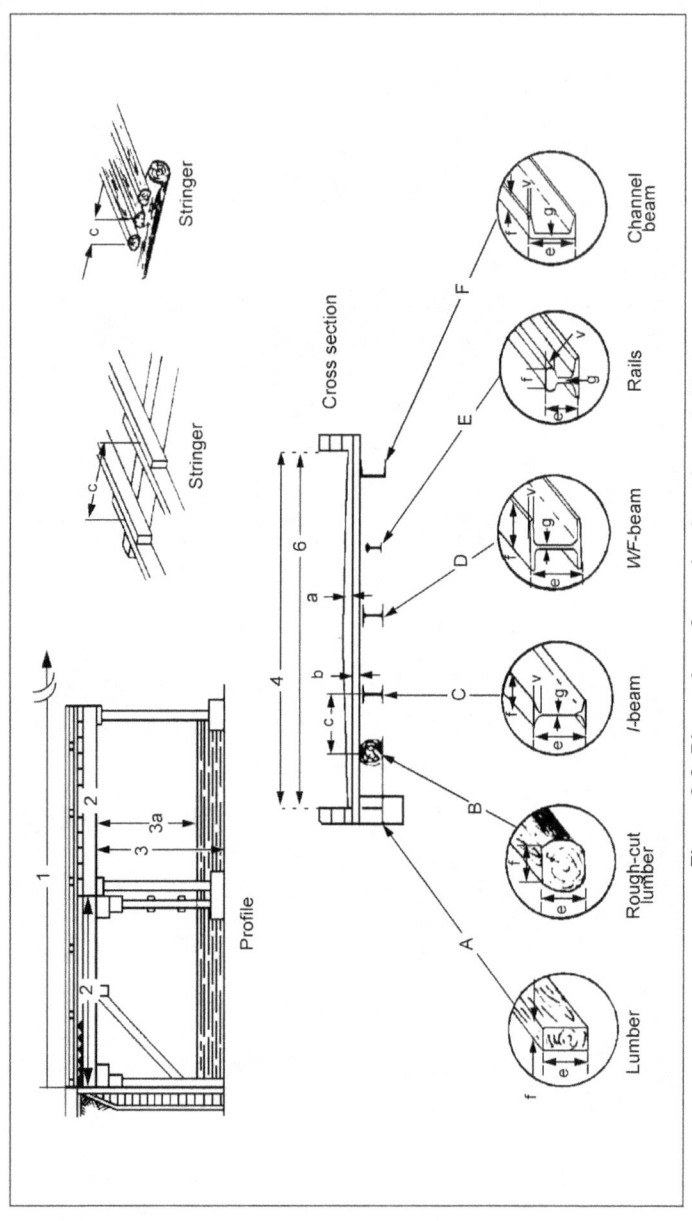

Figure 3-8. Dimensions for a simple stringer bridge

3-10 Reconnaissance

FM 5-34

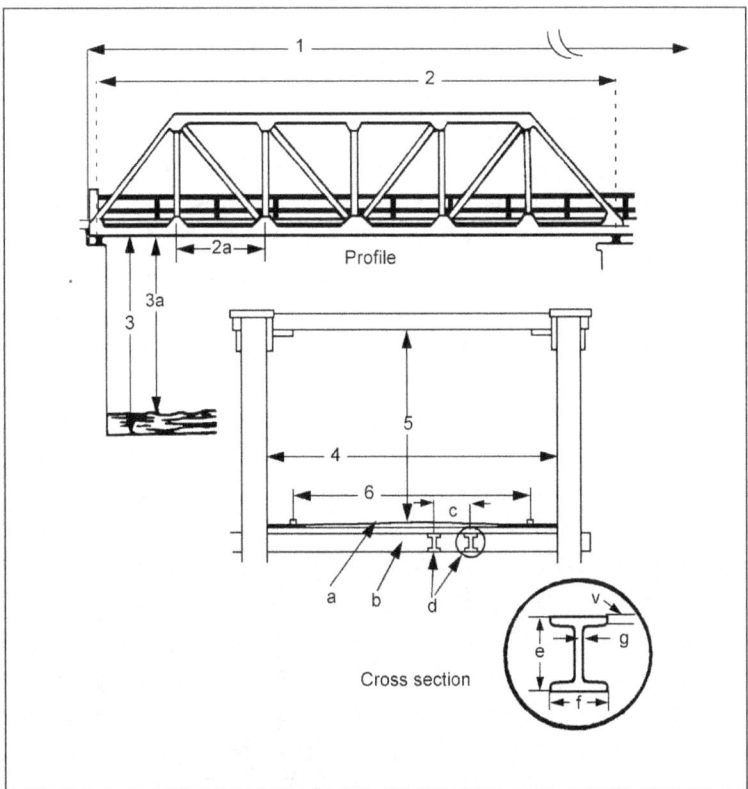

Figure 3-9. Dimensions for steel-truss bridges

Reconnaissance 3-11

FM 5-34

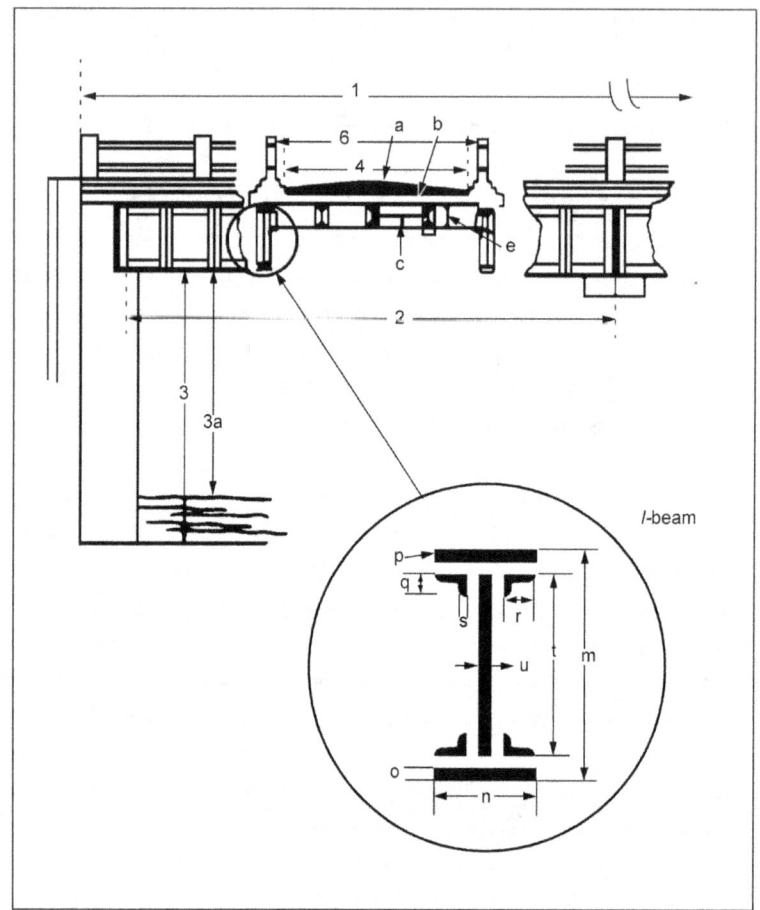

Figure 3-10. Dimensions for plate-girder bridges

3-12 Reconnaissance

FM 5-34

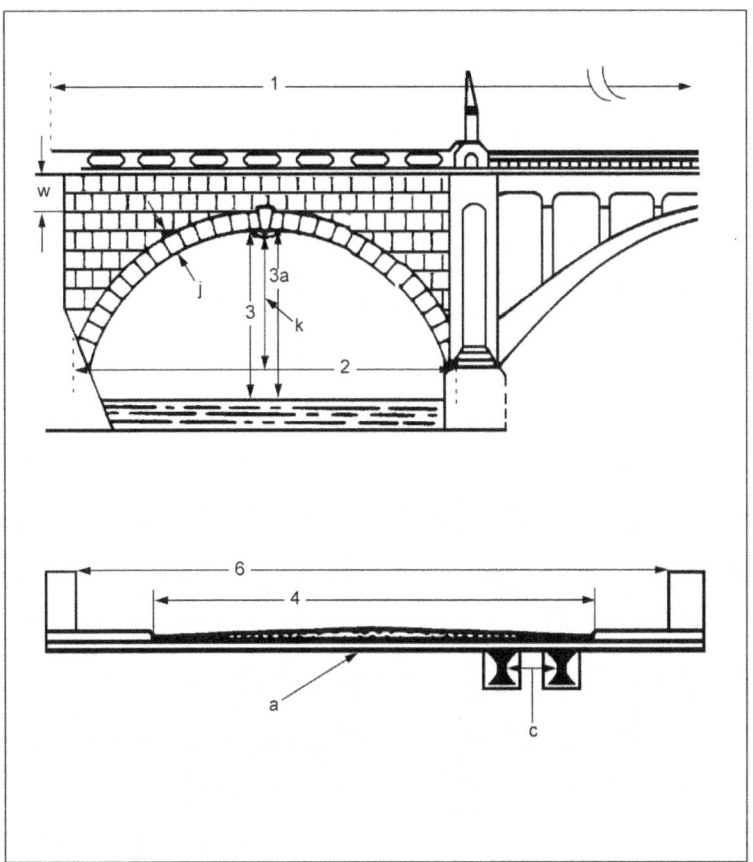

Figure 3-11. Dimensions for arch bridges

Reconnaissance 3-13

FM 5-34

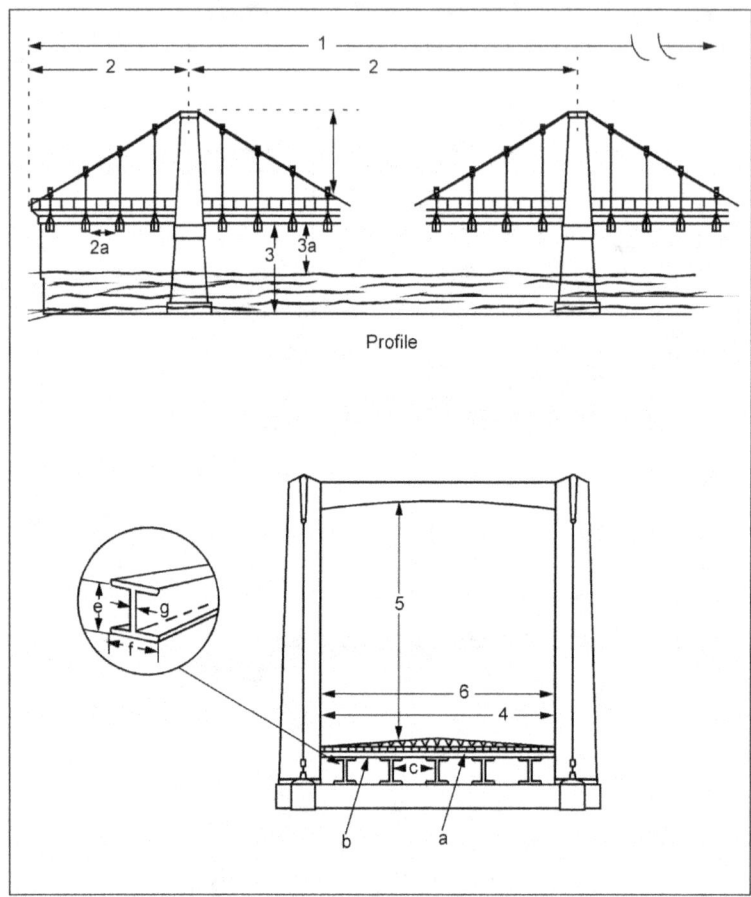

Figure 3-12. Dimensions for suspension bridges

3-14 Reconnaissance

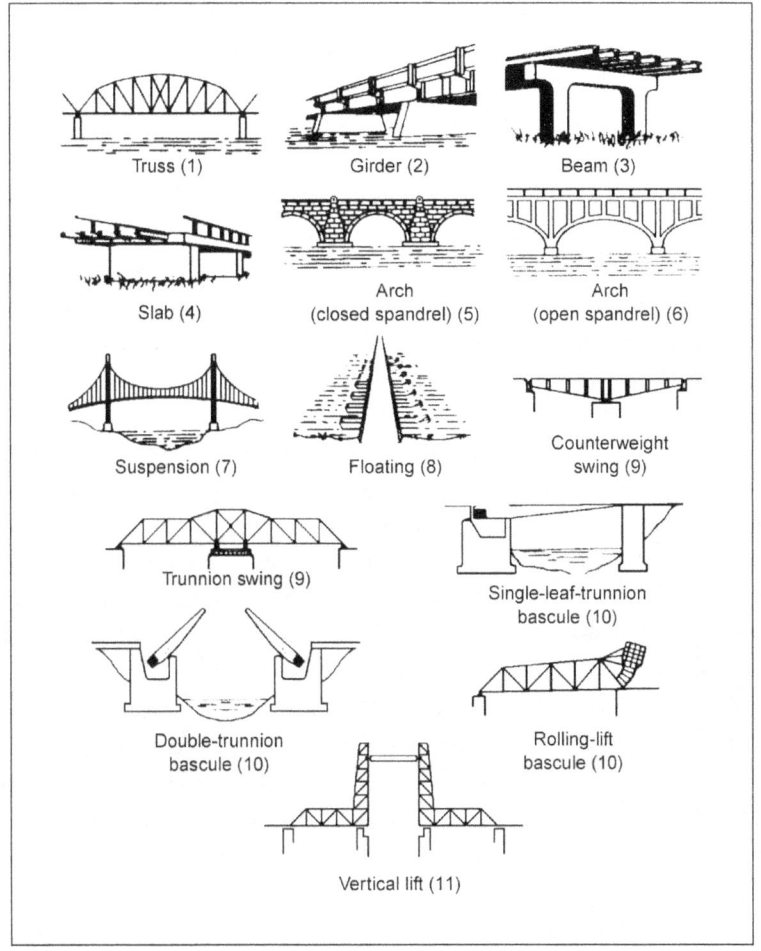

Figure 3-13. Span types

FM 5-34

Table 3-4. Dimensions required on the seven basic bridges

Entries for Front of DA Form 1249 (Figure number reference is in parenthesis under bridge type)

No from Figure	Dimension Data	Simple Stringer (3-7)	Slab (3-8)	T-Beam (3-8)	Truss (3-9)	Girder (3-10)	Arch (3-11)	Suspension (3-12)
1	Overall length	X	X	X	X	X	X	X
2	Number of spans	X	X	X	X	X	X	X
2a	Span length	X	X	X	X	X	X	X
3	Panel length				X			
3a	Height above estimated normal water level	X	X	X	X	X	X	X
4	Traveled-way width	X	X	X	X	X	X	X
5	Overhead clearance				X			
6	Horizontal clearance	X	X	X	X	X	X	X

Entries for Back Side of DA Form 1249

Letter	Capacity(a) Dimension Data	Simple Stringer	Slab	T-Beam	Truss	Girder	Arch	Suspension
a	Wearing-surface thickness	X	X	X	X	X	X	X
b	Flooring and deck thickness or depth of fill at crown	X	X	X	X	X	X	X

		Timber		Steel			
Letter	Capacity Dimension Data	Rectangle	Log	I-Beam	Channel	Rail	
c	Distance, c-to-c, between T-beams, stringers, or floor beams	X	X	X	X	X	
d	No. of T-beams or stringers	X	(b)	X	X	(c)	
e	T-beam or stringer depth (ea)	X		X	X	(c)	
f	T-beam or stringer width (ea)	X		(c)	(c)	(c)	
g	Web thickness of I-beams, WF-beams, channels, or rails			(c)	(c)	(c)	

3-16 Reconnaissance

FM 5-34

Table 3-4. Dimensions required on the seven basic bridges (continued)

Entries for Back Side of DA Form 1249 (continued)

Letter	Capacity[a] Dimension Data	Simple Stringer	Slab	T-Beam	Truss	Girder	Arch	Suspension
h	Sag of cable							X
i	No. of each cable size							X
j	Arch-ring thickness						X	
k	Rise of arch						X	
l	Diameter of each cable size							X
m	Plate girder depth					X		
n	Width of flange plates					X		
o	Thickness of flange plates					X		
p	Number of flange plates					X		
q	Flange-angle depth					X		
r	Flange-angle width					X		
s	Flange-angle thickness					X		
t	Web-plate depth					X		
u	Web-plate thickness					X		
v	Average thickness of flange	X						
w	Cover depth						X	

(a) Capacity is computed by using formulas and data in FM 5-446.
(b) Diameter
(c) Width of flange

Reconnaissance 3-17

FM 5-34

BRIDGE-RECONNAISSANCE REPORTS

To send bridge-reconnaissance information, complete a DA Form 1249 (Figures 3-14 and 3-15). Use Table 3-5, page 3-20, to ensure that you include all the necessary information.

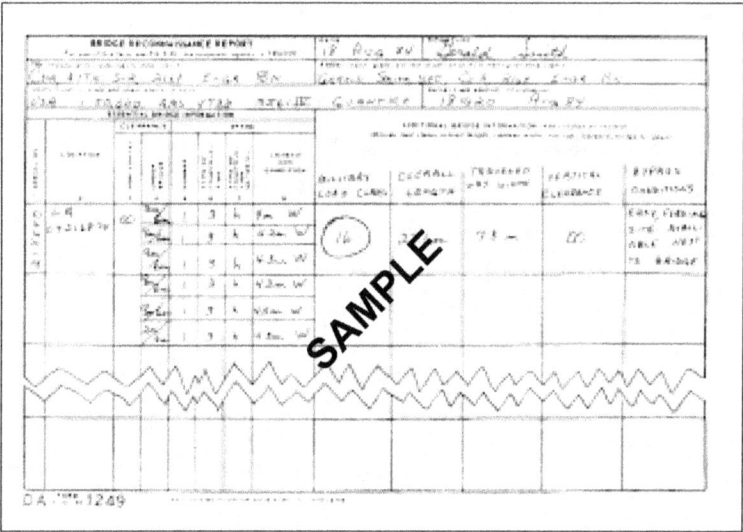

Figure 3-14. Sample, DA Form 1249 (front)

TUNNEL RECONNAISSANCE

Overhead clearances less than 4.3 meters are classified as obstructions. Complete DA Form 1250 the same as DA Form 1249. Figure 3-16, page 3-21, shows a typical sketch of a tunnel with minimum required dimensions. Use Table 3-1, page 3-1, for roadway width requirements.

WATER-CROSSING RECONNAISSANCE

All water-crossing reconnaissance, such as swim, ford, raft, bridge, and ferry, includes the following factors:

- Road network, which should support the largest vehicles and have good drainage facilities.

- Avenues to and from the river, which should be straight for at least 150 meters, have a 10 percent maximum grade, have two lanes with a turnaround, and have an all-weather surface whenever possible.

3-18 Reconnaissance

FM 5-34

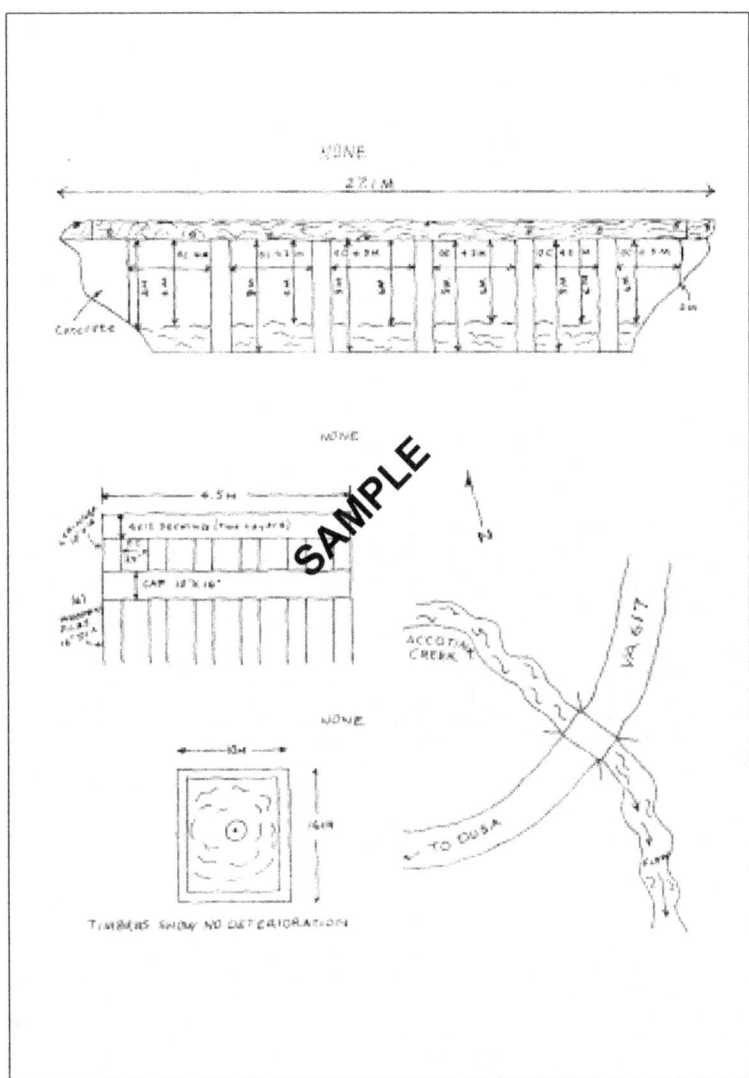

Figure 3-15. Sample, DA Form 1249 (back)

Reconnaissance 3-19

FM 5-34

Table 3-5. Engineer-reconnaissance checklist

——**Roads**. Classify using symbols.
——**Bridges, Fords, and Ferries**. Classify using symbols and include possible bypass for existing crossings.
——**Obstacles to Movement**. Report natural and artificial obstacles including demolitions, mines, and booby traps.
——**Terrain**. Report general nature, ridge system, drainage system including fordability, forests, swamps, and areas suitable for mechanized operations.
——**Engineer Materials**. Report road material, bridge timbers, lumber, steel, and explosives.
——**Engineer Equipment**. Record data on rock crushers, sawmills, garages, machine shops, blacksmith shops, or other facilities or equipment.
——**Errors and Omissions on Maps Used**.
——**Water Points**. Recommend locations.
——**Restrictions on Enemy Movement**. Describe natural or artificial obstacles and sites for construction of improvements (work estimates).
——**Streams**. Give a general description of width, depth, banks, approaches, character of bottom, navigability, and possible ways to cross.
——**Defensive Positions**.
——**Bivouac Areas**. Give data on entrances, soil, drainage, sanitation, and concealment.
——**Petroleum Storage and Equipment**.
——**Utilities**. Report water, sewage, electricity, and gas utilities available.
——**Ports**. Show wharves, sunken obstacles, cargo handling facilities, storage facilities, and transportation routes.
——**Construction Sites**. Report drainage, water supply, power source, earthwork, access, acreage, and soil conditions.
——**Any Other Information of Importance**.

NOTE: Give work estimates as required.

- Riverbanks, which should have stability, slope, and height (see Figure 3-17, page 3-22.)
- Widths, which you can measure by using a string or tape across the river, scaled off the map, or as shown in Figure 3-18, page 3-23.
- Depths, which you record every 3 meters by using a measured pole/rod or weighted ropes/strings.
- Sites, which are assembly areas and other needed areas that should be spacious, provide good concealment, and have easy access routes.
- Velocity, which you measure by using the procedures in Figure 3-19, page 3-23.
- Obstructions, which can be sandbars, floating debris, and other water obstacles or restrictions.
- Drainage, which should be adequate.
- Soil stability, which should be adequate for anchoring. Check the banks and river bottoms for stability.

3-20 Reconnaissance

FM 5-34

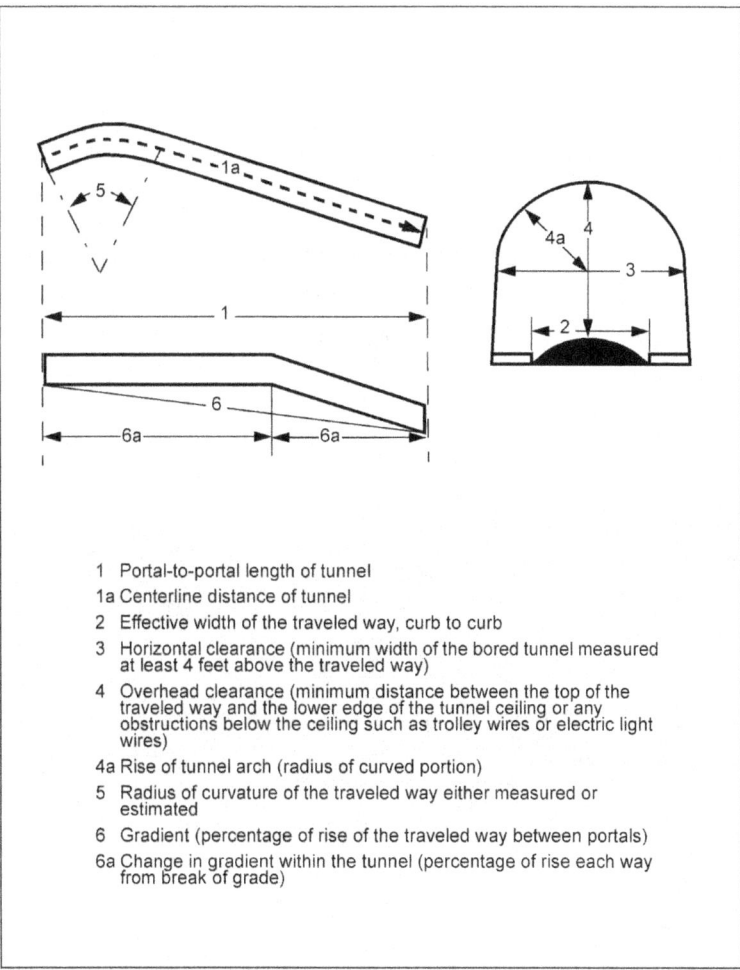

1 Portal-to-portal length of tunnel
1a Centerline distance of tunnel
2 Effective width of the traveled way, curb to curb
3 Horizontal clearance (minimum width of the bored tunnel measured at least 4 feet above the traveled way)
4 Overhead clearance (minimum distance between the top of the traveled way and the lower edge of the tunnel ceiling or any obstructions below the ceiling such as trolley wires or electric light wires)
4a Rise of tunnel arch (radius of curved portion)
5 Radius of curvature of the traveled way either measured or estimated
6 Gradient (percentage of rise of the traveled way between portals)
6a Change in gradient within the tunnel (percentage of rise each way from break of grade)

Figure 3-16. Tunnel sketch with required measurements

Reconnaissance 3-21

FM 5-34

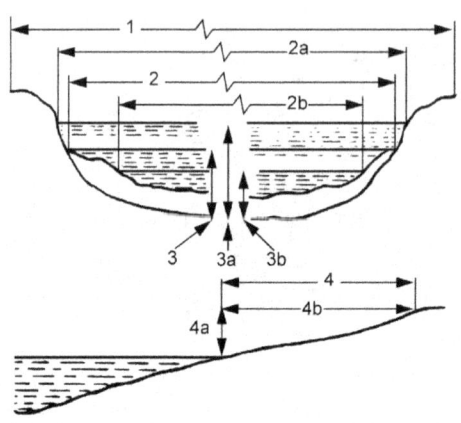

1. The width of streambed from bank to bank
2. The actual width of the water, measured at normal stage (maximum width 2a and minimum width 2b are estimated, based on local observations or records of high water and low water)
3. The actual depth of the stream at normal water level
3a Estimated maximum water depth based on local observations or records
3b /Estimated minimum water depth based on local observations (watermarks) or records
4. The slope of the approaches, which is the slope of the stream banks through that the approach roads are cut

$$\% \text{ slope} = \frac{4a}{4b} \times 100$$

4a = Approach elevation
4b = Approach distance

Figure 3-17. River or stream measurements

3-22 Reconnaissance

FM 5-34

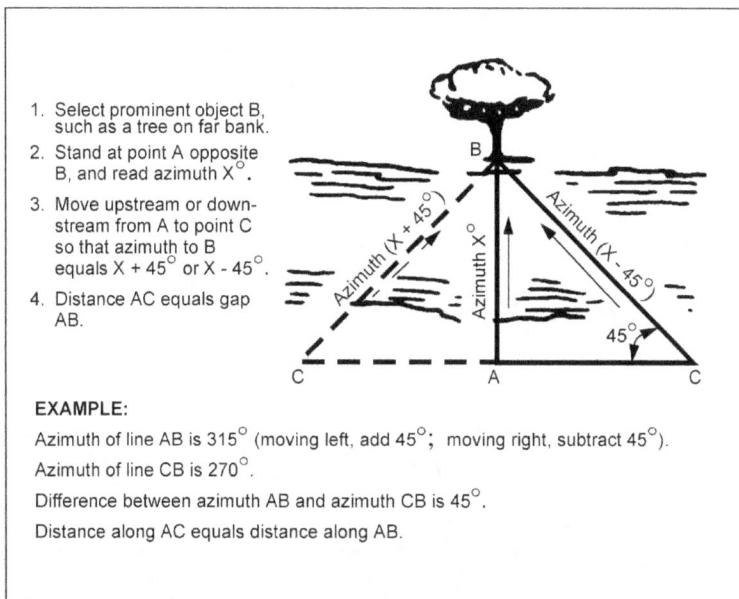

1. Select prominent object B, such as a tree on far bank.
2. Stand at point A opposite B, and read azimuth X°.
3. Move upstream or downstream from A to point C so that azimuth to B equals X + 45° or X - 45°.
4. Distance AC equals gap AB.

EXAMPLE:

Azimuth of line AB is 315° (moving left, add 45°; moving right, subtract 45°).

Azimuth of line CB is 270°.

Difference between azimuth AB and azimuth CB is 45°.

Distance along AC equals distance along AB.

Figure 3-18. Measuring stream width with a compass

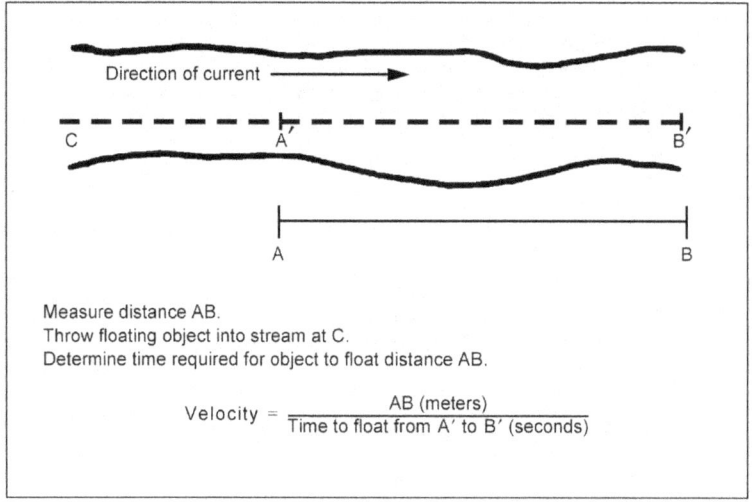

Measure distance AB.
Throw floating object into stream at C.
Determine time required for object to float distance AB.

$$\text{Velocity} = \frac{\text{AB (meters)}}{\text{Time to float from A' to B' (seconds)}}$$

Figure 3-19. Measuring stream velocity

Reconnaissance 3-23

FORD RECONNAISSANCE

Use Table 3-6 to determine trafficability. When DA Form 1251 is used for a swim site, it must specify that the site is for swimming only.

Table 3-6. Ford-site trafficability

Type of Traffic	Shallow, Fordable Depth (m)	Minimum Width (m)	Maximum Percent of Slope or Approaches[1]
Foot	1	1—single file 2—columns of 2	100% 1:1
Trucks and truck-drawn artillery	0.75	3.6	33% 1:3
Tanks	1	4.2	50% 1:2

[1] Based on hard, dry surface

ENGINEER RECONNAISSANCE

An engineer reconnaissance report consists of a completed DA Form 1711-R and an engineer reconnaissance overlay (Figure 3-20 and Figure 3-21, page 3-26). When looking for water-point locations, select sites with running water, if possible. To determine the capacity of the water source, in liters per minute, use the following formula:

$$Q = AV48,000$$

where—

Q = flow, in liters per minute

A = cross section of stream flow, in square meters

V = velocity, in meters per second

48,000 is a conversion and correction factor

FM 5-34

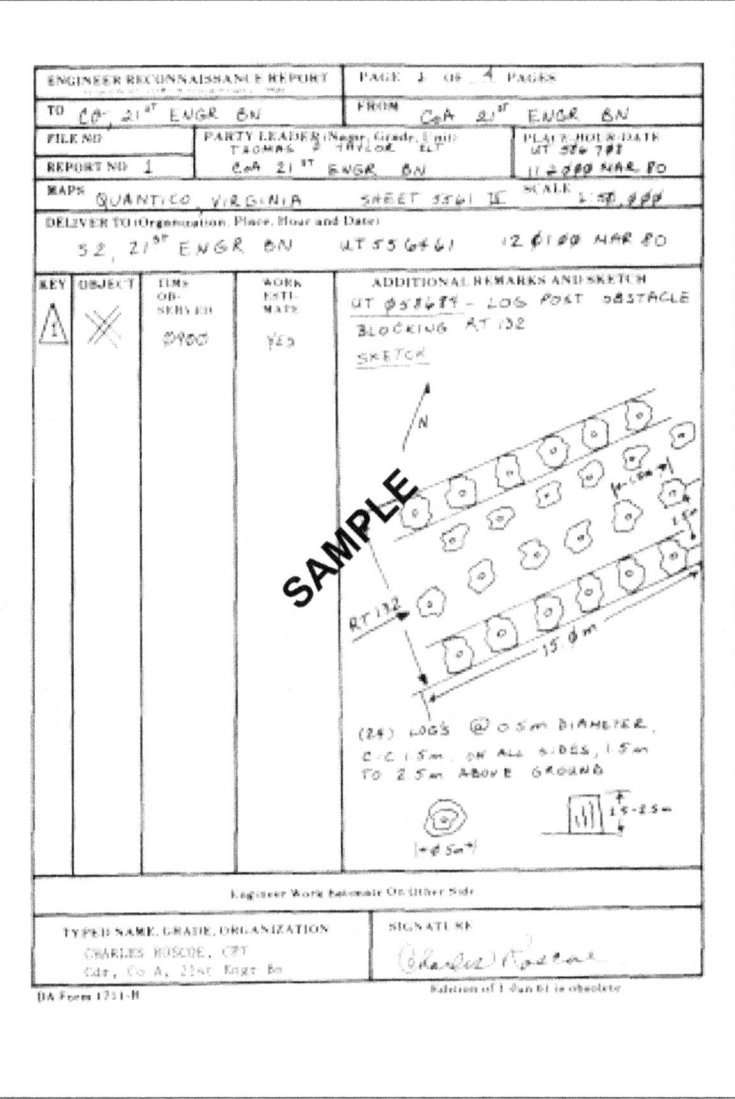

Figure 3-20. Sample, DA Form 1711-R (front)

Reconnaissance 3-25

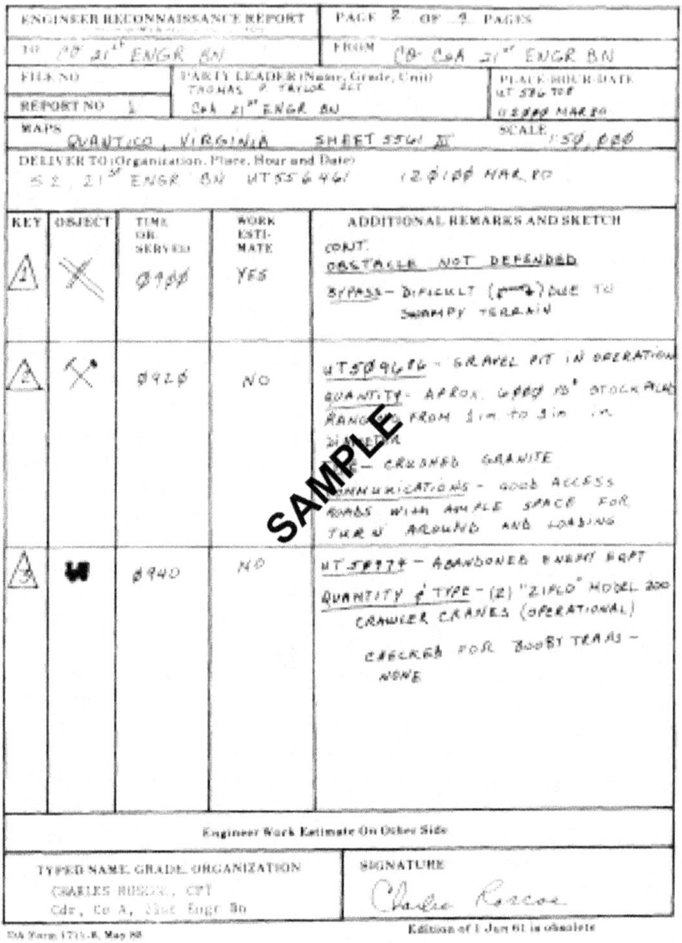

ENGINEER RECONNAISSANCE REPORT | PAGE 2 OF 4 PAGES

TO CO 21st ENGR BN | FROM CO CoA 21st ENGR BN

FILE NO | PARTY LEADER (Name, Grade, Unit) THOMAS P. TAYLOR SGT CoA 21st ENGR BN | PLACE HOUR DATE UT 576 708 0800 MAR 80

REPORT NO 1

MAPS QUANTICO, VIRGINIA SHEET 5561 II | SCALE 1:50,000

DELIVER TO (Organization, Place, Hour and Date) S2, 21st ENGR BN UT556461 1200 MAR 80

KEY	OBJECT	TIME OB-SERVED	WORK ESTI-MATE	ADDITIONAL REMARKS AND SKETCH
1		0900	YES	CONT. OBSTACLE NOT DEFENDED BYPASS – DIFICULT (→) DUE TO SWAMPY TERRAIN
2		0920	NO	UT509686 – GRAVEL PIT IN OPERATION QUANTITY – APPROX. 6000 m³ STOCK PILED RANGING FROM 1 in. to 3 in. in DIAMETER TYPE – CRUSHED GRANITE COMMUNICATIONS – GOOD ACCESS ROADS WITH AMPLE SPACE FOR TURN AROUND AND LOADING
3		0940	NO	UT 58777 – ABANDONED ENEMY EQPT QUANTITY & TYPE – (2) "ZIFCO" MODEL 200 CRAWLER CRANES (OPERATIONAL) CHECKED FOR BOOBY TRAPS – NONE

Engineer Work Estimate On Other Side

TYPED NAME, GRADE, ORGANIZATION CHARLES RUSCOE, CPT Cdr, Co A, 21st Engr Bn | SIGNATURE Charles Ruscoe

DA Form 1711-R, May 88 | Edition of 1 Jan 61 is obsolete

SAMPLE

Figure 3-21. Sample, DA Form 1711-R (back)

Check the color, odor, turbidity, and taste (do not drink) of water. Report any possible pollution such as human or industrial waste or dead fish. Overlay symbols are shown in Figure 3-22; material, facility equipment, and service symbols are shown in Figure 3-23, page 3-29. A reconnaissance checklist is provided in Table 3-5, page 3-20.

FM 5-34

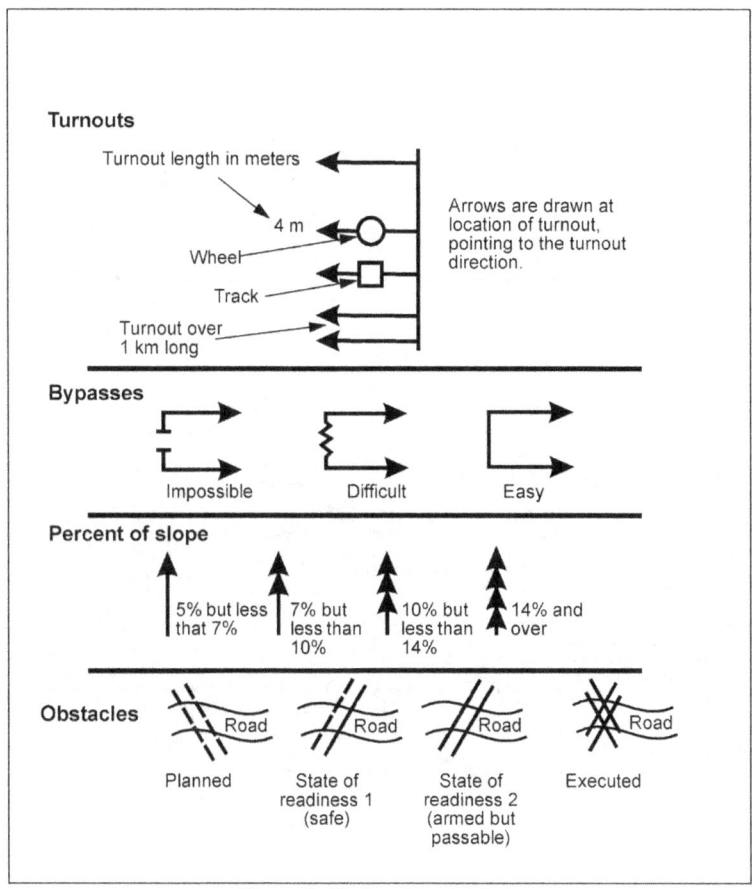

Figure 3-22. Overlay symbols

Reconnaissance 3-27

FM 5-34

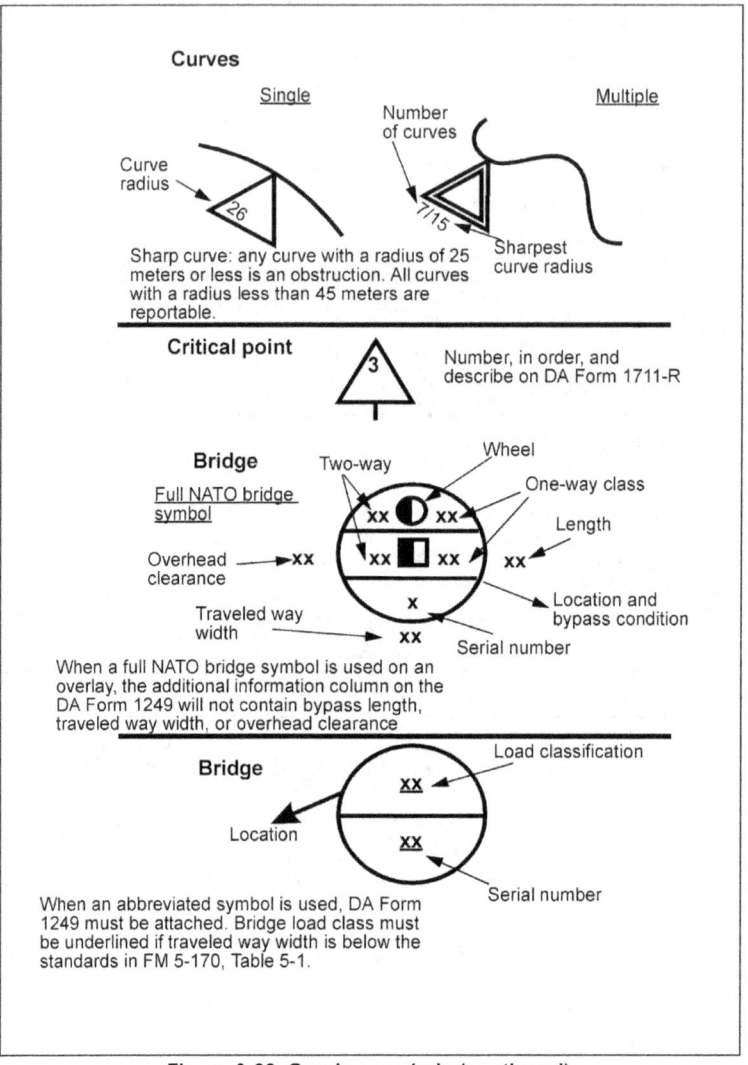

Figure 3-22. Overlay symbols (continued)

3-28 Reconnaissance

FM 5-34

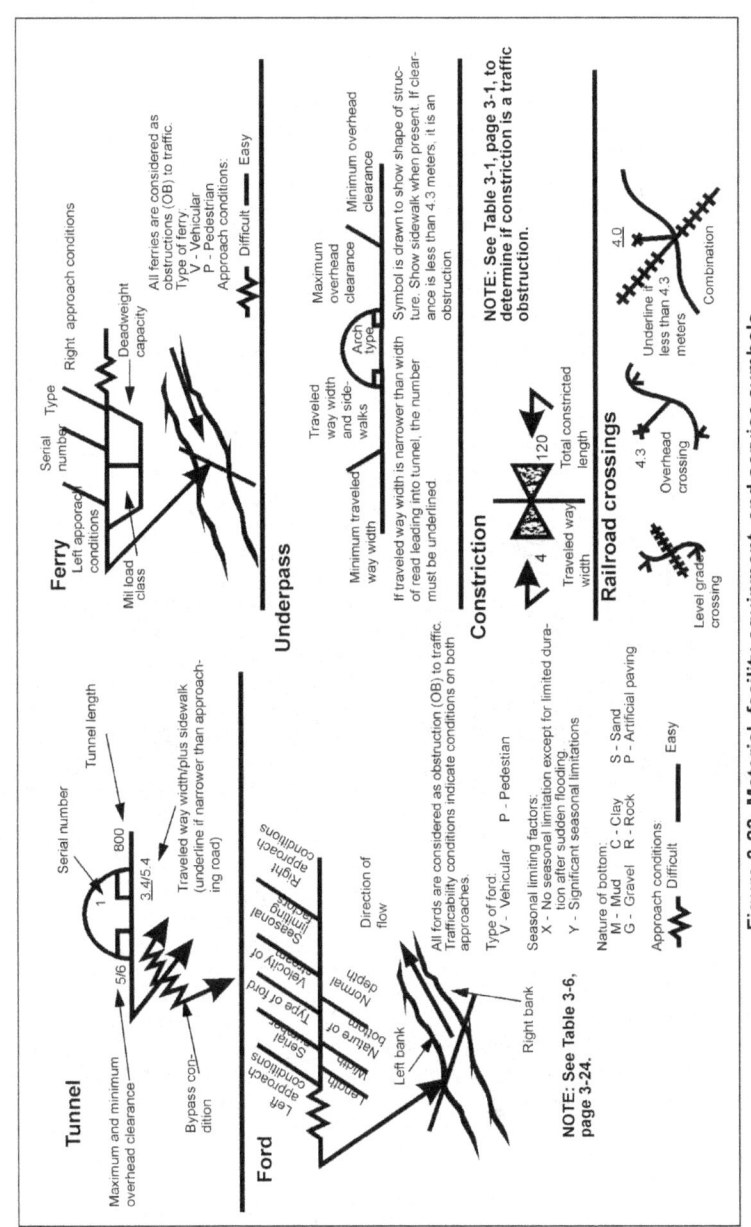

Figure 3-23. Material, facility equipment, and service symbols

Reconnaissance 3-29

FM 5-34

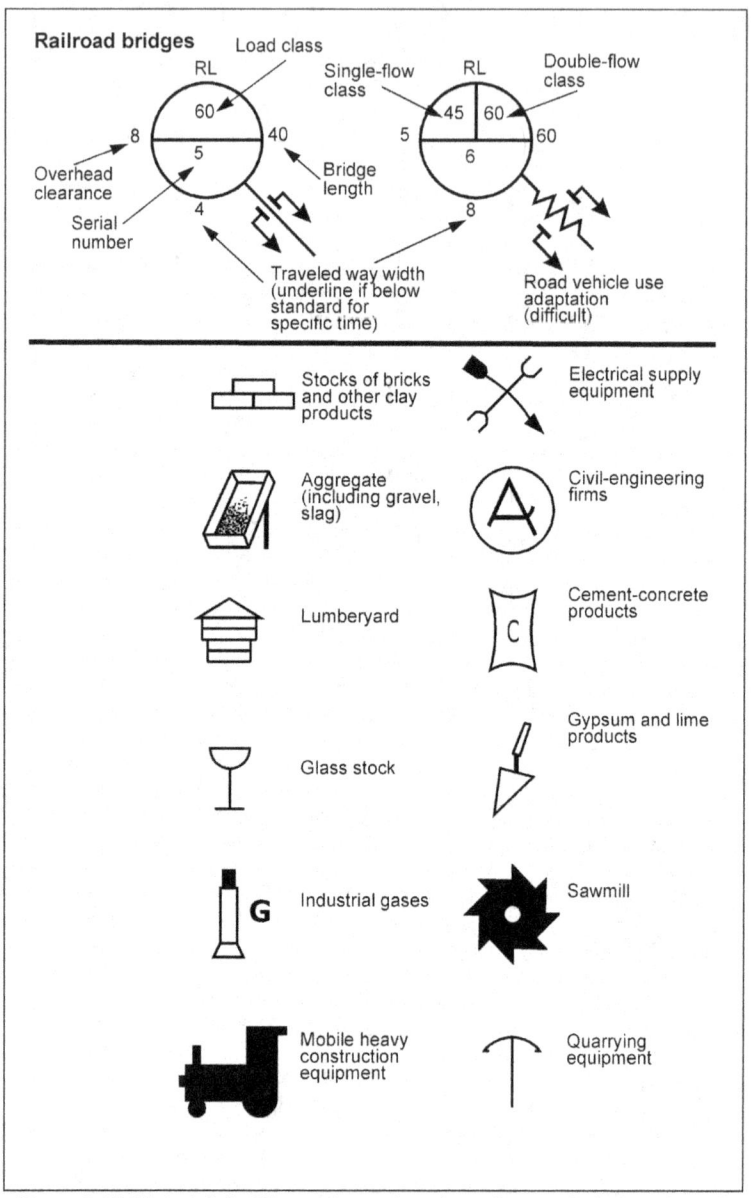

Figure 3-23. Material, facility equipment, and service symbols (continued)

3-30 Reconnaissance

FM 5-34

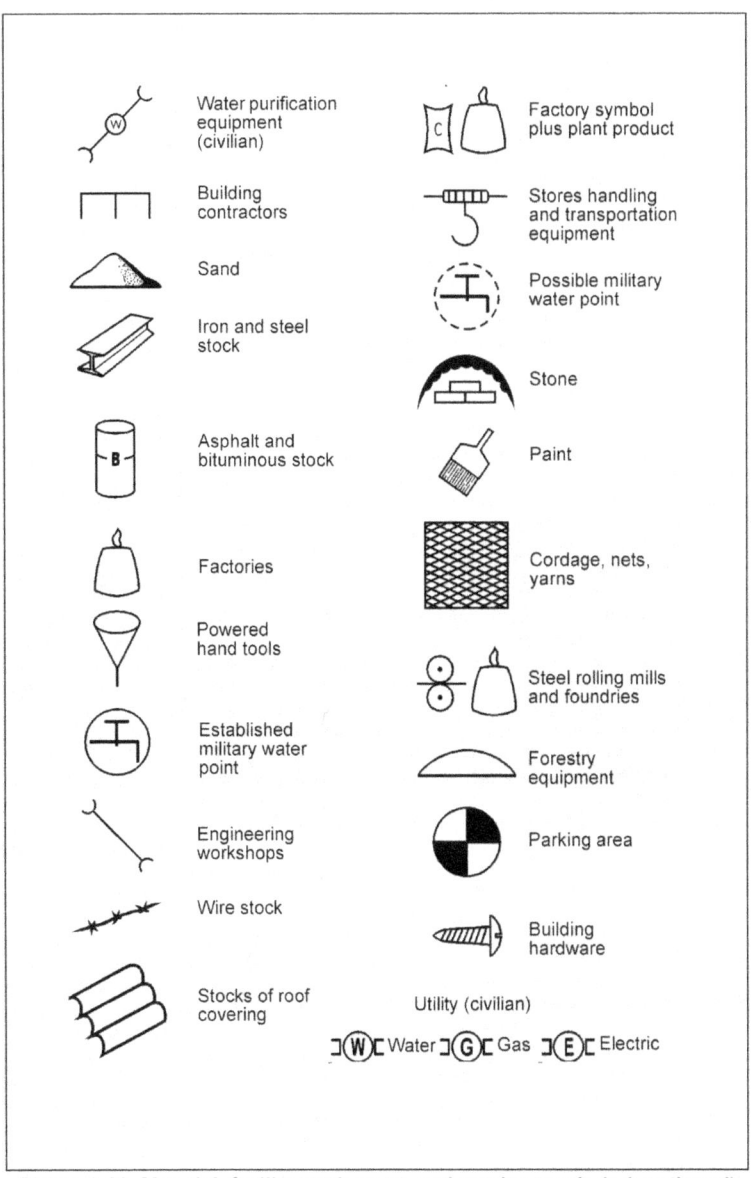

Figure 3-23. Material, facility equipment, and service symbols (continued)

Reconnaissance 3-31

Chapter 4

Mobility

MINE DETECTION

Conduct an analysis by reviewing the terrain, enemy capabilities, and past performances.

MINEFIELD INDICATORS

The following could be indicators of a minefield. (The Thermal Mine Acquisition System [TMAS] can also help you identify mines.)

- Damaged vehicles.
- Dead animals.
- Avoidance by the local populace.
- Signs of digging.
- Signs of concrete removal.
- Holes or grooves in the road.
- Boxes or parcels placed along the road or shoulder.
- Parked vehicles and bicycles without operators.
- Wires on the road surface or extending to the shoulder.
- Metallic devices on the roadway surface.
- Evidence of mine-peculiar supplies (wrenches, shipping plugs, safety collars).
- Disturbances in previous tire tracks.
- Disturbance of road potholes or puddles.
- Disturbance in the cobblestone pattern or missing cobblestone.
- Differences in the amount of moisture or dew on road surface.
- Differences in plant growth (wilting, changing colors, or dead foliage).
- Signs posted on trees that covertly alert the local populace to the presence of mines.

DETECTION AND REMOVAL

Mine-detection and removal methods include visual inspections, probing, using an electronic mine detector, and clearing manually.

VISUAL INSPECTIONS

Check for ground disturbances, posted signs, tripwires, odd features on the ground, and signs of road repairs.

PROBING

Fasten and secure all equipment to your body and remove all metallic objects from your body. Use a slender, nonmetallic probe, in the prone position (as a last resort only, use a bayonet, a screwdriver, a cleaning rod, an antenna, or another sharp object) and probe every 2.5 centimeters across a 1-meter front (gently push the probe into the ground at a 30-degree angle while applying just enough pressure on the probe to sink it slowly into the ground to a depth of at least 3 inches).

ELECTRONIC MINE DETECTOR

Rotate operators at least every 20 minutes.

MANUAL CLEARING

Figures 4-1 and 4-2 and Table 4-1 show an example of a team composition and equipment for a clearing operation. The sweep-team composition is subject to change due to personnel availability and the tactical situation. Figure 4-3, page 4-4, shows a sweep team in echelon. Table 4-2, page 4-5, shows a team organization for a route clearance.

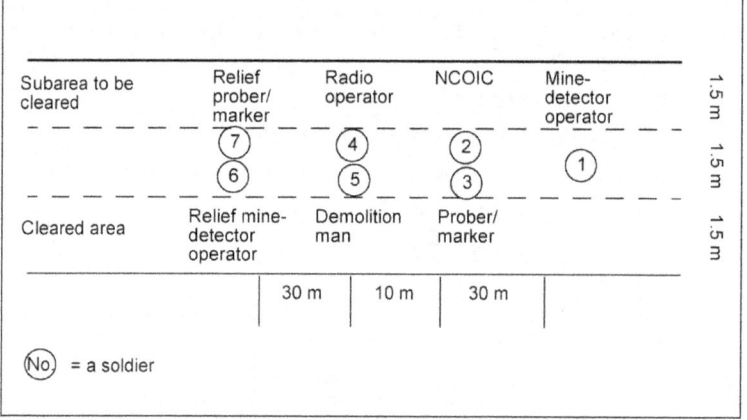

Figure 4-1. Squad-size sweep team

4-2 Mobility

C2, FM 5-34

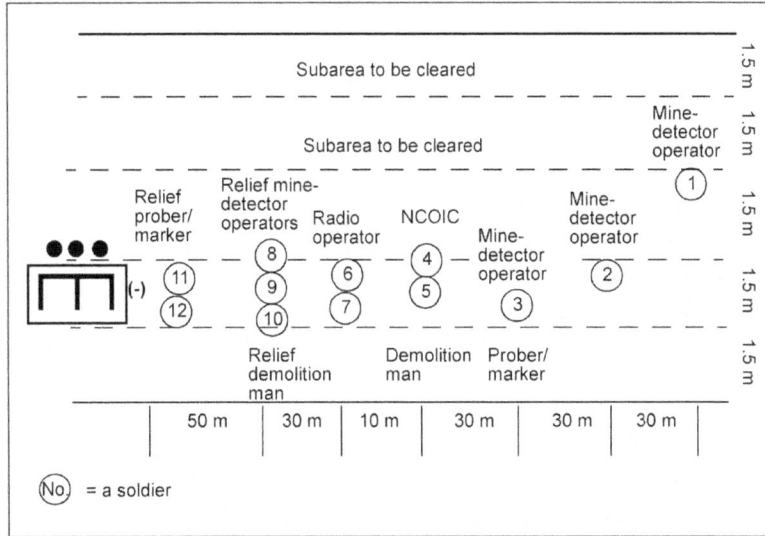

Figure 4-2. Platoon-size sweep team

Table 4-1. Personnel and equipment requirements for a sweep team

Personnel	Support Personnel	Equipment
• NCOIC • Mine-detector operator • Probers/markers • Radio operator • Demolition teams	• Medics • Vehicle operator	• One panel marker • Operational map with required maneuver graphics • Four smoke grenades (minimum) • Six mine detectors (includes three backups) and extra batteries • Two grappling hooks with 60 meters of rope each • One demolition kit or bag for each demolition man • Six probes • Mine-marking material

Mobility 4-3

C2, FM 5-34

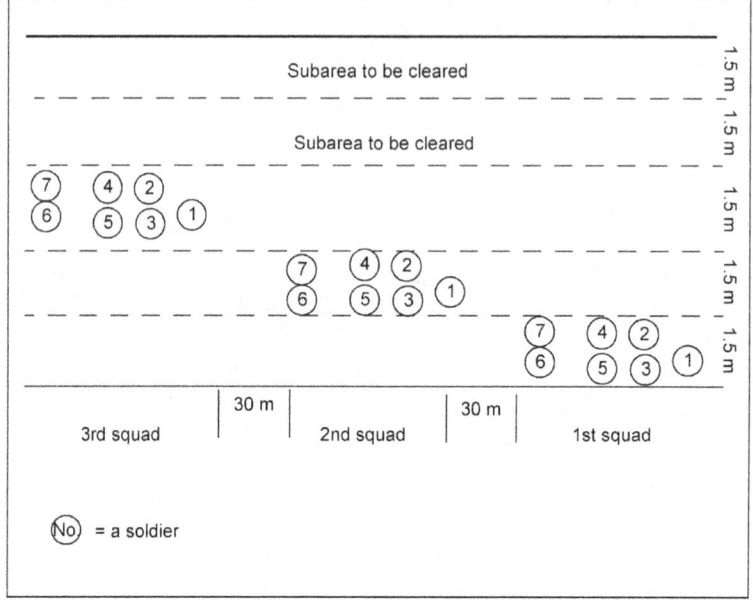

Figure 4-3. Sweep teams in echelon

4-4 Mobility

FM 5-34

Table 4-2. Route-clearance team organization

Team	Support Force	Assault Force	Breach Force
Heavy	• Mechanized infantry platoon with dismount capability • Armor platoon	• Mechanized infantry platoon • Engineer squad • Mortar section • Medical team (two ambulances) • PSYOP team • FIST	• Engineer platoon with organic vehicles • Armor platoon with plows and rollers
Light/heavy	Two infantry platoons (light)	• Bradley platoon with dismount capability • Engineer squad • 60-mm mortar section • Medical team (two ambulances) • PSYOP team • Forward observer	• Engineer platoon with organic vehicles • Armor platoon with plows and rollers
Light	Two infantry platoons (light)	• AT/MP section with M60/MK19 mix • 60-mm mortar section • Medical team (two ambulances) • PSYOP team • Forward observer	• Engineer squad (+) • Infantry platoon (light) • AT/MP section with M60/MK19 mix

Mobility 4-5

FM 5-34

OBSTACLE-BREACHING THEORY

The five breaching tenets are intelligence; breaching fundamentals (suppress, obscure, secure, reduce [SOSR]); breaching organization (support, breach, and assault forces); mass; and synchronization. For more information on obstacle breaching, see FM 90-13-1.

OBSTACLE-REDUCTION TECHNIQUES

The mine-clearing line charge (MICLIC) is a rocket-propelled, explosive line charge used primarily to reduce minefields containing single-pulse, pressure-activated AT mines and mechanically activated AP mines. It clears a path 100 by 14 meters. All pressure-activated mines in this path will be destroyed except the deeply buried mines along a narrow skip zone (see Figure 4-4). Figures 4-5 and 4-6 show other uses of the MICLIC.

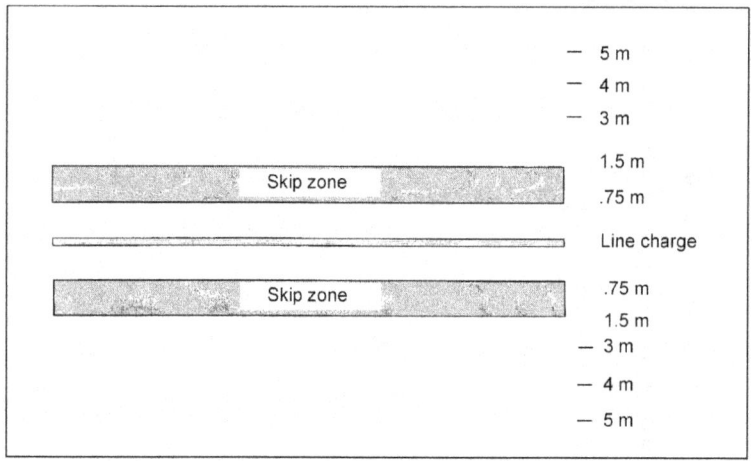

Figure 4-4. MICLIC skip zone

Plows lift and push mines that are surface-laid or buried up to 6 inches deep to the side of the track-width lanes (see Figure 4-7, page 4-8). When plowing, restrict the speed to less than 10 kmph, continue on a straight course through the minefield to prevent damage to the plow, and make sure that the gun tube is traversed to the side. The area you plow must be relatively flat and free of rocks. You should begin plowing about 100 meters from the estimated leading edge of the minefield and 100 meters beyond the estimated far edge. A *dog-bone* assembly between the plows will defeat tilt-rod fused mines. The improved dog-bone assembly (IDA) defeats tilt-rod and magnetically fused mines.

4-6 Mobility

FM 5-34

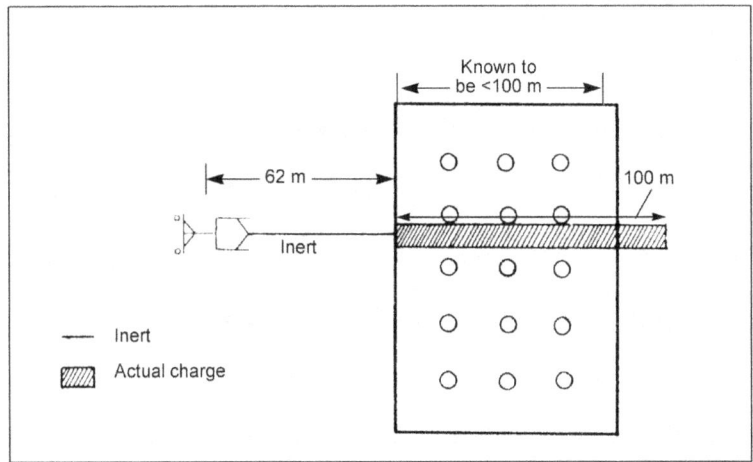

Figure 4-5. Using a MICLIC (depth is less than 100 meters)

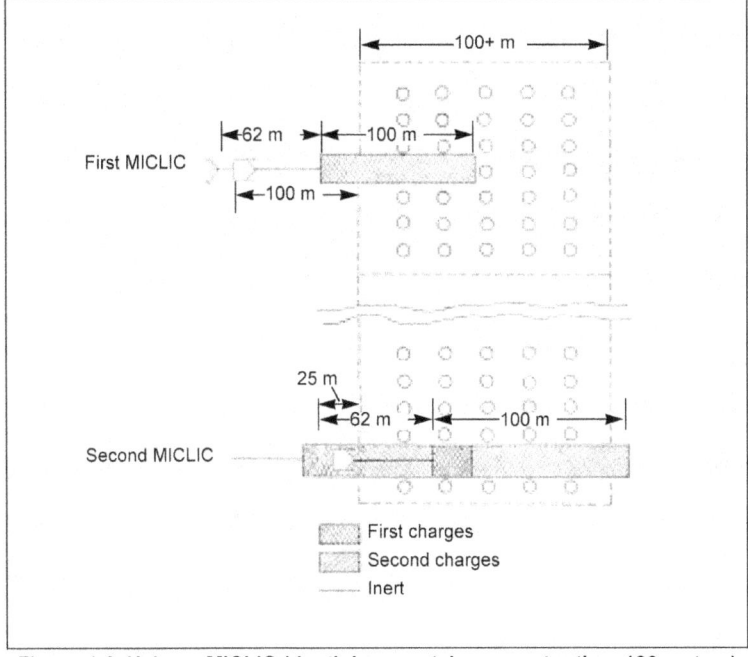

Figure 4-6. Using a MICLIC (depth is uncertain or greater than 100 meters)

Mobility 4-7

FM 5-34

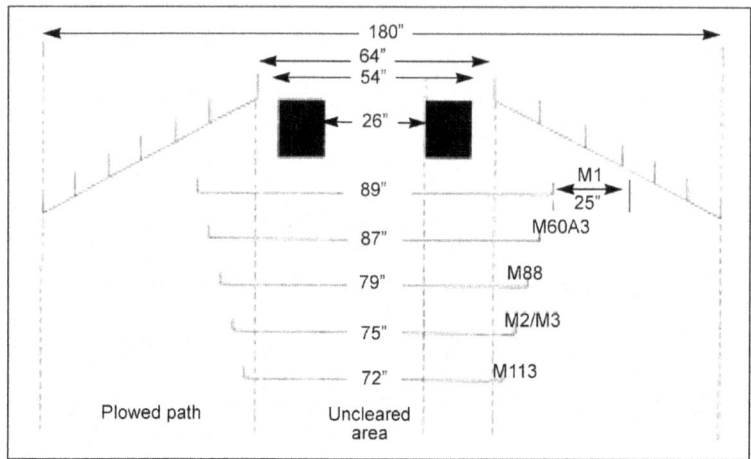

Figure 4-7. Mine-plow width compared to tracked-vehicle widths

Mine rollers are designed to detect minefields. The roller sweeps a 44-inch path in front of each track (see Figure 4-8). You must travel in a relatively straight path, since tight turns may cause the rollers to deviate from the path of the tracks and miss mines. Ensure that the main gun is traversed to the rear or side for a mine encounter. The rollers are designed to defeat most single-pulse, pressure-activated AT and AP mines. A dog-bone and chain assembly between the rollers defeats tilt-rod-fused mines. The IDA can be fitted to the mine roller.

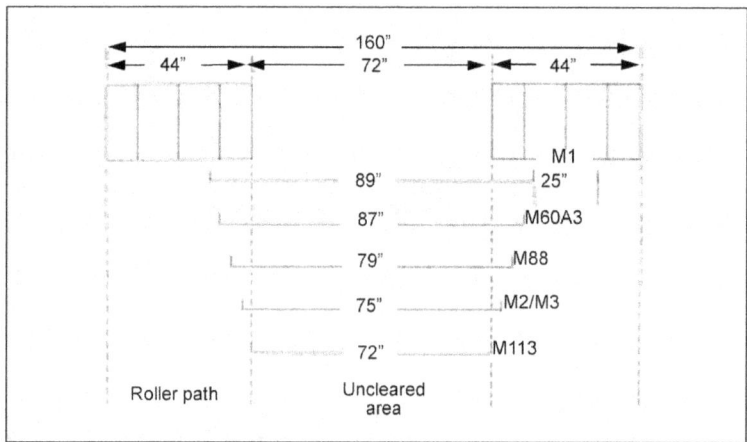

Figure 4-8. Mine-roller width compared to tracked-vehicle widths

4-8 Mobility

You can use the armored combat earthmover. M9 (ACE) as a last resort in a minefield as the blades were not designed for reducing a minefield, thus its use is extremely dangerous to the equipment and crew. However, the blades are ideally suited to break down and reduce earthen gaps such as AT ditches and road craters. ACEs are also effective against wire. Table 4-3 contains information on the ACE and other nonexplosive obstacle breaching equipment. Figure 4-9, page 4-10, shows the blade pattern.

Table 4-3. Nonexplosive obstacle-breaching equipment

Nomen-clature	Load Class	Height (m)	Width (m)	Speed (kmph)	Arma-ment	Mobility Employment
ACE	18	2.3	3.2	48	None	Fills craters and ditches; removes road blocks, trees, and rubble; prepares river and ford accesses; prepares and maintains routes
D7F (dozer)	28	2.4	3.48	10	None	Cuts tactical routes; fills craters and ditches; removes rubble and trees
Loader (2.5)	20	3.7	2.6		None	Fills craters and ditches; removes wire obstacles
AVLB w/bridge w/o bridge	57 37	5	4	48	None	Bridges gaps 18 meters or less; bridges gaps 15 meters or less for Class 70

STANDARD LANE MARKING

The three standard levels of marking breach lanes and bypasses are initial, intermediate, and full.

Marking a breach lane or bypass is a critical subcomponent of obstacle reduction. Effective lane marking allows leaders to project forces through the obstacle quickly, with combat power and command and control (C^2) intact. It provides the assault force a confident, safe lane and helps prevent unnecessary minefield casualties. There are two critical components to any lane marking system:

FM 5-34

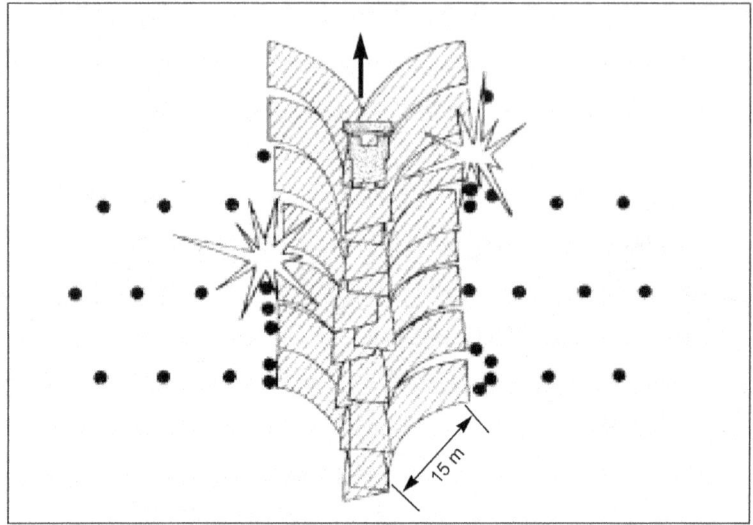

Figure 4-9. Engineer-blade skim pattern

- Lane-marking pattern (the location of markers indicating the entrance, the lane, and the exit).
- Marking device (the type of hardware emplaced to mark the entrance, the lane, and the exit).

INITIAL LANE MARKING

The breach force emplaces the initial lane-marking pattern immediately after the lane is reduced and proofed (see Figure 4-10). The initial lane marking is centered around the minimum markings needed to pass immediate assault forces through the lane to seize the initial foothold on the objective.

INTERMEDIATE LANE MARKING

Upgrading an initial lane marking to an intermediate pattern is triggered by either the commitment of larger combat forces or the rearward passage of sustainment traffic (casualty evacuation [CASEVAC] and vehicle recovery) (see Figure 4-11, page 4-12). Intermediate lane marking has two goals. It—

- Increases the lane signature to assist in the passage of larger, more distant combat forces.
- Provides sufficient marking for two-way, single-lane traffic.

FM 5-34

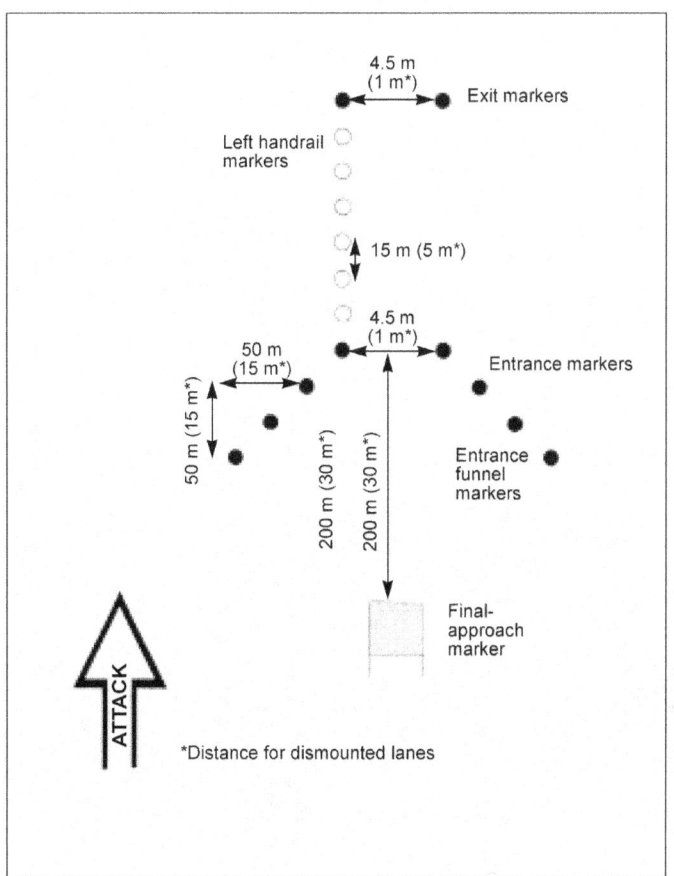

Figure 4-10. Initial lane-marking pattern

Mobility 4-11

FM 5-34

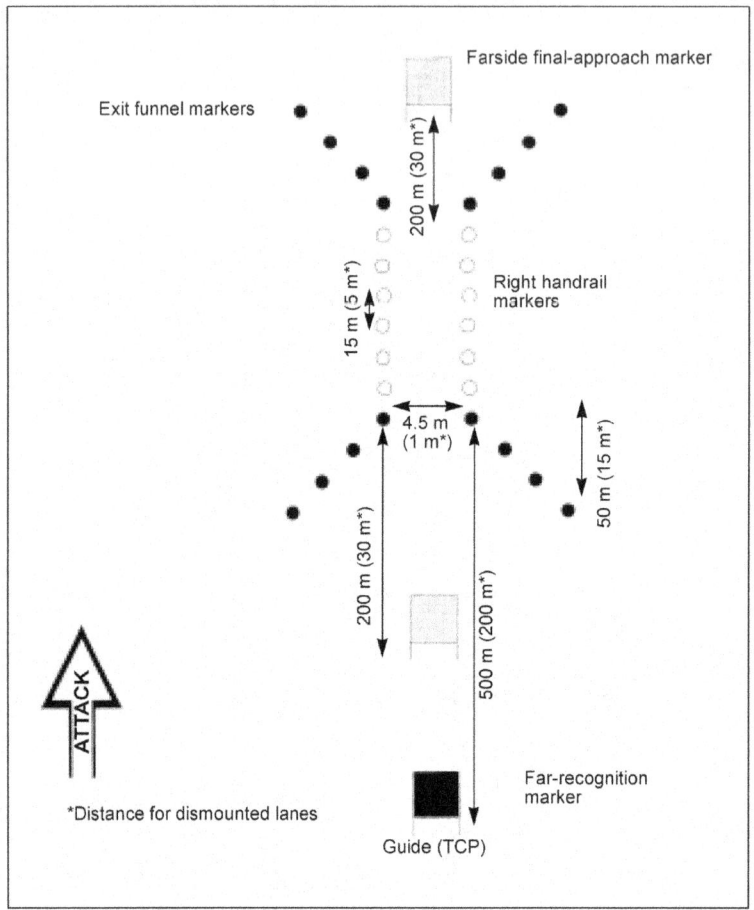

Figure 4-11. Intermediate lane-marking pattern

FULL LANE MARKING

Upgrading to a full lane is normally assigned to follow-on engineer forces. The full lane-marking pattern is also used when marking a lane through friendly obstacles along a major supply route or passage lane (see Figure 4-12). Expanding the breach lane to a full lane involves—

- Expanding the width of the lane to accommodate two-way traffic.
- Modifying the marking pattern to give forces passing forward or rearward the same visual signature.

4-12 Mobility

FM 5-34

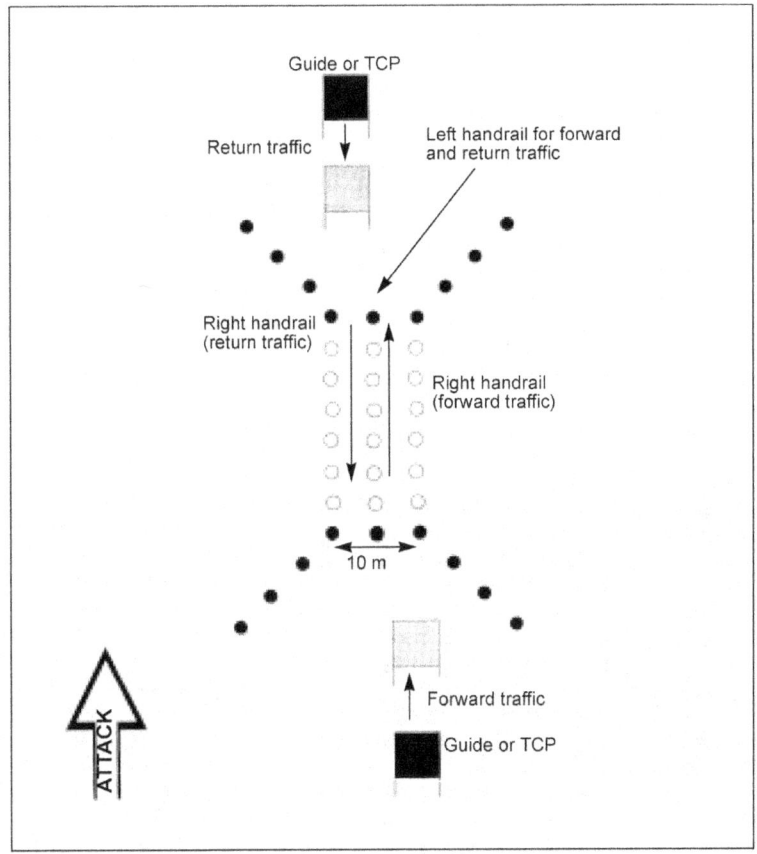

Figure 4-12. Full lane-marking pattern

Table 4-4, page 4-14, lists who is responsible for emplacing each type of lane-marking pattern and what events trigger the emplacement. Table 4-5, page 4-15, lists the requirements for the various components of the lane-marking pattern. Figure 4-13, page 4-16, shows examples of nonstandard marking devices.

Mobility 4-13

FM 5-34

Table 4-4. Lane-marking levels, unit responsibilities, and trigger events

Breach Type	Initial	Intermediate	Full
		Unit Responsibility	
Deliberate	TF breach force	TF breach force	Brigade
Covert	TF breach force	TF breach force	Brigade
In stride	Breach company/team	TF mobility reserve	Brigade
Assault	Assault platoon	TF assault force	NA
		Trigger Events	
	When— • Lanes are reduced • Passing platoon- or company-size forces	When passing— • Battalion- or company-size forces • Forces that cannot see the lane • TF combat trains	When— • Passing brigade- or battalion-size forces • Situation requires uninterrupted sustainment traffic
		Lane Markers	
	Entrance/exit Left handrail Entrance funnel Final approach	The following markers are added: • Right handrail • Exit funnel • Far-side final approach • Far recognition • Guides/TCPs	Lane width is expanded to 10 meters. Existing markers are adjusted. Far-side-recognition markers and guides/TCPs are added.

4-14 Mobility

Table 4-5. Guidelines for lane-marking devices

Marker	Mounted Forces	Dismounted Forces
Handrail and funnel	Visible by TC and driver, buttoned up from 50 meters Quick and easy to emplace minimizing the need to expose soldiers to the outside area	Visible by a dismounted soldier in a prone position from 15 meters Lightweight, quick and easy to emplace; a dismounted soldier should be able to carry enough markers for the lane and still be able to fire and maneuver.
Entrance and exit	Visible by TC, buttoned up from 100 meters Visually different from handrail and funnel markers Quick and easy to emplace; may require soldiers to dismount to emplace; easily manportable	Visible by a dismounted soldier from 50 meters Visually different from the handrail and funnel markers Lightweight, quick and easy to emplace
Final approach and far recognition	Visible by TC, not buttoned up from 500 meters Visually different from each other Visually alterable to facilitate traffic control through multiple lanes	Visible by a dismounted soldier on the march from 100 meters Visually different from each other Visually alterable to facilitate traffic control through multiple lanes

Mobility 4-15

FM 5-34

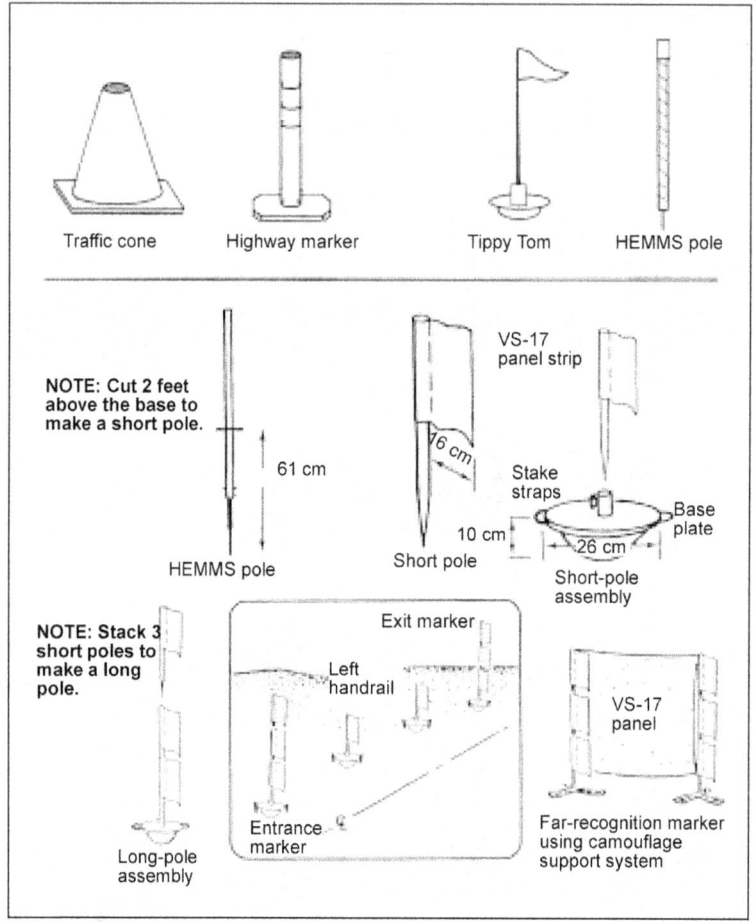

Figure 4-13. Nonstandard marking devices

4-16 Mobility

FM 5-34

NORTH ATLANTIC TREATY ORGANIZATION (NATO) STANDARD MARKING

> This paragraph implements STANAG 2036

To convert intermediate and full lane marking to NATO standards, affix NATO markers to long pickets and replace the existing entrance, exit, funnel, and handrail markers one for one. (See Figure 4-14 for examples of NATO markers.) Since international forces may not be accustomed to using the right lane, place directional arrows to identify lane traffic directions. In addition, lay a barbed wire or concertina fence (one strand minimum) 1 meter above the ground to connect the funnel entrance and handrail markers and the exit pickets.

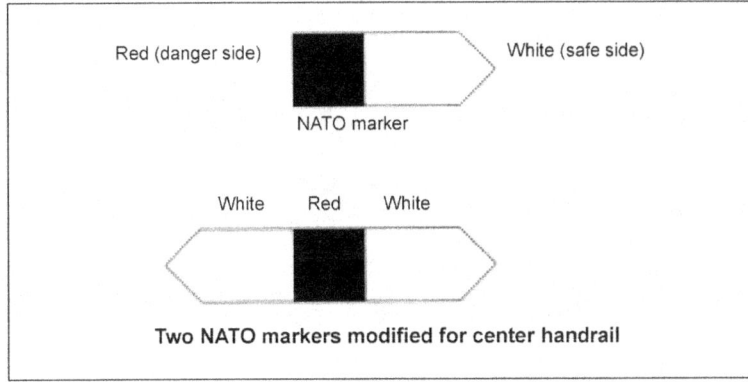

Figure 4-14. NATO standard marker

NATO uses white or green lights to illuminate the markers at night. Entrance and exit markers are marked with either two green or two white lights, placed horizontally, so that the safe and dangerous markings on them are clearly visible. One white or green light will be used on each funnel and handrail marker. The commander may decide whether the light is placed on top of the marker or placed so that it illuminates the marker. Lights should be visible from a minimum of 50 meters under normal conditions.

Mobility 4-17

COMBAT ROADS AND TRAILS

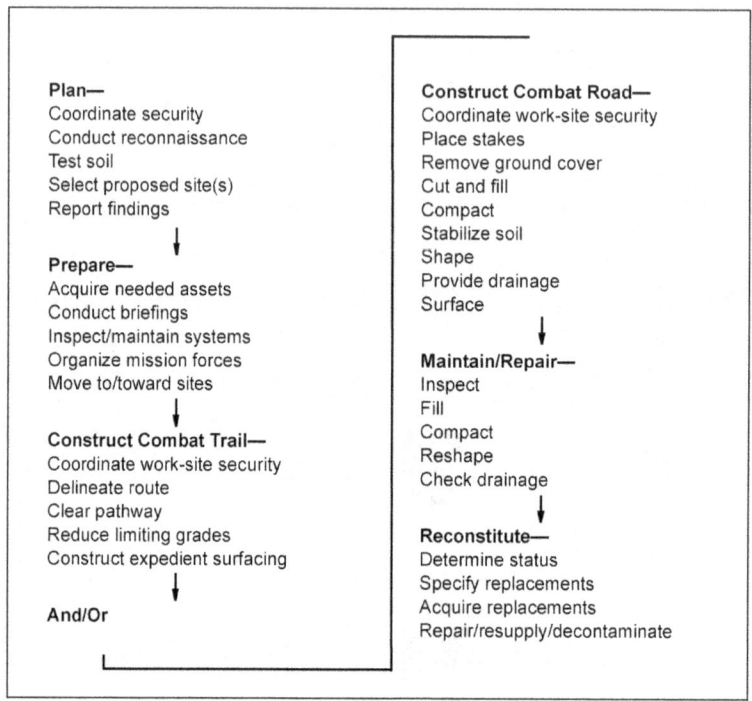

Figure 4-15. Combat roads and trails process

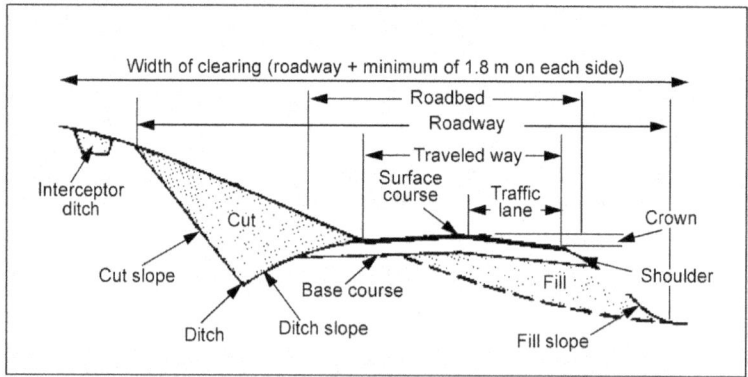

Figure 4-16. Typical cross-section showing road nomenclature

4-18 Mobility

EXPEDIENT SURFACES OVER MUD

CHESPALING MATS

Chespaling mats are made by placing small saplings, 2 meters long and about 3.8 centimeters in diameter, side by side (see Figure 4-17). Wire the saplings together with chicken wire mesh or strands of heavy smooth wire. A chespaling road is constructed by laying mats lengthwise with a 0.3 meter side overlap at the junction of the mats. The resulting surface is 36 meters wide. Unless mats are laid on wet ground, this type of road requires periodic wetting down to retain its springiness and to prevent splitting. Chespaling mats also require extensive maintenance.

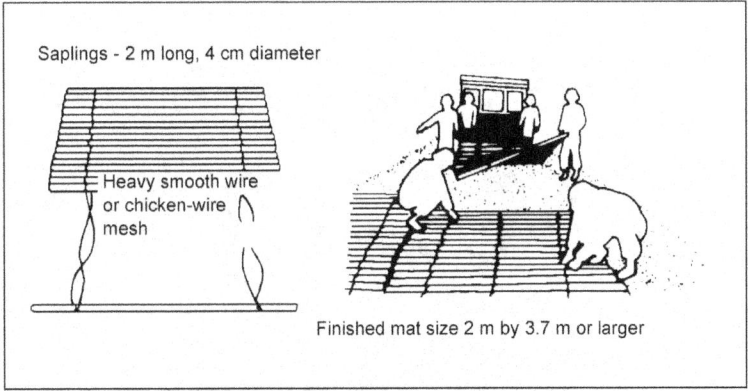

Figure 4-17. Chespaling-surface road construction

CORDUROY

If the road surface is standard corduroy, logs 15 to 20 centimeters in diameter and about 4 meters long are placed adjacent to each other (butt to tip) (see Figure 4-18, page 4-20). Curbs are made by placing 15-centimeter-diameter logs along the edges of the roadway (draft-pinned in place). Pickets about 4 feet long are driven into the ground at regular intervals along the outside edge of the road to hold the road in place. For a smoother surface, the chinks between the logs should be filled with brush, rubble, and twigs. The whole surface is then covered with a layer of gravel or dirt. Side ditches and culverts are constructed as for normal roads.

If the road surface is corduroy with stringers, the corduroy decking is securely pinned to stringers, and then the surface is prepared as standard corduroy (see Figure 4-18). A road surface that is heavy corduroy uses sleepers (heavy logs 25- to 30-inch diameter and long

FM 5-34

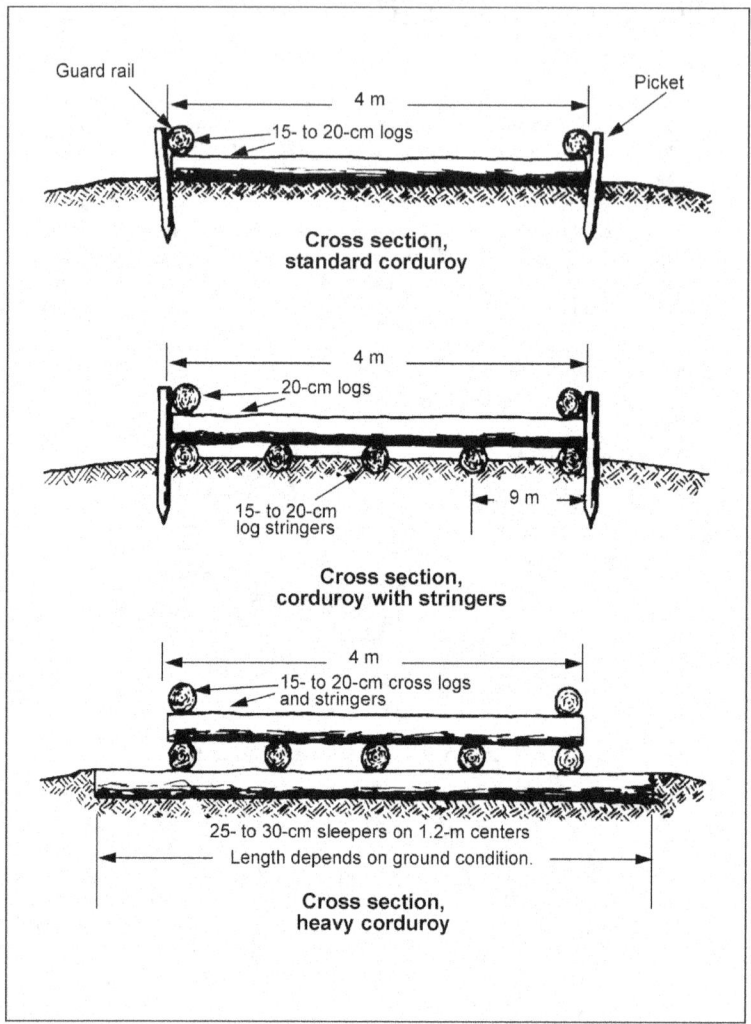

Figure 4-18. Corduroy road surfaces

enough to cover the entire road) placed at right angles to the centerline on 1.2-meter centers (see Figure 4-18). A road surface that is fascine corduroy uses fascine instead of logs for stringers (see Figure 4-19).

4-20 Mobility

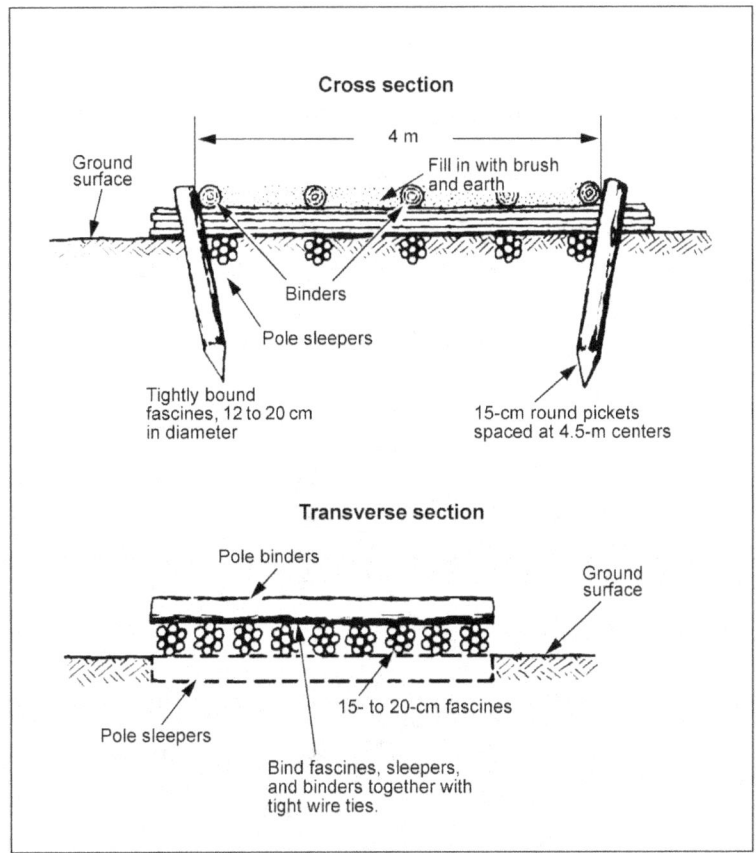

Figure 4-19. Fascine corduroy

TREAD ROADS

Tread roads are made by preparing two narrow parallel treadways of selected material using anything from palm leaves to 4-inch planks. The most common tread road is the plank tread road (see Figure 4-20, page 4-22).

OTHER SURFACES

Surfaces can be constructed from rubble, bricks, concrete blocks, loose aggregate or gravel, and airfield matting. See Figures 4-21 through 4-23, pages 4-22 and 4-23, for other types of road surfaces. For more information, see FM 5-430-00-1, Chapter 9.

FM 5-34

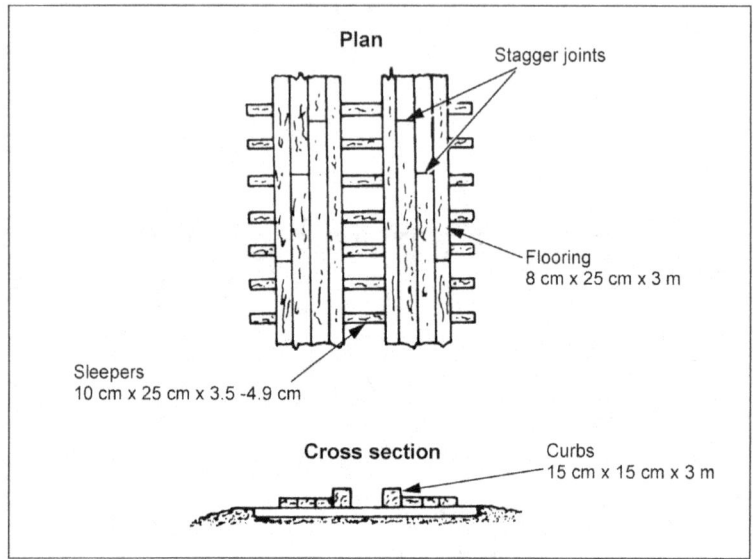

Figure 4-20. Plank tread road

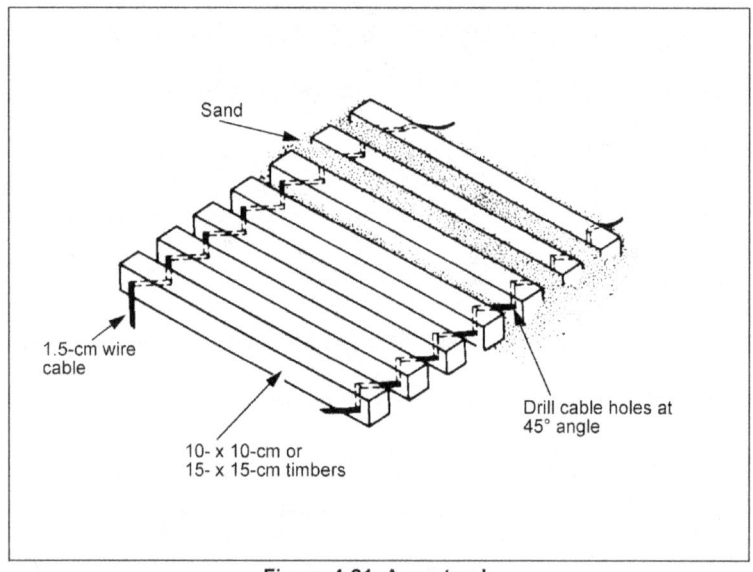

Figure 4-21. Army track

4-22 Mobility

FM 5-34

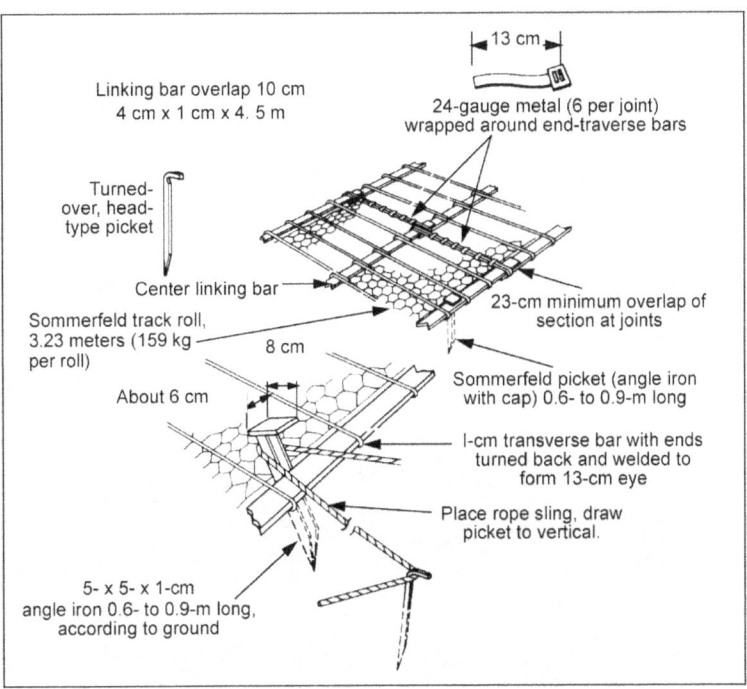

Figure 4-22. Component parts of a Sommerfield truck

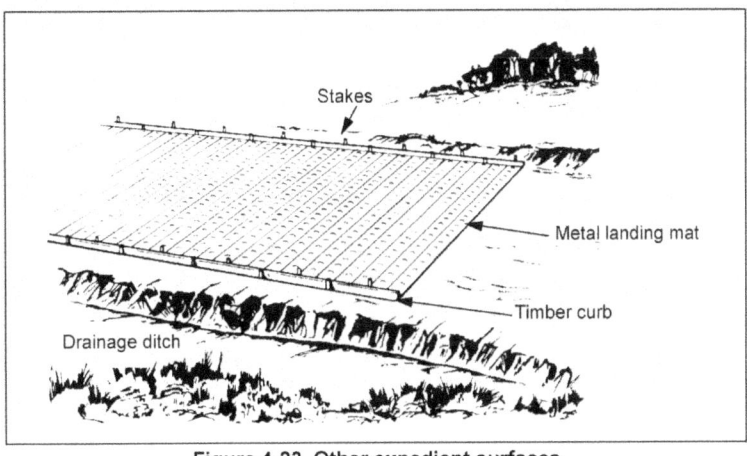

Figure 4-23. Other expedient surfaces

Mobility 4-23

FM 5-34

EXPEDIENT SURFACES OVER SAND

WIRE MESH

Chicken wire, expanded metal lath, or chain-link wire mesh (cyclone fence) may be used for expedient surfaces over sand. Adding a layer of burlap or similar material underneath the wire mesh helps confine the sand. The edges of the wire-mesh road must be picketed at 0.9- to 1.2-meter intervals. Diagonal wires that cross the centerline at 45° angles and are securely attached to buried pickets fortify the lighter meshes. The more layers used, the more durable the pad will become (Figure 4-24).

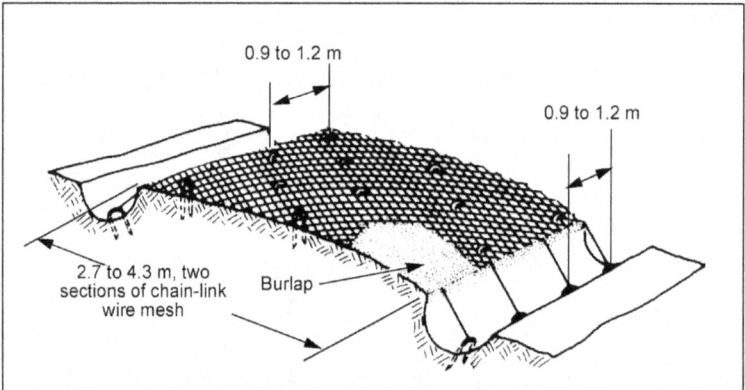

Figure 4-24. Chain-link wire-mesh road

SAND GRID

In a sand grid, each grid section expands to cover an area 2.4 meters x 6 meters x 20 centimeters (see Figure 4-25). Use pickets or place sand on the corners and sides to maintain placement. You may use a bucket loader to fill in the grids. Use hand shovels to fill each grid completely. A full grid section will hold the weight of a bucket loader. To compact the surface, you can use a rubber-tire or steel-wheel roller. You may apply a sand asphalt surface of about one gallon of RC-250 asphalt per square yard.

Place the initial 8-inch layer of sand grid in the crater, parallel to the centerline of the runway or roadway. Expand and place the sand grid, using shovels to fill the grid-edge sections to hold it in place. Cover the entire bottom of the crater with sand grid in this manner. Place grid so it conforms to the shape of the crater, curving or cutting as necessary. Sections that do not fully expand only add strength to this base.

4-24 Mobility

FM 5-34

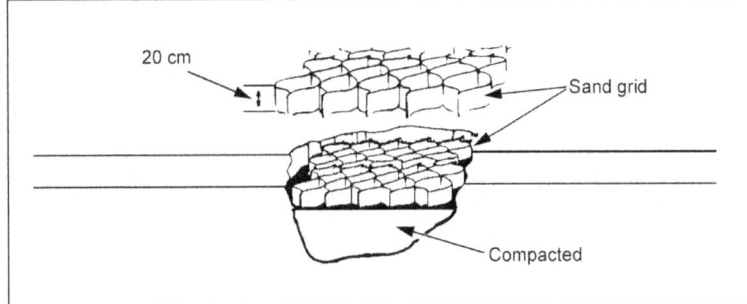

Figure 4-25. Sand grid

After laying the entire first layer, use a bucket loader to begin filling the grid, working from one edge of the crater towards the center. As the grids are filled, the loader can drive forward onto the grids to expand their range. Use shovels and rakes to spread the material to overfill the cells by a uniform layer of 2.5 to 5.0 centimeters. The loader should provide sufficient compaction; however, you could use the towed vib or plate tampers. Over compaction will damage the sand grid; your repair will fail. After compacting the first layer, repeat the process, starting with placing a layer of membrane.

Place the top layer of sand grid with its long dimension stretched 90 degrees to the first, which makes it perpendicular to the runway centerline for added strength.

Overfill the grid cells by 2.4 to 5.0 centimeters. Compact the cells level with the pavement surface. If excess material is present after compaction, remove it.

FORWARD AVIATION

ARMY AIRCRAFT AND HELICOPTER CHARACTERISTICS

Table 4-6, page 4-26, lists information on Army helicopters; Table 4-7, page 4-27, lists the requirements for combat airfields.

FM 5-34

Table 4-6. Army helicopter characteristics

Nomenclature	Name	Length (m)	Rotor Diameter (m)	Height (m)	Basic Weight (kips)	External Load Capacity* (1,000 kg)	Cargo Winch Capacity (1,000 kg)
OH-6A	Cayuse	11.98	8.02	2.67	1.16	--	--
OH-58 A/C	Kiowa	12.85	10.66	3.91	1.90	--	--
CH47B	Chinook	30.14	18.28	5.76	19.59	9,072	1,361
CH-47C	Chinook	30.14	18.28	5.66	20.48	9,072	1,361
CH-47D	Chinook	30.14	18.28	5.68	22.50	11,794	1,361
CH-54A/B	Crane	26.97	21.94	7.74	21.20	9,072/11,640	6,804/11,340
UH 1 H/V/	Iroquois	17.39	14.63	4.42	5.13	1,815	--
UH-60A	Black Hawk	19.76	16.35	5.00	11.04	3,629	--
AH-64	Apache	17.75	14.63	4.67	14.66	2,722	--
AH-1S	Cobra	16.15	13.41	3.53	6.60	454	--

*Maximum lifting capability

FM 5-34

Table 4-7. Combat-area airfield requirements

Airfield Type	Anticipated Service Life (months)	Possible Using Aircraft (US type)	Takeoff Ground Run at Sea Level at 51°F (m)	Minimum Runway Length (m)	Minimum Runway Width (m)	Minimum Shoulder Width (m)
Close battle area	0-6	C-17 C-130*+	914.4	1,066.8	27.43 18.29	3.05
Support area	0-6	C-17 C-130*+	914.4	1,920.24	27.43 18.29	12.19
Rear area	6-24	C-5* C-141+ C-17 C-130 KC-135 B-1 KC-135 F-117 E-3 F-4/F-15 A-7	3,505.20 2,712.72 2,316.48 1,219.20 2,042.16 3,018.96 2,042.16 2,926.08 1,993.39 1,219.20 1,944.62	4,389.12 F-111 F-16	45.72 1645.92 883.92	15.24

*Controlling aircraft for runway length
+Controlling aircraft for required surface depth

Mobility 4-27

CONSTRUCTION OF FORWARD LANDING ZONE OR AIRSTRIP

You can use membrane or available timber to construct an expedient hardened landing-pad surface. Mark all obstacles in the landing zone or airstrip. Sprinkling water, lime, lime solutions, or oils will provide temporary dust control (see Table 4-8). Table 4-9 and Figures 4-26 through 4-29, pages 4-30 and 4-31, show geometric requirements for landing zones and helipads.

Table 4-8. Dust-control requirements for heliports

Helicopter	Diameter of Dust Proofing Area When Parked or on Taxiway (m)	Diameter of Dust Proofing for Landing or Takeoff (m)
AH-64	45.72	91.44
OH-58	45.72	48.77
UH-60	45.72	80.47
CH-47	91.44	179.83
CH-54	91.44	131.67

MAINTENANCE AND REPAIR

Coordinate maintenance and repair operations with tactical operations. Try to do your work at night. Do not leave hazardous equipment on the landing zone. Clearly mark all the areas that are under construction or repair them. Ensure that mud is continuously removed. Remove all debris away from the traffic and landing areas for repair of all mats and membrane surfaces. Replace damaged timber and levels accordingly.

FM 5-34

Table 4-9. Minimum geometric requirements for landing zones in close battle areas

Item No.	Specifications	OH-58	UH-60	AH-64	CH-47	CH-54
	Landing Pad					
1	Length, (m)	3.7	12.2	15.2	15.2	15.2
2	Width (m)	3.7	7.0	7.6	7.6	15.2
3	Landing-pad grade in direction of approach or departure (%)	3	3	3	3	3
4	Shoulder width (ft)	--	--	--	--	--
5	Grade of shoulder in direction of approach or departure (%)	--	--	--	--	--
6	Traverse grade of shoulder (%)	--	--	--	--	--
7	Grade of clear area, maximum (%)	10	10	10	10	10
	Landing Area					
8	Length (m)	21.9	36.6	32.0	45.7	45.7
9	Width (m)	21.9	33.8	30.5	38.1	45.7
	Approach/Departure Zone					
10	Approach, departure surface ratio	10:1	10:1	10:1	10:1	10:1
11	Length (m)	457.2	457.2	457.2	457.2	457.2
12a	Width at end of landing area (m)	21.9	33.8	30.5	38.1	45.7
12b	Width at outer end (m)	152.4	169.2	152.4	152.4	152.4
	Takeoff Safety Zone					
13	Length (m)	152.4	152.4	152.4	152.4	152.4
14a	Width at end of landing area (m)	21.9	33.8	30.5	38.4	45.7
14b	Width at outer end (m)	65.5	73.4	71.0	76.2	81.4

Mobility 4-29

FM 5-34

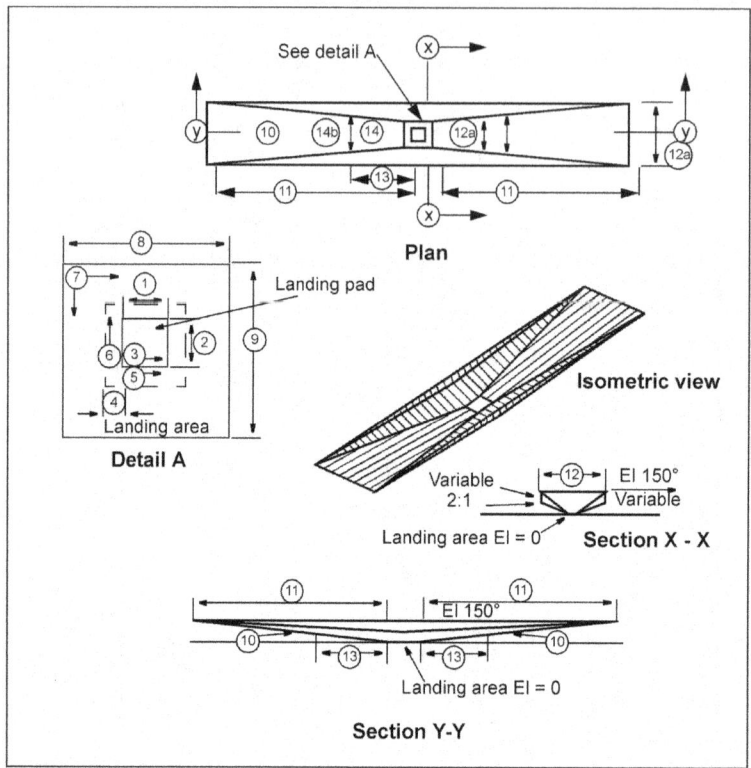

Figure 4-26. Geometric layout of landing zones

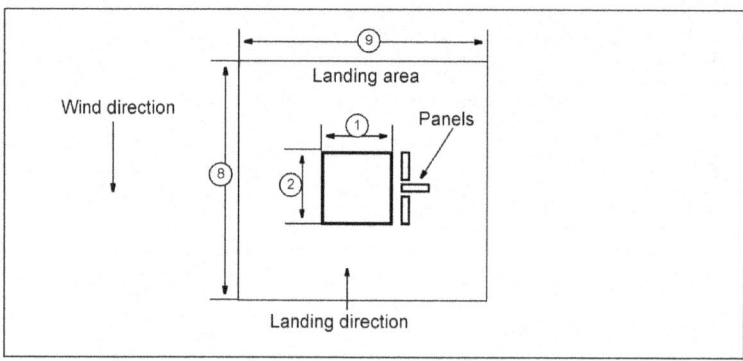

Figure 4-27. Panel layout of landing zones

4-30 Mobility

FM 5-34

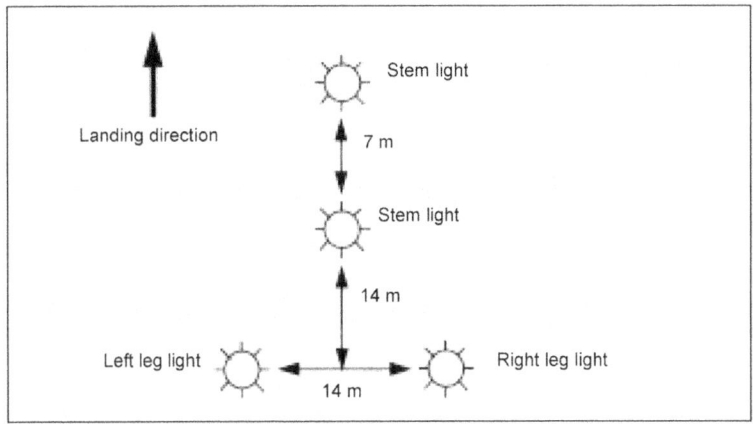

Figure 4-28. Inverted Y

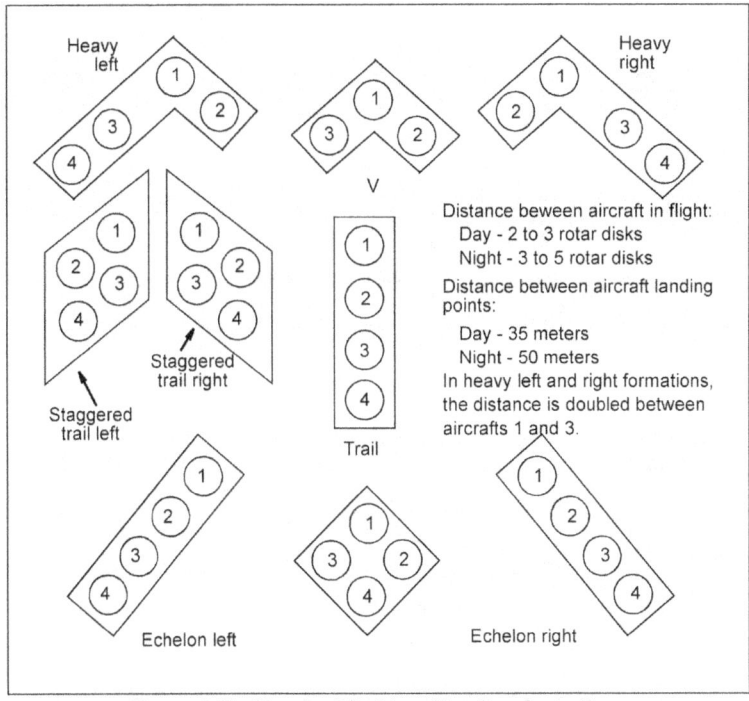

Figure 4-29. Standard flight and landing formations

Mobility 4-31

Chapter 5

Defensive Operations and Obstacle Integration Framework

PLANNING

PROCEDURES

- Integrate into the intelligence preparation of the battlefield (IPB) process with the Intelligence Officer (US Army) (S2), and identify enemy engineer capabilities.
- Analyze the mission, maneuver commander's intent, AAs, engagement areas (EAs), battle positions (BPs), and key weapon-system emplacement/ranges.
- Conduct a leader's reconnaissance with the task force (TF)/company commanders.
- Obtain the commander's obstacle priorities.
- Determine the possible obstacle locations/types.
- Calculate the maximum squad and blade hours available.
- Determine the availability of Class IV/V items.
- Finalize the obstacle plan.
- Develop the obstacle-emplacement priorities.
- Develop a survivability-execution matrix.
- Calculate the Class III and additional operators that will be required for sustainment.
- Coordinate with the TF/company staff.
- Initiate/process reports.
- Supervise obstacle emplacement, verify locations, report any changes.
- Obtain the mission for engineers during the battle: battle position, assembly area, and so forth.

MANEUVER TF RESPONSIBILITIES

TF Commander

- Provides the TF engineer with specific guidance to include the tasks, priorities, and intent; specifically, identifies where he wants to kill the enemy.
- Conducts a leader's reconnaissance with a TF engineer to identify obstacle-group locations.
- Determines the survivability priorities for blades.
- Determines the Class III, IV, and V responsibilities.
- Obtains authorization from the brigade/division to emplace minefields and scatterable mines (SCATMINEs), cut roads, and detonate demolitions.

Company Commander

- Conducts a leader's reconnaissance with an engineer representative.
- Covers obstacles with fires and observation.
- Guards obstacles to ensure that the enemy does not prebreach.
- Accepts target turnover.
- Provides security for the engineer elements who are emplacing the obstacles (at a mine dump).
- Closes passage lanes.
- Furnishes passage-lane guides.
- Provides a supervisor to direct heavy engineer equipment emplacement of survivability positions to ensure correct emplacement.
- Knows the construction standards of the vehicle-fighting/crew-served positions.
- Provides infantry labor augmentation to assist engineer emplacement of obstacles.
- Secures cache sites.
- Conducts counterreconnaissance to prevent breaching obstacles during limited visibility.
- Coordinates with the senior engineer on emplacing obstacles to ensure proper siting.

TF Operations and Training Officer (US Army) (S3)

- Makes enemy breaching equipment a high-priority target.
- Coordinates air-defense artillery (ADA) coverage of engineer equipment, mine dumps, and Class IV/V supply points.

5-2 Defensive Operations and Obstacle Integration Framework

- Provides a location to hide heavy engineer equipment during a battle.
- Provides an engineer mission during a battle (TF reserve, battle position).

SPECIFIC ENGINEER COORDINATIONS

With a TF Commander

- Leader's reconnaissance.
- Commander's intent, scheme of engineer operations (SOEO).
- Survivability and countermobility priorities.
- Class III, IV, and V responsibilities.
- Intent and location of SCATMINEs.

With a Maneuver Company Commander

- Leader's reconnaissance.
- Verifying direct/indirect-fire coverage of obstacles.
- Obstacle security, infantry squad or platoon forward at night (counterreconnaissance).
- Final obstacle placement.
- Senior equipment representative with infantry commander in chief (CINC) dozer.
- Target turnover.
- Security of mine dumps/Class IV and V supply points.
- Infantry security of engineers emplacing obstacles.
- Labor augmentation for obstacle emplacement.
- Passage-point guides.
- Obstacle siting.
- Passage lane closing.

With a TF S2

- IPB of threat intentions and capabilities.
- Daily/repeated intelligence updates.

With a TF S3

- Location to hide heavy engineer equipment during battle.
- Verifying reporting requirements.
- Verifying division's/brigade's authorization to emplace minefields, cut roads, and detonate demolitions and scattterable mines.
- Engineer's mission during conduct of defense.

Defensive Operations and Obstacle Integration Framework

FM 5-34

With a TF Supply Officer (US Army) (S4)

- Class III needs and distribution plan, if not provided by the parent unit.
- Haul/distribution of engineer heavy equipment, if required.
- Additional haul assets for engineer heavy equipment, if required.
- Helicopter assets to slingload Class IV/V forward, container delivery system (CDS).
- Maintenance priority to engineer heavy equipment.
- Location and manning of Class IV/V points.

With a Fire-Support Officer (FSO)/ADA

- Verifying airborne data and analysis system (area denial artillery munition [ADAM])/remote, antiarmor mine system (RAAM) employment location/trigger.
- Bringing FSO/fire-support element (FSE) out to see obstacles, if possible.
- ADA coverage of all personnel emplacing obstacles, heavy equipment, and supply points.
- Integrating obstacles and indirect fires according to FM 90-7.

OBSTACLES

The basic principles of obstacle employment are to support a maneuver commander's plan; integrate with observed fires, existing obstacles, and other reinforcing obstacles; and employ in-depth and for surprise.

OBSTACLE CLASSIFICATION

The two types of obstacles are existing and reinforcing (see Figure 5-1). For more information on obstacle classification, see FM 90-7.

OBSTACLE COMMAND AND CONTROL

Obstacle C^2 focuses on—

- Obstacle emplacement authority. In a theater of operations (TO), theater commanders have the authority to emplace obstacles. This authority can be delegated to corps, and then division level. Commanders subordinate to corps and division do not have the authority to emplace obstacles unless the higher commander gives them that authority for the current mission. Commanders use control measures and other specific guidance or orders to grant obstacle-emplacement authority to subordinate commanders. Emplacement authority for SCATMINEs depends on the particular system characteristics (see Table 5-1).

5-4 Defensive Operations and Obstacle Integration Framework

FM 5-34

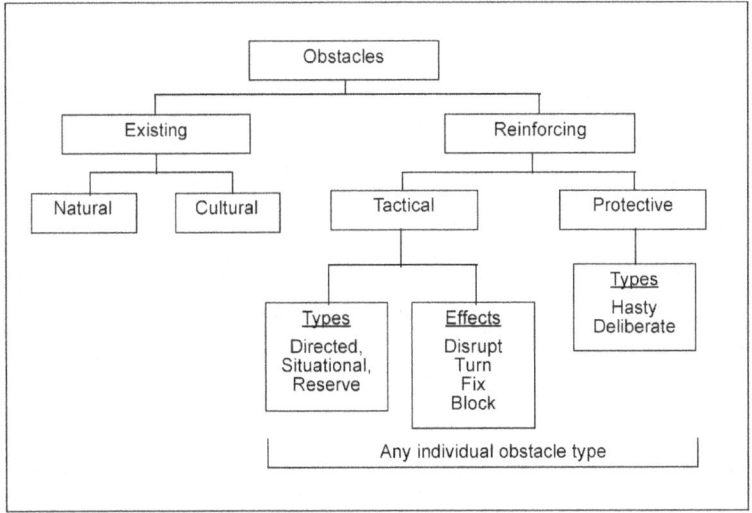

Figure 5-1. Obstacle classification

Table 5-1. SCATMINE emplacement authority

System Characteristics	Emplacement Authority
Ground- or artillery-delivered, with SD time greater than 48 hours (long-duration)	The corps commander may delegate emplacement authority to division level, who may further delegate it to brigade level.
Ground- or artillery-delivered, with SD time of 48 hours or less (short-duration)	The corps commander may delegate emplacement authority to division level, who may further delegate it to brigade level, who may further delegate it to TF level.
Aircraft-delivered (Gator), regardless of SD time	Emplacement authority is normally at corps, theater, or Army command level, depending on who has air-tasking authority.
Helicopter-delivered (Volcano), regardless of SD time	Emplacement authority is normally delegated no lower than the commander who has command authority over the emplacing aircraft.
MOPMS, when used strictly for a protective minefield	Emplacement authority is usually granted to the company, team, or base commander. Commanders at higher levels restrict MOPMS use only as necessary to support their operations.

Defensive Operations and Obstacle Integration Framework 5-5

FM 5-34

- Obstacle control. Obstacle control is the control that commanders exercise to ensure that obstacles support current and future missions. Table 5-2 lists obstacle-control measures and Figure 5-2 shows the obstacle-control-measure graphics and examples of their use.

Table 5-2. Obstacle-control measures

Obstacle-Control Measure	Echelon	Specific Obstacle Effects Assigned	Size of Enemy AA/MC		Planning Guidance
			Armored	Light vs Armored	
Zone	Corps or division	Optional	Division/ brigade	Brigade/ battalion	Requires anticipating belts and intents
Belt	Brigade	Optional but normal	Brigade/ battalion	Battalion/ company	Requires anticipating groups and intents
Group	Corps, brigade, division, or battalion/ TF	Mandatory	Battalion/ TF	Company/ platoon	Based on individual obstacle norms
Restriction	All	NA	NA	NA	Used only when necessary to support the scheme of maneuver

- Obstacle-effect graphics. There are separate graphics for each obstacle effect (see Figure 5-3, page 5-8). Commanders use obstacle-effect graphics to convey the effect they want the obstacles to have on the enemy.

- Battle tracking. The tracking of obstacles from their emplacement or discovery to their recovery or clearance from the battlefield is critical. A 12-character designator is given to each individual obstacle (see Figure 5-4, page 5-9, and Tables 5-3 and 5-4, pages 5-10 and 5-11). Reporting is an essential element of battle tracking. As a minimum, units must report the intent to lay (if required), initiation, completion, and recovery of all obstacles in their AO. Units must report all obstacles by the fastest secure means available, and classify them SECRET when they are completed. The command's SOP should specify the exact format.

FM 5-34

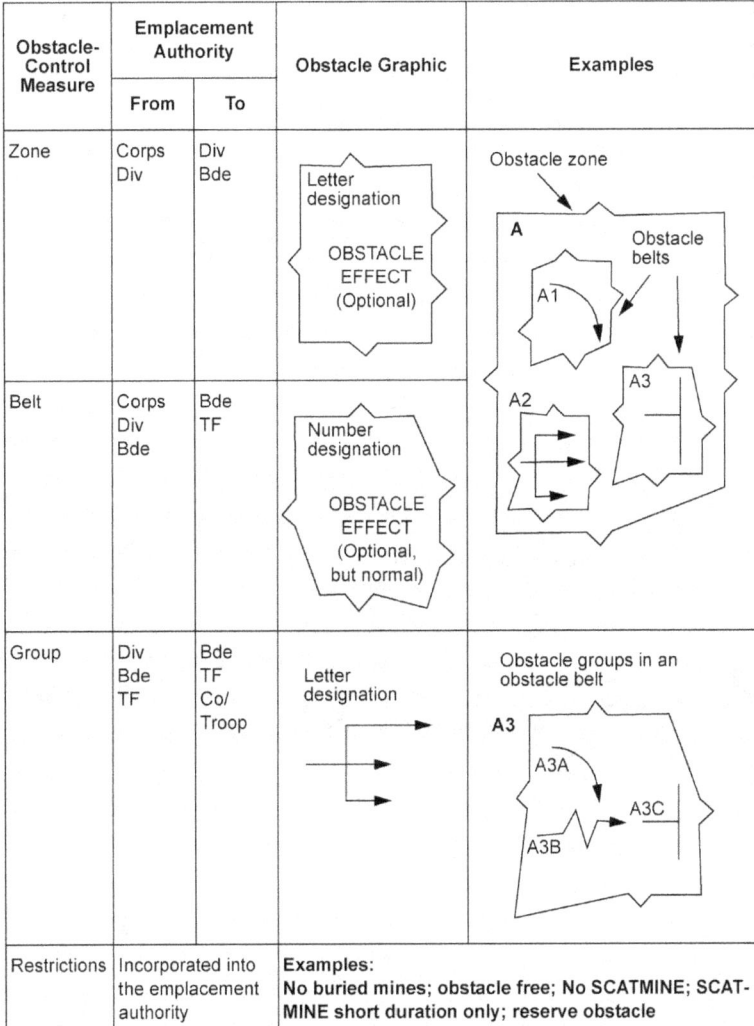

Figure 5-2. Obstacle-control-measure graphics

Defensive Operations and Obstacle Integration Framework 5-7

FM 5-34

Obstacle-Effect Graphics	Application	Examples Conveying Intent
Disrupt 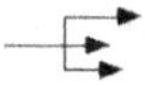	Short arrow indicates where enemy is attacked by obstacles. Long arrows indicate where bypass is allowed and attacked by fires.	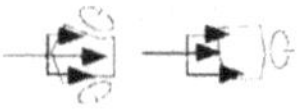
Turn	Heel of arrow is anchor point. Direction of arrow indicates desired direction of turn.	
Fix 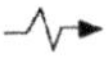	Irregular part of arrow indicates where enemy advance is slowed by obstacles.	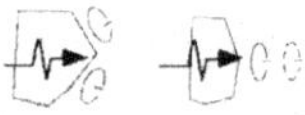
Block	The ends of the vertical line indicate the limit of enemy advance. The ends of the vertical line also indicate where obstacles tie to NO-GO terrain.	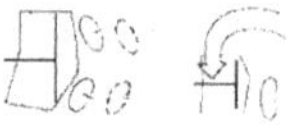
	Direction of Enemy Attack →	

Figure 5-3. Obstacle-effect graphics

5-8 Defensive Operations and Obstacle Integration Framework

FM 5-34

Obstacle No.	Grid Coordinates	Type of Mines	Obstacle-Marking Method	Unit to Clear Obstacle	DTG of Obstacle Clearance	Lane/Grid Marking	Remarks
ENA001MN01/	NK123456-NK125457	SB-MV	Single-strand concertina on all four sides	A/99 EN BN		NK124456-NK124457 Lane marked to full-lane pattern using traffic cones	Obstacle reported by A/1-23 IN (031500JAN97)
ENA001MN02+	NK450200-NK453202	SB-MV	Single-strand concertina on enemy side of minefield			NA	Reported by Engineer Recon Team 1 (NAI 301) (100200JAN97)
ENA001MN03X	NK189765-NK190768	SB-MV	NA	B/99 EN BN	011200JAN97	NA	

As of: 100600JAN97

NOTE: Obstacle numbering system: ENXXXXXXXXX.
- Characters 1-2: EN meaning enemy obstacle.
- Characters 3-6: Alphanumeric description of the headquarters type and numerical designation that reported the obstacle. Character 3 designates the unit type:
  — A, armor division/brigade
  — I, infantry division/brigade
  — C, cavalry division
  — R, cavalry regiment
  — Z, corps
- Characters 7-8: Letters indicating obstacle type (see FM 20-32).
- Characters 9-10: Two numbers indicating obstacle number within the obstacle type.
- Character 11: One of four characters indicating obstacle status:
  — + obstacle reported, no clearance planned
  — / clearance of obstacle planned
  — - clearance of obstacle in progress
  — X clearance of obstacle complete

Figure 5-4. Example of enemy obstacle-tracking chart

Defensive Operations and Obstacle Integration Framework 5-9

FM 5-34

Table 5-3. Obstacle numbers

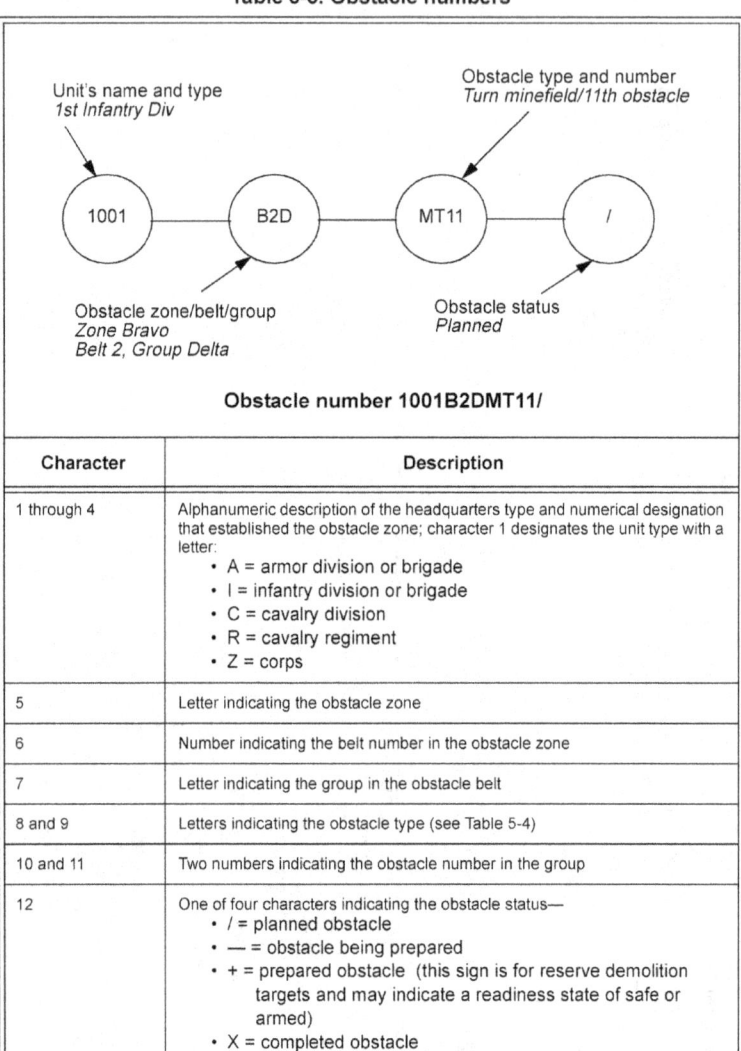

Obstacle number 1001B2DMT11/

Character	Description
1 through 4	Alphanumeric description of the headquarters type and numerical designation that established the obstacle zone; character 1 designates the unit type with a letter: • A = armor division or brigade • I = infantry division or brigade • C = cavalry division • R = cavalry regiment • Z = corps
5	Letter indicating the obstacle zone
6	Number indicating the belt number in the obstacle zone
7	Letter indicating the group in the obstacle belt
8 and 9	Letters indicating the obstacle type (see Table 5-4)
10 and 11	Two numbers indicating the obstacle number in the group
12	One of four characters indicating the obstacle status— • / = planned obstacle • — = obstacle being prepared • + = prepared obstacle (this sign is for reserve demolition targets and may indicate a readiness state of safe or armed) • X = completed obstacle

5-10 Defensive Operations and Obstacle Integration Framework

Table 5-4. Obstacle type abbreviations

M—Minefield/Munition Field		W—Wire Obstacle	
MB	Block	WA	Double-apron
MC	Chemical	WC	Concertina
MD	Disrupt	WF	Tanglefoot
MF	Fix	WG	General-purpose, barbed tape
MH	Hasty protective	WN	Nonstandard/unspecified
MN	Nonstandard	WR	Roadblock
MO	Point	WT	Triple-standard
MP	Protective	**S—Scatterable Minefield/Munition Field**	
MQ	Nuisance	SB	Gator
MS	Standard-pattern	SF	ADAM and RAAM
MT	Turn	SM	MOPMS
MU	Dummy/decoy	SV	Volcano
A—Miscellaneous		SW	Scatterable mines (generic)
AB	Abatis	**H—Hand-Emplaced Munitions**	
AC	Chemical by explosives	HC	Claymore
AD	AT ditch	HH	Hornet/WAM
AF	Thermobaric or flame	HO	Other
AH	Log hurdle	HS	SLAM
AL	Log crib or log obstacle	**I—Improvised Explosive Devices**	
AM	Movable obstacle (car, bus)	ID	Directional, special-purpose explosive hazard
AN	Expedient nonstandard obstacle		
AP	Post obstacle (hedgehog, tetrahedron)	IO	Omnidirectional, special-purpose explosive hazard
AR	Rubble		
AT	AT ditch with AT mines	**B—Bridge Demolition**	
AW	Earthwork (berms, parapets, dunes, pits)	BA	Abutment
		BC	Abutment and span
T—Booby Traps		BS	Span
TA	Booby-trapped area	**R—Road Crater**	
TB	Booby-trapped bodies	RD	Deliberate
TE	Booby-trapped equipment	RH	Hasty
TM	Booby-trapped materiel	RM	Mined
TP	Booby-trapped passage/ confined space	**U—Unexploded Ordnance**	
		UC	Chemical UXO hazard area
TS	Booby-trapped structure	UH	UXO hazard area
TV	Booby-trapped vehicle	UN	Nuclear hazard area

REPORTS

Report all minefields by the fastest secure means available, and classify them SECRET when they are completed. The local command's SOP should specify the exact format.

REPORT OF INTENTION TO LAY

When planing to emplace a minefield, a unit must submit a report of intention (see Table 5-5 for an example). The report doubles as a request when it is initiated below emplacement-authority level.

Table 5-5. Report of intention to lay

Explanation	Letter Designation	Example
Tactical objectives (temporary security road-block or other)	ALPHA	Bridge-construction-site security
Type of minefield	BRAVO	Hasty protective
Estimated number and types of mines and whether surface-laid mines or ones with AHDs	CHARLIE	25 each M16, buried w/trip-wires and no AHD
Location of minefield by coordinates	DELTA	WQ04500359 to WQ04560365
Location and width of minefield lanes and gaps	ECHO	Route Blue at WQ04550363, 16 feet wide
Start and complete DTG (estimate)	FOXTROT	Start: 011000SAUG96 Complete: 011030ZAUG96

REPORT OF INITIATION

A report of initiation is mandatory (see Table 5-6 for an example). It informs higher HQ of an area that is no longer safe for friendly movement.

Table 5-6. Report of initiation

Explanation	Letter Designation	Example
Location of minefield, by coordinates	DELTA	WQ03567843
Start and complete DTG (estimate)	FOXTROT	Start: 011045ZAUG96 Completion: 011130ZAUG96

REPORT OF PROGRESS

During the emplacement process, a commander may require periodic updates. Table 5-7 is an example of a progress report.

Table 5-7. Report of progress

Explanation	Letter Designation	Example
Location of minefield by coordinates: 25%, 50%, 75% or 100% completed.	DELTA	WQ03567843: 75% complete

5-12 Defensive Operations and Obstacle Integration Framework

FM 5-34

REPORT OF COMPLETION

After a completion report (see Table 5-8 for an example), fill out a completed DA Form 1355 or DA Form 1355-1-R for minefields or munitions fields.

Table 5-8. Report of completion of minefield

Explanation	Letter Designation	Example
Changes in information submitted in intention-to-lay report	ALFA	Lane width is 14 feet
Total number and type of AT and AP mines emplaced	BRAVO	M15-299 M21- 865 M16 -203
Date and time of completion	CHARLIE	011145ZAUG96
Method of lay (buried by hand or by machine, surface-laid)	DELTA	Buried by hand
Details of lanes or gaps including marking	ECHO	WD1 wire on centerline azimuth of $165°$; entrance and exit marked with U-shaped pickets and red chemical lights
Details of perimeter marking	FOXTROT	Standard barbed-wire fence
Overlay showing perimeter, lanes, and gaps	GOLF	NA
Laying unit and signature of individual authorizing laying of the field	HOTEL	2d Plt, Co A, 307th Engr Bn

REPORT OF TRANSFER

Use a transfer report when the responsibility for a minefield is transferred between commanders. Both commanders must sign the report. Included must be a certificate stating that the receiving commander was shown or informed of all the mines within the zone of responsibility and that he is responsible for all the mines within the zone. The report is sent to a higher commander who has authority over the relieved and relieving commanders.

REPORT OF CHANGE

A change report is submitted when any alterations are made to a minefield form in which a completion report and record have already been submitted.

Defensive Operations and Obstacle Integration Framework 5-13

Chapter 6
Constructed and Preconstructed Obstacles

WIRE OBSTACLES

Table 6-1 and Table 6-2, page 6-2, list materials and planning factors for wire obstacles.

Table 6-1. Wire and tape obstacle material

Material	Approximate Weight (kg)	Approximate Length (m)	Number That One Soldier can Carry	Approximate Weight of Man Load (kg)
Barbed-wire reel	41.5	400	0.5	21
Bobbin	3.5 to 4.0	30	4-6	14.5 to 24.5
Barbed-tape dispenser	0.77	0.45	20	15.5
Barbed-tape carrying case	14.5	300	1	14.5
Standard barbed-tape concertina	14	15.2	1	25
Standard barbed-wire concertina	25.4	15.2	1	25
GPBTO— • Hand • Vehicular	 15.8 117.9	 20 140	 1 0.25	 15.8 29.5
U-shaped pickets— • Long • Medium • Short	 4.5 2.7 1.8	 1.5 0.81 0.61	 4 6 8	 18.1 16.3 14.5
NOTE: Whenever you use U-shaped pickets, make sure that the open end of the U faces the enemy.				

Table 6-2. Requirements for 300-meter sections of various wire obstacles

Entanglement Type	Pickets			Reels of Barbed Wire[1]	Number of GPBTO	Number of Concertinas	Staples	Man-Hours to Erect[2]	Kg of Materials per Linear Meter of Entanglement[3]
	Long	Medium	Short						
Double apron, 4 and 2 pace	100		200	15-16 (19)[4]				71	4.6 (3.5)[5]
Double apron, 6 and 3 pace	66		132	15-17 (18)[4]				59	3.6 (2.6)[5]
High wire (less guy wires)	198			19-21 (24)[4]				95	5.3 (4.0)[5]
Low wires, 4 and 2 pace		100	200	11				59	3.6 (2.8)[5]
4-strand cattle fence	100		2[7]	6-7 (7)[4]				24	2.2 (1.8)[5]
Triple-standard concertina	160		4[8]	3 (4)[4]		59	317	30	8.2 (7.3)[5]
GPBTO					(8)[6]			(1)[6]	2.7

[1]The lower number of reels applies when you use U-shaped pickets; the higher number applies if you use wooden pickets. If there is only one number, use it for both pickets.
[2]Man-hours are based on the use of driven pickets. Multiply these figures by 0.67 if experienced troops are being used, and by 1.5 for night work.
[3]Average weight when you use any issue metal pickets (1 truckload = 2,268 kg)
[4]Number of barbed-tape carrying cases required if barbed tape is used in place of barbed wire
[5]Kilograms of material required per linear meter of entanglement if barbed tape is used in place of barbed wire and barbed-tape concertina is used in place of standard barbed-tape wire concertina
[6]Based on vehicular emplaced obstacles installed in triple belts
[7]Only 2 required for one belt
[8]Only 4 required for one belt

6-2 Constructed and Preconstructed Obstacles

BARBED-WIRE OBSTACLES

Barbed-wire obstacles are classified according to their use. You can estimate the quantity of concertina required following basic rules of thumb:

- Construct barbed-wired obstacles for conventional deployment along the forward edge of the battle area (FEBA)(see Figure 6-1):

    — Tactical wire = (front) x (1.25) x (number of belts).

    — Protective wire = (front) x (5) x (number of belts).

    — Supplementary wire—

    ° Forward of FEBA = (front) x (1.25) x (number of belts).

    ° Rear of FEBA = (2.5) x (unit depth) x (number of belts).

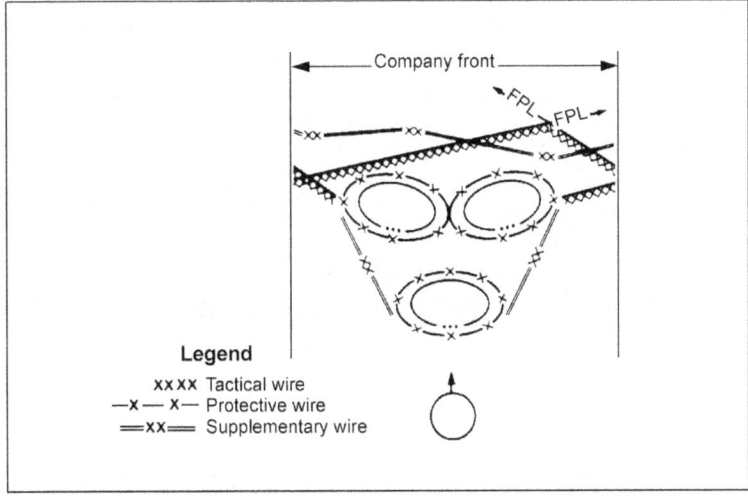

Figure 6-1. Schematic layout of barbed-wire obstacles (defense)

- Construct a base-camp defense along a perimeter (see Figure 6-2, page 6-4):

    — Tactical wire = (mean perimeter) x (1.25) x (number of belts).

    — Protective wire = (perimeter) x (1.10) x (number of belts).

    — Supplementary wire = (mean perimeter) x (1.25) x (number of belts).

Constructed and Preconstructed Obstacles 6-3

FM 5-34

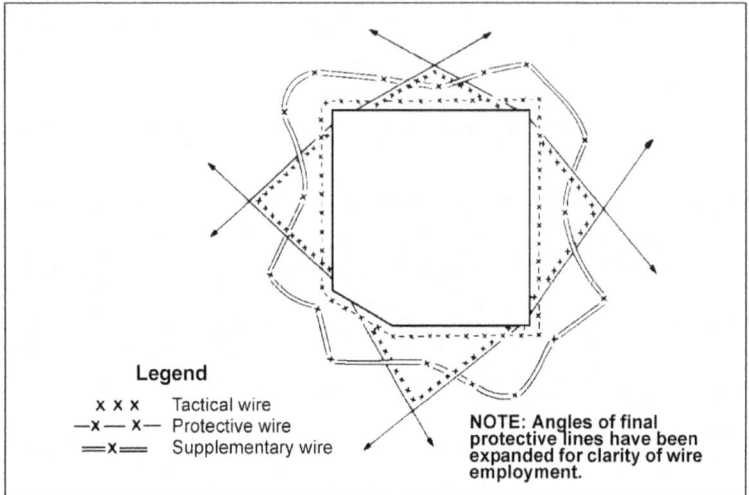

Figure 6-2. Perimeter wire (defense)

- Ensure that the job site is secure.
- Organize a work party into three equal crews. The first two crews lay out pickets, and the third crew installs pickets (open end of *U* toward enemy).
- Reorganize the party into crews of two to four soldiers.
- Install wire in numerical order as shown in Figure 6-3.
- Avoid having any soldier cut off between the enemy and the fence.
- Ensure that the wires are properly secured and tight.

TRIPLE-STANDARD CONCERTINA

When laying out triple-standard concertina, follow the basic rules listed below:

- Ensure that the job site is secure.
- Organize a work party into three crews. The first crew lays pickets (see Figure 6-4, page 6-6). The second crew lays out concertina. It places one roll on the enemy's side at every third picket and two rolls on the friendly's side at every third picket. The third crew installs all the pickets.
- Reorganize the party into four-soldier crews.
- Install the concertina (see Figures 6-5 and 6-6, pages 6-6 and 6-7).

6-4 Constructed and Preconstructed Obstacles

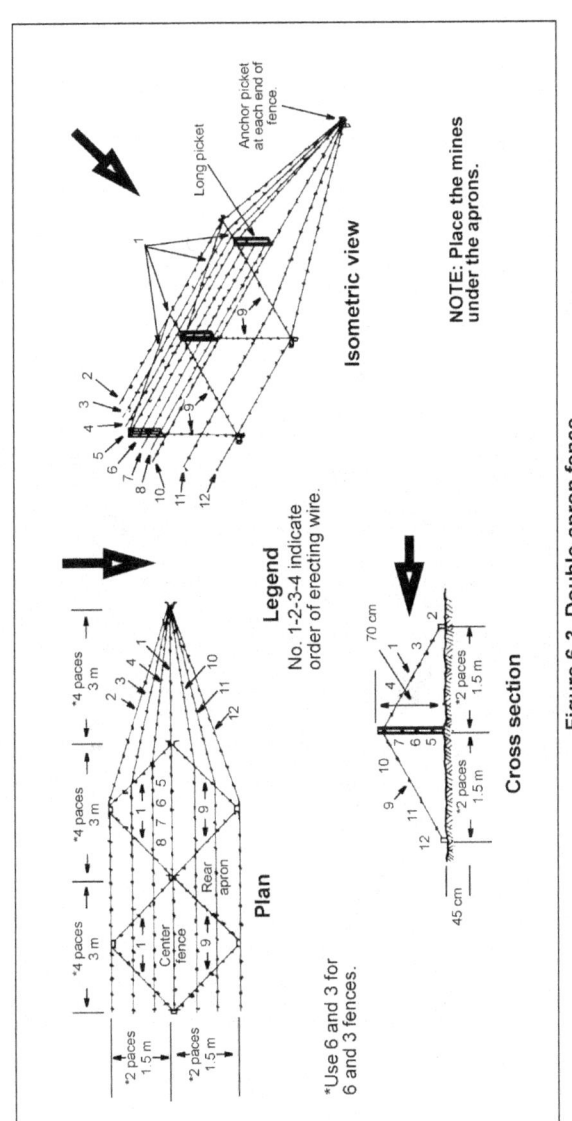

Figure 6-3. Double-apron fence

Constructed and Preconstructed Obstacles 6-5

FM 5-34

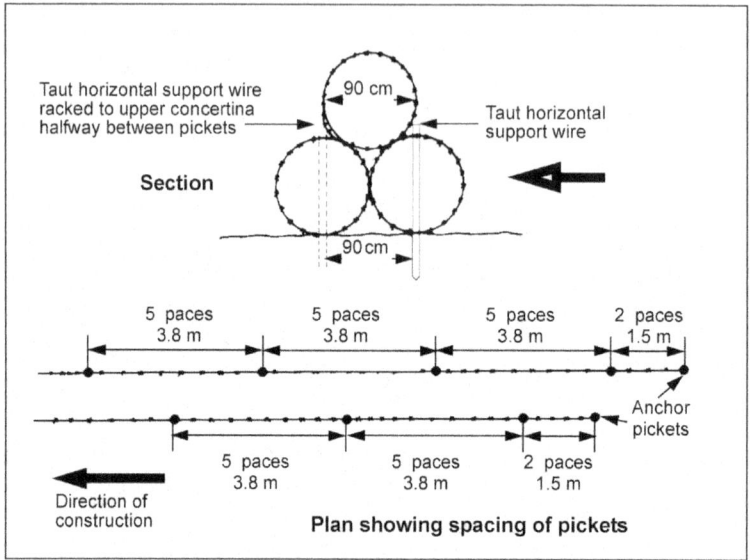

Figure 6-4. Triple-standard concertina fence

- Ensure that the concertina is properly tied and all horizontal wire is properly installed.

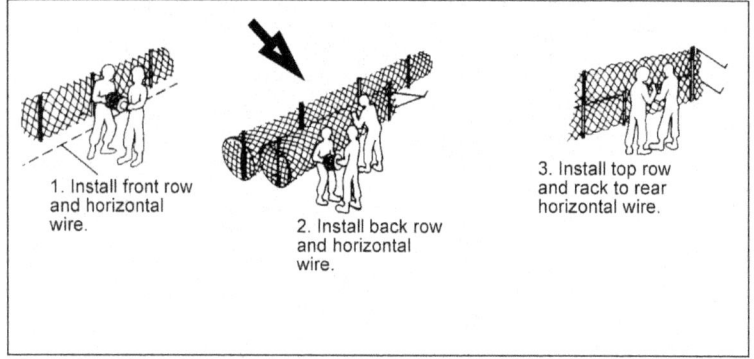

Figure 6-5. Installing concertina

6-6 Constructed and Preconstructed Obstacles

FM 5-34

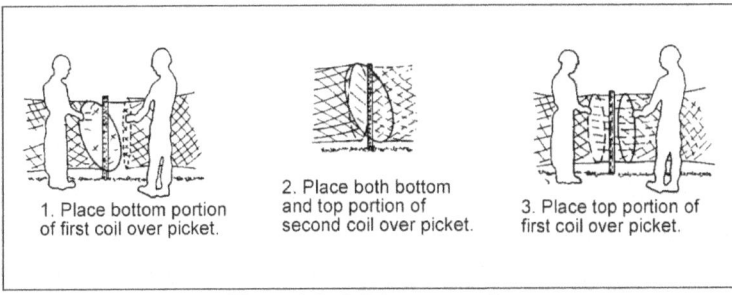

Figure 6-6. Joining concertina

FOUR-STRAND CATTLE FENCE

When laying out four-strand cattle fence (see Figure 6-7), follow the basic rules listed below:

- Ensure that the job site is secure.

- Organize a work party into four-soldier crews. The first crew lays out long pickets 3 meters apart, and the second crew installs them.

- Reorganize the party into two-soldier teams; one team carries the reel and the other team makes the ties.

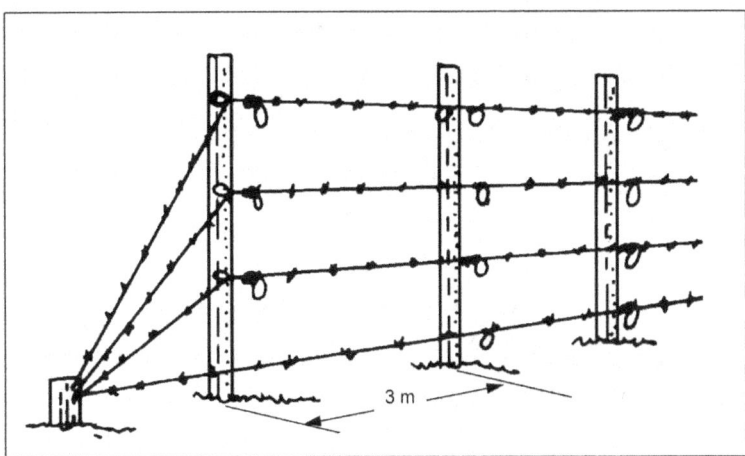

Figure 6-7. Four-strand cattle fence

OTHER WIRE OBSTACLES

Construct other wire obstacles from enemy to friendly side and from bottom up (see Figures 6-8 through 6-11, pages 6-8 and 6-9).

Constructed and Preconstructed Obstacles 6-7

FM 5-34

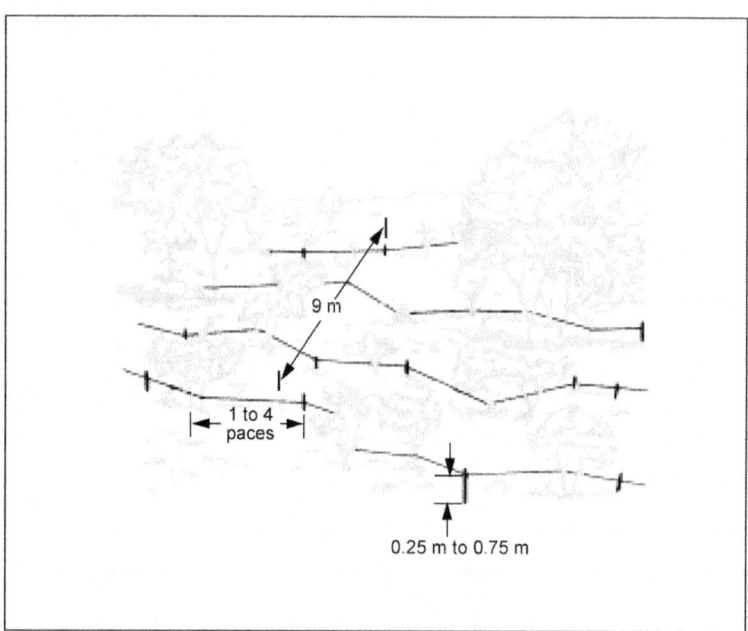

Figure 6-8. Tanglefoot

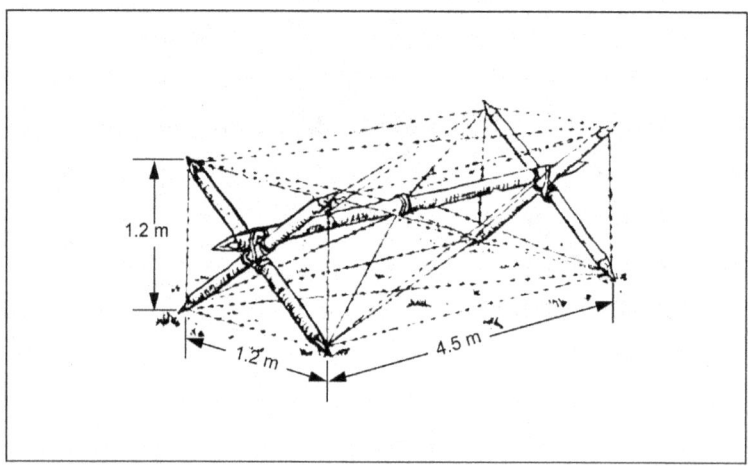

Figure 6-9. Knife rest

6-8 Constructed and Preconstructed Obstacles

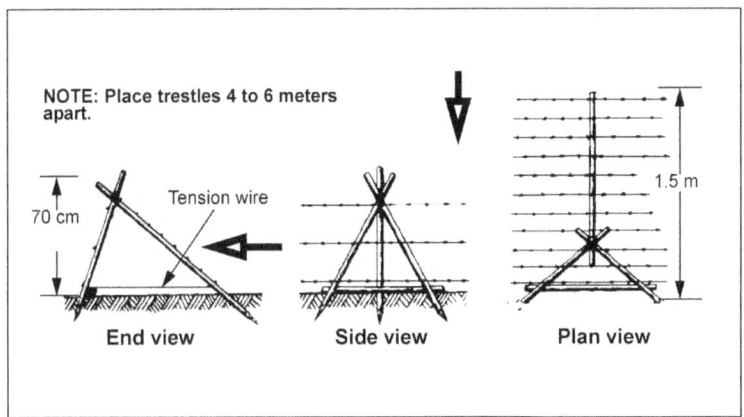

Figure 6-10. Trestle-apron fence

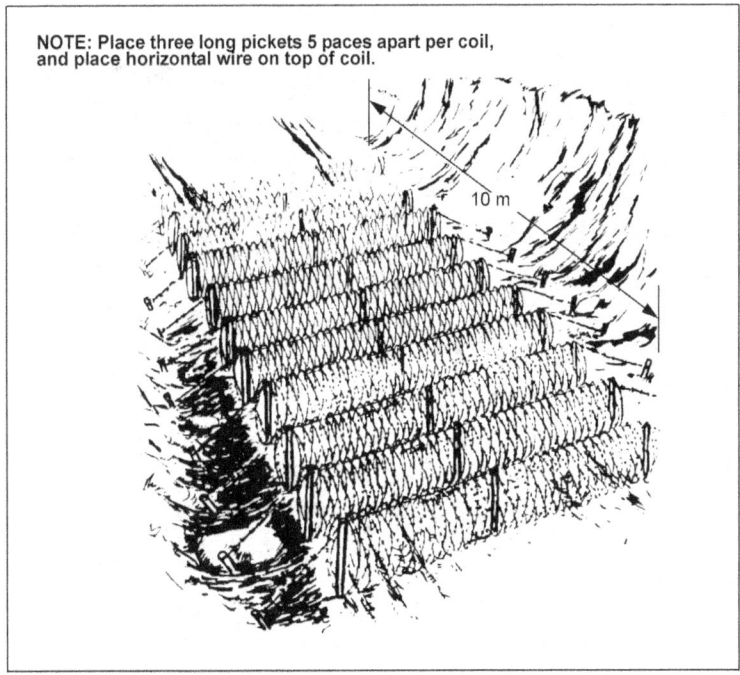

Figure 6-11. Eleven-row antivehicular wire obstacle

FM 5-34

ANTIVEHICULAR OBSTACLES

AT Ditches and Road Craters

Figure 6-12 shows some AT ditches. See Chapter 9 for specific details and construction of road craters.

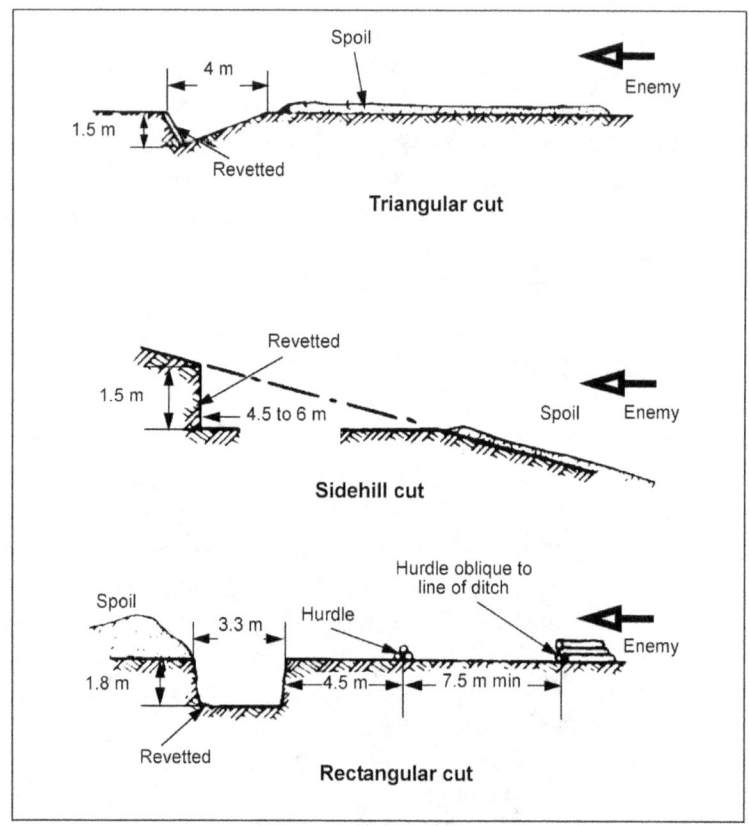

Figure 6-12. AT ditches

Log Cribs

Figures 6-13 and 6-14 show different log-crib designs; Table 6-3, page 6-12, lists post requirements in constructing log cribs. The manpower requirement in constructing log cribs is 4 to 8 engineer-platoon hours, equipped with hand tools, for a 6-meter-wide road.

6-10 Constructed and Preconstructed Obstacles

FM 5-34

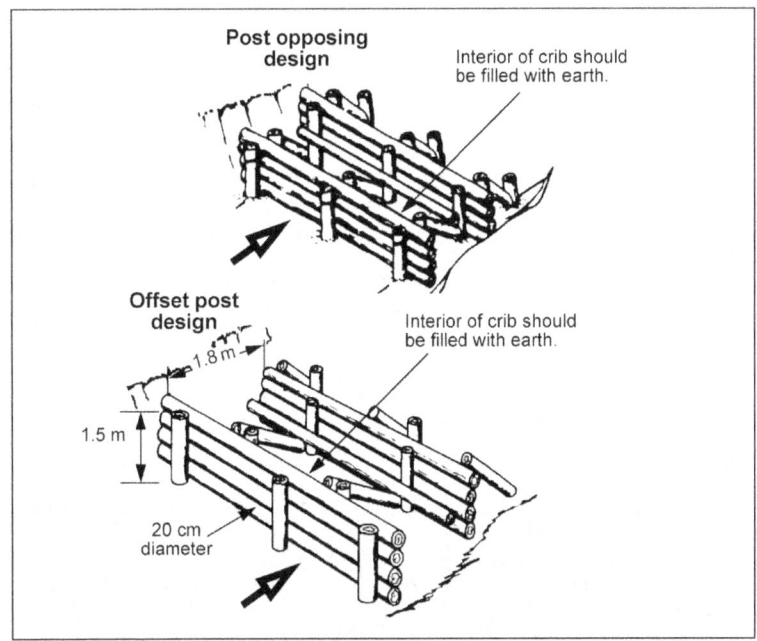

Figure 6-13. Rectangular log-crib design

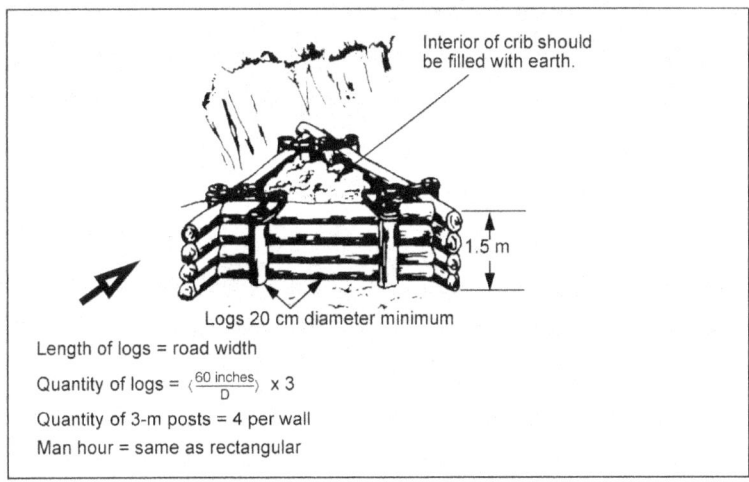

Length of logs = road width

Quantity of logs = $(\frac{60 \text{ inches}}{D}) \times 3$

Quantity of 3-m posts = 4 per wall

Man hour = same as rectangular

Figure 6-14. Triangular log-crib design

Constructed and Preconstructed Obstacles 6-11

Table 6-3. Post requirements (post opposing/offset post)

Posts	Road Width (m)							
	1.8	2.1 to 3.6	3.9 to 5.4	5.8 to 7.3	7.6 to 9.1	9.4 to 10.9	11.3 to 12.8	13.1 to 14.6
Long 3	8	12	16	20	24	28	32	36
Short 2.1	2	3	4	5	6	7	8	9
Braces 2.1	4	6	8	10	12	14	16	18

LOG HURDLES

Make sure that you site log hurdles at the steepest part of a slope (see Figure 6-15).

LOG/STEEL POST OBSTACLE

Figure 6-16 shows a log-post obstacle and instructions on constructing an obstacle.

TETRAHEDRONS, HEDGEHOGS, AND OTHER BARRIERS

Figures 6-17 through 6-21, pages 6-14 and 6-15, show other materials used for barriers and their placements.

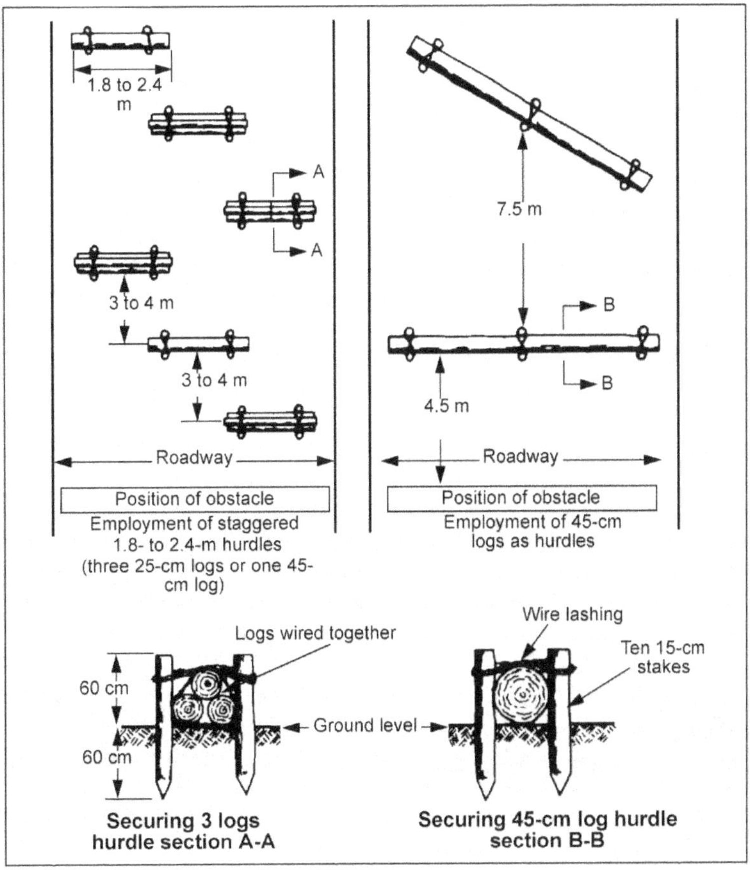

FM 5-34

Figure 6-15. Log hurdles

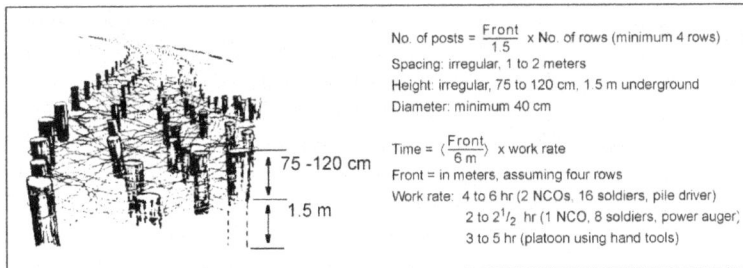

No. of posts = $\frac{Front}{1.5}$ x No. of rows (minimum 4 rows)

Spacing: irregular, 1 to 2 meters

Height: irregular, 75 to 120 cm, 1.5 m underground

Diameter: minimum 40 cm

Time = $(\frac{Front}{6\,m})$ x work rate

Front = in meters, assuming four rows

Work rate: 4 to 6 hr (2 NCOs, 16 soldiers, pile driver)
2 to $2^{1}/_{2}$ hr (1 NCO, 8 soldiers, power auger)
3 to 5 hr (platoon using hand tools)

Figure 6-16. Post obstacles

Constructed and Preconstructed Obstacles 6-13

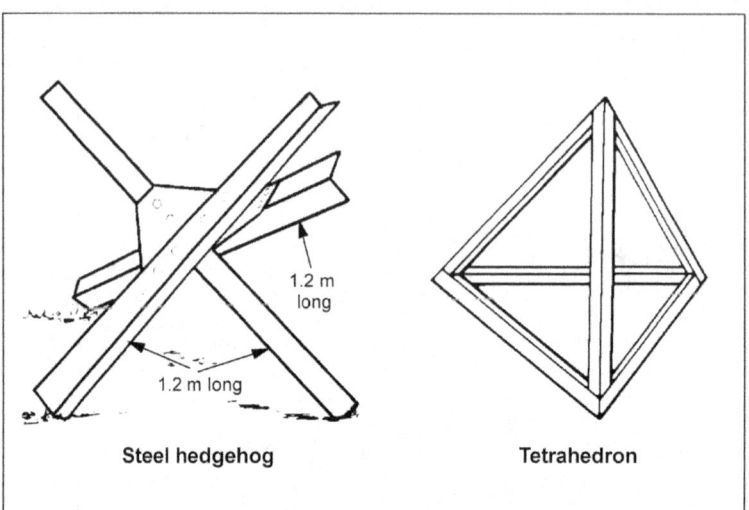

Figure 6-17. Steel hedgehog and tetrahedron

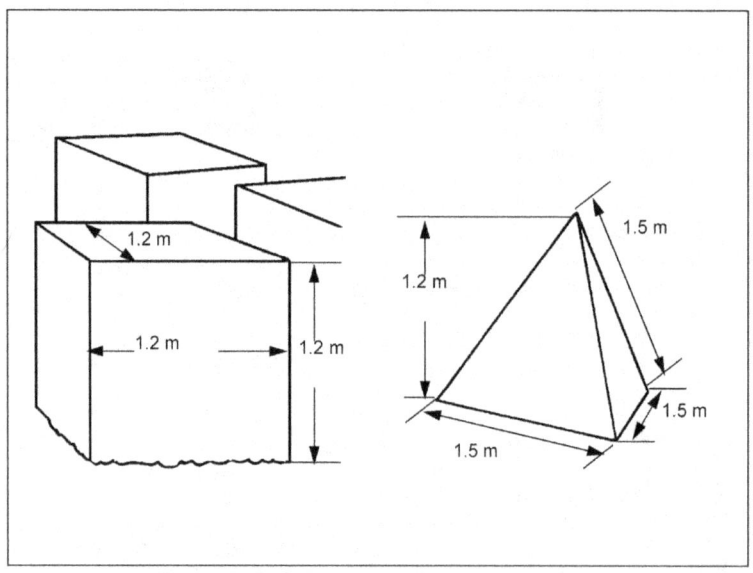

Figure 6-18. Concrete tetrahedron and cubes

6-14 Constructed and Preconstructed Obstacles

FM 5-34

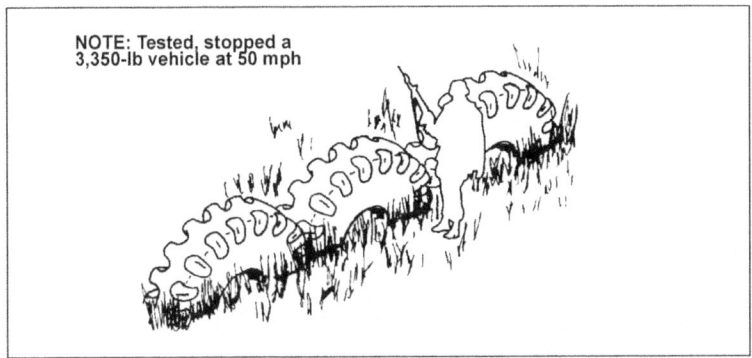

Figure 6-19. Heavy equipment tires

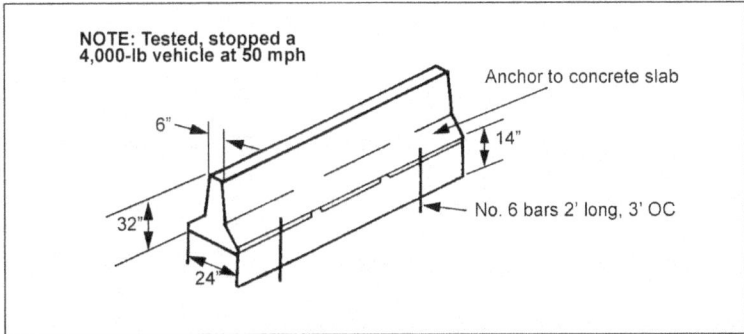

Figure 6-20. Jersey barrier

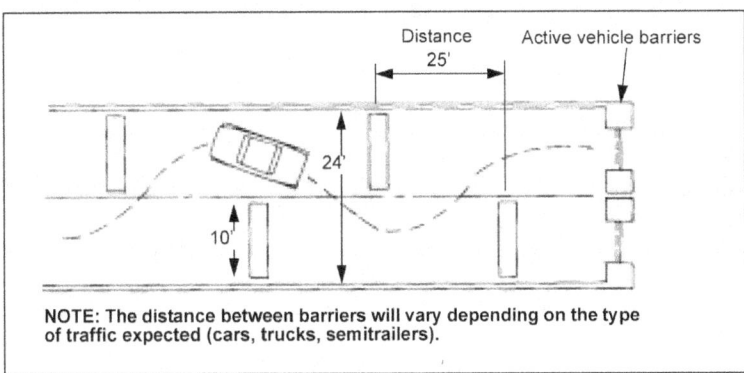

Figure 6-21. Concrete-obstacle placement

Constructed and Preconstructed Obstacles 6-15

Chapter 7
Landmine and Special-Purpose Munition Obstacles

CONVENTIONAL MINEFIELDS

ROW MINING

Table 7-1 lists the characteristics of standard row minefields.

Table 7-1. Standard minefield characteristics

Characteristics	Disrupt	Turn	Fix	Block
Frontage (m)	250	500	250	500
Depth (m)	100	300	120	320
AT, full-width (No. rows)	1	4	1	4
AT, track-width (No. rows)	2	2	2	2
AP (No. rows)*	0	0	0	2
IOE (yes/no)	No	No	Yes	Yes
Platoon hours required to emplace	1.5	3.5	1.5	5.0
AHD	No	No	No	Yes
Total AT mines—				
Full width	42	336	63	378
Track width	84	168	84	168
Total AP mines*				
M16	0	0	0	84
M14	0	0	0	84
Density AT	0.5	1.0	0.6	1.1
Density AP (M16/M14)	0	0	0	0.17/1.0
Symbol	↑↑↑	↶	⟂	⊤
NOTE: Mine spacing is 6 meters for all standard minefields. *Korea only, units have a choice of either the M16 or M14.				

Use the following procedure to calculate the number of mines and minefield rows if you did not use the standard minefields described in Table 7-1:

FM 5-34

Step 1. The number of mines required is equal to the desired density times the minefield front.

$$density(0.5) \times front(400) = 200 \; AT \; mines$$

Step 2. The number of AT mines per row is determined by dividing the minefield front by the spacing interval between the AT mines (normally 6 meters between mines).

$$\frac{400}{6} = 66.6 \; AT \; mines \; per \; row = 67 \; (rounded \; up)$$

Step 3. The number of rows needed in the minefield is equal to the number of AT mines required (step 1) divided by the number of AT mines per row (step 2). (Round up your answer to the next whole number.)

$$\frac{200}{67} = 3 \; rows \; (rounded \; up)$$

Step 4. The number of truckloads required for minefield emplacement depends on the type and quantity of mines and vehicular carrying capacity (see Table 7-2). Multiply the total number of rows by the number of mines per row and round up.

$$67 \times 3 = 201$$

Step 5. Multiply the total number of mines by 1.10 and round up.

$$201 \times 1.10 = 221.1 = 222 \; (rounded \; up)$$

Step 6. The number of truckloads required is equal to the total number of AT mines divided by the truck's capacity. In this example, 5-ton dump trucks are used to haul M15 AT mines.

$$\frac{222}{204} = 1.09 = 2 \; truckloads \; (rounded \; up)$$

7-2 Landmine and Special-Purpose Munition Obstacles

FM 5-34

Table 7-2. Class IV/V haul capacity

Vehicle	Concertina Wire[1]	AT Mine M15	AT Mine M19	AT Mine M21	AP Mine M16	AP Mine M14	MOPMS Mine	Flipper Mine	Volcano Mine	MICLIC Reload[2]	Hornet
HMMWV 1,124 kg, 6 cu m	2	51	34	27	55	56	15	11	1	NA	1
M35 2½-ton truck 2,250 kg, 12.5 cu m	4	102	69	55	111	113	30	23	2	2	2
M1078 2½-ton truck 2,250 kg, 13.4 cu m	4	102	69	55	111	113	30	23	2	2	2
M54 5-ton truck 4,500 kg, 13.6 cu m	7	204	138	109	222	227	61	46	5	3	5
M1083 5-ton truck 4,500 kg, 15.6 cu m	8	204	138	109	222	227	61	46	5	3	5
M930 5-ton dump truck (without sideboards) 4,500 kg, 3.8 cu m	2	112	64	32	168	71	23	39	3	2	2
M930 5-ton dump truck (with sideboards) 4,500 kg, 8.2 cu m	4	204	138	70	222	153	51	46	5	3	4

[1]The number of concertina in bundles; 1 bundle = 40 rolls
[2]Line charge + rocket

Landmine and Special-Purpose Munition Obstacles 7-3

FM 5-34

Table 7-2. Class IV/V haul capacity (continued)

Vehicle	Concertina Wire[1]	AT Mine M15	AT Mine M19	AT Mine M21	AP Mine M16	AP Mine M14	MOPMS Mine	Flipper Mine	Volcano Mine	MICLIC Reload[2]	Hornet
M1090 5-ton dump truck 4,500 kg, 3.8 cu m	2	112	64	32	168	71	23	39	3	2	2
HEMTT truck 9,000 kg, 15 cu m	8	408	277	128	444	317	94	92	10	7	8
12-ton S&T 10,800 kg, 24.5 cu m	13	489	333	208	533	514	148	110	12	9	13
40-ton lowboy 36,000 kg, 49.3 cu m	27	1,466	1,035	419	1,777	1,035	308	368	30	27	27
M548 cargo 5,400 kg, 14.9 cu m	8	244	166	125	266	272	74	55	6	4	6
M1077 PLS flat rack 14,900 kg, 17.6 cu m	9	440	352	164	586	293	110	152	11	9	9
No. of mines per box	NA	1	2	4	4	90	21	40	240	NA	30
Weight per box (kg)	531	22	33	41	21	20	73	97.7	833	1,195	810
Size of box (cu m)	1.8	0.04	0.05	0.12	0.03	0.06	0.16	0.1	1.6	1.8	1.8

[1]The number of concertina in bundles; 1 bundle = 40 rolls
[2]Line charge + rocket

7-4 Landmine and Special-Purpose Munition Obstacles

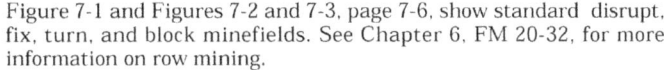

Figure 7-1 and Figures 7-2 and 7-3, page 7-6, show standard disrupt, fix, turn, and block minefields. See Chapter 6, FM 20-32, for more information on row mining.

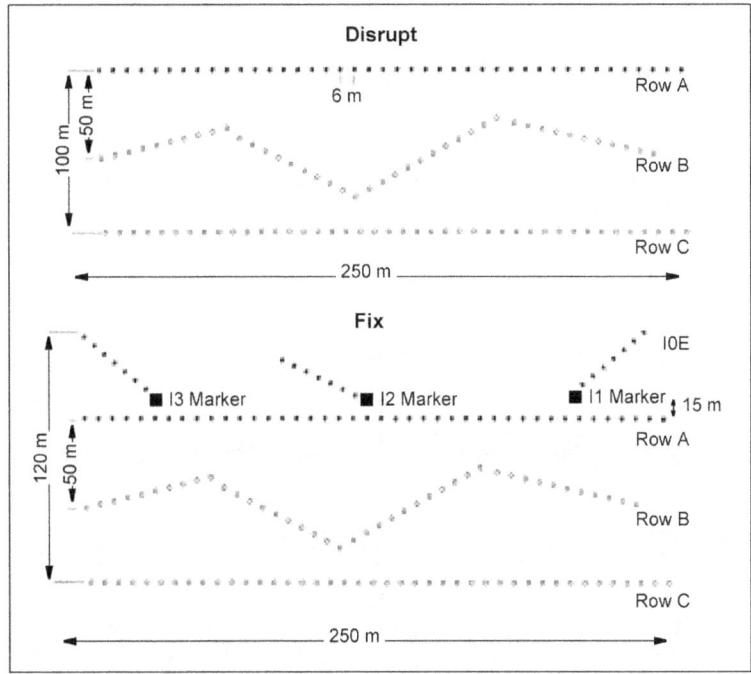

Figure 7-1. Standard disrupt and fix row minefields

Landmine and Special-Purpose Munition Obstacles 7-5

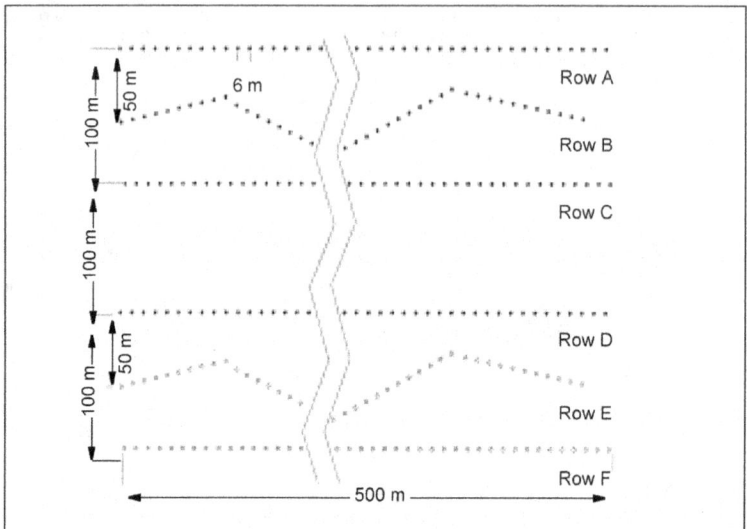

Figure 7-2. Standard turn row minefield

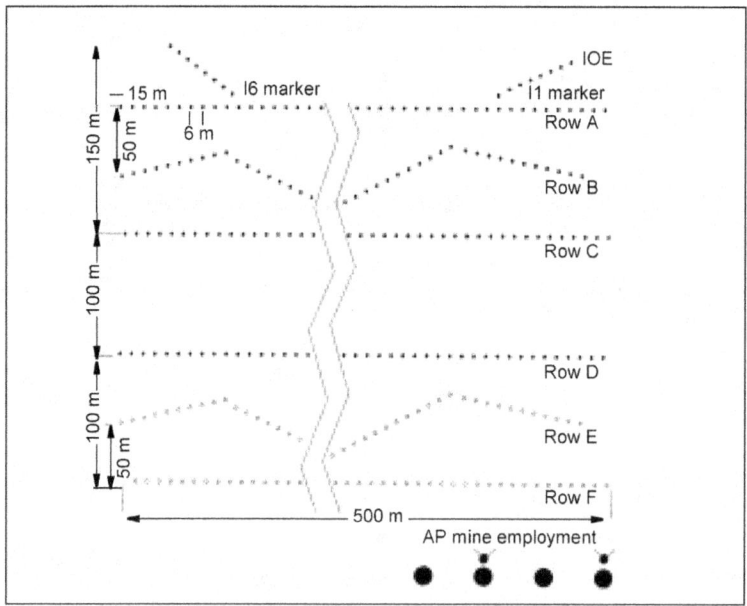

Figure 7-3. Standard block row minefield

7-6 Landmine and Special-Purpose Munition Obstacles

STANDARD-PATTERN MINEFIELDS

See FM 20-32, Chapter 7, for detailed information on standard-pattern minefields. Table 7-3 lists the platoon organization for a row minefield.

Table 7-3. Platoon organization for row mining

Personnel	Officer	NCO	EM	Equipment
Siting and recording party	1		2-3	SITEMP, maneuver and fire support graphics, obstacle overlay, execution matrix, GPS, lensatic compass, minefield record forms, stakes, cones, or pickets, picket pounder, engineer tape on reels, and nails to peg tape (Note 1)
Marking party		1	6-8	Barbed wire or concertina, marking signs, lane signs, wire cutters, gloves, picket pounder, and pickets (Note 2)
Mine dump party		1	6-8	Wire cutters, grease, rags, gloves, NVDs, and pliers (Note 3)
Laying party (Note 4)		1		Strip-feeder reports
Carrier team			2	Vehicle
Sapper team		1	3-4	Wrenches, fuses, row markers
Digging team			1-2	Picks, shovels, sandbags
Total (Note 4)	1	4	20-27	

NOTES:
1. The use of the mini-rehearsal to site an obstacle group during EA development may initially require all of the engineer platoon vehicles to portray the enemy's maneuver through the EA. An alternate solution is to use the engineer platoon leader, maneuver company team 1SG, and other maneuver team headquarters' vehicles.

2. Minefield marking is time- and labor-intensive. Any available soldiers should be placed on this team, especially when marking scatterable minefields.

3. If the unit uses the supply point or tailgate resupply technique, the unit must task-organize to do the mine-dump tasks at the Class IV/V point or at the point where the mines are transferred to the emplacing vehicle. See Table 2-2, FM 20-32, for additional mine-dump planning factors.

4. The unit may employ one or more laying parties. Each laying party emplaces one row of mines at a time.

5. The organization may vary depending on the terrain, soldiers, and material available and proximity of the enemy. This typically requires augmentation by nonengineer soldiers. Nonengineer soldiers can be integrated into any of the parties, but it is simplest to integrate them into the marking and mine-dump parties in squad-size units, or as individuals into the digging team.

Landmine and Special-Purpose Munition Obstacles

FM 5-34

HASTY PROTECTIVE ROW MINEFIELDS

Figure 7-4 shows a hasty protective row minefield record that illustrates a typical layout.

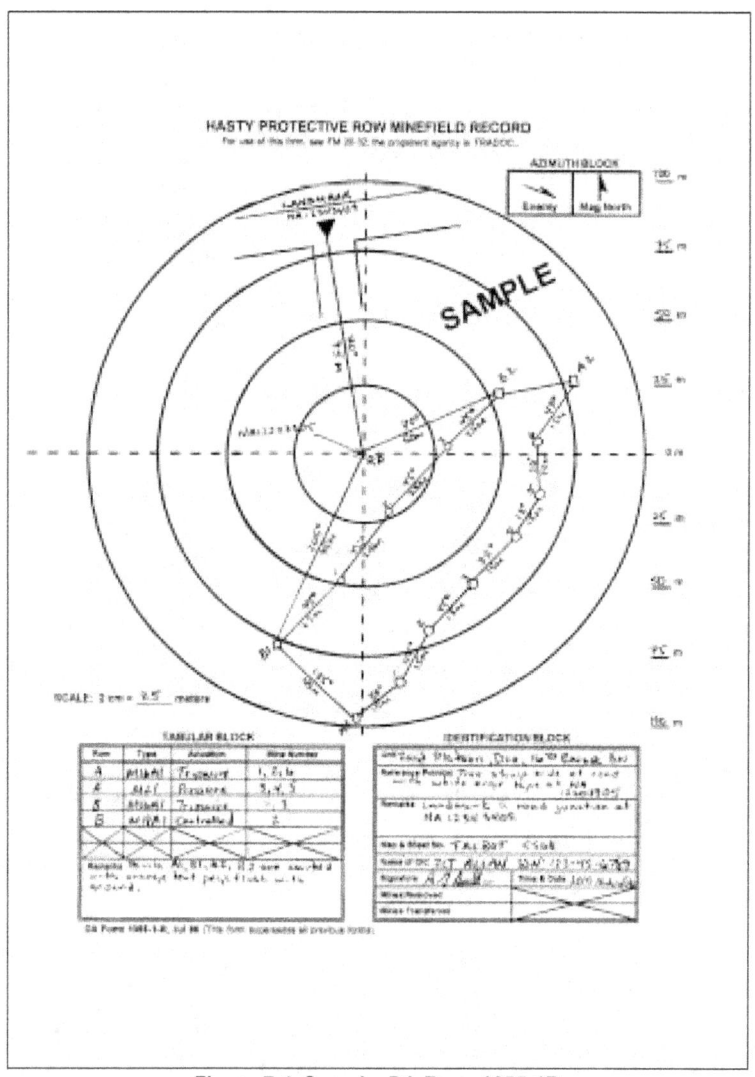

Figure 7-4. Sample, DA Form 1355-1R

7-8 Landmine and Special-Purpose Munition Obstacles

SCATTERABLE MINEFIELDS

Table 7-4. SCATMINEs' sizes and safety zones

Type	Area Density	Minefield Size (m)	Safety Zone
ADAM/RAAM	0.001, 0.002, 0.004 0.001, 0.002, 0.004	200 x 200 400 x 400	1,400 x 1,400 m[1] 1,500 x 1,500 m[1]
Air Volcano	0.0067	1,115 x 140	235 m on all sides
Ground Volcano	0.01	1,110 x 120	235 m on all sides
MOPMS	0.01	35 x 180° semicircle	50 m on all sides
Gator[2]	0.003	650 x 200	275 m on all sides

[1] Maximum size based on maximum error
[2] Based on proper delivery altitude

Table 7-5. SCATMINEs' self-destruct times

Type	Arming Time	Short (hours)	Long	
ADAM/RAAM	2 min/45 sec*	4	48 hr	15 days
Volcano	2 min	4	48 hr	15 days
MOPMS	2 min	4, can recycle up to 4 times		
Gator	2 min	4	48 hr	15 days
PDM	50 sec	4		

NOTE: Mines begin self-destructing at 80 percent of laid life (4 hours x 0.8 = 3 hours 12 minutes). You must recycle before 3 hours to avoid self-destruct. At least 20 percent of mines have internal AHDs.
*Rounds with 45 seconds; PIP designated w/ "A1"

MODULAR PACK MINE SYSTEM (MOPMS)

The MOPMS is a man-portable, 73-kilogram, suitcase-shaped mine dispenser that can be emplaced anytime before dispensing mines. The dispenser contains 21 mines (17 AT and 4 AP) that are dispensed, on command, using an M71 remote-control unit or an electronic initiating device, such as the M34 blasting machine. The

Landmine and Special-Purpose Munition Obstacles 7-9

mines are propelled within a 35-meter distance from the container in a 180° semicircle (see Figure 7-5). Figures 7-6 and 7-7 show other features of the MOPMS.

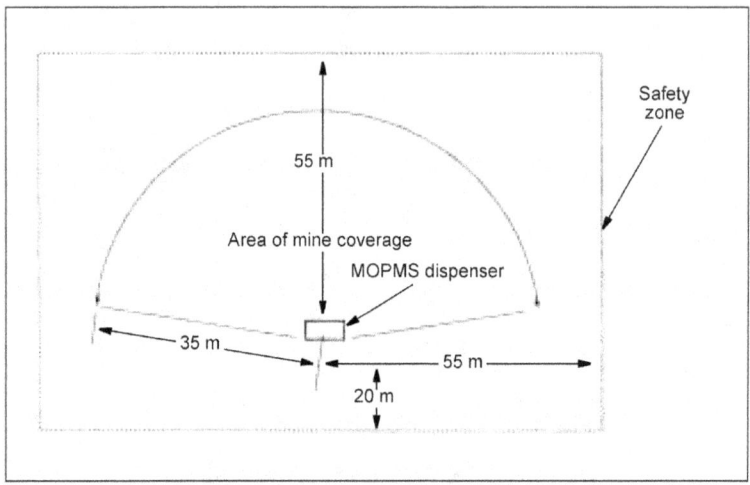

Figure 7-5. MOPMS dispenser emplacement and safety zone

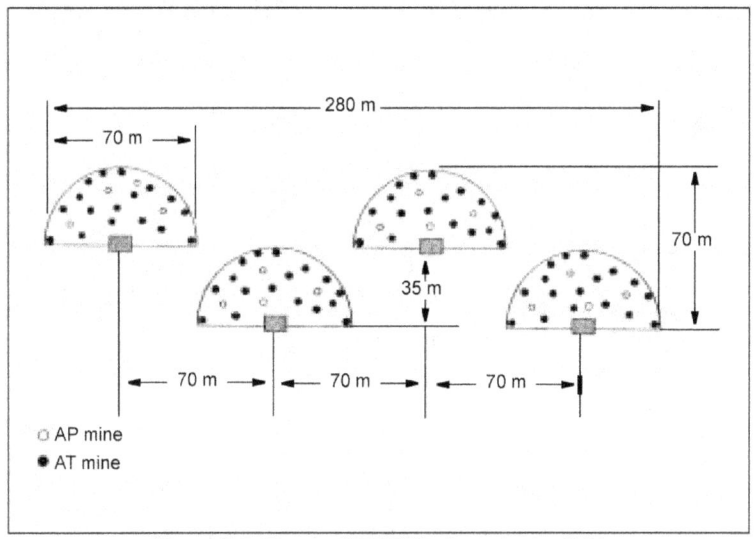

Figure 7-6. Standard MOPMS disrupt minefield

7-10 Landmine and Special-Purpose Munition Obstacles

FM 5-34

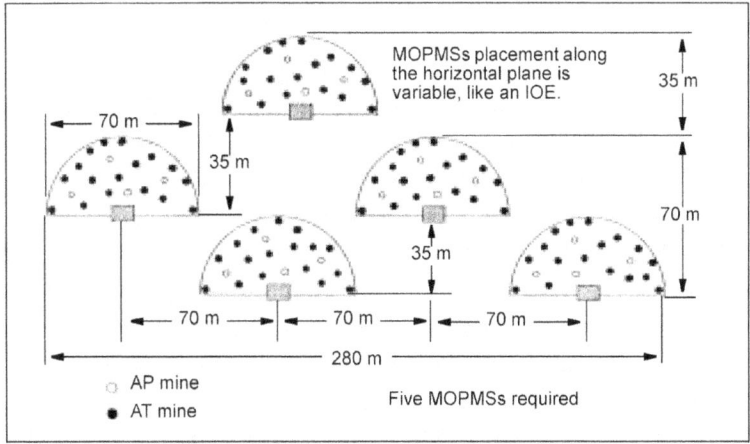

Figure 7-7. Standard MOPMS fixed minefield

VOLCANO

The Volcano is a single-mine delivery system which can be dispensed from the air or ground. It can be mounted on a 5-ton vehicle (heavy, expanded, mobility tactical truck [HEMTT]), an M548 tracked cargo carrier, or a UH-60 Blackhawk helicopter. Up to four Volcano racks can be mounted on each vehicle. Each rack can hold up to 40 canisters, each prepackaged with five AT mines and one AP mine. Figure 7-8 and Figure 7-9, page 7-12, show the disrupt, fix, turn, and block minefields using the ground/air Volcano. Table 7-6, page 7-12, lists characteristics of the Volcano.

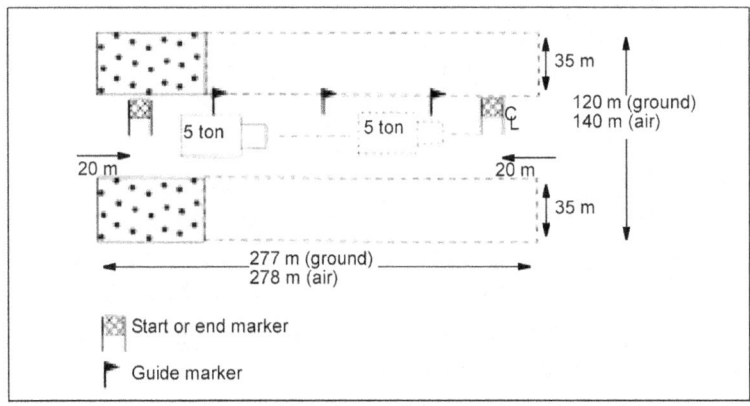

Figure 7-8. Ground/air Volcano disrupt and fixed minefields

Landmine and Special-Purpose Munition Obstacles 7-11

FM 5-34

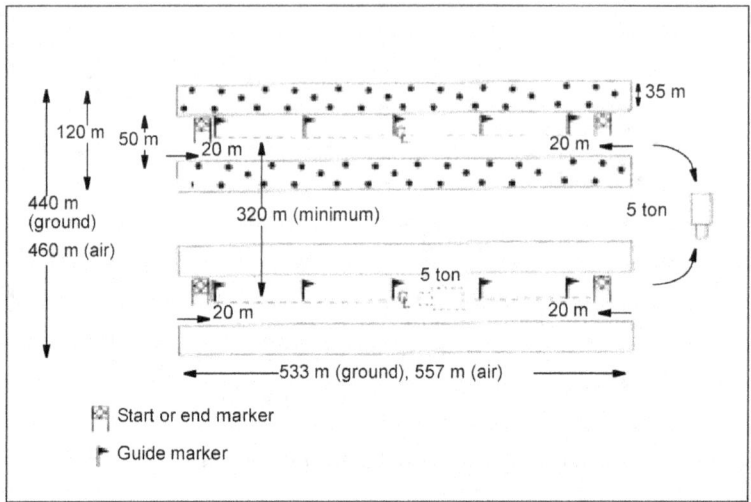

Figure 7-9. Ground/air Volcano turn and block minefields

Table 7-6. Volcano minefield's characteristics

Minefield Type	Depth (m)	Front (m) Ground/Air	Number of Strips	Canisters per Strip	Total Canisters	Minefields per Load
Disrupt	120	277/278	1	40 (20 each side)	40	4
Fix	120	277/278	1	40 (20 each side)	40	4
Turn	320	555/557	2	80 (20 each side)	160	1
Block	320	555/557	2	80 (20 each side)	160	1

ADAM/RAAM

ADAM and RAAM mines are delivered by a 155-millimeter howitzer. (The mines are contained within the 155-millimeter projectile.) Each ADAM projectile contains 36 mines; each RAAM projectile contains 9 mines. ADAM/RAAM minefields have a significant safety zone based on the method of delivery. Table 7-7 lists the minefield's density and sizes and Table 7-8, the safety zones for the ADAM/RAAM.

7-12 Landmine and Special-Purpose Munition Obstacles

Table 7-7. ADAM/RAAM minefield's density and size

Obstacle Effect	Minefield Densities				Width (m)	Depth (m)
	RAAM		ADAM			
	Area[1]	Linear[2]	Area[1]	Linear[2]		
Disrupt	0.001	0.2	0.0005	0.1	200	200
Turn	0.002	0.8	0.001	0.4	400	400
Fix	0.002	0.4	0.0005	0.1	200	200
Block	0.004	1.6	0.002	0.8	400	400

[1]Area density = mines per square meter
[2]Linear density = mines per meter

Table 7-8. ADAM/RAAM minefield's safety zones

Projectile and Trajectory	Range (km)	Meteorological + Velocity-Error/Transfer Technique	Observer-Adjust Technique
RAAM low angle	4	500 x 500	500 x 500
	7	550 x 550	500 x 500
	10	700 x 700	550 x 550
	12	850 x 850	550 x 550
	14	1,000 x 1,000	650 x 650
	16	1,050 x 1,050	650 x 650
	17.5	1,200 x 1,200	650 x 650
ADAM low angle	4	700 x 700	700 x 700
	7	750 x 750	700 x 700
	10	900 x 900	750 x 750
	12	1,050 x 1,050	750 x 750
	14	1,200 x 1,200	850 x 850
	16	1,250 x 1,250	850 x 850
	17.5	1,400 x 1,400	850 x 850
RAAM or ADAM high angle	4	750 x 750	700 x 700
	7	900 x 900	700 x 700
	10	1,050 x 1,050	750 x 750
	12	1,200 x 1,200	750 x 750
	14	1,400 x 1,400	850 x 850
	16	1,500 x 1,500	850 x 850
	17.5	1,400 x 1,400	850 x 850

GATOR

The area of a minefield depends on the speed and altitude of the aircraft. The normal size of a minefield is 650 x 200 meters. Density depends on the number of canisters that are dropped. The Gator system is used primarily for interdiction minefields; somewhat lower

Landmine and Special-Purpose Munition Obstacles 7-13

than normal densities (0.001 mines/meters^2) are usually planned. Each canister (bomblet) contains 72 AT and 22 AP mines. Up to six canisters may be mounted on each aircraft.

SPECIAL-PURPOSE MUNITIONS

M86 Pursuit Deterrent Munition (PDM)

The PDM is similar in configuration and functioning to the ADAM, but must be manually armed (see Figure 7-10).

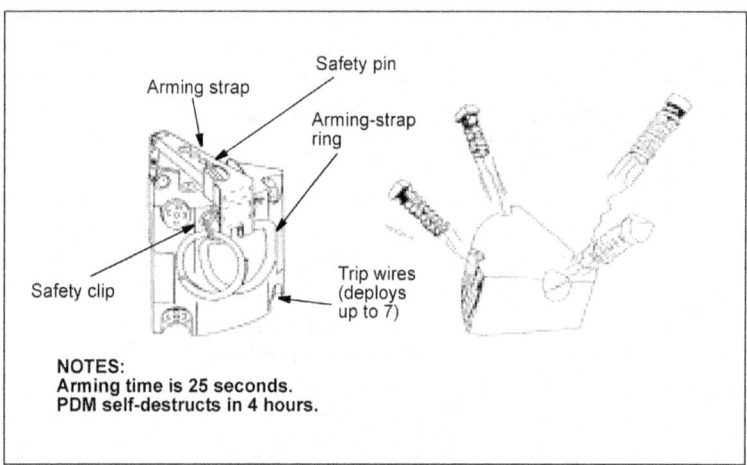

Figure 7-10. M86 PDM

M18A1 Claymore

The M18A1 claymore munition (see Figure 7-11) is a fragmentation munition that contains 700 steel balls and 682 grams of composition C4 explosive. It weighs 1.6 kilograms and can be detonated by command **(Korea only: or by trip wire)**.

When employing the M18A1 claymore with other munitions or mines, separate the munitions by the following minimum distances:

- 50 meters in front of or behind other M18A1s.
- 3 meters between M18A1s that are placed side by side.
- 10 meters from AT or fragmentation AP munitions.
- 2 meters from blast AP munitions.

7-14 Landmine and Special-Purpose Munition Obstacles

FM 5-34

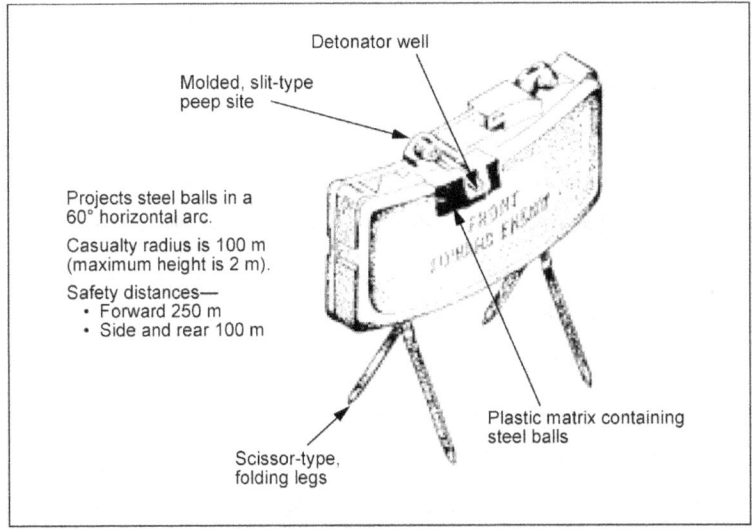

Detonator well
Molded, slit-type peep site
Projects steel balls in a 60° horizontal arc.
Casualty radius is 100 m (maximum height is 2 m).
Safety distances—
• Forward 250 m
• Side and rear 100 m
Plastic matrix containing steel balls
Scissor-type, folding legs

Figure 7-11. M18A1

SELECTABLE LIGHTWEIGHT ATTACK MUNITION (SLAM)

The SLAM (see Figure 7-12, page 7-16) is a multipurpose munition with an antitamper feature.

M93 HORNET

The M93 Hornet (see Figure 7-13, page 7-16) is an AT/antivehicular off-route munition made of lightweight material (35 pounds) that one person can carry and employ. The Hornet is a nonrecoverable munition that is capable of destroying vehicles by using sound and motion detection methods. It will automatically search, detect, recognize, and engage moving targets by using top attack at a standoff distance up to 100 meters from the munition.

Figures 7-14 through 7-18, pages 7-17 through 7-19, show basic emplacement scenarios for the Hornet.

Munitions placed at ground level should be no closer to obstructions than the distances shown in Table 7-9, page 7-19.

Landmine and Special-Purpose Munition Obstacles 7-15

FM 5-34

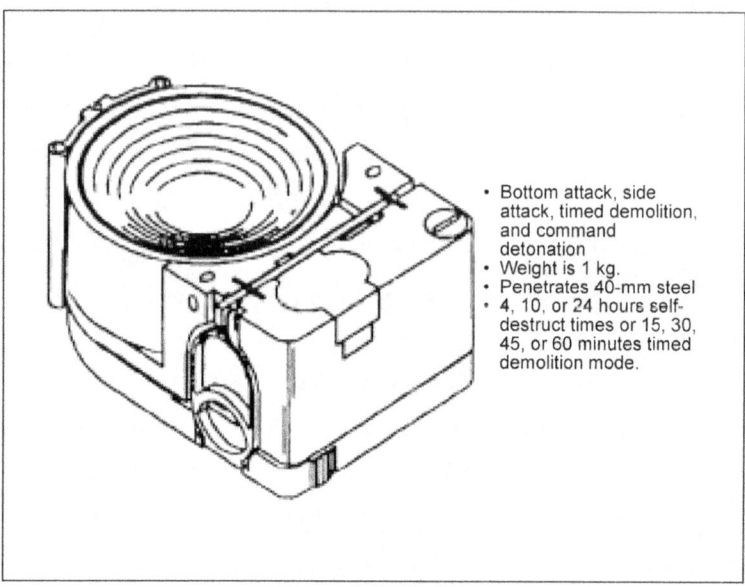

- Bottom attack, side attack, timed demolition, and command detonation
- Weight is 1 kg.
- Penetrates 40-mm steel
- 4, 10, or 24 hours self-destruct times or 15, 30, 45, or 60 minutes timed demolition mode.

Figure 7-12. SLAM

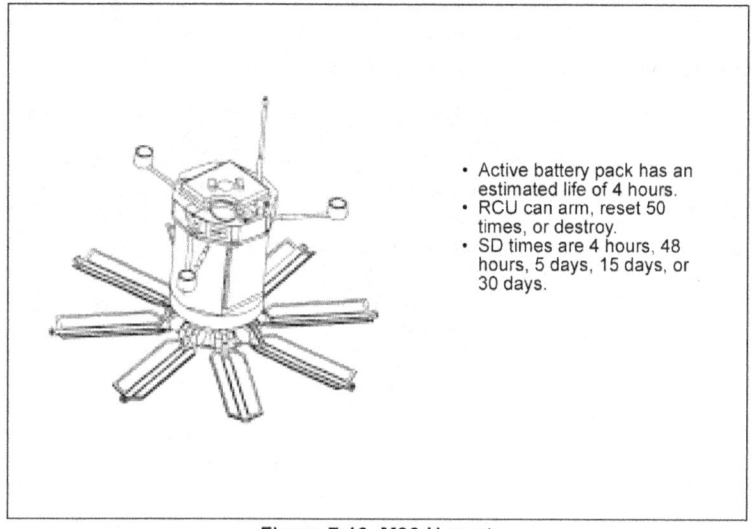

- Active battery pack has an estimated life of 4 hours.
- RCU can arm, reset 50 times, or destroy.
- SD times are 4 hours, 48 hours, 5 days, 15 days, or 30 days.

Figure 7-13. M93 Hornet

7-16 Landmine and Special-Purpose Munition Obstacles

FM 5-34

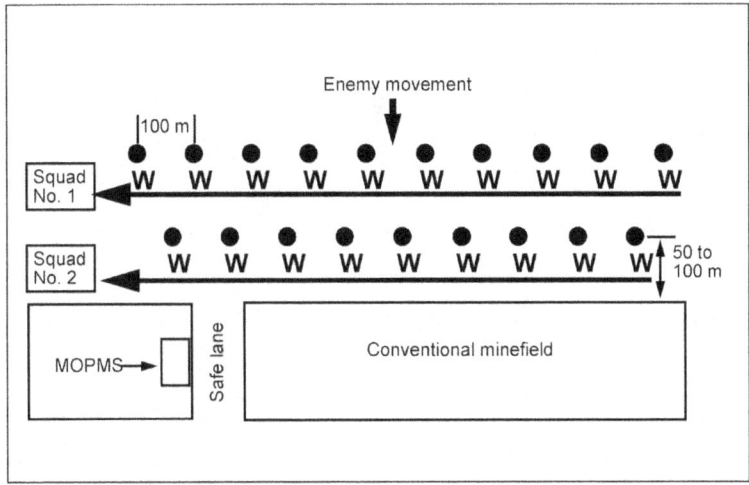

Figure 7-14. Hornet reinforcing a conventional minefield

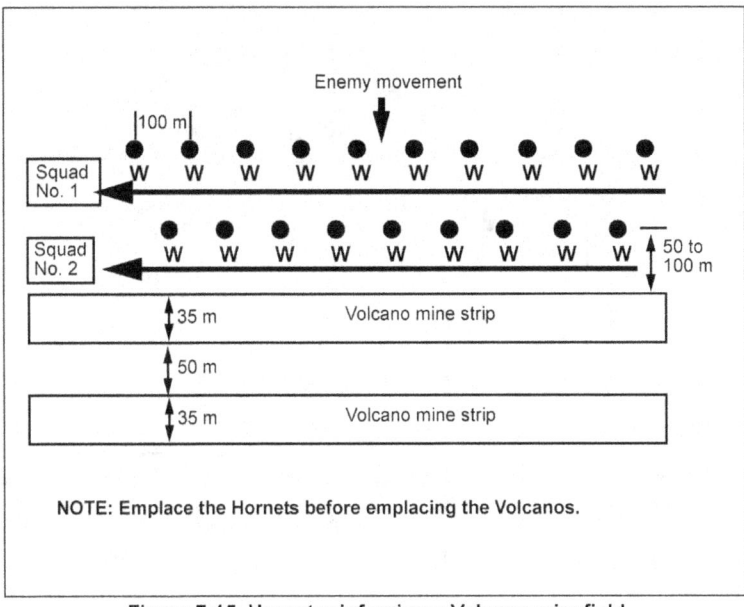

Figure 7-15. Hornet reinforcing a Volcano minefield

Landmine and Special-Purpose Munition Obstacles 7-17

FM 5-34

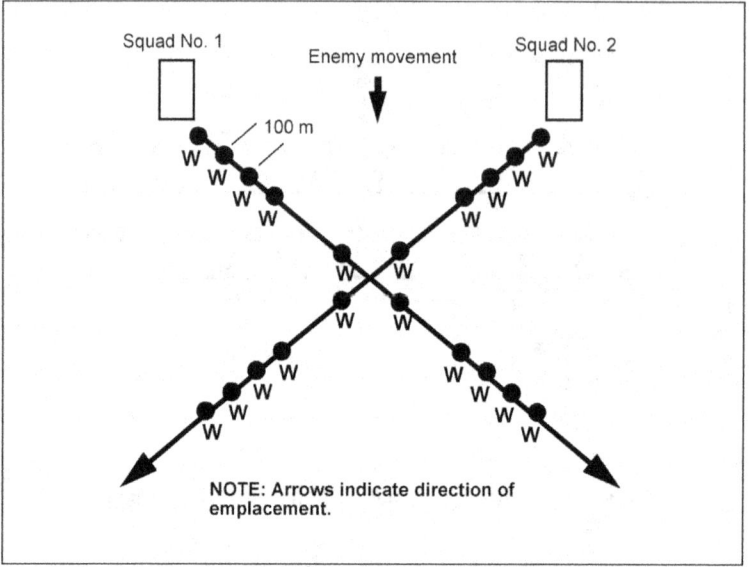

Figure 7-16. Hornet area-disruption obstacle

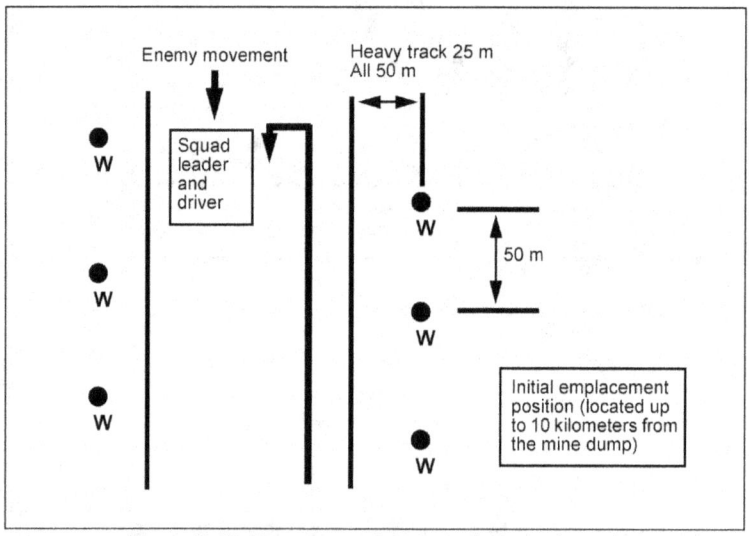

Figure 7-17. Hornet gauntlet obstacle (one cluster)

7-18 Landmine and Special-Purpose Munition Obstacles

FM 5-34

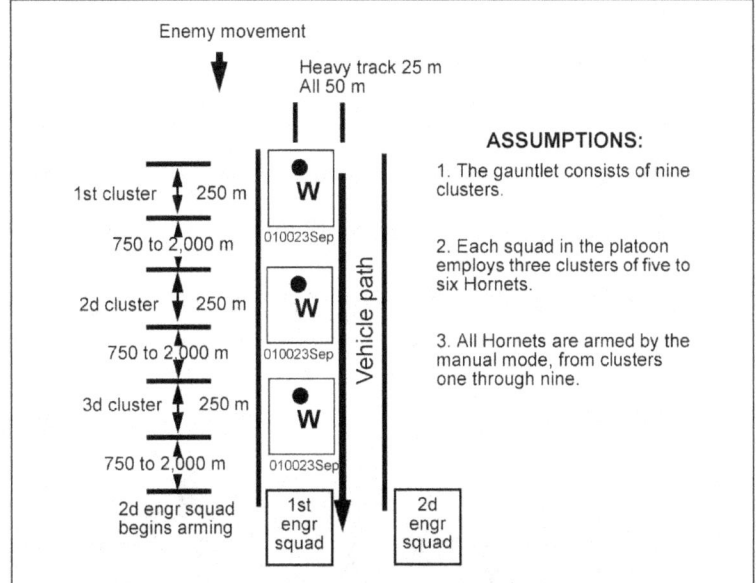

Figure 7-18. Hornet gauntlet obstacle (platoon)

Table 7-9. Hornet minimum emplacement distances

Maximum Obstruction Height	Minimum Employment Distance from Obstruction
1 m	3 m
2.4 m	5 m
6.5 m	15 m
25 m	25 m

RAPTOR INTELLIGENT COMBAT OUTPOST

Raptor introduces an entirely new concept to the combined-arms team. A combined system involving the hand emplaced-wide area munition product improvement plan (HE-WAM PIP) also known as "Hornet-PIP", advanced acoustic sensors, electronic gateways for Hornet coordination, and an overall electronic control station. The Raptor detects, classifies, and engages heavy and light tracked and wheeled vehicles (see Figure 7-19, page 7-20). It is capable of being ordered or programmed to develop coordinated attacks with other minefields and/or direct and indirect fire weapons. Once activated,

Landmine and Special-Purpose Munition Obstacles 7-19

FM 5-34

the Raptor can be inactivated, allowing freedom of maneuver through the munitions while still providing near-real-time intelligence and situational awareness.

A hand-emplaced, top attack system, the Raptor can be used as a stand-alone tactical obstacle or integrated with other conventional or situational obstacles. It can communicate with its employing unit for remote on/off/on programming and reporting of battle-space intelligence. Intelligence data may include target descriptions, numbers, and the direction and rate of movement.

Because a soldier can arm the Hornet-PIP, a planning consideration should be that the soldier moves to a safe separation distance (500 meters) within 5 minutes of arming. This prevents the Hornet-PIP from accidentally engaging the emplacing unit during obstacle construction.

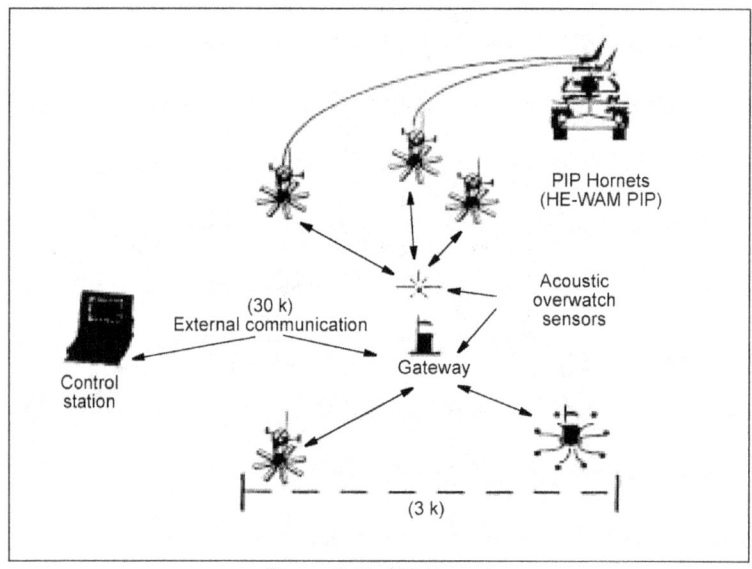

Figure 7-19. PIP Hornet

RECORDING

Use DA Form 1355 to record data on most conventional minefields/munition fields and DA Form 1355-1-R for hasty protective row or munition fields (see Figures 7-20 through 7-23, pages 7-21 through 7-24). Figure 7-4, page 7-8, shows an example of a hasty protective row minefield record.

7-20 Landmine and Special-Purpose Munition Obstacles

FM 5-34

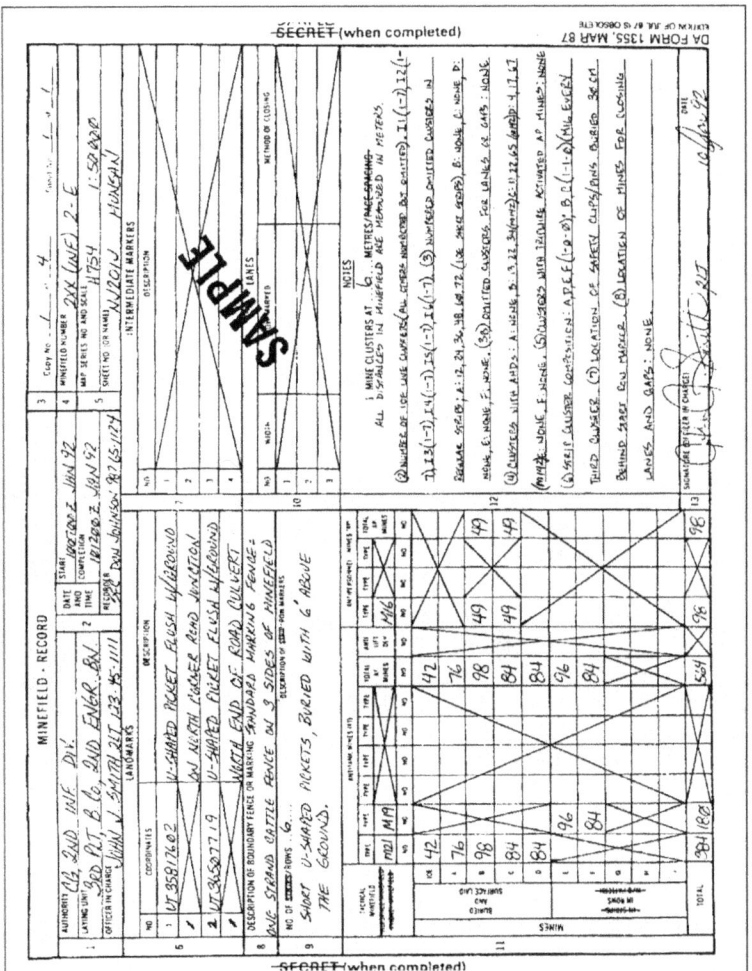

Figure 7-20. Sample, DA Form 1355 (front) (standard-pattern minefield)

Landmine and Special-Purpose Munition Obstacles 7-21

FM 5-34

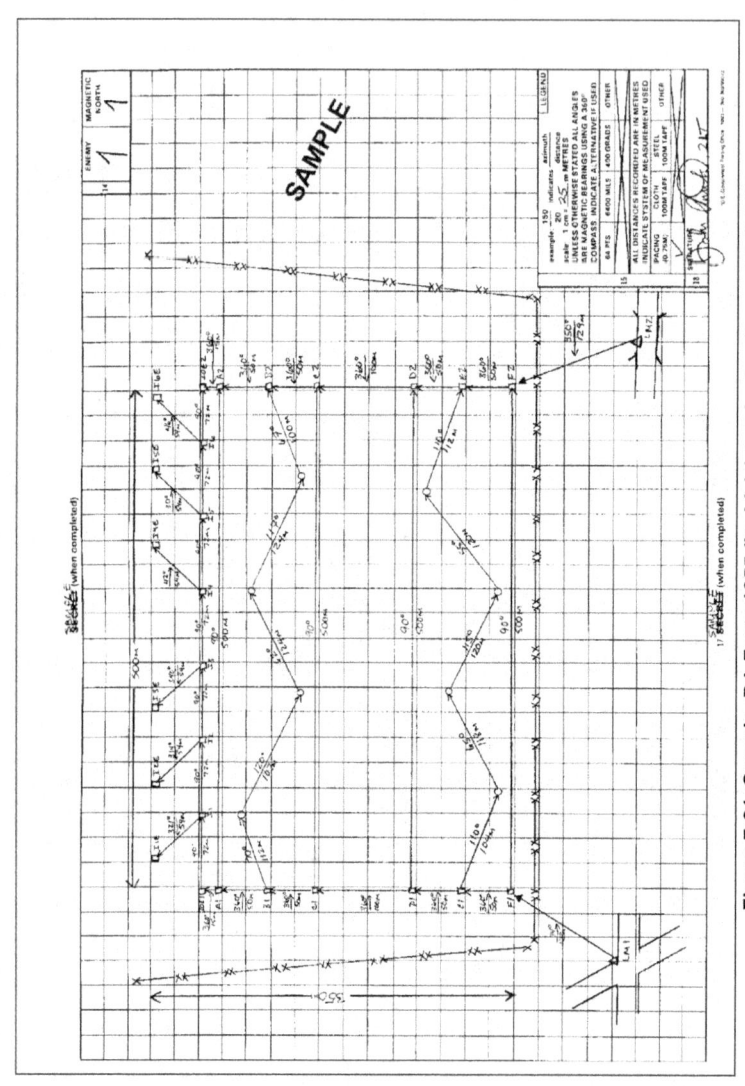

Figure 7-21. Sample, DA Form 1355 (inside) (standard-pattern minefield)

7-22 Landmine and Special-Purpose Munition Obstacles

FM 5-34

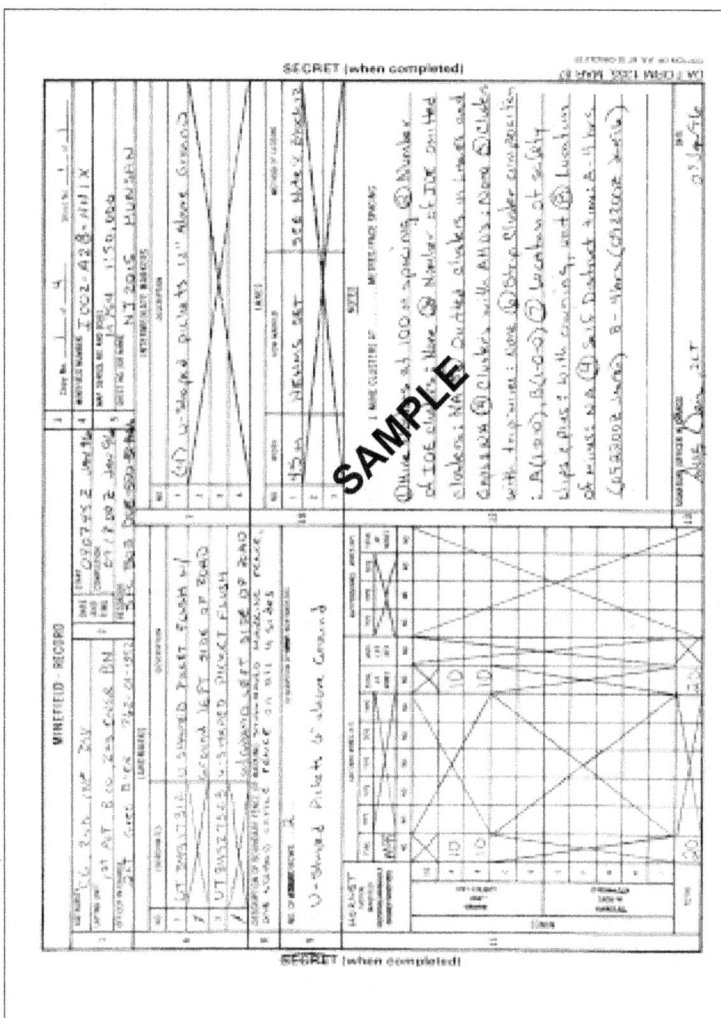

Figure 7-22. Sample DA Form 1355 (front side) for a Hornet minefield/munition field

Landmine and Special-Purpose Munition Obstacles 7-23

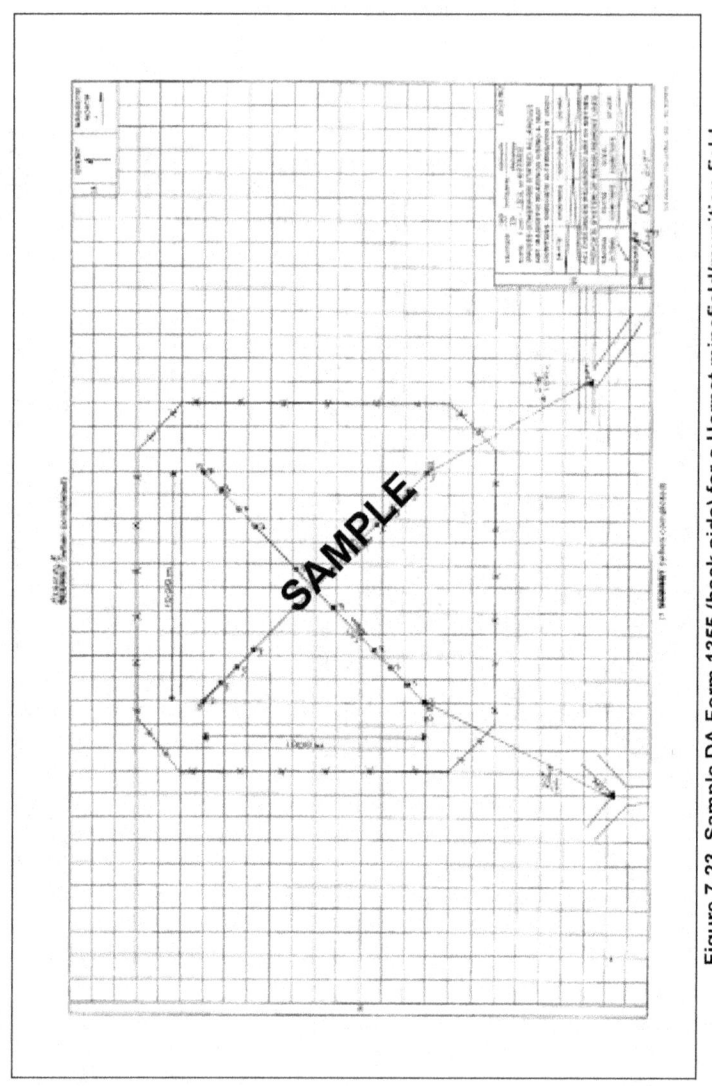

Figure 7-23. Sample DA Form 1355 (back side) for a Hornet minefield/munition field

7-24 Landmine and Special-Purpose Munition Obstacles

To facilitate reporting (discussed at the beginning of this chapter) and recording of scatterable minefields/munition fields, a simple, uniform procedure is used. This procedure combines the report and the record into one document (Table 7-10) that is applicable for all delivery systems. Table 7-10 and Table 7-11, page 7-26, deal with scatterable minefields.

Table 7-10. Scatterable minefield's report and record

Line	Instructions	Example
1	Approving authority	CDR, 3AD
2	Target/obstacle number: If the minefield is part of an obstacle plan, enter the obstacle number. If the minefield is not part of an obstacle plan or does not have a number, leave blank or enter N/A.	I001C3BSV04X
3	Type of emplacing system: Enter the type system that emplaced the minefield (MOPMS, Volcano).	Ground Volcano
4	Mine type: Enter AP or AT. If both, enter AP/AT	AP/AT
5	Life cycle: Enter the DTG the minefield was emplaced until the last mine self-destructs.	021005ZAUG96 - 041005ZAUG96
6-14	Aim point/corner points of the minefield: If the system used to emplace the minefield uses a single aim point to deliver the mines, enter that aim point. If the system has distinct corner points, as the Volcano does, enter those corner points.	MB 17955490 MB 18604860 MB 18504895 MB 17804850
15	Size of safety zone from aim point: If an aim point is given in line 6, enter the size of the safety zone from that aim point.	N/A
16	Unit emplacing mines and report number: Reports should be numbered consecutively. This would be the fourth minefield that the company has emplaced.	HHC, 307th Engr Bn, 4
17	Person completing the report	CPT Zimmerman
18	DTG of report	021015ZAUG96
19	Remarks: Any other items that the reporting unit deems important	Centerline generally follows east-west route, Route Blue.

Landmine and Special-Purpose Munition Obstacles 7-25

Table 7-11. Scatterable minefield warning (SCATMINEWARN) report

Line	Message	Example
ALPHA	Emplacing system	Arty
BRAVO	AT mines (Yes or No)	Yes
CHARLIE	AP mines (Yes or No)	Yes
DELTA	Number of aim points or corner points	One
ECHO	Grid coordinates of aim points or corner points and size of safety zone	WQ03574598 500-m
FOXTROT	DTG of life cycle	020615ZAUG96- 061015ZAUG96

In addition to the scatterable minefield/munition field report and record, the (SCATMINWARN) (see the example in Table 7-11) notifies effected units that SCATMINEs will be emplaced.

MINEFIELD MARKINGS

MARKING SETS

A hand-emplaced minefield marking set (HEMMS) is normally used for temporary markings and can mark 70 to 1,000 meters. A US Number 2 minefield marking set can mark about 400 meters per set and replaces the HEMMS if a minefield is left in place for more than 15 days.

MARKING PROCEDURES

A minefield is normally marked to prevent friendly personnel from accidentally entering it. Figures 7-24 and 7-25 show typical markings and marked minefield perimeters and lanes. Scatterable minefields will be marked to the maximum extent possible to protect friendly troops. The same marking procedures for a conventional minefield will be used. Table 7-12, page 7-28, lists marking requirements.

FM 5-34

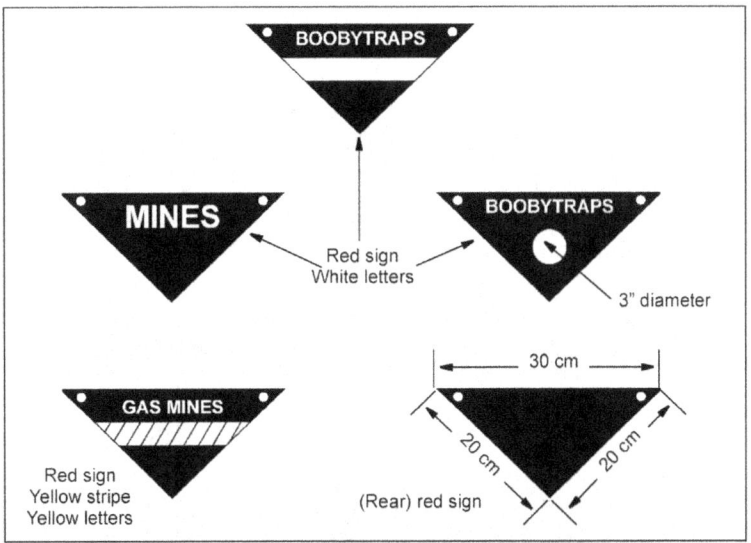

Figure 7-24. Standard marking signs

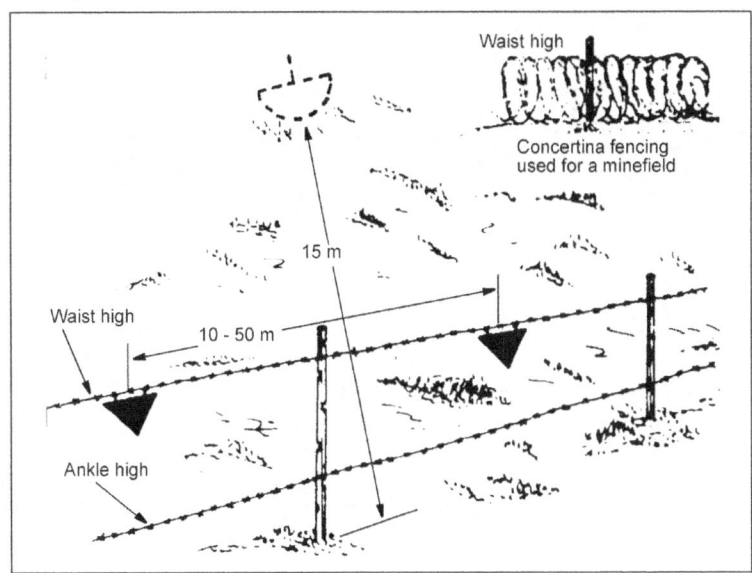

Figure 7-25. Minefield marking fence

Landmine and Special-Purpose Munition Obstacles 7-27

FM 5-34

Table 7-12. Scatterable minefield's marking requirements

Minefield Location	Marking
Enemy's forward area	Not marked
Friendly's forward area	Sides and rear marked
Friendly's rear area	All sides marked

US MINES AND FUSES

Figures 7-26 through 7-28, pages 7-29 through 7-35, show and describe AT and AP mines and firing devices and trip flares. Figure 7-29 and Table 7-13, page 7-36, show characteristics of AP SCATMINEs. Figure 7-30 and Table 7-14, page 7-37, show characteristics of AT SCATMINEs.

FM 5-34

Munition	Packing	Arming Procedures			Disarming
 M14 blast AP mine	Carton contains: 90 mines 90 detonators 6 or 9 wrenches Dimensions (cm): Length: 20 Width: 44 Height: 22 Total wt: 21 kg	1.  Unscrew shipping plug from bottom of mine. Turn pressure plate to ARMED position with arming tool.	2. 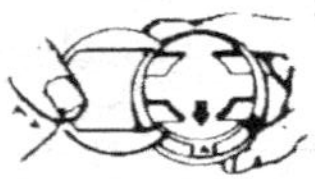 Remove safety clip and check for malfunctioning.	3.  Replace safety clip.	TO DISARM: Insert safety clip and remove detonator
Wt .... 3 1/3 oz. Explosive ..... 1 oz. TETRYL Fuse .... integral (with Belleville Spring) Functioning 20 to 35 lb Penetrate boot and foot		4.  Screw detonator into detonator well.	5.  Bury mine and remove safety clip.		CAUTION: Repeated turning of arming dial may cause excessive wear. TO BURY: Pressure plate should be slightly above ground level.
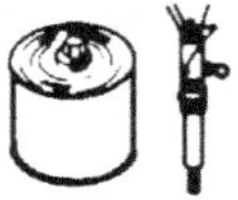 M16A1 bounding AP mines	Wooden box: 4 mines per box 4 fuses per box 1 arming wrench 4 trip wires Dimensions (cm): Length: 41 Width: 28 Height: 22 Total wt: 20 kg	1.  Remove shipping plug and screw in fuse.	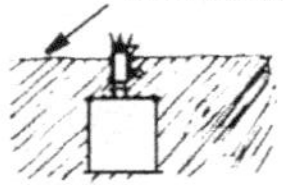 2. Ground level	3.  Trip wire installation	TO DISARM: Reverse arming procedures
Wt .... 8.25 lb Projectiles ..... steel Fuse .... M605 (Combination) Functioning: Pressure ...... 8 to 20 lb Pull .... 3 to 10 lb Bounding height ... 6–1.2 m Casualty radius .... 30 m		4. 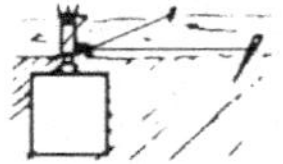 Attach trip wires—first to anchor, then to pull ring.	5.  Remove locking safety pin first. The interlocking pins should fall free. Then remove positive safety.		M16A2 is similar to M16A1/M16, but fuse well is not centered on mine.

Figure 7-26. AP mines (Korea only)

Landmine and Special-Purpose Munition Obstacles 7-29

FM 5-34

Munition	Packing	Arming Procedures	Disarming
M18A1 fragmentation*	Wooden box: 6 mines with accessories. Dimensions (in): Length: 20 Width: 11.5 Height: 9.75 Total wt: 33 lb	1. TEST CIRCUIT: Mate firing device, circuit tester, and blasting cap. Depress handle. Light should show in window. Separate test components. 2. AIMING: In aiming the M18A1, when using the slit-type peep sight, aim the mine at an individual's head when standing 45 m from the mine. When using the knife edge sight, aim the mine at an individual's feet when standing 50 m from the mine. 3. Remove shipping plug-priming adapter, insert blasting cap, and screw into either cap well.	TO DISARM: Reverse arming procedures.
Weight .... 3.5 lb Explosive .... 1.5 lb C4 Projectiles .... 700 (steel balls) Equipment: One electric cap. 30-m firing wire per mine One electric firing device per mine One tester per 6 mines		4. Unroll firing wire and connect directly to firing device with safety engaged. 5. Direction of arm 60° Dangerous out to 250 m 50 m 100 m FIRING POSITION: Minimum of 15 meters from rear of mine to fighting position; friendly troops at side and rear should be under cover at a minimum of 100 meters.	TO FIRE: Disengage safety bail and depress handle.

* US policy regarding the use and employment of APLs outlined in this FM is subject to the convention on Certain Conventional Weapons and Executive Orders. Current US policy limits the use of non-self-destructing APLs to (1) defending the US and its allies from armed aggression across the Korean demilitarized zone and (2) training personnel engaged in demining and countermine operations. The use of the M18A1 claymore in the command-detonation mode is not restricted under international law or Executive Order.

Figure 7-26. AP mines (Korea only) (continued)

7-30 Landmine and Special-Purpose Munition Obstacles

C2, FM 5-34

Munition	Packing	Arming Procedures	Disarming
M15 heavy AT mine Wt .... 30 lb Explosive .... .22 lb Fuse ...... M603, M624 Secondary fuse wells ... 2 Functioning ... 300 to 400 lb	Individual crate: 1 mine with fuse 1 activator Dimensions (in): Length: 18 Width: 15.3 Height: 7.5 Total wt: 49 lb	1. Remove plug and inspect fuse well.* 2. Inspect fuse and remove safety 5. Replace plug with dial in SAFE position. 6. Turn dial to ARMED. 4. Insert fuse TO BURY: Put mine in hole with pressure plate at or slightly above ground level.	TO DISARM: Reverse arming procedures.
M15 AT mine used with M608 fuse 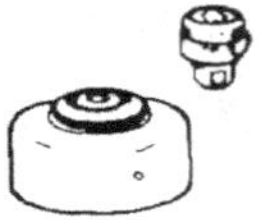 Functioning ... 200-350 lb for 250-450 milliseconds; resistant to blast-type countermeasures	Same as above	1.  Remove plug and inspect fuse well. Ensure fuse is in SAFE position. Thread fuse into mine. HAND TIGHT. Locking ring Base fuse Hold fuse to prevent rotating, turn locking ring down until it locks against pressure plate. 2.  Turn dial from SAFE to ARMED. Place mine in hole and remove pull pin from fuse.	 3. TO DISARM: Reverse arming procedures except DO NOT replace pull pin.

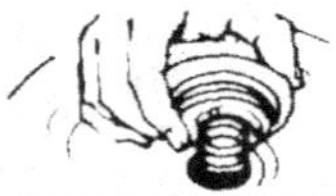

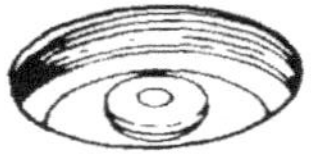

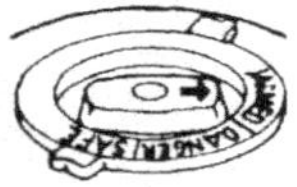

*Inspect secondary fuze wells for corrosion. Do not fit an M1 activator into a corroded fuze well. In training, return a mine with a corroded fuze well to the ammunition supply point as unserviceable.

Figure 7-27. AT mines

Landmine and Special-Purpose Munition Obstacles 7-31

FM 5-34

Munition	Packing	Arming Procedures	Disarming
M21 metallic (killer) AT mine Wt .... 18 lb Explosive .... 10.5 lb Fuse .... M607 Functioning ..... 290 lb (Pressure or pressure ring or 20° deflection of tilt rod)	Wooden box: 4 mines 2 wrenches Dimensions (in): Length: 22.2 Width: 20.2 Height: 16 Total wt: 90.8 lb	1. Remove closing plug, insert M120 booster in bottom, and replace closing plug. 2. Remove closure assembly from fuse. 3. Remove shipping plug from mine and screw in fuse, then screw in tilt-rod extension. 4. Bury mine. 5. For pressure-type mine bury with fuse cap flush with ground surface. Tilt-rod mines should be seated firmly in snug-fitting hole. Most effective in tall brush or grass.	TO DISARM: Reverse arming procedures.
M19 plastic heavy AT mine Wt .... 28 lb Explosive .... 21 lb Fuse .... M606 integral (with pressure plate) Secondary fuse wells .. 2 Functioning .. 350 to 500 lb	Wooden box: 4 mines 4 fuses 1 arming wrench Dimensions (in): Length: 16.8 Width: 10.8 Height: 16 Total wt: 71.8 lb	1. Remove pressure-plate fuse. 2. Remove shipping plug; check position of striker (off-set). Remove safety fork, then turn dial to ARMED position. Check position of striker (center). Turn to SAFE and replace safety fork. 3. Screw threaded detonator into detonator well. 4. Place mine in hole, remove safety fork, and turn dial to ARMED. 5. Complete camouflage.	TO DISARM: Reverse arming positions. TO BURY: Put mine in hole with pressure plate at or slightly above ground level.

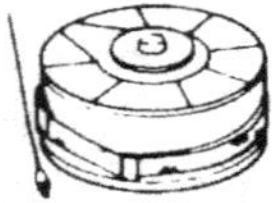

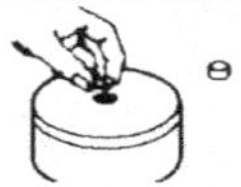

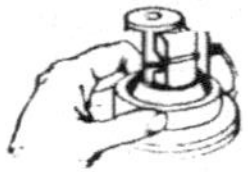

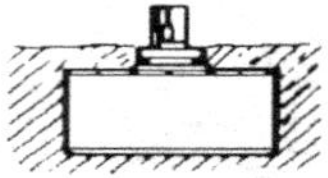

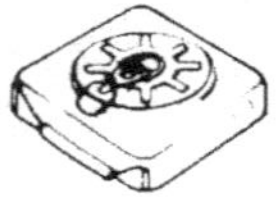

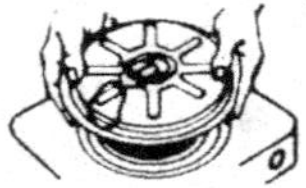

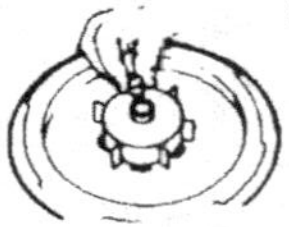

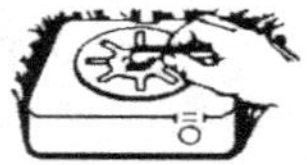

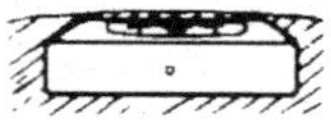

Figure 7-27. AT mines (continued)

7-32 Landmine and Special-Purpose Munition Obstacles

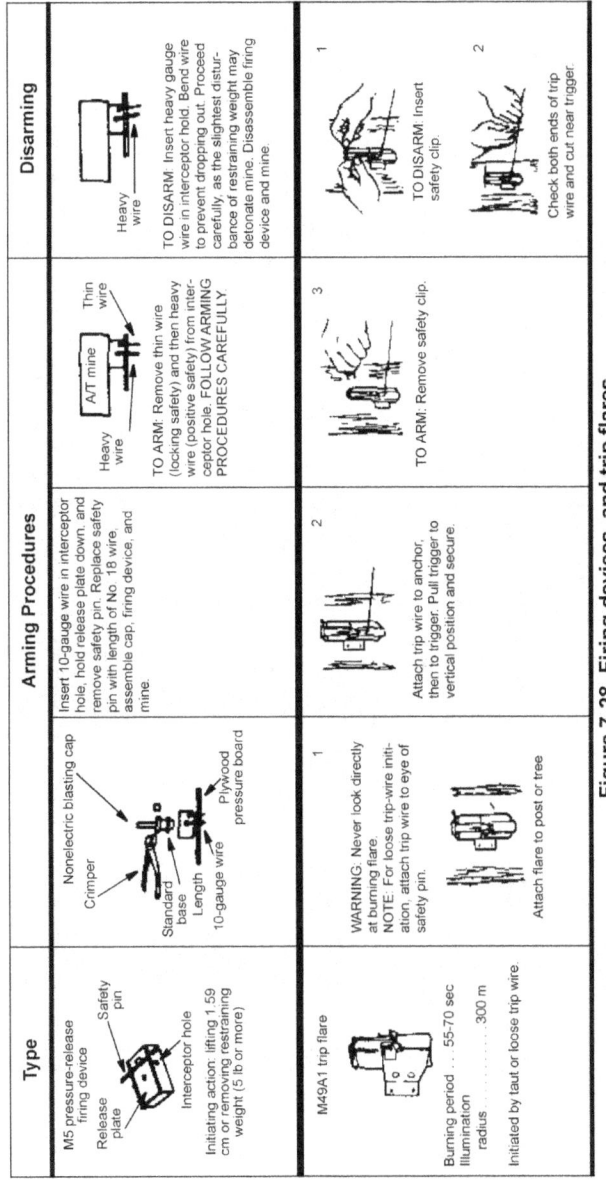

Figure 7-28. Firing devices and trip flares

FM 5-34

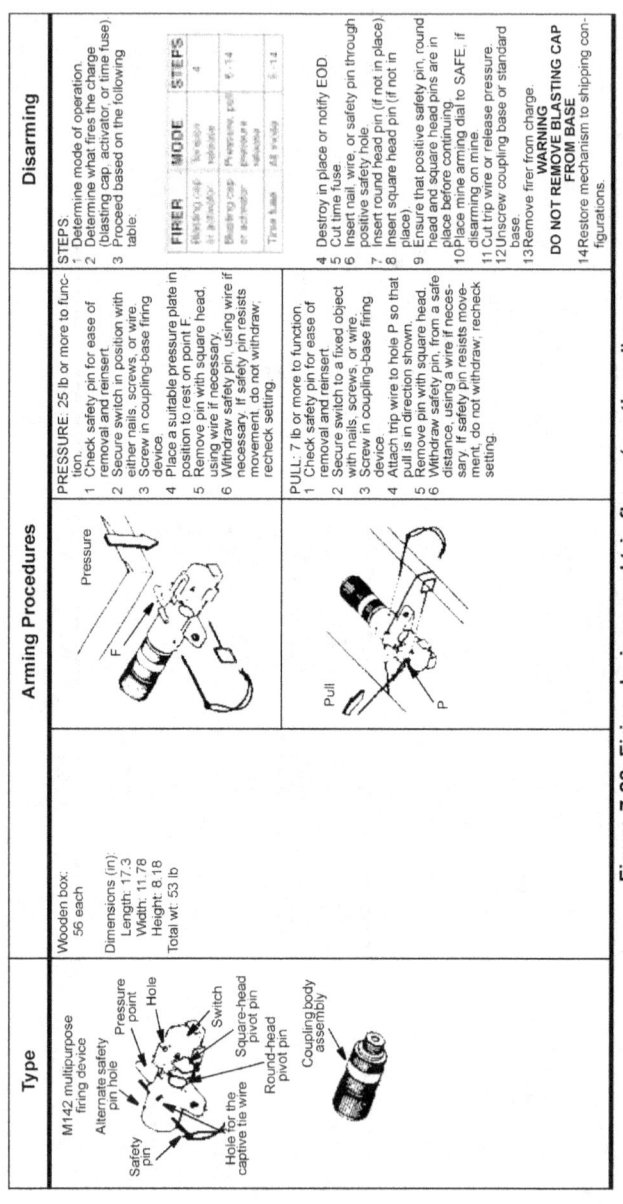

Figure 7-28. Firing devices and trip flares (continued)

7-34 Landmine and Special-Purpose Munition Obstacles

FM 5-34

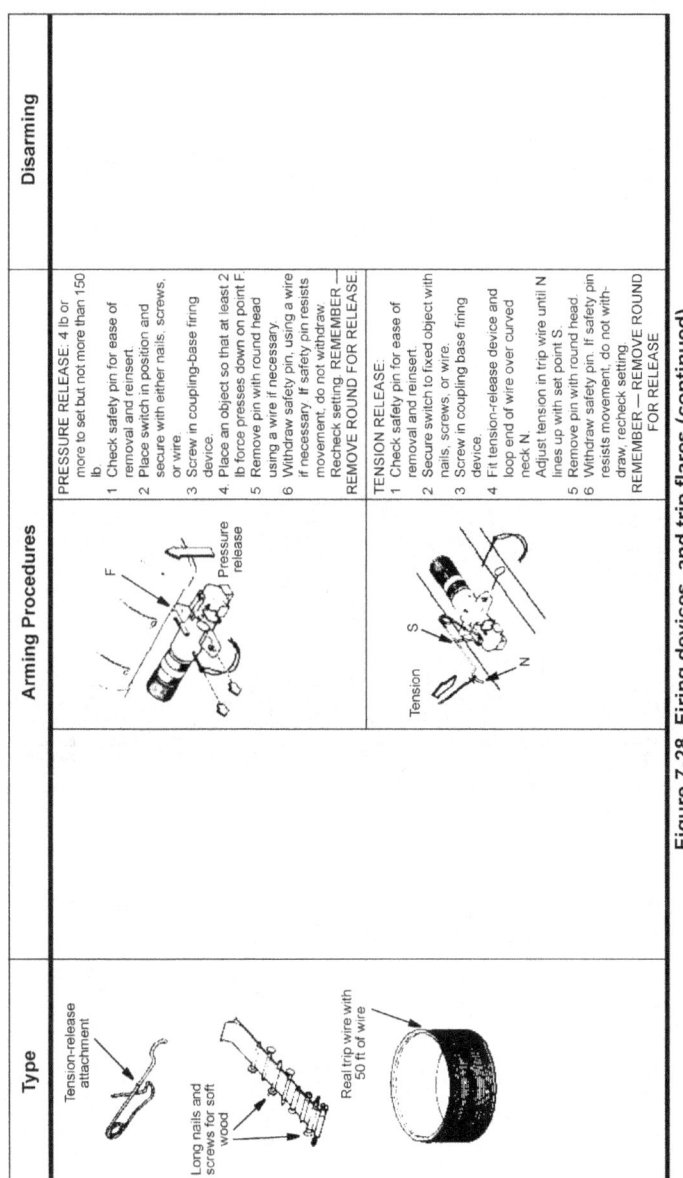

Figure 7-28. Firing devices and trip flares (continued)

Landmine and Special-Purpose Munition Obstacles 7-35

FM 5-34

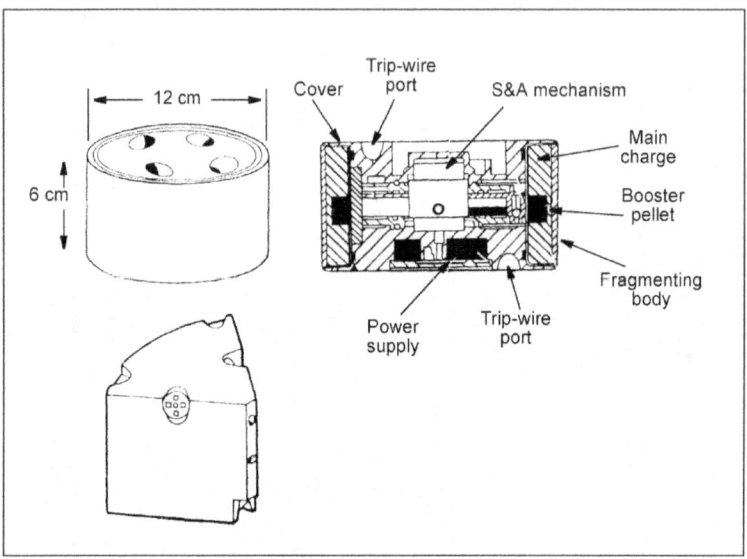

Figure 7-29. AP SCATMINEs

Table 7-13. Characteristics of AP SCATMINEs

Munition	Delivery System	DODIC	Arming Time	Fuse	War-head	AHD	SD Time	Explosive Weight	Munition Weight	Number of Mines
M67	155-mm artillery (ADAM)	D502	45 sec 2 min	Trip wire	Bounding frag	20%	4 hr	21 g Comp A5	540 g	36 per M731 projectile
M72	155-mm artillery (ADAM)	D501	45 sec 2 min	Trip wire	Bounding frag	20%	48 hr	21 g Comp A5	540 g	36 per M692 projectile
M74	Flipper	K151	45 min	Trip wire	Blast frag	20%	5 days 15 days	540 g Comp B4	1.44 kg	5 per sleeve
BLU 92/B	USAF (Gator)	K291 K292 K293	2 min	Trip wire	Blast frag	100%	4 hr 48 hr 15 days	540 g Comp B4	1.44 kg	22 per CBU 89/B dispenser
M77	MOPMS	K022	2 min	Trip wire	Blast frag	0%	4 hr (recycle up to 3 times)	540 g Comp B4	1.44 kg	4 per M131 dispenser
Volcano	Ground/ air	K045	4 min	Trip wire	Blast frag	0%	4 hr 48 hr 15 days	540 g Comp B4	1.44 kg	1 per M87 canister

7-36 Landmine and Special-Purpose Munition Obstacles

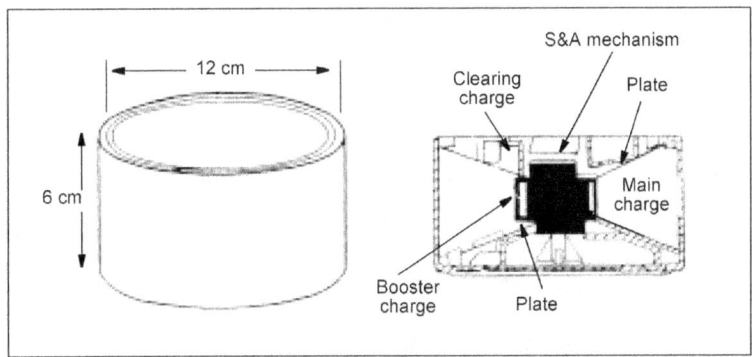

Figure 7-30. AT SCATMINE

Table 7-14. Characteristics of AT SCATMINEs

Mine	Delivery System	DODIC	Arming Time	Fuse	Warhead	AHD	SD Time	Explosive Weight	Mine Weight	Number of Mines
M73	155-mm artillery (RAAM)	D503	45 sec 2 min	Magnetic	M-S plate	20%	48 hr	585 g RDX	1.7 kg	9 per M718 projectile
M70	155-mm artillery (RAAM)	D509	45 sec 2 min	Magnetic	M-S plate	20%	4 hr	585 g RDX	1.7 kg	9 per M741 projectile
M75	Flipper	K184	45 min	Magnetic	M-S plate	20%	5 days 15 days	585 g RDX	1.7 kg	5 per sleeve
BLU 91/ B	USAF (Gator)	K291 K292 K293	2 min	Magnetic	M-S plate	NA	4 hr 48 hr 15 days	585 g RDX	1.7 kg	72 per CBU 89/B dispenser
M76	MOPMS	K022	2 min	Magnetic	M-S plate	NA	4 hr (recycle up to 3 times)	585 g RDX	1.7 kg	17 per M131 dispenser
Volcano	Ground/ air	K045	2 min 30 sec	Magnetic	M-S plate	NA	4 hr 48 hr 15 days	585 g RDX	1.7 kg	5 per M87 canister; 6 per M87A1 canister

FOREIGN MINES

Unless directed differently, all foreign mines will be destroyed in place rather than removed/disarmed.

Landmine and Special-Purpose Munition Obstacles 7-37

FM 5-34

Mine	Description	Sketch
Former Soviet Union		

TM 60	Plastic
	Total weight: 11.4 kg
	Weight of explosive: 9.9 kg
	Diameter: 300 mm
	Fuze: Two available
	  a. Nonmetallic chemical
	  b. Mechanical pressure

TMS-B TMB1 TMB2	Tar-impregnated cardboard, glass blug over fuze well Diameter: 275 mm Total weight: 6.9 kg Weight of explosive: 5.0 kg Fuze: MV-5K	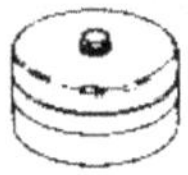
TM46 TMN46 TM41	Metal Total weight: 8.7 kg Weight of explosive: 5.7 kg Diameter: 300 mm Fuze: MVM	
TM57	Material (metal) Total weight: 9-12 kg Diameter: 310 mm Fuze: Pressure, tilt rod, or pneumatic	
LMG	Rocket Total weight: 10 kg Weight of explosive: 3.2 kg Fuze: Pull (MUV)	
MZD series	Wood, field fabricated Total weight: Variable Weight of explosive: 0.4 - 4.0 kg Fuze: Vibration, electric	
TMD B TMD44	Wood Total weight: 7.7 - 10 kg Weight of explosive: 5 - 6.8 kg Fuze: Pressure (MV-5)	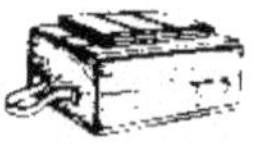
YAM 5/10 TMD41	Wood Total weight: 7.7 kg Weight of explosive: 5.8 kg Fuze: pull (MUV)	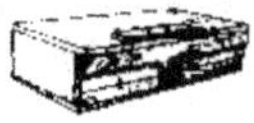
TMK2	Metal Total weight: 12.5 kg Fuze: Tilt rod (adjustable)	

Figure 7-31. Foreign AT mines

7-38 Landmine and Special-Purpose Munition Obstacles

FM 5-34

Mine	Description	Sketch
Former Czechoslovakia		
PT Mi Ba PT Mi Ba 53	Plastic Total weight: 7.6 kg Weight of explosive: 5.6 kg Diameter: 310 mm Fuze: Pressure	
PT Mi Ba II/III	Plastic Total weight: 9.9 kg Weight of explosive: 6 kg Fuze: Pressure	
PT Mi K	Metal Total weight: 7.1 kg Weight of explosive: 5 kg Fuze: Pressure	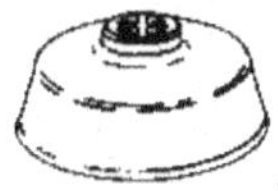
PT Mi D/II/III	Wood Total weight: 9 kg+ Weight of explosive: 6.2 kg	
Former East Germany		
PM 60	Similar to TM60 (Russian)	
K1	Plastic Total weight: 11 kg Weight of explosive: 7 kg Fuze: Pressure	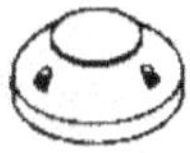
Hungary		
Shape-charge mine	Cardboard and plywood Total weight: 5.4 kg Fuze: Pressure	
Denmark		
M/47-1	Metal Total weight: 10 kg Weight of explosive: 6.3 kg Fuze: Pressure or antidisturbance	
M/52	Plastic Total weight: 10.7 kg Weight of explosive: 8.3 kg Fuze: Pressure-chemical	

Figure 7-31. Foreign AT mines (continued)

Landmine and Special-Purpose Munition Obstacles 7-39

FM 5-34

Mine	Description	Sketch
Model 1951 Nonmetallic	Has no case, cast TNT Total weight: 7 kg Diameter: 300 mm Fuze: Pressure chemical 1950 or pressure friction 1952	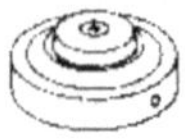
Model 1947 Nonmetallic	Bakelite case Total weight: 11 kg Diameter: 330 mm Fuze: Pressure chemical 1950 or pressure friction 1952	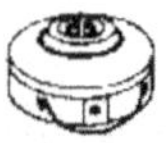
Model 1948	Metal Total weight: 9 kg Diameter: 310 mm Fuze: Main and two secondary fuze wells	
Italy		
CS 42/2 CS 42/3	Wood Total weight: 6.9 kg Weight of explosive: 5 kg Fuze: Pressure	
SH-55	Plastic Total weight: 7.3 kg Diameter: 265 mm Fuze: Integral pneumatic pressure	

"Saci" 54/7	Plastic case but metal striker detectable
	Diameter: 265 mm
	Total weight:Two models
	  a. Light — 6.2 g
	  b. Heavy — 10.2 kg
	Fuze: Three pressure

Japan		
Type 63	Nonmetallic Total weight: 15 kg Weight of explosive: 11 kg	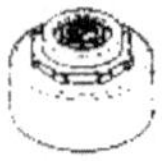
Netherlands		
MIRJAM River mine	Employs normal antitank mine, such as Model 26 (Serial 6) Total weight: 18 kg Length: 605 mm	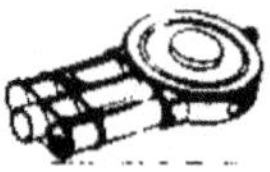
Model 26 Undetectable	Plastic reinforced with glass wool Total weight: 9 kg Diameter: 300 mm Fuze: Pressure friction with shear collar control, two secondary fuze wells for antilift devices	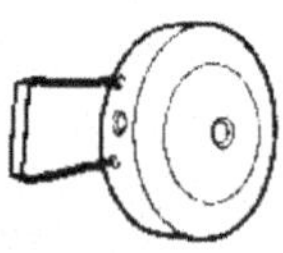

Figure 7-31. Foreign AT mines (continued)

7-40 Landmine and Special-Purpose Munition Obstacles

FM 5-34

Mine	Description	Sketch
Model 25	Metal Total weight: 12.8 kg Diameter: 309 mm Fuze: Pressure with two secondary fuze wells for antihandling devices	
T40	Metal Total weight: 6 kg Diameter: 280 mm Fuze: Pressure	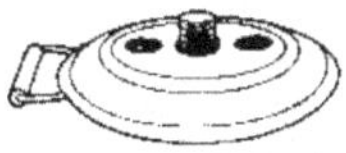
Spain		
C.E.T.M.E.	Nonmetallic Total weight: 9.9 kg Weight of explosive: 5.2 kg Fuze: Chemical or mechanical	
Sweden		
Model 52	Wood and fabrics Total weight: 8.9 kg Weight of explosive: 7.4 kg Fuze: Pressure	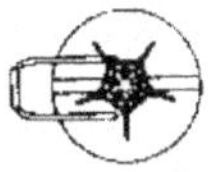
M1 101	Nonmetallic Total weight: 12.4 kg Weight of explosive: 11 kg Fuze: No data	
Model 41-47 and 47	Metallic Weight of explosive: 5 kg Fuze: Pressure	
United Kingdom		
L9A1	Nonmetallic Total weight: 11 kg Length: 1.2 m	
MK7	Metallic Total weight: 14.7 kg Weight of explosive: 8.8 kg Diameter: 330 mm Fuze: Pressure	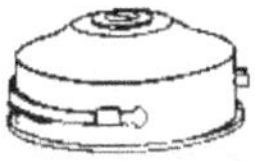
L3A1	Plastic with removable detector ring Total weight: 7.7 kg Diameter: 266 mm	

Figure 7-31. Foreign AT mines (continued)

Landmine and Special-Purpose Munition Obstacles 7-41

FM 5-34

Mine	Description	Sketch
L14A1	Off-road Total weight: 13 kg Maximum range: 80 m Height: 330 mm Length: 260 mm Fuze: Actuated by break wire across kill zone	
Former West Germany		
DM 11	Plastic Total weight: 7.4 kg Weight of explosive: 7 kg Diameter: 300 mm Fuze: DM 46 pressure	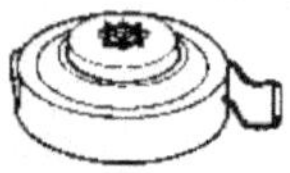
DM 39	Plastic Total weight: 0.50 kg Weight of explosive: 0.31 kg Diameter: 118 mm Fuze: Antilift device with pressure-release fuze	
DM 49	Plastic Total weight: 0.50 kg Weight of explosive: 0.20 kg Diameter: 90 mm Fuze: Antilift device with pressure-release fuze	

Figure 7-31. Foreign AT mines (continued)

7-42 Landmine and Special-Purpose Munition Obstacles

Mine	Description	Sketch
Former Soviet Union		
POM 2-2M	Cast iron case Total weight: 1.7 kg Weight of explosive: 0.75kg Diameter: 60 mm Fuze: MUV-2	
OZM-3 OZM-4	Steel Total weight: 4.54 kg Weight of explosive: 0.75 kg Diameter: 77 mm Fuze: MUV or MUV-2	

MON 100
MON 200
	Metal
	Total weight:
	  MON 100: 5 kg
	  MON 200: 25 kg
	Weight of explosive:
	  MON 100: 2 kg
	  MON 200: 12 kg
	Diameter:
	  MON 100: 220 mm
	  MON 200: 520 mm
	Fuze: Electric command or tripwire

PMN	Phenolic body with rubber cover Total weight: 0.60 kg Weight of explosive: 0.216 kg Diameter: 100 mm Fuze: Integral with mine	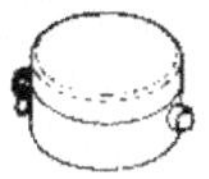
PMD6 PMD7	Wood Total weight: 398 gm Weight of explosive: 200 gm Fuze: Pull (MUV)	
Former Czechoslovakia		
PP Mi S6	Concrete case Total weight: 2.1 kg Weight of explosive: 0.075 kg Diameter: 75 mm Fuze: R01 pull or R08 pressure	
PP Mi Sr	Steel Total weight: 3.25 kg Weight of explosive: 0.325 kg Diameter: 100 mm Fuze: R01 pull or R08 pressure	
PP Mi ST-46	Cast-iron case	

Figure 7-32. Foreign AP mines

Landmine and Special-Purpose Munition Obstacles

FM 5-34

Mine	Description	Sketch
Hungary		
Ramp mine	Metal Total weight: 1.4 kg Weight of explosive: 0.8 kg Fuze: Pull	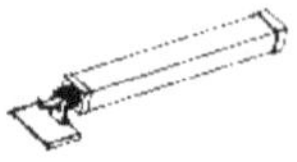
M62	Plastic Total weight: 386 gm Weight of explosive: 74 gm Fuze: Pull (MUV)	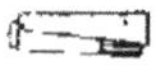
Bounding	Metal case Total weight: 3.6 kg Weight of explosive: 0.8 kg Fuze: Pull	
Former East Germany		
K-2	Plastic with metal Total weight: 4 kg Weight of explosive: 3 kg Fuze: Pressure	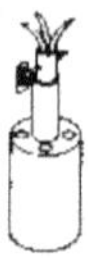
France		
Model 1948	Nonmetallic Total weight: 0.56 kg Weight of explosive: 170 gm	
Model 1951 nonmetallic	Plastic Total weight: 0.85 kg Diameter: 70 mm Fuze: Integral pressure friction	
Model 1951/55 bounding	Metal Total weight: 4.5 kg Diameter: 110 mm Fuze: Model 1952 tilt rod	
DV 56 Model 1956 Nonmetallic	Plastic Total weight: 0.16 kg Fuze: Friction pressure Diameter: 70 mm	

Figure 7-32. Foreign AP mines (continued)

7-44 Landmine and Special-Purpose Munition Obstacles

FM 5-34

Mine	Description	Sketch
Italy		
Minelba Type A	Metal Total weight: 0.17 kg Diameter: 110 mm Fuze: Integral pneumatic	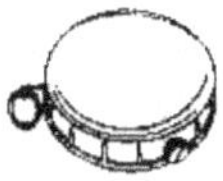
Minelba Type B	Similar in outer appearance to Type A but is made of plastic and has no safety pin hole and no safety device Diameter: 110 mm	
AUS 50/5	Plastic Total weight: 1.4 kg Diameter: 125 mm Fuze: Pressure/pull	
Type R	Wood Total weight: 0.5 kg Fuze: Pressre/pull	
Valmara	Metallic Total weight: 3.2 kg Weight of explosive: 0.54 kg Fuze: Pressure/pull	
Netherlands		
Model 22 Nonmetallic	Plastic Total weight: 0.85 kg Fuze: Integral pressure friction with shear collar control	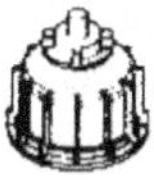
Model 15	Plastic Total weight: 0.6 kg Fuze: Pressure igniter Length: 114 mm Width: 100 mm	
Spain		
FAMD	Plastic Total weight: 97 gm Weight of explosive: 48 gm Fuze: Pressure	
Sweden		
M49 M49B	Cardboard Total weight: 0.23 kg Fuze: Pressure	

Figure 7-32. Foreign AP mines (continued)

Landmine and Special-Purpose Munition Obstacles 7-45

FM 5-34

Mine	Description	Sketch
M48	Fragmentation Total weight: 2.9 kg Weight of explosive: 0.23 kg Fuse: Pull	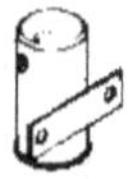
Model 43 Model 43 (T)	Concrete Total weight: 5.8 kg Weight of explosive: 0.6 kg Fuze: Pull	
M/43 T	Cardboard Total weight: 0.23 kg Weight of explosive: 0.14 kg Fuze: Pressure	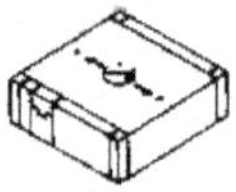
M41	Wood Total weight: 0.35 kg Weight of explosive: 0.12 kg Fuze: Pressure pin withdrawal	
Switzerland		
M3	Nonmetallic Total weight: 93 gm Weight of explosive: 68 gm	
P59	Plastic Weight of explosive: 60 gm Fuze: None	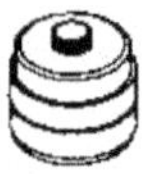
United Kingdom		
AP No. 6 (i)	Plastic mine with metal-detector ring Length: 203 mm	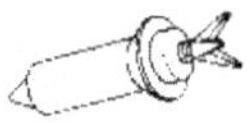
AP No. 7 (Dingbat)	Small metal mine, actuated by a load of 3.20 kg Total weight: 0.11 kg Diameter: 63 mm	
AP C3 (Elsie)	Nonmetallic Small plastic mine with removable detector ring Total weight: 0.08 kg Length: 76 mm	

Figure 7-32. Foreign AP mines (continued)

7-46 Landmine and Special-Purpose Munition Obstacles

FM 5-34

Mine	Description	Sketch
Former West Germany		
DM 11	Plastic Total weight: 200 gm Weight of explosive: 114 gm Diameter: 80 mm	
DM 31	Steel Total weight: 4 kg Weight of explosive: 0.53 kg Diameter: 102 mm Fuze: DM56	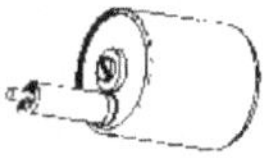

Figure 7-32. Foreign AP mines (continued)

Landmine and Special-Purpose Munition Obstacles 7-47

Chapter 8
Survivability

WEAPONS FIGHTING POSITIONS

These positions may be hasty or deliberate, depending on the time and material availability. They may be dug by hand or mechanically, using a small emplacement excavator (SEE). Table 8-1 shows the required thicknesses for protection against direct and indirect fires. In training, support conservation and safety efforts by backfilling the positions and returning the top soil to the upper layer.

Table 8-1. Material thickness for protection against direct and indirect fires

Material (cm)	Direct Fire			Indirect Fire		
	Small Caliber (7.62)	85-mm HE Shaped charge (RPG7)	107- to 120-mm HE Shaped Charge (RCLR) (Sagger)	82-mm Mortar	120- to 122-mm Mortar/Rocket/ HE Shell	152-mm Mortar/Rocket/ HE Shell
Concrete (not reinforcing)	30	76	91	10	13	15
Gravel, small rocks, bricks, rubble	51	61	91	25	46	51
Soil, sand	107	198	244	30	51	76
Dry timber	91	229	274	20	30	36
Snow (tamped)	183	396	None	152	152	152

INDIVIDUAL FIGHTING POSITIONS

Soldiers must construct fighting positions that protect them and allow them to fire into their assigned sectors. Fighting positions protect soldiers by providing cover through sturdy construction and by providing concealment through positioning and proper camouflage. The enemy must not be able to identify a fighting position until it is too late and he has been effectively engaged. When possible, soldiers should site positions in nonobvious places, such as behind natural cover, and in an easy-to-camouflage location. The most important step in preparing a fighting position is to make sure that it cannot be seen. In constructing a fighting position, soldiers should always—

- Dig the position armpit deep.
- Fill the sandbags about 75 percent full.
- Revet excavations in sandy soil.
- Check stabilization of wall bases.
- Inspect and test the position daily, after heavy rain, and after receiving direct or indirect fires.
- Maintain, repair, and improve positions as required.
- Use proper material, and use it correctly.

> **NOTE: In sandy soil, soldiers should not drive vehicles within 6 feet of the position.**

SITING TO ENGAGE THE ENEMY

Soldiers must be able to engage the enemy within their assigned fire sectors. They should be able to fire out to the maximum effective range of their weapons with maximum grazing fire and minimal dead space. Soldiers and leaders must be able to identify the best position locations to meet this criteria. Leaders must also ensure that fighting positions provide interlocking fires. This siting process allows the soldiers in the fighting positions to cover the platoon's sector and provides a basis for final protective fires (FPFs).

PREPARING BY STAGES

Leaders must ensure that their soldiers understand when and how to prepare fighting positions based on the situation. Soldiers normally prepare hasty fighting positions everytime the platoon halts (except for short security halts) and when only half of the platoon digs in while the other half maintains security. They prepare positions in stages and require a leader to inspect the position before moving on to the next stage. The example below explains the stages:

- Stage 1. The leader checks the fields of fire from the prone position and has a soldier emplace sector stakes (see Figure 8-1).
- Stage 2. The soldiers prepare the retaining walls for the parapets. There must be at least a one-helmet distance from the edge of the hole to the beginning of the front, flanks, and rear cover (see Figure 8-2, page 8-4).
- Stage 3. The soldiers dig the position, throw the dirt forward of the parapet's retaining walls, and pack the dirt down hard (see Figure 8-3, page 8-5).

FM 5-34

• Stage 4. The soldiers prepare the overhead cover (see Figure 8-4, page 8-5). They should camouflage the position so it blends with the surrounding terrain. (The position should not be dectable at a distance of 35 meters.)

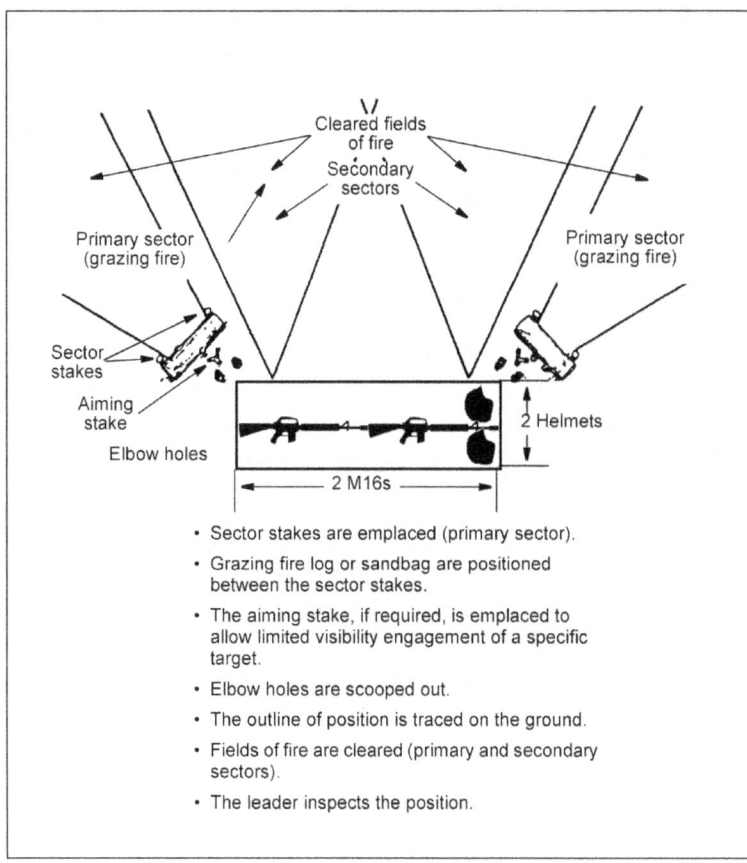

Figure 8-1. Stage 1, preparing a fighting position

Table 8-2, page 8-6, and Figures 8-5 through 8-8, pages 8-7 through 8-8, show details and characteristics of different individual positions. The AT-4 may be fired from any of these positions; however, a back-blast area must be cleared before firing.

Survivability 8-3

FM 5-34

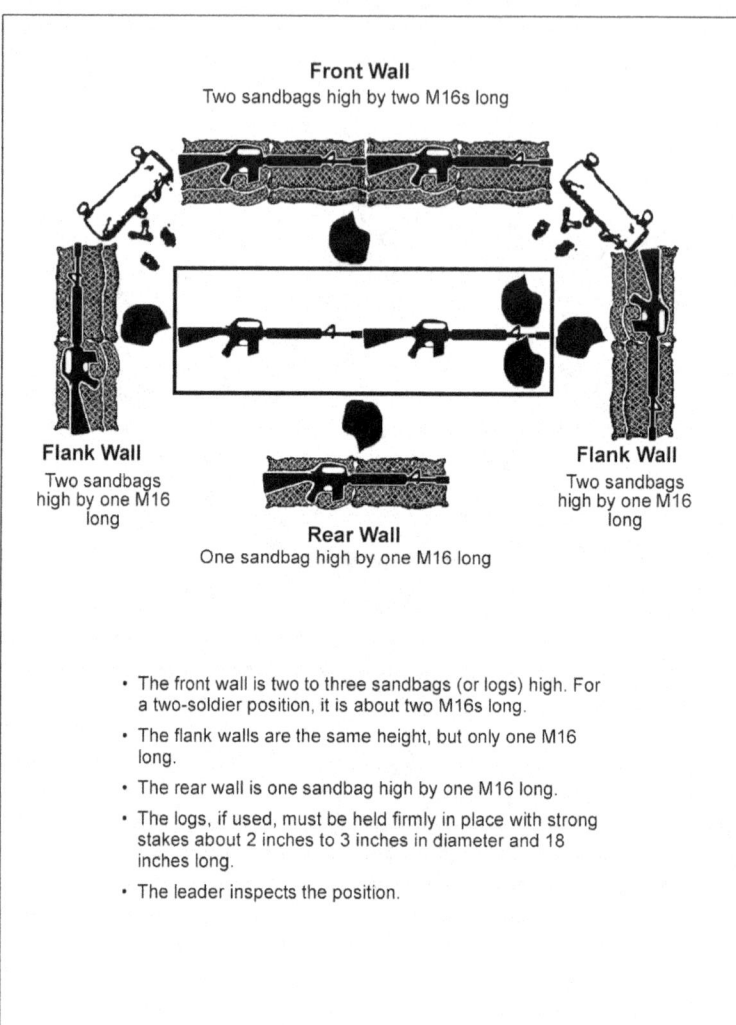

- The front wall is two to three sandbags (or logs) high. For a two-soldier position, it is about two M16s long.
- The flank walls are the same height, but only one M16 long.
- The rear wall is one sandbag high by one M16 long.
- The logs, if used, must be held firmly in place with strong stakes about 2 inches to 3 inches in diameter and 18 inches long.
- The leader inspects the position.

Figure 8-2. Stage 2, preparing a fighting position

8-4 Survivability

FM 5-34

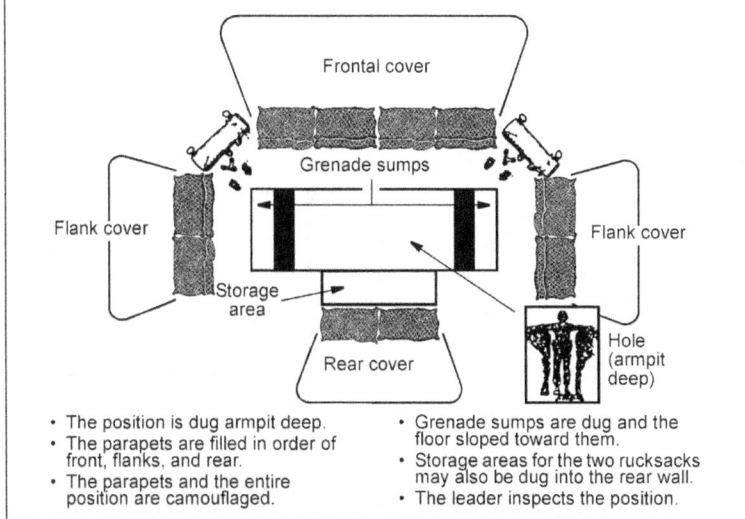

- The position is dug armpit deep.
- The parapets are filled in order of front, flanks, and rear.
- The parapets and the entire position are camouflaged.
- Grenade sumps are dug and the floor sloped toward them.
- Storage areas for the two rucksacks may also be dug into the rear wall.
- The leader inspects the position.

Figure 8-3. Stage 3, preparing a fighting position

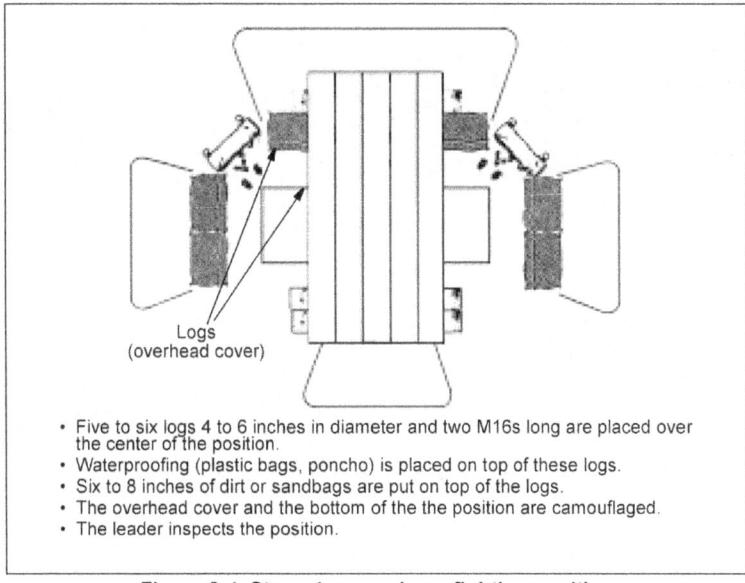

- Five to six logs 4 to 6 inches in diameter and two M16s long are placed over the center of the position.
- Waterproofing (plastic bags, poncho) is placed on top of these logs.
- Six to 8 inches of dirt or sandbags are put on top of the logs.
- The overhead cover and the bottom of the the position are camouflaged.
- The leader inspects the position.

Figure 8-4. Stage 4, preparing a fighting position

Survivability 8-5

FM 5-34

Table 8-2. Characteristics of individual fighting positions

Position Type	Construction Time w/Hand Tools (man-hours, estimate)	Direct-Fire Protection	Indirect-Fire Protection
Hasty Positions			
Crater	0.2	Up to 7.62 mm	Better than in the open, no overhead protection
Skirmisher's trench	0.5	Up to 7.62 mm	Better than in the open, no overhead protection
Prone position	1.0	Up to 7.62 mm	Better than in the open, no overhead protection
Deliberate Positions			
One soldier	3.0	Up to 12.7 mm	Up to medium artillery, no closer than 30 ft, no overhead protection
One soldier w/18 in of overhead cover	8.0	Up to 12.7 mm	Up to medium artillery, no closer than 30 ft
Two soldiers	6.0	Up to 12.7 mm	Up to medium artillery, no closer than 30 ft, no overhead protection
Two soldiers w/18 in of overhead cover	11.0	Up to 12.7 mm	Up to medium artillery, no closer than 30 ft
Three soldiers	9.0	Up to 12.7 mm	Up to medium artillery, no closer than 30 ft, no overhead protection
Three soldiers w/18 in of overhead cover	14.0	Up to 12.7 mm	Up to medium artillery, no closer than 30 ft

NOTES:
1. Positions do not provide protection from indirect-fire blasts or direct hits from indirect fire.
2. Shell sizes are small, 82-mm mortar and 105-mm artillery, and medium, 120-mm mortar and 152-mm artillery.

8-6 Survivability

FM 5-34

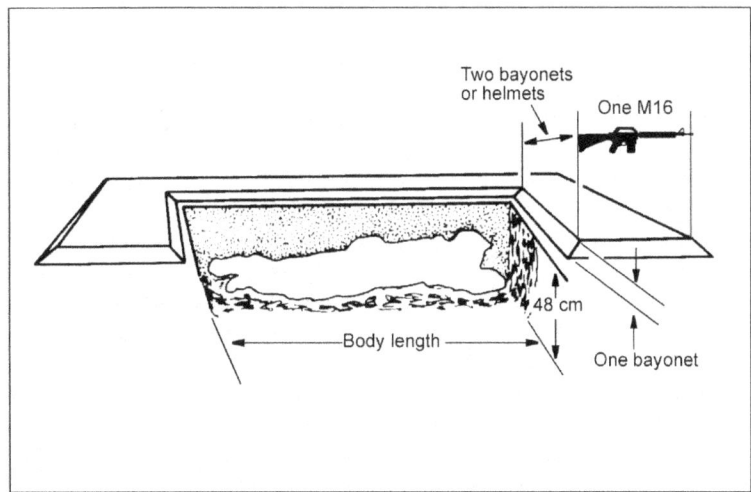

Figure 8-5. Hasty prone position

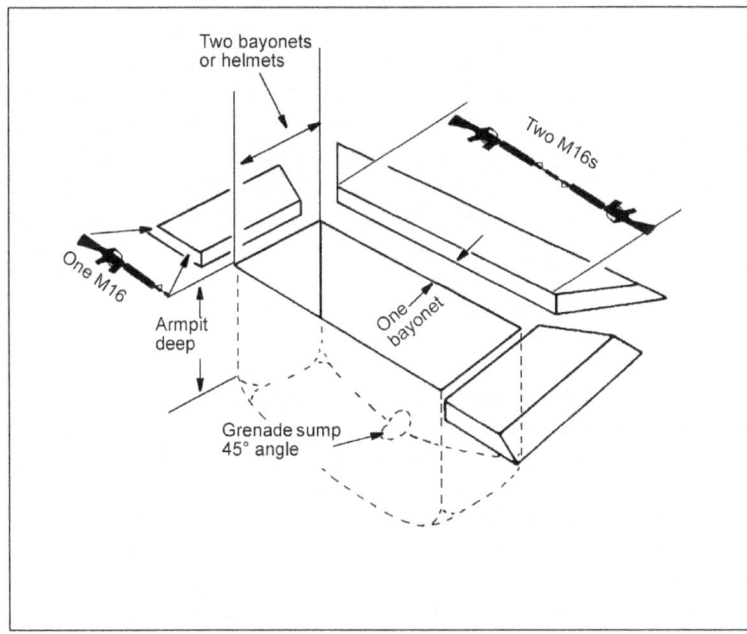

Figure 8-6. Two-soldier fighting position

Survivability 8-7

FM 5-34

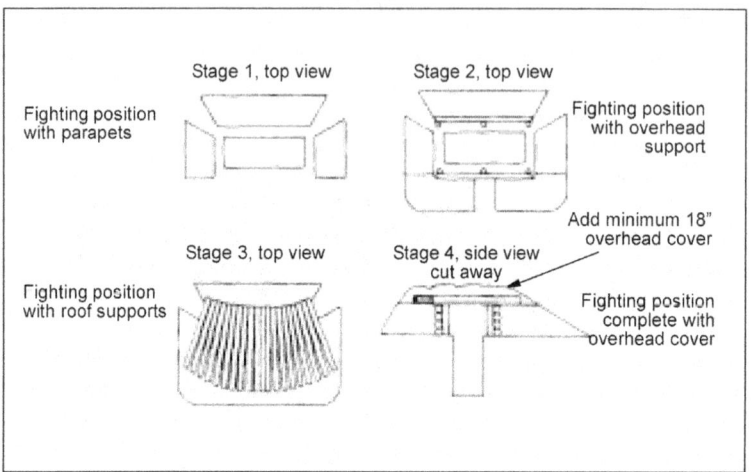

Figure 8-7. Two-soldier fighting position development

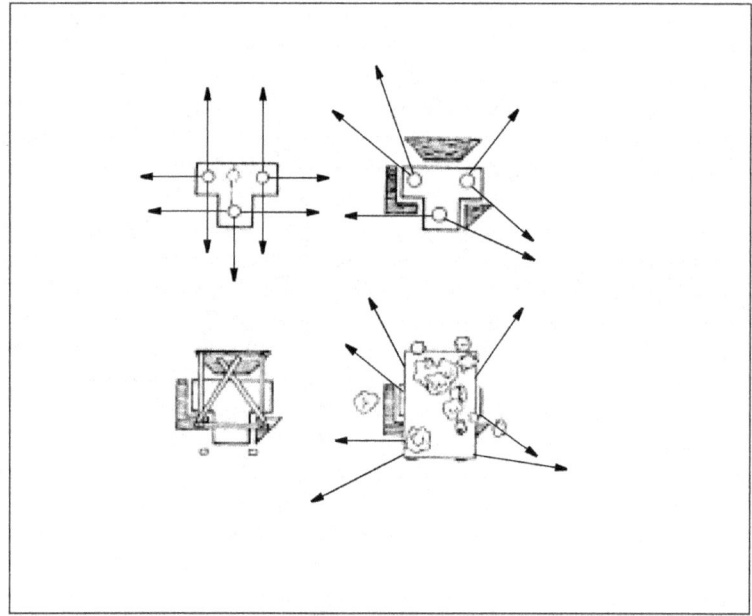

Figure 8-8. Three-soldier *T*-position

8-8 Survivability

FM 5-34

CREW-SERVED-WEAPONS FIGHTING POSITIONS

Table 8-3. Characteristics of crew-served-weapons fighting positions

Position Type	Construction Time w/Hand Tools (man-hours, estimate)	Direct-Fire Protection	Indirect-Fire Protection
AT-4, LAW	4.0	Up to 12.7 mm	Up to medium artillery, no closer than 30 ft, no overhead protection
Dragon, M47	6.0	Up to 12.7 mm	Up to medium artillery, no closer than 30 ft, no overhead protection
Machine gun	7.0	Up to 12.7 mm	Up to medium artillery, no closer than 30 ft, no overhead protection
Machine gun w/ 18 in of overhead cover	12.0	Up to 12.7 mm	Up to medium artillery, no closer than 30 ft
Dismounted	110	Up to 12.7 mm	Up to medium artillery, no closer than 30 ft, no overhead protection
TOW Mortar	14.0	Up to 12.7 mm	Up to medium artillery, no closer than 30 ft, no overhead protection

NOTES:
1. Positions do not provide protection from indirect-fire blasts or direct hits from indirect fire.
2. Shell sizes are small, 82-mm mortar and 105-mm artillery, and medium, 120-mm mortar and 152-mm artillery.

After being assigned a sector of fire with a final protective line (FPL) or a principle direction of fire (PDF), the gun crew starts constructing a fighting position. The crew—

- Positions a tripod and marks it so that the weapon will be pointed in the general direction of the target area.
- Draws a preliminary sketched range card to show the sector's limits.
- Outlines the shape of the platform and hole, to include the area for the frontal cover in the ground (see Figure 8-9).

Figure 8-9. Planning the fighting position

Survivability 8-9

FM 5-34

- Starts digging out the platform.
- Puts the machine gun in place, after digging about 4 to 6 inches, to cover the primary sector of fire until construction is complete.

When assigned an FPL, the gun crew—

- Emplaces the gun by locking the traversing slide to the extreme left or right of the traversing bar, depending on which side of the primary sector the FPL is on.
- Aligns the barrel on the FPL by shifting the tripod.

   NOTE: The crew does not fill in the direction entry in the data section of the range card for the FPL.

When assigned a PDF, the gun crew—

- Emplaces the gun by locking the traversing slide at the center of the traversing bar.
- Shifts the tripod and gun until the barrel is aimed at the center of the sector.
- Checks coverage of the sector limits by traversing the gun fully left and right.

   NOTE: In the data section of the range card, the crew records the directions and elevation data of the PDF and the sector limits from the traverse and elevation mechanism (see Figure 8-10).

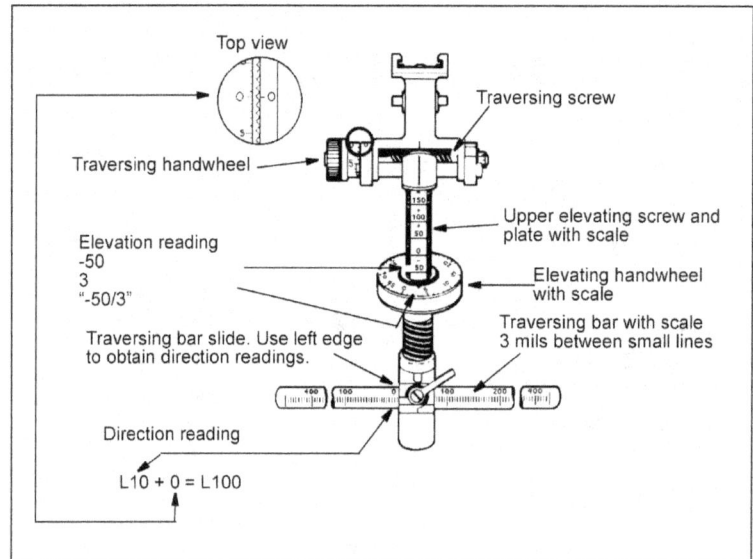

Figure 8-10. Traverse and elevation mechanism

8-10 Survivability

The crew digging the hole—

- Uses the dug-up dirt to build up the cover—first for frontal cover and then for flank and rear cover.
- Digs the hole deep enough so that the crew is protected and the gunner can shoot with comfort (usually about armpit deep) (see Figure 8-11).

Figure 8-11. Digging the fighting position

- Fixes the tripod legs in place by digging, sandbagging, or staking them down. Doing so ensures that the gun does not shift during firing, which would render the range card useless.
- Digs three trench-shaped grenade sumps at various points so that the crew can kick grenades into them (see Figure 8-12, page 8-12).
- Digs only half a position when a position does not have a secondary sector of fire (see Figure 8-13, page 8-12).
- Prepares two firing platforms when a position has both a primary and a secondary sector. The crew prepares overhead cover for a machine gun's position the same as a two-man, small-arms fighting position. Time and material permitting, overhead cover should extend to cover the firing platforms (see Figure 8-14, page 8-13). If the crew improperly constructs the overhead cover, it can result in reduced fields of fire, inability to mount night-vision devices (NVDs), or reloading problems. Properly constructing the overhead cover is critical to survival.

When a three-man crew is available for the machine gun, the ammunition bearer digs a one-man fighting position to the flank of

FM 5-34

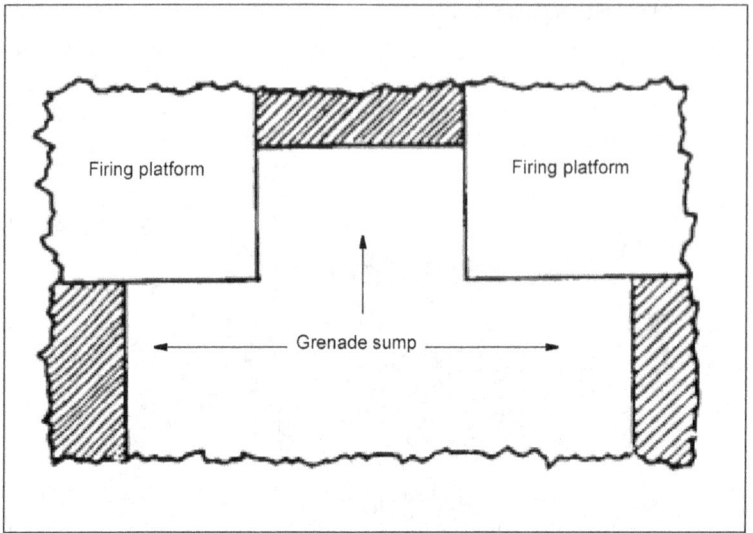

Figure 8-12. Digging grenade sumps

Figure 8-13. Half of a position

8-12 Survivability

FM 5-34

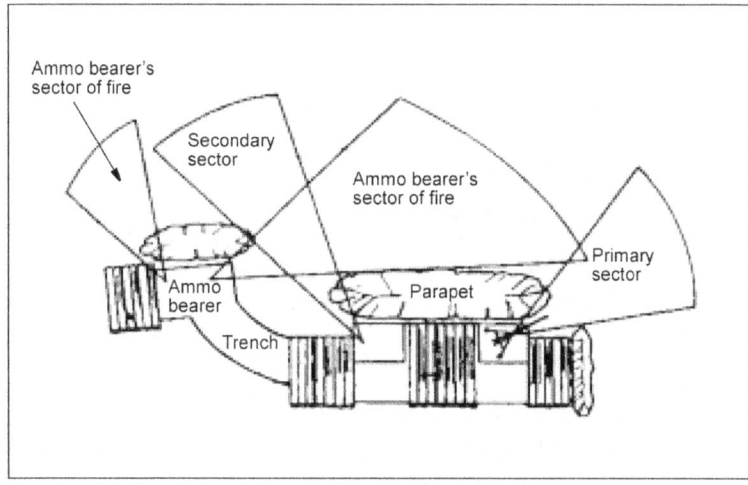

Figure 8-14. Two firing platforms with overhead cover

the gun position so that he can see and shoot to the oblique. In this position, he can cover the front of the machine gun's position (see Figure 8-15).

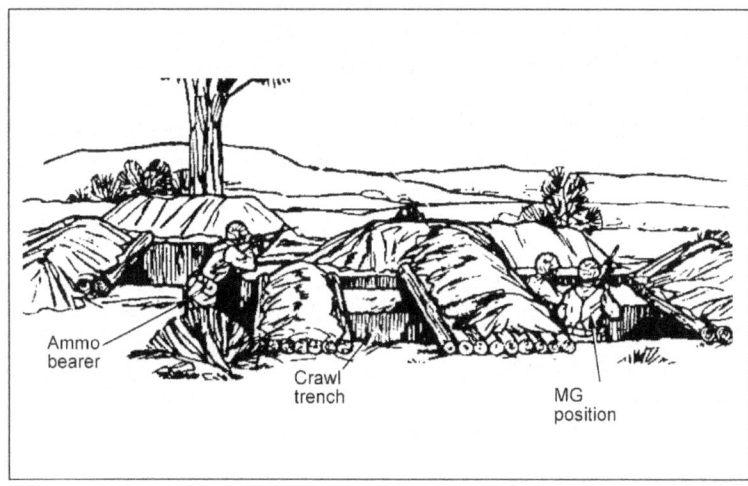

Figure 8-15. Ammo bearer covering front

Survivability 8-13

Figure 8-16 shows a dug-in position for the MK-19.

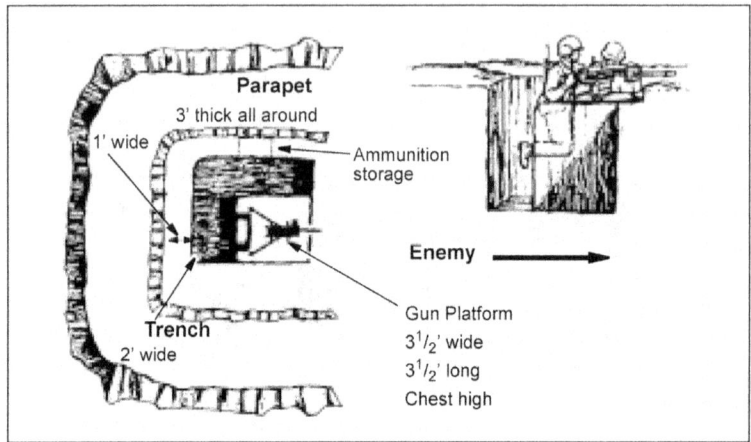

Figure 8-16. Dug-in position for an MK19

RANGE CARD

A range card contains a sketch of a sector that a direct-fire-weapon system is assigned to cover. (See FM 7-8 for a reproducible copy of DA Form 5517-R.) Information on the card—

- Aids in planning and controlling fires.
- Aids the crew in acquiring targets during limited visibility.
- Helps replacement personnel, platoons, or squads move into position and orient on their sector.

During good visibility, a gunner should not have problems staying oriented within his sector. During poor visibility, he may not be able to detect lateral limits. If the gunner does become disoriented and cannot find or locate reference points or sector limit markers, he can use the information on the range card to locate these limits. He should assess the terrain in his sector and update the range card as necessary.

DETAILS

To prepare a range card, a gunner must know the following information:

Sectors of Fire

A sector of fire is the part of the battlefield for which a gunner is responsible. He may be assigned a primary and a secondary sector.

8-14 Survivability

Leaders use sectors of fire to ensure that fires are distributed across the platoon's area of responsibility.

Leaders assign sectors of fire to cover possible enemy AAs. When assigning sectors, leaders should overlap them to provide the best use of overlapping fire and to cover areas that cannot be engaged by a single-weapon system. Leaders assign left and right limits of a sector using prominent terrain features or easily recognizable objects (rocks, telephone poles, fences, emplaced stakes).

Target Reference Points (TRPs)/Reference Points (RPs)

Leaders designate natural or man-made terrain features as RPs. The gunner uses these points in target acquisition and the range-determination process during limited visibility. There will also be predesignated TRPs, which must be useful as such or as indirect-fire targets.

A commander or platoon leader designates indirect-fire targets used as TRPs so that target numbers can be assigned. If TRPs are within the sector of fire, the squad leader points them out and tells the gunner their designated reference numbers. Normally, a gunner has at least one TRP but should not have more than four. The range card should show only pertinent data for reference points and TRPs.

Dead Space

Dead space is any area that cannot be observed or covered by direct-fire systems within the sector of fire. All dead space within the sector of fire must be identified to allow the squad and platoon leaders to plan the use of fires (mortars, artillery) to cover that area. The crew must walk the EA so that the gunners can detect dead spaces through the integrated sight units (ISUs).

Maximum Engagement Line (MEL)

The depth of the sector is normally limited to the maximum effective engagement range of the vehicle's weapon systems; however, it can be less if there are objects that prevent the gunner from engaging targets at the maximum effective engagement range. To assist in determining the distance to each MEL, the gunner or squad leader should use a map to make sure that the MELs are shown correctly on the range card. MEL identification assists in decreasing the ammunition used on an engagement.

Weapon Reference Point (WRP)

The WRP is an easily recognizable terrain feature on the map. It is used to assist leaders in plotting a vehicle's position and to assist replacement personnel in finding a vehicle's position.

FM 5-34

PREPARATION PROCEDURES

A gunner prepares two copies of the range card. If alternate and supplementary firing positions are assigned, the gunner prepares two copies for those positions. He keeps one copy in the vehicle and gives the other to the platoon leader for his sketch. To complete a range card, a gunner should do the following:

Step 1. Draw the weapon symbol in the center of the small circle. Draw two lines from the position of the vehicle, extending left and right to show the limits of the sector (see Figure 8-17).

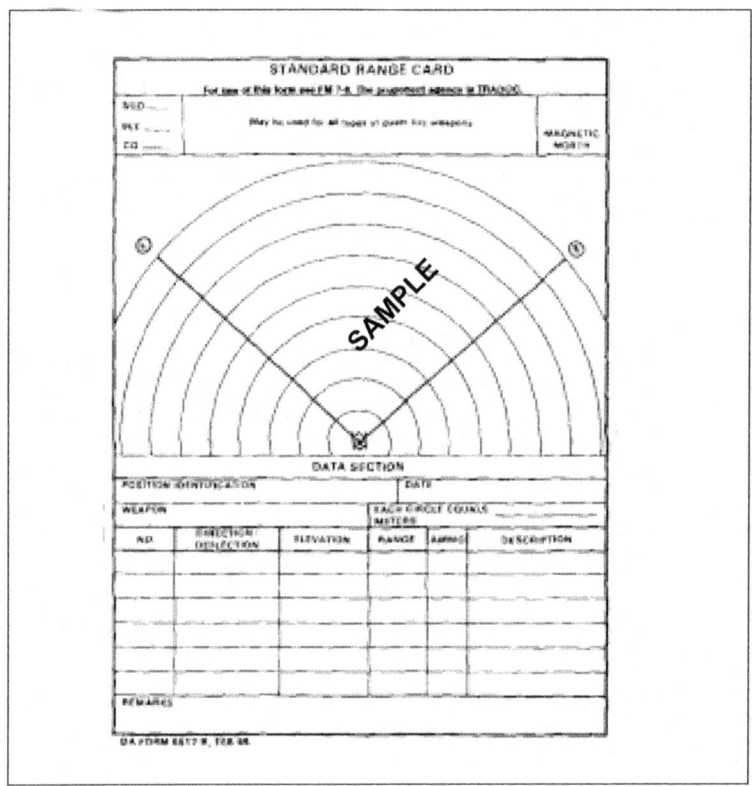

Figure 8-17. Placement of weapon symbol and left and right limits

Step 2. Determine the value of each circle by finding a terrain feature that is farthest from the position within the weapon system's capability. Determine the distance to the terrain feature; round off to the next even hundredth, if necessary. Determine the

8-16 Survivability

maximum number of circles that will divide evenly into the distance. The result is the value of each circle. Draw the terrain feature on the appropriate circle on the range card. Mark the increment clearly for each circle across the area where DATA SECTION is written. For example, Figure 8-18 shows a hilltop at 3,145 meters (rounded to 3,200 meters). The 3,200 is divided by 8, which equals 400, the value of each circle. Figure 8-19, page 8-18, shows a farmhouse at 2,000 meters on the left limit. The right limit is noted by the wood line at 2,600 meters.

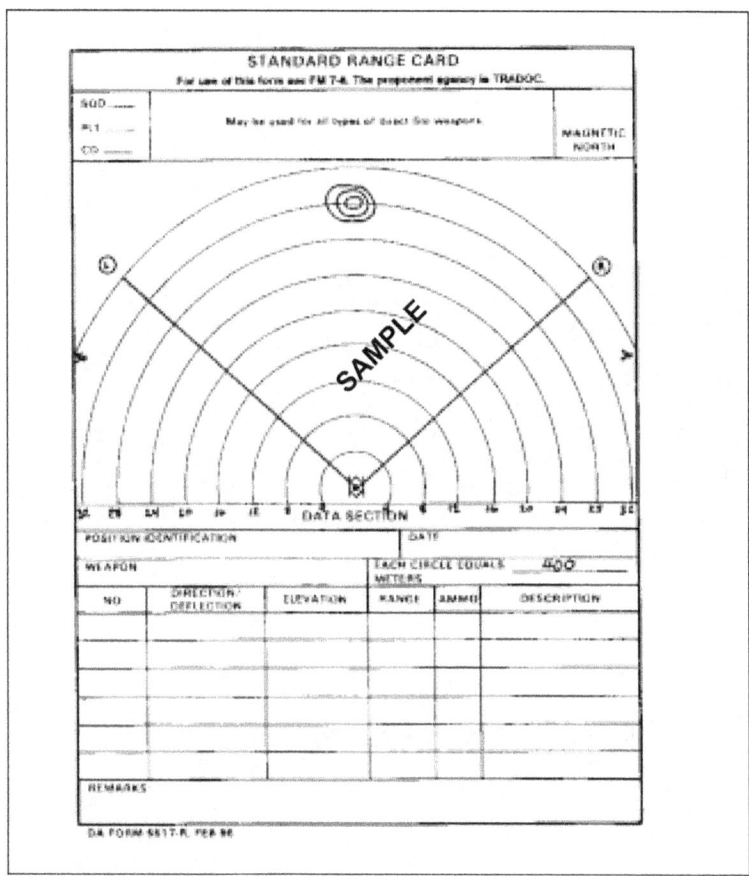

Figure 8-18. Circle value

Survivability 8-17

FM 5-34

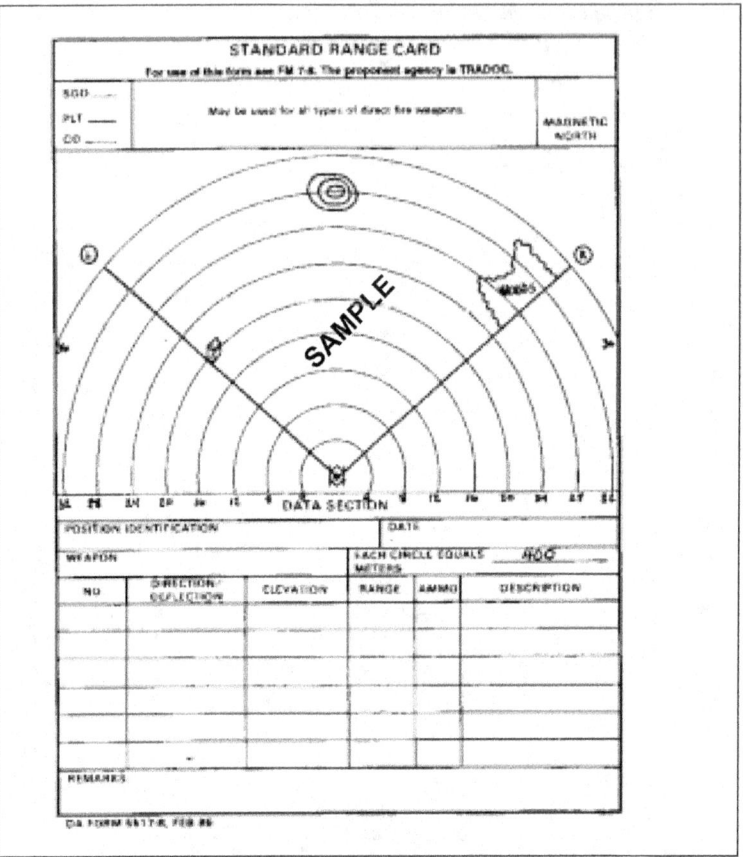

Figure 8-19. Terrain features for left and right limits

Step 3. Determine the distance to these features using a map or a hand-held laser range finder.

Step 4. Draw all TRPs and RPs in the sector. Mark each of these with a circled number beginning with 1. Figure 8-20 shows the hilltop as RP 1 and the road junctions as RP 2 and RP 3. Sometimes, a TRP and an RP are the same point, as RP 2 and RP 3 are in the figure. This occurs when a TRP is used for target acquisition and range determination. Mark the TRP with the first designated number in the upper right quadrant and the RP in the lower left quadrant of the cross. Draw road junctions by determining the range to the junction, drawing it, and then drawing the connecting roads from the road junction.

8-18 Survivability

FM 5-34

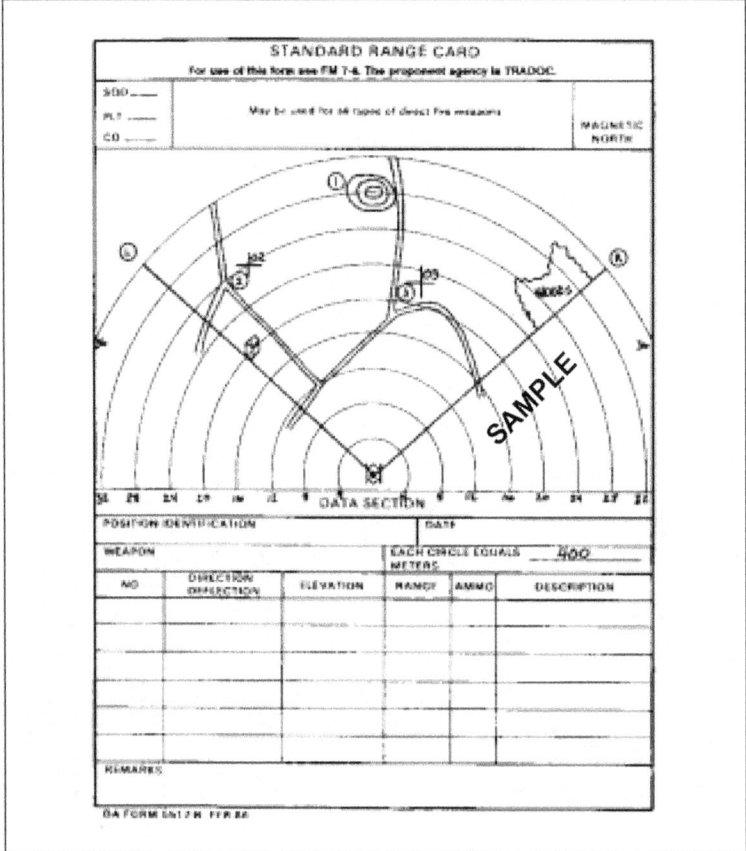

Figure 8-20. Target reference points/reference points

Step 5. Show dead space in an irregular circle with diagonal lines drawn inside (see Figure 8-21, page 8-20). Any object that prohibits observation or coverage with direct fire will have the circle and diagonal lines extend out to the farthest maximum engagement line. If you can engage the area beyond the dead space, close the circle. For example, an area of lower elevation will have a closed circle because you can engage the area beyond it.

Step 6. Draw MELs at the maximum effective engagement range per weapon, if there is no dead space to limit their range capabilities (see Figure 8-22, page 8-21). Note how the MEL for HE extends beyond the dead space in Figure 8-21. This indicates a higher elevation

Survivability 8-19

FM 5-34

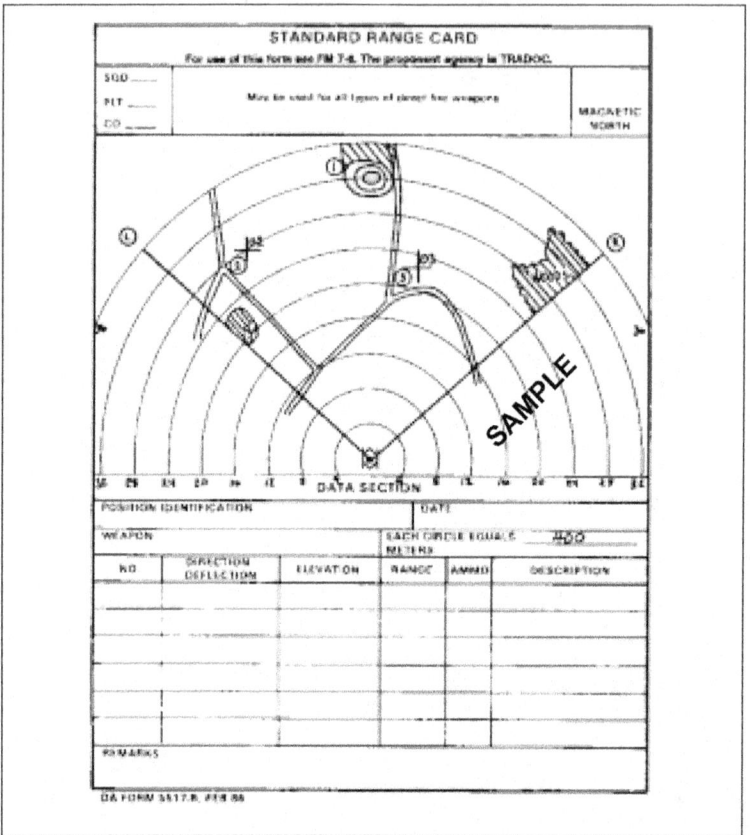

Figure 8-21. Dead space

where HE area suppression is possible. Do not draw MELs through dead space. The maximum effective ranges for the Bradley weapon systems are as follows:

— COAX: 900 meters (tracer burnout).

— APDS-T: 1,700 meters (tracer burnout).

— NEI-T/TOW (basic): 3,000 meters (impact).

— TOW 2: 3,750 meters (impact).

The WRP in Figure 8-23, page 8-22, is shown as a line with a series of arrows extending from a known terrain feature and pointing in the direction of the Bradley symbol. This feature is numbered last. The WRP location is given a six-digit grid. When

8-20 Survivability

FM 5-34

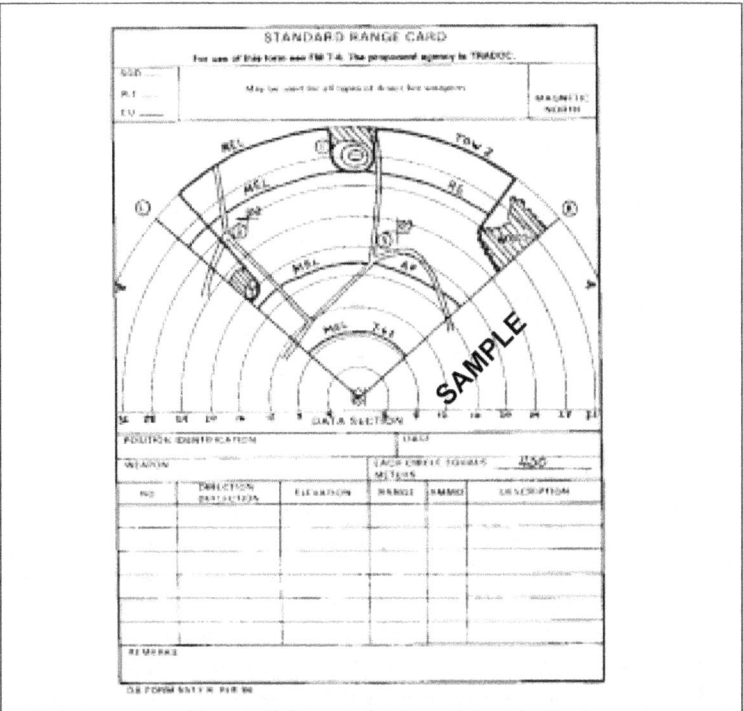

Figure 8-22. Maximum engagement lines

there is no terrain feature to be designated as the WRP, the vehicle's location is shown as an eight-digit grid coordinate in the remarks block of the range card (see Figure 8-24, page 8-23).

NOTE: When you cannot drawn the WRP precisely on the card because of vehicle location, draw it to the left or right nearest the actual direction.

Step 7. Complete the data section as follows (see Figure 8-24):

— Position identification: list either primary, alternate, or supplementary. You must clearly identify alternate and supplemental positions.

— Date: show the date and time that you completed the range card. Because these cards are constantly updated, the date and time are vital in determining current data.

— Weapon: This block indicates M2 and the vehicle's bumper number.

Survivability 8-21

FM 5-34

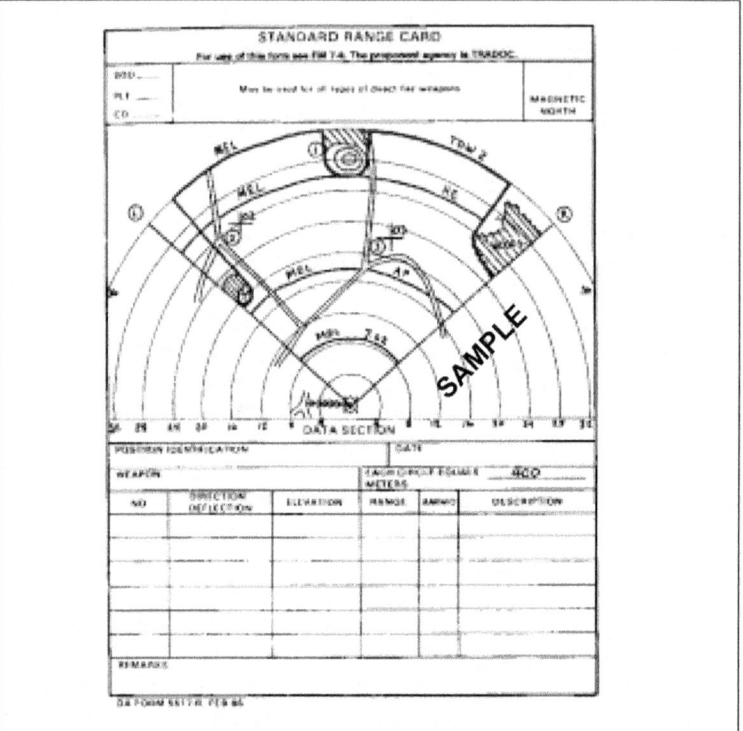

Figure 8-23. Weapon reference point

— Each circle equals_____meters: write in the distance between circles.

— Number (No.): start with L and R limts; then list TRPs and RPs in numerical order.

— Direction deflection: list the direction, in degrees, taken from a lensatic compass. The most accurate technique is to have the gunner aim at the terrain feature and the driver dismount and align himself with the gun barrel and the terrain feature to measure the azimuth.

— Elevation: show the gun-elevation reading in tens or hundreds of mils. The smallest increment of measure on the elevation scale is tens of mils. Any number other than 0 is preceded by a plus or minus sign to show whether the gun needs to be elevated or depressed. Ammunition and range must be indexed to have an accurate elevation reading.

8-22 Survivability

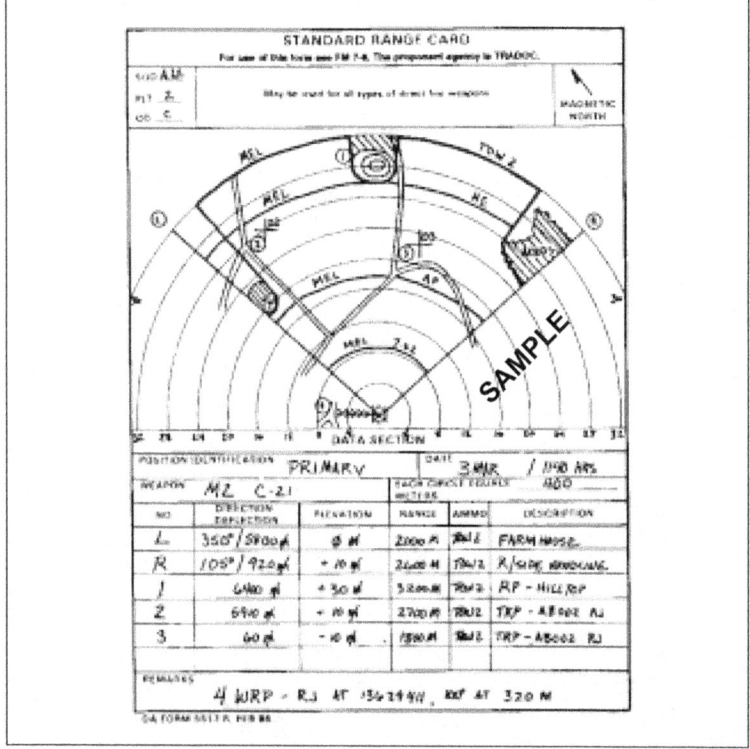

Figure 8-24. Example of a completed range card

- Range: record the distance, in meters, from the vehicle's position to L and R limits and TRPs and RPs.

- Ammo: list the types of ammunition used.

- Description: list the name of the object; for example, farmhouse, wood line, hilltop.

- Remarks: enter the WRP data. As a minimum, WRP data include a description of what the WRP is, a six-digit or eight-digit grid coordinate of the WRP, the magnetic azimuth, and the distance from the WRP to the vehicle's position.

VEHICLE POSITIONS

Positions may be fighting or protective, hasty or deliberate. See Table 8-4, page 8-24, for estimated survivability positions for maneuver units. See Tables 8-5 through 8-8, pages 8-25 through 8-36, for the number of turret-defilade positions (TDPs) and hull-defilade positions (HDPs) that can be constructed, based on the availability of time and assets.

> **WARNING**
>
> Ensure that engineer equipment bowls on scoop loaders, ACEs, SEEs, scrapers, and so on are periodically emptied. Never allow them to remain filled overnight, especially during cold weather.

Table 8-4. Standard survivability estimates for maneuver units

Level	Description of Recommended Priority of Survivability Support		Number of Hull-Down Positions to be Provided per Battle Position			
			Armor Bn	Mech Inf Bn	Armor Co	Mech Inf Co
1	TOWs Tanks APC (Pit and Co HQ only) TOC	- P - P - 50% P - P	80	100	15	15
2	TOWs Tanks APC (Pit and Co HQ only) TOC	- P and A - P - P - P	85	175	15	15
3	TOWs Tanks APC (Pit and Co HQ only) TOC Combat support	- P and A - P and A - P - P - P	150	180	30	25
4	TOWs Tanks APC (all) TOC Combat support Combat train	- P and A - P and A - P - P - P - 50% P	160	190	30	30
5	TOWs Tanks, APC (all) TOC Combat support Combat train	- P, A, and S - P and A - P - P - P	185	295	45	40
6	TOWs, tanks, and APC (all) TOC Combat support Combat train	- P, A, and S - P and A - P and A - p	265	330	45	45

NOTES:
1. P = Primary, A = Aternate, S = Supplementary hull-down positions.
2. Numbers are rounded to the nearest 5.
3. Combat support vehicles comprise mortars and ADA.
4. Platoon and Co. HQ only. Allows for four APCs per platoon and two per Co HQ to be dug in.

FM 5-34

Table 8-5. Dozer team TDP calculations

Real hr avail	Blade Teams Available																		
	1.0	1.5	2.0	2.5	3.0	3.5	4.0	4.5	5.0	5.5	6.0	6.5	7.0	7.5	8.0	8.5	9.0	9.5	10.0
1	0.3	0.4	0.5	0.6	0.8	0.9	1.0	1.1	1.3	1.4	1.5	1.6	1.8	1.9	2.0	2.1	2.3	2.4	2.5
2	0.5	0.8	1.0	1.3	1.5	1.8	2.0	2.3	2.5	2.8	3.0	3.3	3.5	3.8	4.0	4.3	4.5	4.8	5.0
3	0.8	1.1	1.5	1.9	2.3	2.6	3.0	3.4	3.8	4.1	4.5	4.9	5.3	5.6	6.0	6.4	6.8	7.1	7.5
4	1.0	1.5	2.0	2.5	3.0	3.5	4.0	4.5	5.0	5.5	6.0	6.5	7.0	7.5	8.0	8.5	9.0	9.5	10.0
5	1.3	1.9	2.5	3.1	3.8	4.4	5.0	5.6	6.3	6.9	7.5	8.1	8.8	9.4	10.0	10.6	11.3	11.9	12.5
6	1.5	2.3	3.0	3.8	4.5	5.3	6.0	6.8	7.5	8.3	9.0	9.8	10.5	11.3	12.0	12.8	13.5	14.3	15.0
7	1.8	2.6	3.5	4.4	5.3	6.1	7.0	7.9	8.8	9.6	10.5	11.4	12.3	13.1	14.0	14.9	15.8	16.6	17.5
8	2.0	3.0	4.0	5.0	6.0	7.0	8.0	9.0	10.0	11.0	12.0	13.0	14.0	15.0	16.0	17.0	18.0	19.0	20.0
9	2.3	3.4	4.5	5.6	6.8	7.9	9.0	10.1	11.3	12.4	13.5	14.6	15.8	16.9	18.0	19.1	20.3	21.4	22.5
10	2.5	3.8	5.0	6.3	7.5	8.8	10.0	11.3	12.5	13.8	15.0	16.3	17.5	18.8	20.0	21.3	22.5	23.8	25.0
11	2.8	4.1	5.5	6.9	8.3	9.6	11.0	12.4	13.8	15.1	16.5	17.9	19.3	20.6	22.0	23.4	24.8	26.1	27.5
12	3.0	4.5	6.0	7.5	9.0	10.5	12.0	13.5	15.0	16.5	18.0	19.5	21.0	22.5	24.0	25.5	27.0	28.5	30.0
13	3.3	4.9	6.5	8.1	9.8	11.4	13.0	14.6	16.3	17.9	19.5	21.1	22.8	24.4	26.0	27.6	29.3	30.9	32.5
14	3.5	5.3	7.0	8.8	10.5	12.3	14.0	15.8	17.5	19.3	21.0	22.8	24.5	26.3	28.0	29.8	31.5	33.3	35.0
15	3.8	5.6	7.5	9.4	11.3	13.1	15.0	16.9	18.8	20.6	22.5	24.4	26.3	28.1	30.0	31.9	33.8	35.6	37.5
16	4.0	6.0	8.0	10.0	12.0	14.0	16.0	18.0	20.0	22.0	24.0	26.0	28.0	30.0	32.0	34.0	36.0	38.0	40.0
17	4.3	6.4	8.5	10.6	12.8	14.9	17.0	19.1	21.3	23.4	25.5	27.6	29.8	31.9	34.0	36.1	38.3	40.4	42.5
18	4.5	6.8	9.0	11.3	13.5	15.8	18.0	20.3	22.5	24.8	27.0	29.3	31.5	33.8	36.0	38.3	40.5	42.8	45.0
19	4.8	7.1	9.5	11.9	14.3	16.6	19.0	21.4	23.8	26.1	28.5	30.9	33.3	35.6	38.0	40.4	42.8	45.1	47.5
20	5.0	7.5	10.0	12.5	15.0	17.5	20.0	22.5	25.0	27.5	30.0	32.5	35.0	37.5	40.0	42.5	45.0	47.5	50.0

Legend: Ar/Mech Platoon Ar/Mech Company Ar/Mech Battalion

Survivability 8-25

FM 5-34

Table 8-5. Dozer team TDP calculations (continued)

Real hr avail	Blade Teams Available																		
	1.0	1.5	2.0	2.5	3.0	3.5	4.0	4.5	5.0	5.5	6.0	6.5	7.0	7.5	8.5	9.0	9.5	10.0
21	5.3	7.9	10.5	13.1	15.8	18.4	21.0	23.6	26.3	28.9	31.5	34.1	36.8	39.4	42.0	44.6	47.3	49.9	52.5
22	5.5	8.3	11.0	13.8	16.5	19.3	22.0	24.8	27.5	30.3	33.0	35.8	38.5	41.3	44.0	46.8	49.5	52.3	55.0
23	5.8	8.6	11.5	14.4	17.3	20.1	23.0	25.9	28.8	31.6	34.5	37.4	40.3	43.1	46.0	48.9	51.8	54.6	57.5
24	6.0	9.0	12.0	15.0	18.0	21.0	24.0	27.0	30.0	33.0	36.0	39.0	42.0	45.0	48.0	51.0	54.0	57.0	60.0
25	6.3	9.4	12.5	15.6	18.8	21.9	25.0	28.1	31.3	34.4	37.5	40.6	43.8	46.9	50.0	53.1	56.3	59.4	62.5
26	6.5	9.8	13.0	16.3	19.5	22.8	26.0	29.3	32.5	35.8	39.0	42.3	45.5	48.8	52.0	55.3	58.5	61.8	65.0
27	6.8	10.1	13.5	16.9	20.3	23.6	27.0	30.4	33.8	37.1	40.5	43.9	47.3	50.6	54.0	57.4	60.8	64.1	67.5
28	7.0	10.5	14.0	17.5	21.0	24.5	28.0	31.5	35.0	38.5	42.0	45.5	49.0	52.5	56.0	59.5	63.0	66.5	70.0
29	7.3	10.9	14.5	18.1	21.8	25.4	29.0	32.6	36.3	39.9	43.5	47.1	50.8	54.4	58.0	61.6	65.3	68.9	72.5
30	7.5	11.3	15.0	18.8	22.5	26.3	30.0	33.8	37.5	41.3	45.0	48.8	52.5	56.3	60.0	63.8	67.5	71.3	75.0
31	7.8	11.6	15.5	19.4	23.3	27.1	31.0	34.9	38.8	42.6	46.5	50.4	54.3	58.1	62.0	65.9	69.8	73.6	77.5
32	8.0	12.0	16.0	20.0	24.0	28.0	32.0	36.0	40.0	44.0	48.0	52.0	56.0	60.0	64.0	68.0	72.0	76.0	80.0
33	8.3	12.4	16.5	20.6	24.8	28.9	33.0	37.1	41.3	45.4	49.5	53.6	57.8	61.9	66.0	70.1	74.3	78.4	82.5
34	8.5	12.8	17.0	21.3	25.5	29.8	34.0	38.3	42.5	46.8	51.0	55.3	59.5	63.8	68.0	72.3	76.5	80.8	85.0
35	8.8	13.1	17.5	21.9	26.3	30.6	35.0	39.4	43.8	48.1	52.5	56.9	61.3	65.6	70.0	74.4	78.8	83.1	87.5
36	9.0	13.5	18.0	22.5	27.0	31.5	36.0	40.5	45.0	49.5	54.0	58.5	63.0	67.5	72.0	76.5	81.0	85.5	90.0
37	9.3	13.9	18.5	23.1	27.8	32.4	37.0	41.6	46.3	50.9	55.5	60.1	64.8	69.4	74.0	78.6	83.3	87.9	92.5
38	9.5	14.3	19.0	23.8	28.5	33.3	38.0	42.8	47.5	52.3	57.0	61.8	66.5	71.3	76.0	80.8	85.5	90.3	95.0
39	9.8	14.6	19.5	24.4	29.3	34.1	39.0	43.9	48.8	53.6	58.5	63.4	68.3	73.1	78.0	82.9	87.8	92.6	97.5
40	10.0	15.0	20.0	25.0	30.0	35.0	40.0	45.0	50.0	55.0	60.0	65.0	70.0	75.0	80.0	85.0	90.0	95.0	100.0

Legend: Ar/Mech Platoon | Ar/Mech Company | Ar/Mech Battalion

8-26 Survivability

FM 5-34

Table 8-5. Dozer team TDP calculations (continued)

Real hr avail	Blade Teams Available																		
	1.0	1.5	2.0	2.5	3.0	3.5	4.0	4.5	5.0	5.5	6.0	6.5	7.0	7.5	8.0	8.5	9.0	9.5	10.0
41	10.3	15.4	20.5	25.6	30.8	35.9	41.0	46.1	51.3	56.4	61.5	66.6	71.8	76.9	82.0	87.1	92.3	97.4	102.5
42	10.5	15.8	21.0	26.3	31.5	36.8	42.0	47.3	52.5	57.8	63.0	68.3	73.5	78.8	84.0	89.3	94.5	99.8	105.0
43	10.8	16.1	21.5	26.9	32.3	37.6	43.0	48.4	53.8	59.1	64.5	69.9	75.3	80.6	86.0	91.4	96.8	102.1	107.5
44	11.0	16.5	22.0	27.5	33.0	38.5	44.0	49.5	55.0	60.5	66.0	71.5	77.0	82.5	88.0	93.5	99.0	104.5	110.0
45	11.3	16.9	22.5	28.1	33.8	39.4	45.0	50.6	56.3	61.9	67.5	73.1	78.8	84.4	90.0	95.6	101.3	106.9	112.5
46	11.5	17.3	23.0	28.8	34.5	40.3	46.0	51.8	57.5	63.3	69.0	74.8	80.5	86.3	92.0	97.8	103.5	109.3	115.0
47	11.8	17.6	23.5	29.4	35.3	41.1	47.0	52.9	58.8	64.6	70.5	76.4	82.3	88.1	94.0	99.9	105.8	111.6	117.5
48	12.0	18.0	24.0	30.0	36.0	42.0	48.0	54.0	60.0	66.0	72.0	78.0	84.0	90.0	96.0	102.0	108.0	114.0	120.0
49	12.3	18.4	24.5	30.6	36.8	42.9	49.0	55.1	61.3	67.4	73.5	79.6	85.8	91.9	98.0	104.1	110.3	116.4	122.5
50	12.5	18.8	25.0	31.3	37.5	43.8	50.0	56.3	62.5	68.8	75.0	81.3	87.5	93.8	100.0	106.3	112.5	118.8	125.0
51	12.8	19.1	25.5	31.9	38.3	44.6	51.0	57.4	63.8	70.1	76.5	82.9	89.3	95.6	102.0	108.4	114.8	121.1	127.5
52	13.0	19.5	26.0	32.5	39.0	45.5	52.0	58.5	65.0	71.5	78.0	84.5	91.0	97.5	104.0	110.5	117.0	123.5	130.0
53	13.3	19.9	26.5	33.1	39.8	46.4	53.0	59.6	66.3	72.9	79.5	86.1	92.8	99.4	106.0	112.6	119.3	125.9	132.5
54	13.5	20.3	27.0	33.8	40.5	47.3	54.0	60.8	67.5	74.3	81.0	87.8	94.5	101.3	108.0	114.8	121.5	128.3	135.0
55	13.8	20.6	27.5	34.4	41.3	48.1	55.0	61.9	68.8	75.6	82.5	89.4	96.3	103.1	110.0	116.9	123.8	130.6	137.5
56	14.0	21.0	28.0	35.0	42.0	49.0	56.0	63.0	70.0	77.0	84.0	91.0	98.0	105.0	112.0	119.0	126.0	133.0	140.0
57	14.3	21.4	28.5	35.6	42.8	49.9	57.0	64.1	71.3	78.4	85.5	92.6	99.8	106.9	114.0	121.1	128.3	135.4	142.5
58	14.5	21.8	29.0	36.3	43.5	50.8	58.0	65.3	72.5	79.8	87.0	94.3	101.5	108.8	116.0	123.3	130.5	137.8	145.0
59	14.8	22.1	29.5	36.9	44.3	51.6	59.0	66.4	73.8	81.1	88.5	95.9	103.3	110.6	118.0	125.4	132.8	140.1	147.5
60	15.0	22.5	30.0	37.5	45.0	52.5	60.0	67.5	75.0	82.5	90.0	97.5	105.0	112.5	120.0	127.5	135.0	142.5	150.0

Legend: ▨ Ar/Mech Platoon ☐ Ar/Mech Company ☐ Ar/Mech Battalion

Survivability 8-27

FM 5-34

Table 8-6. Dozer team HDP calculations

Real hrs avail	Blade Teams Available																		
	1.0	1.5	2.0	2.5	3.0	3.5	4.0	4.5	5.0	5.5	6.0	6.5	7.0	7.5	8.0	8.5	9.0	9.5	10.0
1	0.6	0.9	1.3	1.6	1.9	2.2	2.5	2.8	3.1	3.4	3.8	4.1	4.4	4.7	5.0	5.3	5.6	5.9	6.3
2	1.3	1.9	2.5	3.1	3.8	4.4	5.0	5.6	6.3	6.9	7.5	8.1	8.8	9.4	10.0	10.6	11.3	11.9	12.5
3	1.9	2.8	3.8	4.7	5.6	6.6	7.5	8.4	9.4	10.3	11.3	12.2	13.1	14.1	15.0	15.9	16.9	17.8	18.8
4	2.5	3.8	5.0	6.3	7.5	8.8	10.0	11.3	12.5	13.8	15.0	16.3	17.5	18.8	20.0	21.3	22.5	23.8	25.0
5	3.1	4.7	6.3	7.8	9.4	10.9	12.5	14.1	15.6	17.2	18.8	20.3	21.9	23.4	25.0	26.6	28.1	29.7	31.3
6	3.8	5.6	7.5	9.4	11.3	13.1	15.0	16.9	18.8	20.6	22.5	24.4	26.3	28.1	30.0	31.9	33.8	35.6	37.5
7	4.4	6.6	8.8	10.9	13.1	15.3	17.5	19.7	21.9	24.1	26.3	28.4	30.6	32.8	35.0	37.2	39.4	41.6	43.8
8	5.0	7.5	10.0	12.5	15.0	17.5	20.0	22.5	25.0	27.5	30.0	32.5	35.0	37.5	40.0	42.5	45.0	47.5	50.0
9	5.6	8.4	11.3	14.1	16.9	19.7	22.5	25.3	28.1	30.9	33.8	36.6	39.4	42.2	45.0	47.8	50.6	53.4	56.3
10	6.3	9.4	12.5	15.6	18.8	21.9	25.0	28.1	31.3	34.4	37.5	40.6	43.8	46.9	50.0	53.1	56.3	59.4	62.5
11	6.9	10.3	13.8	17.2	20.6	24.1	27.5	30.9	34.4	37.8	41.3	44.7	48.1	51.6	55.0	58.4	61.9	65.3	68.8
12	7.5	11.3	15.0	18.8	22.5	26.3	30.0	33.8	37.5	41.3	45.0	48.8	52.5	56.3	60.0	63.8	67.5	71.3	75.0
13	8.1	12.2	16.3	20.3	24.4	28.4	32.5	36.6	40.6	44.7	48.8	52.8	56.9	60.9	65.0	69.1	73.1	77.2	81.3
14	8.8	13.1	17.5	21.9	26.3	30.6	35.0	39.4	43.8	48.1	52.5	56.9	61.3	65.6	70.0	74.4	78.8	83.1	87.5
15	9.4	14.1	18.8	23.4	28.1	32.8	37.5	42.2	46.9	51.6	56.3	60.9	65.6	70.3	75.0	79.7	84.4	89.1	93.8
16	10.0	15.0	20.0	25.0	30.0	35.0	40.0	45.0	50.0	55.0	60.0	65.0	70.0	75.0	80.0	85.0	90.0	95.0	100.0
17	10.6	15.9	21.3	26.6	31.9	37.2	42.5	47.8	53.1	58.4	63.8	69.1	74.4	79.7	85.0	90.3	95.6	100.9	106.3
18	11.3	16.9	22.5	28.1	33.8	39.4	45.0	50.6	56.3	61.9	67.5	73.1	78.8	84.4	90.0	95.6	101.3	106.9	112.5
19	11.9	17.8	23.8	29.7	35.6	41.6	47.5	53.4	59.4	65.3	71.3	77.2	83.1	89.1	95.0	100.9	106.9	112.8	118.8
20	12.5	18.8	25.0	31.3	37.5	43.8	50.0	56.3	62.5	68.8	75.0	81.3	87.5	93.8	100.0	106.3	112.5	118.8	125.0

Legend: Ar/Mech Platoon | Ar/Mech Company | Ar/Mech Battalion

8-28 Survivability

FM 5-34

Table 8-6. Dozer team HDP calculations (continued)

Real hrs avail	Blade Teams Available																		
	1.0	1.5	2.0	2.5	3.0	3.5	4.0	4.5	5.0	5.5	6.0	6.5	7.0	7.5	8.0	8.5	9.0	9.5	10.0
21	13.1	19.7	26.3	32.8	39.4	45.9	52.5	59.1	65.6	72.2	78.8	85.3	91.9	98.4	105.0	111.6	118.1	124.7	131.3
22	13.8	20.6	27.5	34.4	41.3	48.1	55.0	61.9	68.8	75.6	82.5	89.4	96.3	103.1	110.0	116.9	123.8	130.6	137.5
23	14.4	21.6	28.8	35.9	43.1	50.3	57.5	64.7	71.9	79.1	86.3	93.4	100.6	107.8	115.0	122.2	129.4	136.6	143.8
24	15.0	22.5	30.0	37.5	45.0	52.5	60.0	67.5	75.0	82.5	90.0	97.5	105.0	112.5	120.0	127.5	135.0	142.5	150.0
25	15.6	23.4	31.3	39.1	46.9	54.7	62.5	70.3	78.1	85.9	93.8	101.6	109.4	117.2	125.0	132.8	140.6	148.4	156.3
26	16.3	24.4	32.5	40.6	48.8	56.9	65.0	73.1	81.3	89.4	97.5	105.6	113.8	121.9	130.0	138.1	146.3	154.4	162.5
27	16.9	25.3	33.8	42.2	50.6	59.1	67.5	75.9	84.4	92.8	101.3	109.7	118.1	126.6	135.0	143.4	151.9	160.3	168.8
28	17.5	26.3	35.0	43.8	52.5	61.3	70.0	78.8	87.5	96.3	105.0	113.8	122.5	131.3	140.0	148.8	157.5	166.3	175.0
29	18.1	27.2	36.3	45.3	54.4	63.4	72.5	81.6	90.6	99.7	108.8	117.8	126.9	135.9	145.0	154.1	163.1	172.2	181.3
30	18.8	28.1	37.5	46.9	56.3	65.6	75.0	84.4	93.8	103.1	112.5	121.9	131.3	140.6	150.0	159.4	168.8	178.1	187.5
31	19.4	29.1	38.8	48.4	58.1	67.8	77.5	87.2	96.9	106.6	116.3	125.9	135.6	145.3	155.0	164.7	174.4	184.1	193.8
32	20.0	30.0	40.0	50.0	60.0	70.0	80.0	90.0	100.0	110.0	120.0	130.0	140.0	150.0	160.0	170.0	180.0	190.0	200.0
33	20.6	30.9	41.3	51.6	61.9	72.2	82.5	92.8	103.1	113.4	123.8	134.1	144.4	154.7	165.0	175.3	185.6	195.9	206.3
34	21.3	31.9	42.5	53.1	63.8	74.4	85.0	95.6	106.3	116.9	127.5	138.1	148.8	159.4	170.0	180.6	191.3	201.9	212.5
35	21.9	32.8	43.8	54.7	65.6	76.6	87.5	98.4	109.4	120.3	131.3	142.2	153.1	164.1	175.0	185.9	196.9	207.8	218.8
36	22.5	33.8	45.0	56.3	67.5	78.8	90.0	101.3	112.5	123.8	135.0	146.3	157.5	168.8	180.0	191.3	202.5	213.8	225.0
37	23.1	34.7	46.3	57.8	69.4	80.9	92.5	104.1	115.6	127.2	138.8	150.3	161.9	173.4	185.0	196.6	208.1	219.7	231.3
38	23.8	35.6	47.5	59.4	71.3	83.1	95.0	106.9	118.8	130.6	142.5	154.4	166.3	178.1	190.0	201.9	213.8	225.6	237.5
39	24.4	36.6	48.8	60.9	73.1	85.3	97.5	109.7	121.9	134.1	146.3	158.4	170.6	182.8	195.0	207.2	219.4	231.6	243.8
40	25.0	37.5	50.0	62.5	75.0	87.5	100.0	112.5	125.0	137.5	150.0	162.5	175.0	187.5	200.0	212.5	225.0	237.5	250.0

Legend: Ar/Mech Platoon Ar/Mech Company Ar/Mech Battalion

Survivability 8-29

FM 5-34

Table 8-6. Dozer team HDP calculations (continued)

Real hrs avail	1.0	1.5	2.0	2.5	3.0	3.5	4.0	4.5	5.0	5.5	6.0	6.5	7.0	7.5	8.0	8.5	9.0	9.5	10.0
41	25.6	38.4	51.3	64.1	76.9	89.7	102.5	115.3	128.1	140.9	153.8	166.6	179.4	192.2	205.0	217.8	230.6	243.4	256.3
42	26.3	39.4	52.5	65.6	78.8	91.9	105.0	118.1	131.3	144.4	157.5	170.6	183.8	196.9	210.0	223.1	236.3	249.4	262.5
43	26.9	40.3	53.8	67.2	80.6	94.1	107.5	120.9	134.4	147.8	161.3	174.7	188.1	201.6	215.0	228.4	241.9	255.3	268.8
44	27.5	41.3	55.0	68.8	82.5	96.3	110.0	123.8	137.5	151.3	165.0	178.8	192.5	206.3	220.0	233.8	247.5	261.3	275.0
45	28.1	42.2	56.3	70.3	84.4	98.4	112.5	126.6	140.6	154.7	168.8	182.8	196.9	210.9	225.0	239.1	253.1	267.2	281.3
46	28.8	43.1	57.5	71.9	86.3	100.6	115.0	129.4	143.8	158.1	172.5	186.9	201.3	215.6	230.0	244.4	258.8	273.1	287.5
47	29.4	44.1	58.8	73.4	88.1	102.8	117.5	132.2	146.9	161.6	176.3	190.9	205.6	220.3	235.0	249.7	264.4	279.1	293.8
48	30.0	45.0	60.0	75.0	90.0	105.0	120.0	135.0	150.0	165.0	180.0	195.0	210.0	225.0	240.0	255.0	270.0	285.0	300.0
49	30.6	45.9	61.3	76.6	91.9	107.2	122.5	137.8	153.1	168.4	183.8	199.1	214.4	229.7	245.0	260.3	275.6	290.9	306.3
50	31.3	46.9	62.5	78.1	93.8	109.4	125.0	140.6	156.3	171.9	187.5	203.1	218.8	234.4	250.0	265.6	281.3	296.9	312.5
51	31.9	47.8	63.8	79.7	95.6	111.6	127.5	143.4	159.4	175.3	191.3	207.2	223.1	239.1	255.0	270.9	286.9	302.8	318.8
52	32.5	48.8	65.0	81.3	97.5	113.8	130.0	146.3	162.5	178.8	195.0	211.3	227.5	243.8	260.0	276.3	292.5	308.8	325.0
53	33.1	49.7	66.3	82.8	99.4	115.9	132.5	149.1	165.6	182.2	198.8	215.3	231.9	248.4	265.0	281.6	298.1	314.7	331.3
54	33.8	50.6	67.5	84.4	101.3	118.1	135.0	151.9	168.8	185.6	202.5	219.4	236.3	253.1	270.0	286.9	303.8	320.6	337.5
55	34.4	51.6	68.8	85.9	103.1	120.3	137.5	154.7	171.9	189.1	206.3	223.4	240.6	257.8	275.0	292.2	309.4	326.6	343.8
56	35.0	52.5	70.0	87.5	105.0	122.5	140.0	157.5	175.0	192.5	210.0	227.5	245.0	262.5	280.0	297.5	315.0	332.5	350.0
57	35.6	53.4	71.3	89.1	106.9	124.7	142.5	160.3	178.1	195.9	213.8	231.6	249.4	267.2	285.0	302.8	320.6	338.4	356.3
58	36.3	54.4	72.5	90.6	108.8	126.9	145.0	163.1	181.3	199.4	217.5	235.6	253.8	271.9	290.0	308.1	326.3	344.4	362.5
59	36.9	55.3	73.8	92.2	110.6	129.1	147.5	165.9	184.4	202.8	221.3	239.7	258.1	276.6	295.0	313.4	331.9	350.3	368.8
60	37.5	56.3	75.0	93.8	112.5	131.3	150.0	168.8	187.5	206.3	225.0	243.8	262.5	281.3	300.0	318.8	337.5	356.3	375.0

Legend: Ar/Mech Platoon   Ar/Mech Company   Ar/Mech Battalion

8-30 Survivability

FM 5-34

Table 8-7. ACE/ACE team TDP calculations

Real hrs avail	Blade Teams Available																		
	1.0	1.5	2.0	2.5	3.0	3.5	4.0	4.5	5.0	5.5	6.0	6.5	7.0	7.5	8.0	8.5	9.0	9.5	10.0
1	0.2	0.3	0.4	0.4	0.5	0.6	0.7	0.8	0.9	1.0	1.1	1.2	1.3	1.3	1.4	1.5	1.6	1.7	1.8
2	0.4	0.5	0.7	0.9	1.1	1.3	1.4	1.6	1.8	2.0	2.1	2.3	2.5	2.7	2.9	3.0	3.2	3.4	3.6
3	0.5	0.8	1.1	1.3	1.6	1.9	2.1	2.4	2.7	2.9	3.2	3.5	3.8	4.0	4.3	4.6	4.8	5.1	5.4
4	0.7	1.1	1.4	1.8	2.1	2.5	2.9	3.2	3.6	3.9	4.3	4.6	5.0	5.4	5.7	6.1	6.4	6.8	7.1
5	0.9	1.3	1.8	2.2	2.7	3.1	3.6	4.0	4.5	4.9	5.4	5.8	6.3	6.7	7.1	7.6	8.0	8.5	8.9
6	1.1	1.6	2.1	2.7	3.2	3.8	4.3	4.8	5.4	5.9	6.4	7.0	7.5	8.0	8.6	9.1	9.6	10.2	10.7
7	1.3	1.9	2.5	3.1	3.8	4.4	5.0	5.6	6.3	6.9	7.5	8.1	8.8	9.4	10.0	10.6	11.3	11.9	12.5
8	1.4	2.1	2.9	3.6	4.3	5.0	5.7	6.4	7.1	7.9	8.6	9.3	10.0	10.7	11.4	12.1	12.9	13.6	14.3
9	1.6	2.4	3.2	4.0	4.8	5.6	6.4	7.2	8.0	8.8	9.6	10.4	11.3	12.1	12.9	13.7	14.5	15.3	16.1
10	1.8	2.7	3.6	4.5	5.4	6.3	7.1	8.0	8.9	9.8	10.7	11.6	12.5	13.4	14.3	15.2	16.1	17.0	17.9
11	2.0	2.9	3.9	4.9	5.9	6.9	7.9	8.8	9.8	10.8	11.8	12.8	13.8	14.7	15.7	16.7	17.7	18.7	19.6
12	2.1	3.2	4.3	5.4	6.4	7.5	8.6	9.6	10.7	11.8	12.9	13.9	15.0	16.1	17.1	18.2	19.3	20.4	21.4
13	2.3	3.5	4.6	5.8	7.0	8.1	9.3	10.4	11.6	12.8	13.9	15.1	16.3	17.4	18.6	19.7	20.9	22.1	23.2
14	2.5	3.8	5.0	6.3	7.5	8.8	10.0	11.3	12.5	13.8	15.0	16.3	17.5	18.8	20.0	21.3	22.5	23.8	25.0
15	2.7	4.0	5.4	6.7	8.0	9.4	10.7	12.1	13.4	14.7	16.1	17.4	18.8	20.1	21.4	22.8	24.1	25.4	26.8
16	2.9	4.3	5.7	7.1	8.6	10.0	11.4	12.9	14.3	15.7	17.1	18.6	20.0	21.4	22.9	24.3	25.7	27.1	28.6
17	3.0	4.6	6.1	7.6	9.1	10.6	12.1	13.7	15.2	16.7	18.2	19.7	21.3	22.8	24.3	25.8	27.3	28.8	30.4
18	3.2	4.8	6.4	8.0	9.6	11.3	12.9	14.5	16.1	17.7	19.3	20.9	22.5	24.1	25.7	27.3	28.9	30.5	32.1
19	3.4	5.1	6.8	8.5	10.2	11.9	13.6	15.3	17.0	18.7	20.4	22.1	23.8	25.4	27.1	28.8	30.5	32.2	33.9
20	3.6	5.4	7.1	8.9	10.7	12.5	14.3	16.1	17.9	19.6	21.4	23.2	25.0	26.8	28.6	30.4	32.1	33.9	35.7

Legend: ▓ Ar/Mech Platoon ☐ Ar/Mech Company ☐ Ar/Mech Battalion

Survivability 8-31

FM 5-34

Table 8-7. ACE/ACE team TDP calculations (continued)

Real hrs avail	Blade Teams Available																		
	1.0	1.5	2.0	2.5	3.0	3.5	4.0	4.5	5.0	5.5	6.0	6.5	7.0	7.5	8.0	8.5	9.0	9.5	10.0
21	3.8	5.6	7.5	9.4	11.3	13.1	15.0	16.9	18.8	20.6	22.5	24.4	26.3	28.1	30.0	31.9	33.8	35.6	37.5
22	3.9	5.9	7.9	9.8	11.8	13.8	15.7	17.7	19.6	21.6	23.6	25.5	27.5	29.5	31.4	33.4	35.4	37.3	39.3
23	4.1	6.2	8.2	10.3	12.3	14.4	16.4	18.5	20.5	22.6	24.6	26.7	28.8	30.8	32.9	34.9	37.0	39.0	41.1
24	4.3	6.4	8.6	10.7	12.9	15.0	17.1	19.3	21.4	23.6	25.7	27.9	30.0	32.1	34.3	36.4	38.6	40.7	42.9
25	4.5	6.7	8.9	11.2	13.4	15.6	17.9	20.1	22.3	24.6	26.8	29.0	31.3	33.5	35.7	37.9	40.2	42.4	44.6
26	4.6	7.0	9.3	11.6	13.9	16.3	18.6	20.9	23.2	25.5	27.9	30.2	32.5	34.8	37.1	39.5	41.8	44.1	46.4
27	4.8	7.2	9.6	12.1	14.5	16.9	19.3	21.7	24.1	26.5	28.9	31.3	33.8	36.2	38.6	41.0	43.4	45.8	48.2
28	5.0	7.5	10.0	12.5	15.0	17.5	20.0	22.5	25.0	27.5	30.0	32.5	35.0	37.5	40.0	42.5	45.0	47.5	50.0
29	5.2	7.8	10.4	12.9	15.5	18.1	20.7	23.3	25.9	28.5	31.1	33.7	36.3	38.8	41.4	44.0	46.6	49.2	51.8
30	5.4	8.0	10.7	13.4	16.1	18.8	21.4	24.1	26.8	29.5	32.1	34.8	37.5	40.2	42.9	45.5	48.2	50.9	53.6
31	5.5	8.3	11.1	13.8	16.6	19.4	22.1	24.9	27.7	30.4	33.2	36.0	38.8	41.5	44.3	47.1	49.8	52.6	55.4
32	5.7	8.6	11.4	14.3	17.1	20.0	22.9	25.7	28.6	31.4	34.3	37.1	40.0	42.9	45.7	48.6	51.4	54.3	57.1
33	5.9	8.8	11.8	14.7	17.7	20.6	23.6	26.5	29.5	32.4	35.4	38.3	41.3	44.2	47.1	50.1	53.0	56.0	58.9
34	6.1	9.1	12.1	15.2	18.2	21.3	24.3	27.3	30.4	33.4	36.4	39.5	42.5	45.5	48.6	51.6	54.6	57.7	60.7
35	6.3	9.4	12.5	15.6	18.8	21.9	25.0	28.1	31.3	34.4	37.5	40.6	43.8	46.9	50.0	53.1	56.3	59.4	62.5
36	6.4	9.6	12.9	16.1	19.3	22.5	25.7	28.9	32.1	35.4	38.6	41.8	45.0	48.2	51.4	54.6	57.9	61.1	64.3
37	6.6	9.9	13.2	16.5	19.8	23.1	26.4	29.7	33.0	36.3	39.6	42.9	46.3	49.6	52.9	56.2	59.5	62.8	66.1
38	6.8	10.2	13.6	17.0	20.4	23.8	27.1	30.5	33.9	37.3	40.7	44.1	47.5	50.9	54.3	57.7	61.1	64.5	67.9
39	7.0	10.4	13.9	17.4	20.9	24.4	27.9	31.3	34.8	38.3	41.8	45.3	48.8	52.2	55.7	59.2	62.7	66.2	69.6
40	7.1	10.7	14.3	17.9	21.4	25.0	28.6	32.1	35.7	39.3	42.9	46.4	50.0	53.6	57.1	60.7	64.3	67.9	71.4

Legend: ☐ Ar/Mech Platoon   ☐ Ar/Mech Company   ☐ Ar/Mech Battalion

8-32 Survivability

FM 5-34

Table 8-7. ACE/ACE team TDP calculations (continued)

Real hrs avail	Blade Teams Available																			
	1.0	1.5	2.0	2.5	3.0	3.5	4.0	4.5	5.0	5.5	6.0	6.5	7.0	7.5	8.0	8.5	9.0	9.5	10.0
41	7.3	11.0	14.6	18.3	22.0	25.6	29.3	32.9	36.6	40.3	43.9	47.6	51.3	54.9	**58.6**	62.2	65.9	69.6	73.2
42	7.5	11.3	15.0	18.8	22.5	26.3	30.0	33.8	37.5	41.3	45.0	48.8	52.5	56.3	60.0	63.8	67.5	71.3	75.0
43	7.7	11.5	15.4	19.2	23.0	26.9	30.7	34.6	38.4	42.2	46.1	49.9	53.8	57.6	61.4	65.3	69.1	72.9	76.8
44	7.9	11.8	15.7	19.6	23.6	27.5	31.4	35.4	39.3	43.2	47.1	51.1	55.0	**58.9**	62.9	66.8	70.7	74.6	78.6
45	8.0	12.1	16.1	20.1	24.1	28.1	32.1	36.2	40.2	44.2	48.2	52.2	56.3	60.3	64.3	68.3	72.3	76.3	80.4
46	8.2	12.3	16.4	20.5	24.6	28.8	32.9	37.0	41.1	45.2	49.3	53.4	57.5	61.6	65.7	69.8	73.9	78.0	82.1
47	8.4	12.6	16.8	21.0	25.2	29.4	33.6	37.8	42.0	46.2	50.4	54.6	**58.8**	62.9	67.1	71.3	75.5	79.7	83.9
48	8.6	12.9	17.1	21.4	25.7	30.0	34.3	38.6	42.9	47.1	51.4	55.7	60.0	64.3	68.6	72.9	77.1	81.4	85.7
49	8.8	13.1	17.5	21.9	26.3	30.6	35.0	39.4	43.8	48.1	52.5	56.9	61.3	65.6	70.0	74.4	78.8	83.1	87.5
50	8.9	13.4	17.9	22.3	26.8	31.3	35.7	40.2	44.6	49.1	53.6	**58.0**	62.5	67.0	71.4	75.9	80.4	84.8	89.3
51	9.1	13.7	18.2	22.8	27.3	31.9	36.4	41.0	45.5	50.1	54.6	59.2	63.8	68.3	72.9	77.4	82.0	86.5	91.1
52	9.3	13.9	18.6	23.2	27.9	32.5	37.1	41.8	46.4	51.1	55.7	60.4	65.0	69.6	74.3	78.9	83.6	88.2	92.9
53	9.5	**14.2**	18.9	23.7	28.4	33.1	37.9	42.6	47.3	52.1	56.8	61.5	66.3	71.0	75.7	80.4	85.2	89.9	94.6
54	9.6	14.5	19.3	24.1	28.9	33.8	38.6	43.4	48.2	53.0	57.9	62.7	67.5	72.3	77.1	82.0	86.8	91.6	96.4
55	9.8	14.7	19.6	24.6	29.5	34.4	39.3	44.2	49.1	54.0	**58.9**	63.8	68.8	73.7	78.6	83.5	88.4	93.3	98.2
56	10.0	15.0	20.0	25.0	30.0	35.0	40.0	45.0	50.0	55.0	60.0	65.0	70.0	75.0	80.0	85.0	90.0	95.0	100.0
57	10.2	15.3	20.4	25.4	30.5	35.6	40.7	45.8	50.9	56.0	61.1	66.2	71.3	76.3	81.4	86.5	91.6	96.7	101.8
58	10.4	15.5	20.7	25.9	31.1	36.3	41.4	46.6	51.8	57.0	62.1	67.3	72.5	77.7	82.9	88.0	93.2	98.4	103.6
59	10.5	15.8	21.1	26.3	31.6	36.9	42.1	47.4	52.7	57.9	63.2	68.5	73.8	79.0	84.3	89.6	94.8	100.1	105.4
60	10.7	16.1	21.4	26.8	32.1	37.5	42.9	48.2	53.6	**58.9**	64.3	69.6	75.0	80.4	85.7	91.1	96.4	101.8	107.1

Legend: ☐ Ar/Mech Platoon ☐ Ar/Mech Company ☐ Ar/Mech Battalion

Survivability 8-33

FM 5-34

Table 8-8. ACE/ACE team HDP calculations

Real hrs avail	\multicolumn{19}{c}{Blade Teams Available}
	1.0	1.5	2.0	2.5	3.0	3.5	4.0	4.5	5.0	5.5	6.0	6.5	7.0	7.5	8.0	8.5	9.0	9.5	10.0
1	0.4	0.6	0.8	1.0	1.3	1.5	1.7	1.9	2.1	2.3	2.5	2.7	2.9	3.1	3.3	3.5	3.8	**4.0**	**4.2**
2	0.8	1.3	1.7	2.1	2.5	2.9	3.3	3.8	**4.2**	**4.6**	**5.0**	**5.4**	**5.8**	**6.3**	**6.7**	**7.1**	**7.5**	7.9	8.3
3	1.3	1.9	2.5	3.1	3.8	4.4	**5.0**	**5.6**	6.3	6.9	7.5	8.1	8.8	9.4	10.0	10.6	11.3	11.9	12.5
4	1.7	2.5	3.3	**4.2**	**5.0**	5.8	6.7	7.5	8.3	9.2	10.0	10.8	11.7	12.5	13.3	14.2	**15.0**	**15.8**	**16.7**
5	2.1	3.1	**4.2**	5.2	6.3	7.3	8.3	9.4	10.4	11.5	12.5	13.5	**14.6**	**15.6**	**16.7**	17.7	18.8	19.8	20.8
6	2.5	3.8	5.0	6.3	7.5	8.8	10.0	11.3	12.5	13.8	**15.0**	**16.3**	17.5	18.8	20.0	21.3	22.5	23.8	25.0
7	2.9	**4.4**	5.8	7.3	8.8	10.2	11.7	13.1	**14.6**	**16.0**	17.5	19.0	20.4	21.9	23.3	24.8	26.3	27.7	29.2
8	3.3	5.0	6.7	8.3	10.0	11.7	13.3	**15.0**	**16.7**	18.3	20.0	21.7	23.3	25.0	26.7	28.3	30.0	31.7	33.3
9	3.8	5.6	7.5	9.4	11.3	13.1	**15.0**	16.9	18.8	20.6	22.5	24.4	26.3	28.1	30.0	31.9	33.8	35.6	37.5
10	**4.2**	6.3	8.3	10.4	12.5	**14.6**	**16.7**	18.8	20.8	22.9	25.0	27.1	29.2	31.3	33.3	35.4	37.5	39.6	41.7
11	4.6	6.9	9.2	11.5	13.8	16.0	18.3	20.6	22.9	25.2	27.5	29.8	32.1	34.4	36.7	39.0	41.3	43.5	45.8
12	5.0	7.5	10.0	12.5	**15.0**	17.5	20.0	22.5	25.0	27.5	30.0	32.5	35.0	37.5	40.0	42.5	45.0	47.5	50.0
13	5.4	8.1	10.8	13.5	16.3	19.0	21.7	24.4	27.1	29.8	32.5	35.2	37.9	40.6	43.3	46.0	48.8	51.5	54.2
14	5.8	8.8	11.7	**14.6**	17.5	20.4	23.3	26.3	29.2	32.1	35.0	37.9	40.8	43.8	46.7	49.6	52.5	55.4	**58.3**
15	6.3	9.4	12.5	15.6	18.8	21.9	25.0	28.1	31.3	34.4	37.5	40.6	43.8	46.9	50.0	53.1	56.3	**59.4**	62.5
16	6.7	10.0	13.3	16.7	20.0	23.3	26.7	30.0	33.3	36.7	40.0	43.3	46.7	50.0	53.3	56.7	**60.0**	63.3	66.7
17	7.1	10.6	**14.2**	17.7	21.3	24.8	28.3	31.9	35.4	39.0	42.5	46.0	49.6	53.1	56.7	**60.2**	63.8	67.3	70.8
18	7.5	11.3	15.0	18.8	22.5	26.3	30.0	33.8	37.5	41.3	45.0	48.8	52.5	56.3	**60.0**	63.8	67.5	71.3	75.0
19	7.9	11.9	15.8	19.8	23.8	27.7	31.7	35.6	39.6	43.5	47.5	51.5	55.4	**59.4**	63.3	67.3	71.3	75.2	79.2
20	8.3	12.5	16.7	20.8	25.0	29.2	33.3	37.5	41.7	45.8	50.0	54.2	**58.3**	62.5	66.7	70.8	75.0	79.2	83.3

Legend: ▨ Ar/Mech Platoon ▨ Ar/Mech Company ▨ Ar/Mech Battalion

8-34 Survivability

FM 5-34

Table 8-8. ACE/ACE team HDP calculations (continued)

Real hrs avail	Blade Teams Available																	
	1.0	1.5	2.0	2.5	3.0	3.5	4.0	4.5	5.0	5.5	6.0	6.5	7.0	7.5	8.0	8.5	9.0	9.5	10.0
21	8.8	13.1	17.5	21.9	26.3	30.6	35.0	39.4	43.8	48.1	52.5	56.9	61.3	65.6	70.0	74.4	78.8	83.1	87.5
22	9.2	13.8	18.3	22.9	27.5	32.1	36.7	41.3	45.8	50.4	55.0	59.6	64.2	68.8	73.3	77.9	82.5	87.1	91.7
23	9.6	14.4	19.2	24.0	28.8	33.5	38.3	43.1	47.9	52.7	57.5	62.3	67.1	71.9	76.7	81.5	86.3	91.0	95.8
24	10.0	15.0	20.0	25.0	30.0	35.0	40.0	45.0	50.0	55.0	60.0	65.0	70.0	75.0	80.0	85.0	90.0	95.0	100.0
25	10.4	15.6	20.8	26.0	31.3	36.5	41.7	46.9	52.1	57.3	62.5	67.7	72.9	78.1	83.3	88.5	93.8	99.0	104.2
26	10.8	16.3	21.7	27.1	32.5	37.9	43.3	48.8	54.2	59.6	65.0	70.4	75.8	81.3	86.7	92.1	97.5	102.9	108.3
27	11.3	16.9	22.5	28.1	33.8	39.4	45.0	50.6	56.3	61.9	67.5	73.1	78.8	84.4	90.0	95.6	101.3	106.9	112.5
28	11.7	17.5	23.3	29.2	35.0	40.8	46.7	52.5	58.3	64.2	70.0	75.8	81.7	87.5	93.3	99.2	105.0	110.8	116.7
29	12.1	18.1	24.2	30.2	36.3	42.3	48.3	54.4	60.4	66.5	72.5	78.5	84.6	90.6	96.7	102.7	108.8	114.8	120.8
30	12.5	18.8	25.0	31.3	37.5	43.8	50.0	56.3	62.5	68.8	75.0	81.3	87.5	93.8	100.0	106.3	112.5	118.8	125.0
31	12.9	19.4	25.8	32.3	38.8	45.2	51.7	58.1	64.6	71.0	77.5	84.0	90.4	96.9	103.3	109.8	116.3	122.7	129.2
32	13.3	20.0	26.7	33.3	40.0	46.7	53.3	60.0	66.7	73.3	80.0	86.7	93.3	100.0	106.7	113.3	120.0	126.7	133.3
33	13.8	20.6	27.5	34.4	41.3	48.1	55.0	61.9	68.8	75.6	82.5	89.4	96.3	103.1	110.0	116.9	123.8	130.6	137.5
34	14.2	21.3	28.3	35.4	42.5	49.6	56.7	63.8	70.8	77.9	85.0	92.1	99.2	106.3	113.3	120.4	127.5	134.6	141.7
35	14.6	21.9	29.2	36.5	43.8	51.0	58.3	65.6	72.9	80.2	87.5	94.8	102.1	109.4	116.7	124.0	131.3	138.5	145.8
36	15.0	22.5	30.0	37.5	45.0	52.5	60.0	67.5	75.0	82.5	90.0	97.5	105.0	112.5	120.0	127.5	135.0	142.5	150.0
37	15.4	23.1	30.8	38.5	46.3	54.0	61.7	69.4	77.1	84.8	92.5	100.2	107.9	115.6	123.3	131.0	138.8	146.5	154.2
38	15.8	23.8	31.7	39.6	47.5	55.4	63.3	71.3	79.2	87.1	95.0	102.9	110.8	118.8	126.7	134.6	142.5	150.4	158.3
39	16.3	24.4	32.5	40.6	48.8	56.9	65.0	73.1	81.3	89.4	97.5	105.6	113.8	121.9	130.0	138.1	146.3	154.4	162.5
40	16.7	25.0	33.3	41.7	50.0	58.3	66.7	75.0	83.3	91.7	100.0	108.3	116.7	125.0	133.3	141.7	150.0	158.3	166.7

Legend: ☐ Ar/Mech Platoon   ☐ Ar/Mech Company   ☐ Ar/Mech Battalion

Survivability 8-35

FM 5-34

Table 8-8. ACE/ACE team HDP calculations (continued)

Real hrs avail	\\	Blade Teams Available																		
	1.0	1.5	2.0	2.5	3.0	3.5	4.0	4.5	5.0	5.5	6.0	6.5	7.0	7.5	8.0	8.5	9.0	9.5	10.0
41	17.1	25.6	34.2	42.7	51.3	59.8	68.3	76.9	85.4	94.0	102.5	111.0	119.6	128.1	136.7	145.2	153.8	162.3	170.8
42	17.5	26.3	35.0	43.8	52.5	61.3	70.0	78.8	87.5	96.3	105.0	113.8	122.5	131.3	140.0	148.8	157.5	166.3	175.0
43	17.9	26.9	35.8	44.8	53.8	62.7	71.7	80.6	89.6	98.5	107.5	116.5	125.4	134.4	143.3	152.3	161.3	170.2	179.2
44	18.3	27.5	36.7	45.8	55.0	64.2	73.3	82.5	91.7	100.8	110.0	119.2	128.3	137.5	146.7	155.8	165.0	174.2	183.3
45	18.8	28.1	37.5	46.9	56.3	65.6	75.0	84.4	93.8	103.1	112.5	121.9	131.3	140.6	150.0	159.4	168.8	178.1	187.5
46	19.2	28.8	38.3	47.9	57.5	67.1	76.7	86.3	95.8	105.4	115.0	124.6	134.2	143.8	153.3	162.9	172.5	182.1	191.7
47	19.6	29.4	39.2	49.0	58.8	68.5	78.3	88.1	97.9	107.7	117.5	127.3	137.1	146.9	156.7	166.5	176.3	186.0	195.8
48	20.0	30.0	40.0	50.0	60.0	70.0	80.0	90.0	100.0	110.0	120.0	130.0	140.0	150.0	160.0	170.0	180.0	190.0	200.0
49	20.4	30.6	40.8	51.0	61.3	71.5	81.7	91.9	102.1	112.3	122.5	132.7	142.9	153.1	163.3	173.5	183.8	194.0	204.2
50	20.8	31.3	41.7	52.1	62.5	72.9	83.3	93.8	104.2	114.6	125.0	135.4	145.8	156.3	166.7	177.1	187.5	197.9	208.3
51	21.3	31.9	42.5	53.1	63.8	74.4	85.0	95.6	106.3	116.9	127.5	138.1	148.8	159.4	170.0	180.6	191.3	201.9	212.5
52	21.7	32.5	43.3	54.2	65.0	75.8	86.7	97.5	108.3	119.2	130.0	140.8	151.7	162.5	173.3	184.2	195.0	205.8	216.7
53	22.1	33.1	44.2	55.2	66.3	77.3	88.3	99.4	110.4	121.5	132.5	143.5	154.6	165.6	176.7	187.7	198.8	209.8	220.8
54	22.5	33.8	45.0	56.3	67.5	78.8	90.0	101.3	112.5	123.8	135.0	146.3	157.5	168.8	180.0	191.3	202.5	213.8	225.0
55	22.9	34.4	45.8	57.3	68.8	80.2	91.7	103.1	114.6	126.0	137.5	149.0	160.4	171.9	183.3	194.8	206.3	217.7	229.2
56	23.3	35.0	46.7	58.3	70.0	81.7	93.3	105.0	116.7	128.3	140.0	151.7	163.3	175.0	186.7	198.3	210.0	221.7	233.3
57	23.8	35.6	47.5	59.4	71.3	83.1	95.0	106.9	118.8	130.6	142.5	154.4	166.3	178.1	190.0	201.9	213.8	225.6	237.5
58	24.2	36.3	48.3	60.4	72.5	84.6	96.7	108.8	120.8	132.9	145.0	157.1	169.2	181.3	193.3	205.4	217.5	229.6	241.7
59	24.6	36.9	49.2	61.5	73.8	86.0	98.3	110.6	122.9	135.2	147.5	159.8	172.1	184.4	196.7	209.0	221.3	233.5	245.8
60	25.0	37.5	50.0	62.5	75.0	87.5	100.0	112.5	125.0	137.5	150.0	162.5	175.0	187.5	200.0	212.5	225.0	237.5	250.0

Legend: ☐ Ar/Mech Platoon ☐ Ar/Mech Company ☐ Ar/Mech Battalion

FM 5-34

HASTY FIGHTING POSITIONS

Figure 8-25 shows hasty fighting positions for combat vehicles. Remember, berms will not protect vehicles from enemy armor fire.

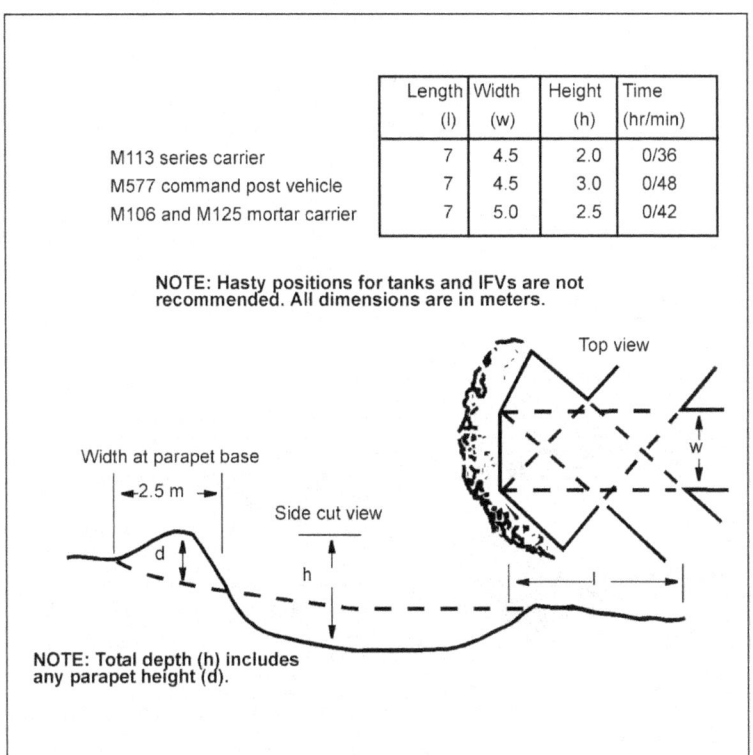

Figure 8-25. Hasty fighting positions for combat vehicles

MODIFIED FIGHTING POSITIONS

Figures 8-26 and 8-27, pages 8-38 and 8-39, show various modified fighting positions.

Survivability 8-37

FM 5-34

	A Length (ft)	B Width (ft)	C Turret Depth (ft)
M2	28	16	10
M1	32	18	9

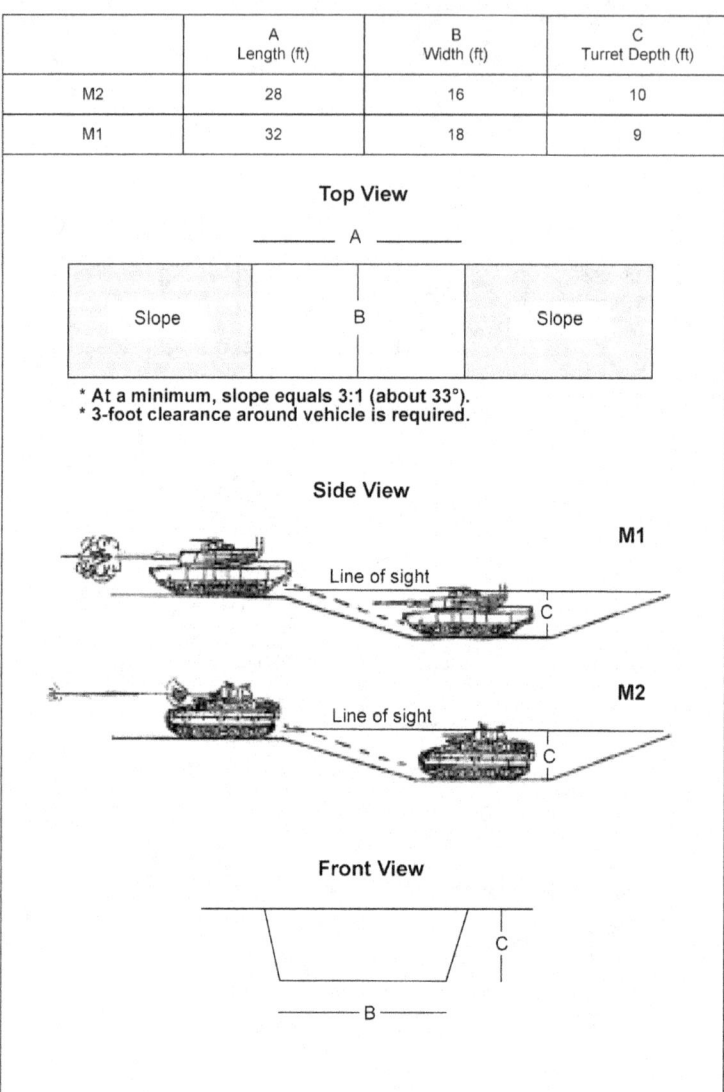

Figure 8-26. Modified, two-tiered hiding position

8-38 Survivability

FM 5-34

	A Length (ft)	B Width (ft)	C Hull depth (ft)
M109/M548	109	18	5

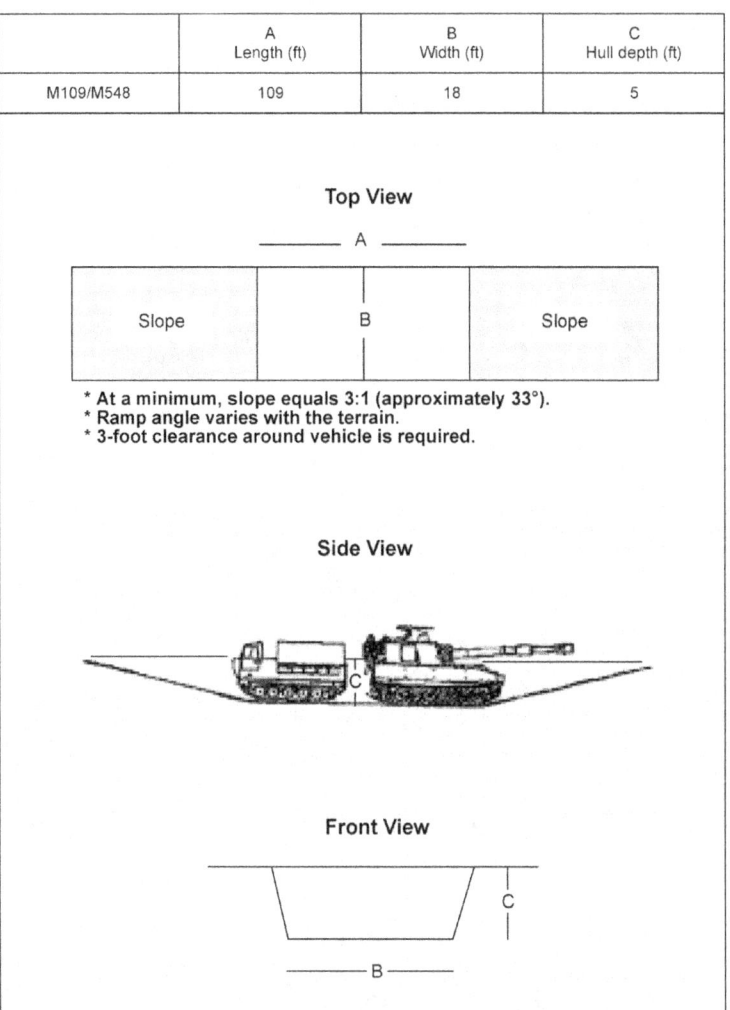

* At a minimum, slope equals 3:1 (approximately 33°).
* Ramp angle varies with the terrain.
* 3-foot clearance around vehicle is required.

Figure 8-27. Modified, two-tiered artillery position

Survivability 8-39

FM 5-34

DELIBERATE FIGHTING POSITIONS

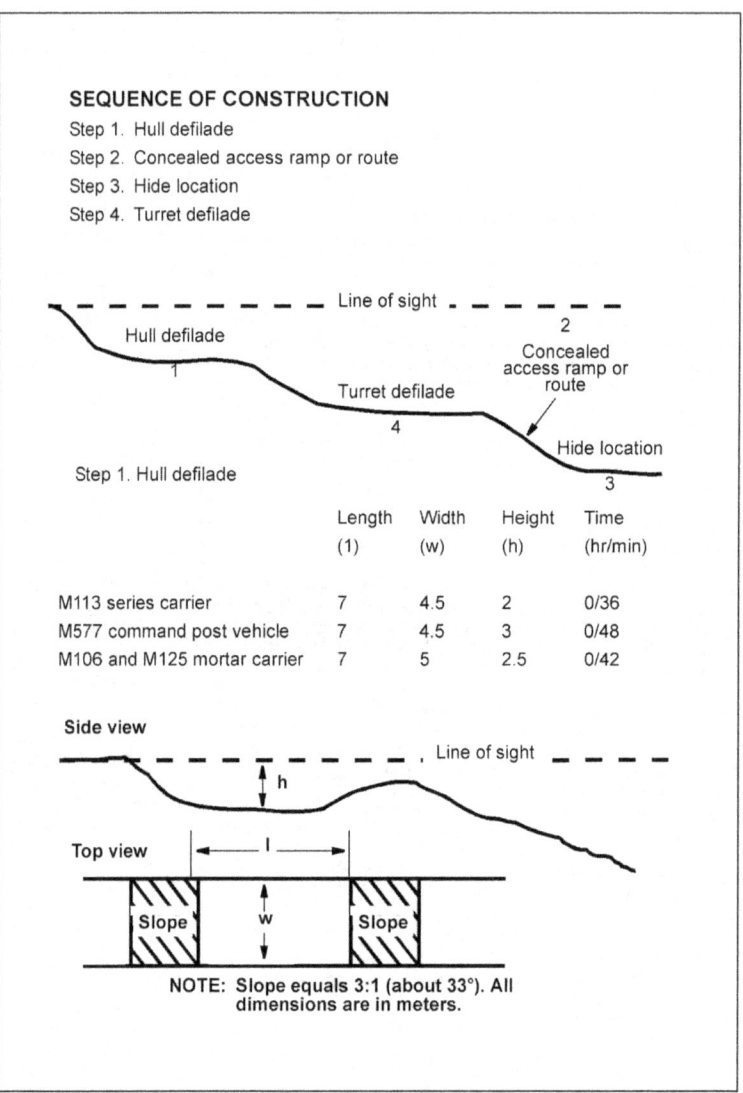

Figure 8-28. Deliberate fighting positions for fighting vehicles

FM 5-34

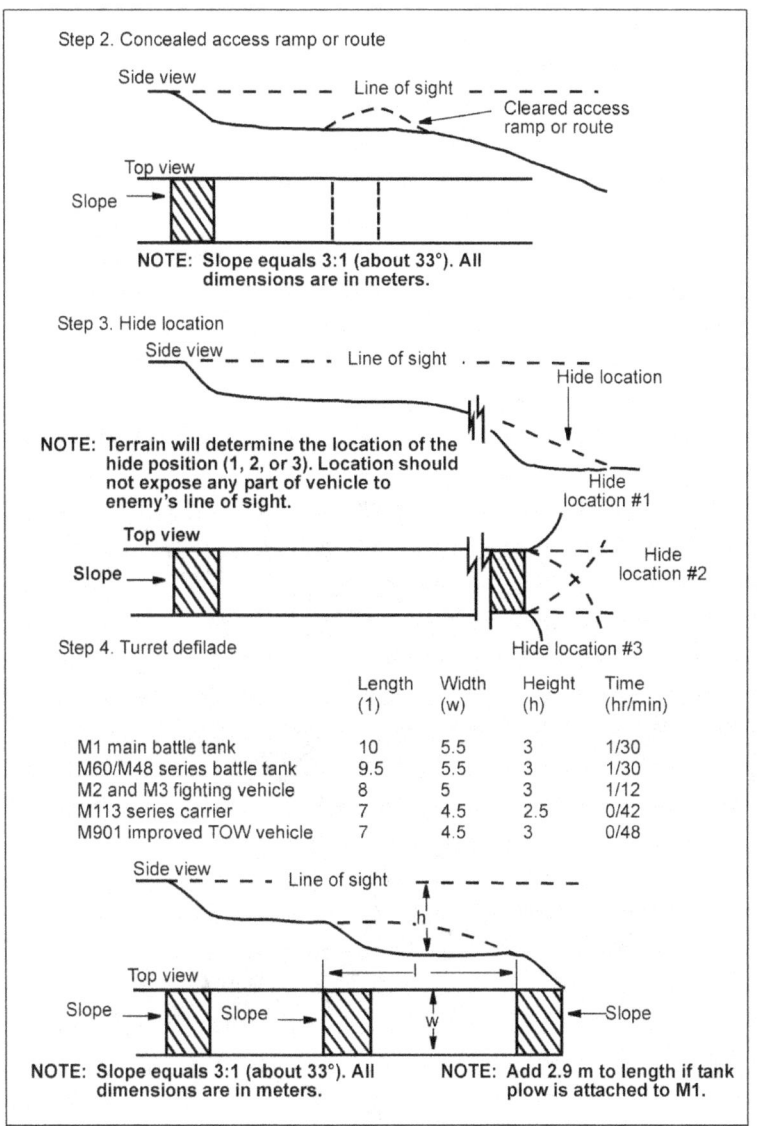

Figure 8-28. Deliberate fighting positions for fighting vehicles (continued)

Survivability 8-41

Protective Fighting Positions

Artillery and Parapet

Figure 8-29 shows a parapet position. Table 8-9 shows the dimensions of field artillery vehicle positions.

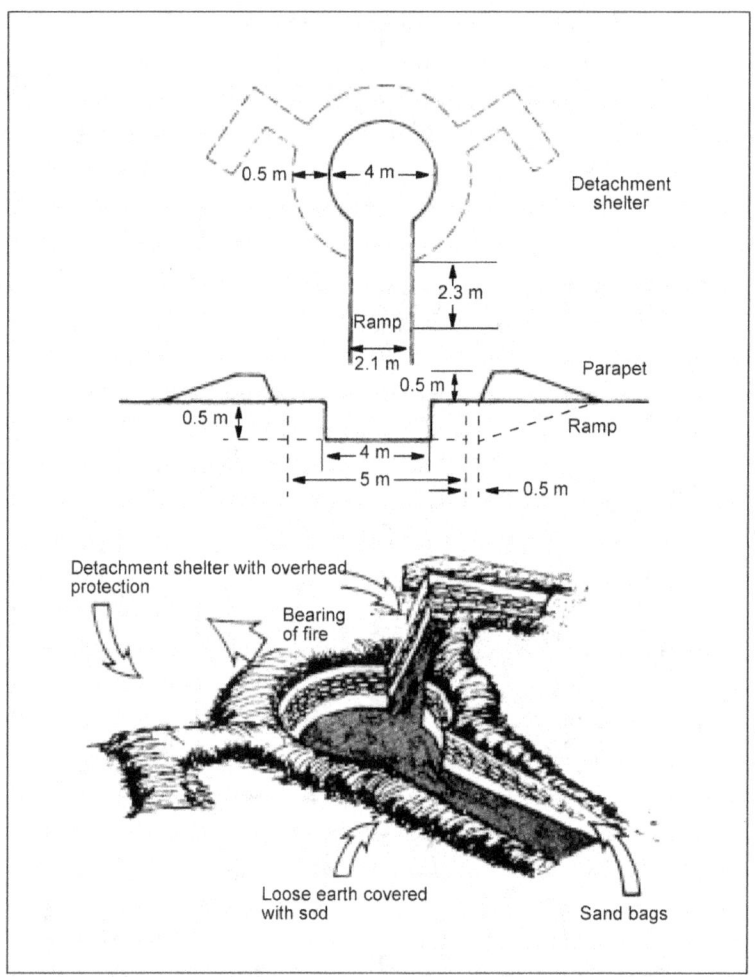

Figure 8-29. 105-mm parapet-position construction detail

8-42 Survivability

FM 5-34

Table 8-9. Dimensions of field artillery vehicle positions

Vehicle Type	Dimension[1]			Equipment Hours[3] (D7 Dozer/ ACE)	Minimum Parapet Thickness at Base (m)	Remarks
	Length (m)	Width (m)	Depth[2,4] (m)			
General-support rocket launcher	8	5.1	0.9			
155-mm self-propelled howitzer (M109)	32	5.4	1.5			*
175-mm self-propelled gun (M107)	31.5	4.8	1.5			*
8-in self-propelled howitzer (M110)	32.4	5	1.5			*

*Length accommodates ammunition supply vehicles.
[1]Position dimensions provide an approximate 0.9 m clearance around the vehicle for movement and maintenance but do not include the ramp(s).
[2]Total depth includes any parapet height.
[3]Production is at a rate of 100 bank cubic yards per 0.75 hour. Divide the construction time by 0.85 for rocky or hard soil, night conditions, or closed hatch operations (ACE). Using natural terrain features will reduce construction time.
[4]All depths are approximate and will need adjustment for surrounding terrain and fields of fire.

Deep-Cut

Figure 8-30 shows a deep-cut position; Table 8-10, page 8-44, shows the dimensions of deep-cut positions.

Figure 8-30. Deep-cut position

Survivability 8-43

Table 8-10. Dimensions of typical deep-cut positions

Vehicle Type	Dimensions[1]			Equipment Hours[3] (D7 Dozer/ACE)	Remarks
	Length (m)	Width (m)	Depth[2,4] (m)		
1 1/4-ton truck/ HMMWV	6	3.9	2.7	0.7	Add 2.7 m to length for cargo trailer.
2 1/2-ton cargo truck	8.7	3.9	3	1.1	Add 4.2 m to length for cargo or water trailer.
2 1/2-ton shop van	8.4	4.2	3.6	1.3	
HEMTT					
5-ton cargo	11.4	4.2	3	1.5	
5-ton shop van	10.8	4.2	3.6	1.7	
10-ton cargo truck	10.2	4.8	3.6	1.9	
10-ton tractor w/ van semitrailer	15.9	4.8	3.6	2.9	Dimensions shown are for trailer length of 9.3 m; for other trailers, add 6.9 m to actual trailer length.

[1] Position dimensions provide an approximate 0.9 m of clearance around a vehicle for movement and maintenance but do not include ramp(s).
[2] Production rate is 100 bank cubic yards per 0.75 hour. Divide the construction time by 0.85 for rocky or hard soil, night conditions, or closed-hatch operations (ACE). Using natural terrain features will reduce construction time.
[3] Ensure that drainage is provided.
[4] See Table 8-11 for maximum slope-cut ratios.
In training, deep cuts cause significant impacts to root systems, natural-drainage patterns, and training-area foliage. Coordinate with the environmental office regarding deep-cut employment.

Table 8-11. Recommended requirements for slope ratios in cuts and fills

Unified Soil Classification System	Slopes not Subject to Saturation		Slopes Subject to Saturation	
	Maximum Height of Earth Face	Maximum Slope Ratio	Maximum Height of Earth Face	Maximum Slope Ratio
GW, GP, GMd, SW, SP, SMd	Not critical	1.5:1	Not critical	2:1
GMu, GC, SMu, SC, ML, MH, CL, CH	Less than 50 feet	2:1	Less than 50 feet	3:1
OL, OH, PT	Generally not suitable for construction			

NOTES:
1. The recommended slopes are valid only in homogeneous soils that have either an in-place or compacted density equaling or exceeding 95 percent CE55 maximum dry density. For nonhomogeneous soils, or soils at a lower densities, a deliberate slope stability analysis is required.
2. Backslopes that cut into loess soil will seek to maintain a near-vertical cleavage. Do not apply loading above this cut face. Expect sloughing to occur.
3. Chapter 11 contains more information on the Unified Soil Classification System (USCS).

TRENCHES, REVETMENTS, BUNKERS, AND SHELTERS

TRENCHES

Construct trenches to connect fighting positions and provide protection and concealment for personnel moving between positions. They may be open, with an overhead cover, or a combination (see Figure 8-31, page 8-46).

REVETMENTS

Retaining Wall

Materials that you can use for a retaining wall are sandbags, sod blocks (20 x 45 centimeters), lumber, timber, or corrugated metal. When using sandbags, fill them 3/4 full with one part cement to 10 parts earth. Place a bottom row as a header at about 15 centimeters below floor level. Alternate rows as header and stretcher (see Figure 8-32, page 8-46). Ensure that the wall slopes forward of the revetted face at a 1:4 slope ratio. See Figure 8-33, page 8-47, for an anchoring method.

FM 5-34

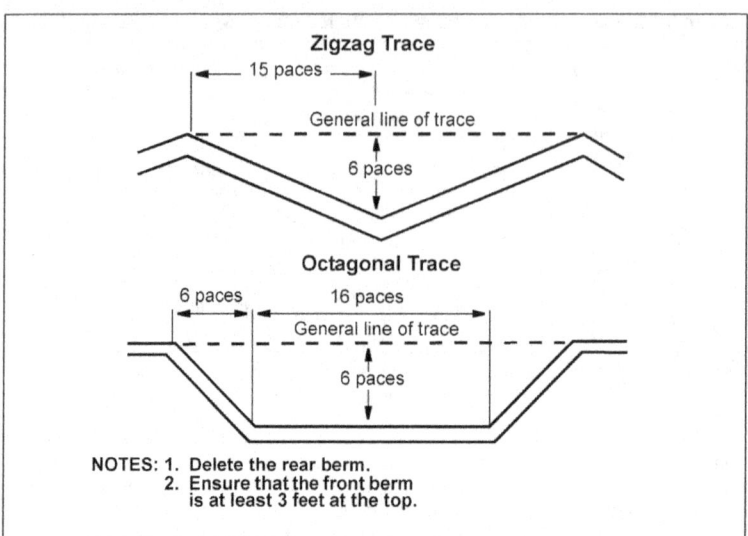

Figure 8-31. Standard trench traces

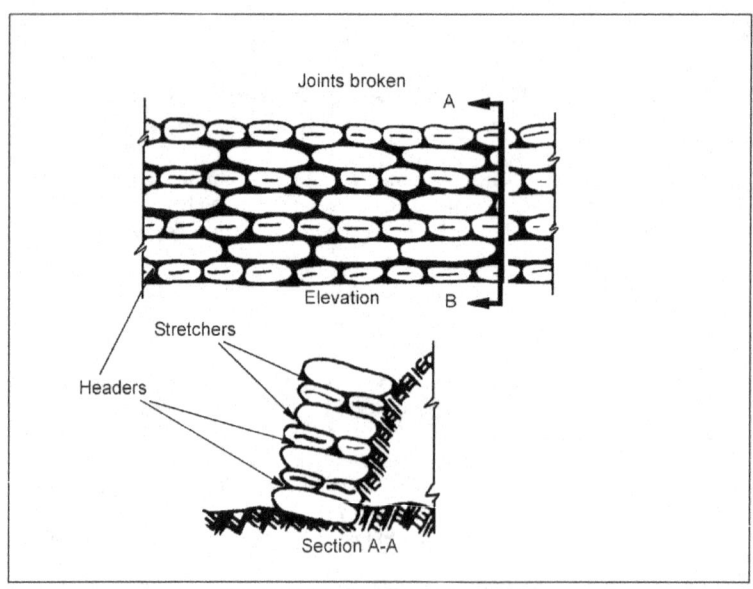

Figure 8-32. Sandbag revetment

8-46 Survivability

FM 5-34

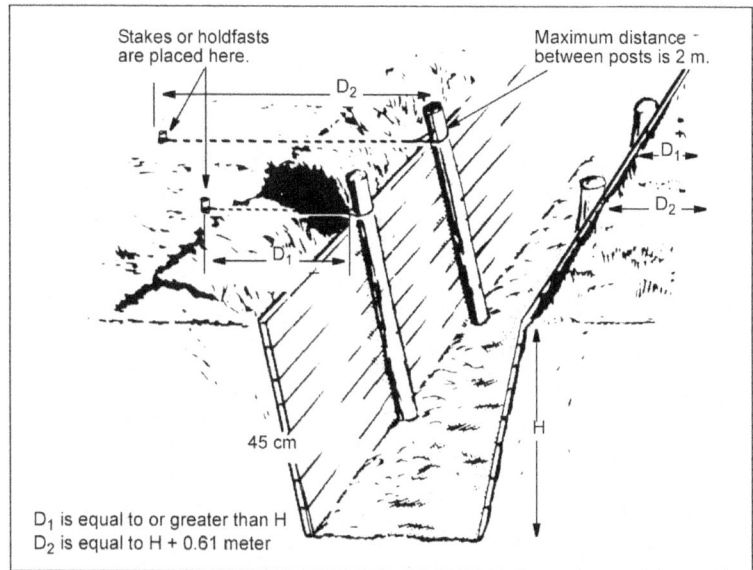

Figure 8-33. Retaining-wall anchoring method

Facing Revetments

You use facing revetments mainly to protect a surface from weather and damage from occupation. You can use brushwood hurdles (Figure 8-34), continuous brush, pole and dimensional timbers,

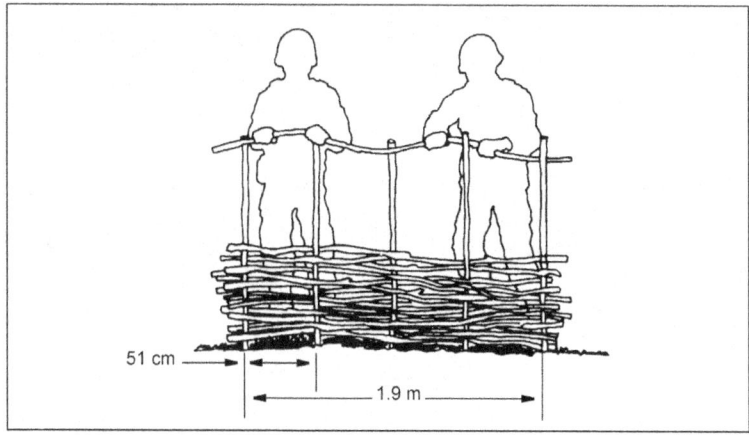

Figure 8-34. Brushwood hurdle

Survivability 8-47

corrugated metal or burlap, and chicken wire for construction material. To emplace a facing revetment, pickets should be 8 centimeters or larger in diameter and at least 1.75 meters apart. Drive the pickets into the ground at least 0.5 meter and anchor at the top as shown in Figure 8-34, page 8-47.

BUNKERS

When designing a bunker, consider its purpose (command post or fighting position) and the degree of protection desired (small arms, mortars, bombs) (see Table 8-12). A bunker can be constructed wholly or partly underground. Prefabricated bunker assemblies (wall and roof) afford rapid construction and placement flexibility. When using timber, avoid notching construction timber. Figures 8-35 and 8-36, pages 8-50 and 8-51, show common field bunkers.

SHELTERS

The most effective shelters are cut and cover. Figures 8-37 and 8-38, pages 8-51 and 8-52, show some typical shelters.

CAMOUFLAGE

The purpose of camouflage is to alter or eliminate recognition (shape, shadow, color, texture, position, and movement). Materials for camouflaging can be natural or man-made. Natural materials include vegetation (growing, cut, or dead), inert substances of the earth (soil and mud), and debris.

NOTE: In training, avoid obstruction of natural vegetation by using man-made and/or inert camouflage.

Man-made materials are divided into three groups: hiding and screening (net sets, wire netting, snow fencing, tarpaulins, and smoke); garnishing and texturing (gravel, cinders, sawdust, fabric strips, feather, and Spanish moss); and coloring (paints, oil, and grease). Table 8-13, page 8-52, shows expedient paints that you can make in the field.

POSITION DEVELOPMENT STAGES

- Planning. Consider the unit's mission, access routes, existing concealment, and area size.
- Occupation. Carefully control traffic to avoid unnecessary movement and disruption of existing concealment. Mark trails and paths and avoid vehicle spacing less than 30 meters apart. Disperse the main congested areas (kitchen, CP, and maintenance).
- Camouflage maintenance. Inspect the area frequently and upgrade as needed. Maintain light and noise discipline to include equipment blackout. Do not create additional paths or trails.
- Evacuation. Leave the area as undisturbed as possible.

FM 5-34

Table 8-12. Center-to-center spacing for wood-supporting soil cover to defeat various contact bursts

Nominal Stringer Size (in)	Depth of Soil (d) (m)	Span Length (L) (m)				
		0.6	1.2	1.8	2.4	3
		Center-to-center stringer spacing (H), in cm				
		82-mm contact bursts				
2 x 4	0.6 0.9 1.2	7.6 46 46	10 30 36	10 20 18	10 13 10	8 8 8
2 x 6	0.6 0.9 1.2	10 46 46	18 46 46	20 41 46	20 30 28	15 20 18
4 x 4	0.6 0.9 1.2	18 46 46	25 46 46	25 46 46	22 30 25	18 20 18
4 x 8	0.5 0.6 0.9	10 36 46	13 46 46	18 46 46	20 46 46	20 46 46
		120- and 122-mm contact bursts				
4 x 8	1.2 1.5 1.8	9 30 46	10 30 46	13 30 46	13 28 41	15 25 30
6 x 6	1.2 1.5 1.8	 36 46	 36 46	14 33 46	15 30 41	15 25 30
6 x 8	1.2 1.5	14 46	15 46	20 46	23 46	25 46
8 x 8	1.2 1.5	1.2 1.5	19 46	23 46	28 46	33 46
		152-mm contact bursts				
4 x 8	1.2 1.5 1.8 2.1	 15 43 46	 15 41 46	 18 36 46	 18 30 38	9 18 25 28
6 x 6	1.5 1.8 2.1	18 46 46	20 46 46	20 38 46	20 30 38	18 25 28
6 x 8	1.2 1.5 1.8	 25 46	 30 46	 30 46	 30 46	15 30 43
8 x 8	1.2 1.5 1.8	 36 46	 38 46	 41 46	 43 46	20 41 46

NOTE: The maximum beam spacing listed in the table is 46 cm. This is to preclude further design for roof material placed over the stringers to hold the earth cover. Use a maximum of 1-inch wood or plywood over stringers to support the earth cover for 82-mm bursts; use 2-inch wood or plywood for 120-mm, 122-mm, and 152-mm bursts.

Survivability 8-49

FM 5-34

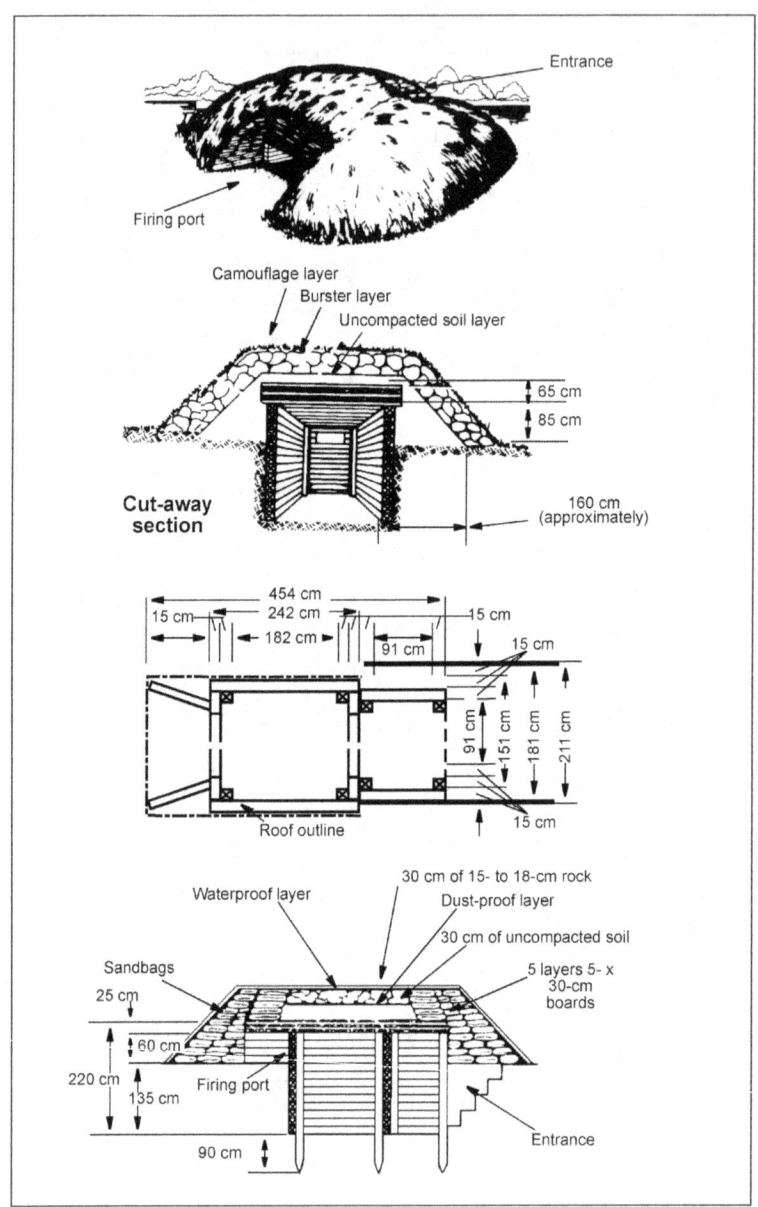

Figure 8-35. Typical bunker

8-50 Survivability

FM 5-34

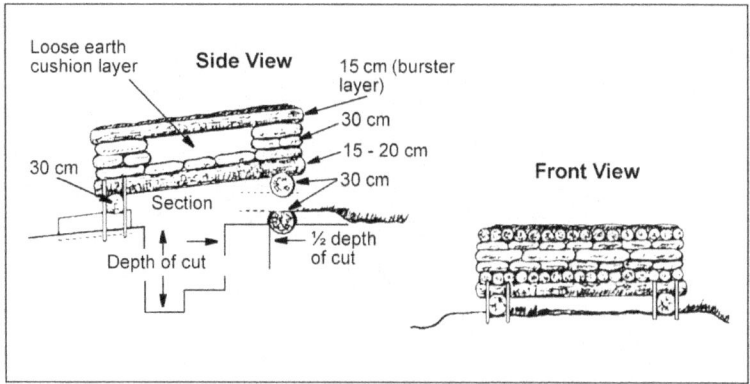

Figure 8-36. Log fighting bunker with overhead cover

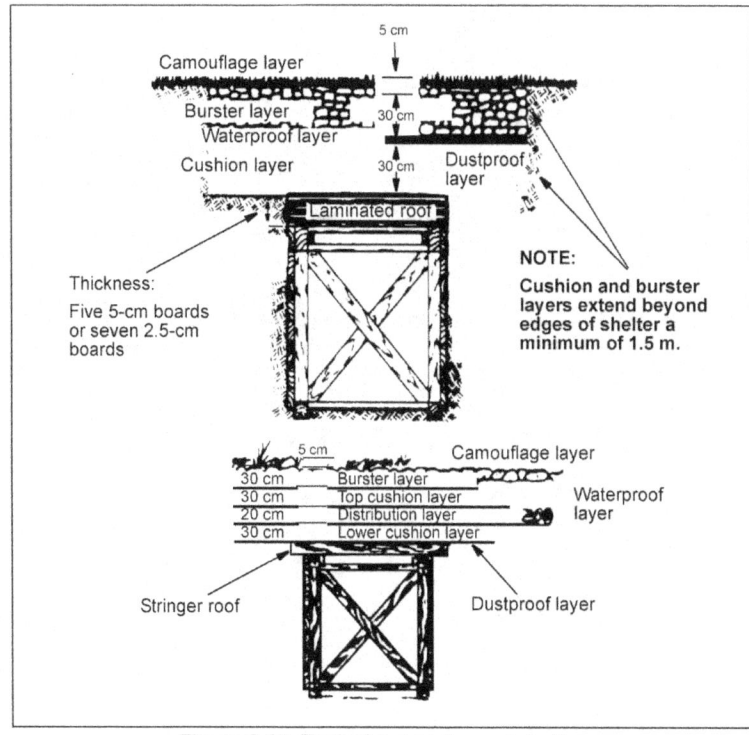

Figure 8-37. Typical cut-and-cover shelter

Survivability 8-51

FM 5-34

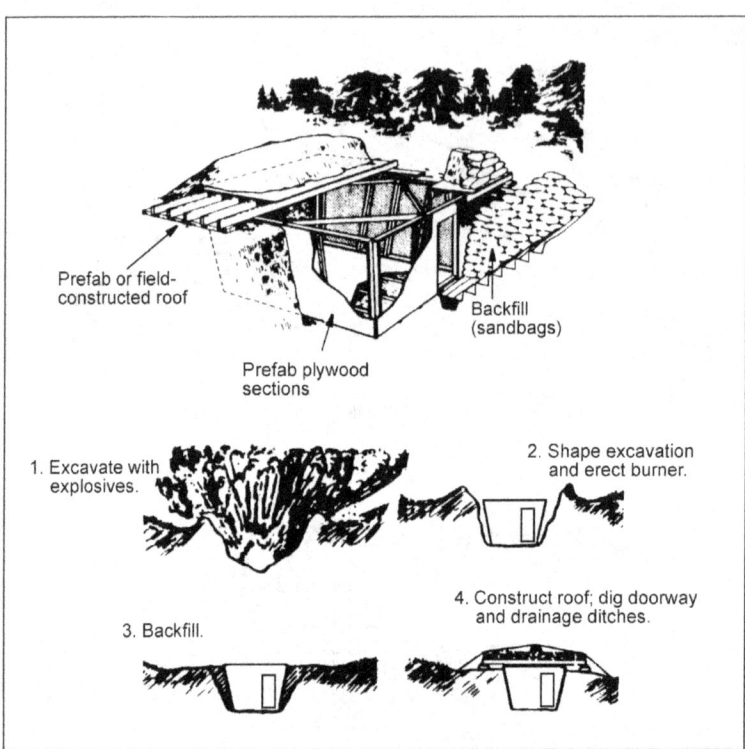

Figure 8-38. Air-transportable prefab shelter

Table 8-13. Expedient paints

Paint Materials	Mixing	Color	Finish
No. 1: local earth, GI soap, water, soot, paraffin	Mix soot with paraffin; add to 8 gal of water and 1/2-lb soap solution; stir in earth.	Dark gray	Flat, lusterless
No. 2: oil, ground clay, water, gasoline, earth	Mix 2 gal of water with 1 gal. oil and 1/4 to 1/2 gal of clay, add earth; thin with gasoline or water.	Depends on earth colors	Glossy on metal; otherwise dull
No. 3: oil, clay, GI soap water, earth	Mix 1 1/2 bars GI soap with 3 gal of water; add 1 gal of oil; stir in 1 gal of clay; add earth for color.	Depends on earth colors	Glossy on metal; otherwise dull
NOTE: You can use canned milk or powdered eggs to increase binding properties of either issue of field-expedient paints.			

8-52 Survivability

LIGHTWEIGHT CAMOUFLAGE SCREEN

Estimation

Use Figure 8-39 to determine the number of screen modules you need to camouflage vehicles and equipment.

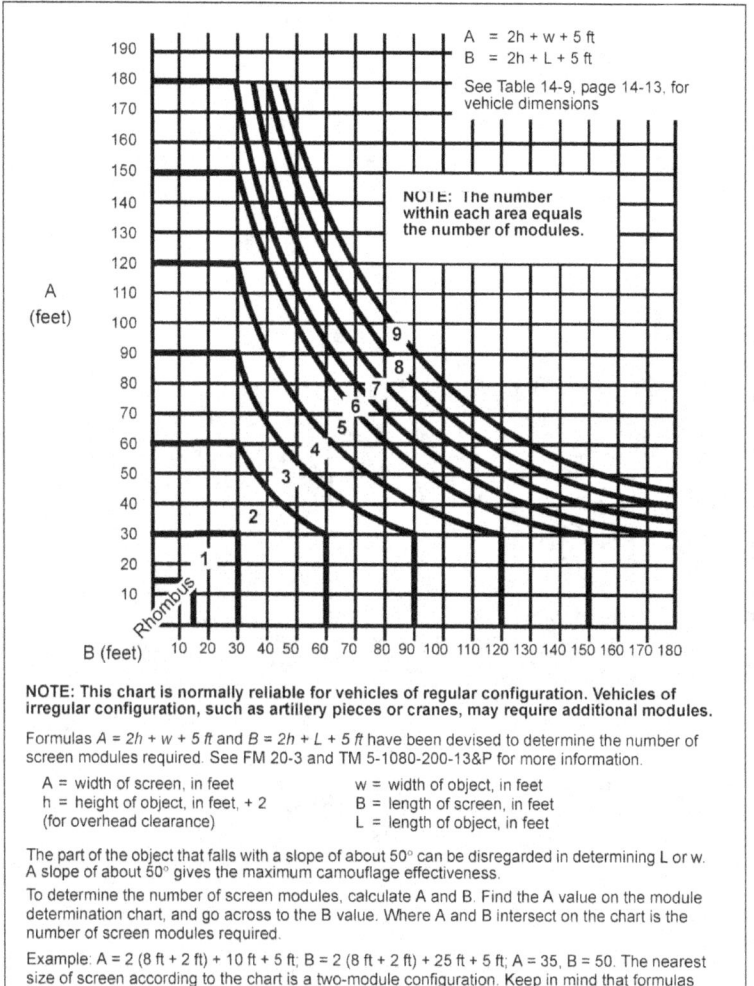

NOTE: This chart is normally reliable for vehicles of regular configuration. Vehicles of irregular configuration, such as artillery pieces or cranes, may require additional modules.

Formulas $A = 2h + w + 5\,ft$ and $B = 2h + L + 5\,ft$ have been devised to determine the number of screen modules required. See FM 20-3 and TM 5-1080-200-13&P for more information.

A = width of screen, in feet
h = height of object, in feet, + 2 (for overhead clearance)
w = width of object, in feet
B = length of screen, in feet
L = length of object, in feet

The part of the object that falls with a slope of about 50° can be disregarded in determining L or w. A slope of about 50° gives the maximum camouflage effectiveness.

To determine the number of screen modules, calculate A and B. Find the A value on the module determination chart, and go across to the B value. Where A and B intersect on the chart is the number of screen modules required.

Example: $A = 2\,(8\,ft + 2\,ft) + 10\,ft + 5\,ft$; $B = 2\,(8\,ft + 2\,ft) + 25\,ft + 5\,ft$; $A = 35$, $B = 50$. The nearest size of screen according to the chart is a two-module configuration. Keep in mind that formulas and charts are guides and that the selection of screens and combinations must be tailored to the individual situation.

Figure 8-39. Hasty module determination chart

FM 5-34

Emplacement

Assemble modules into one net (see Figure 8-40) and place it over the vehicle. Keep the screen away from all hot surfaces and exhaust systems. Ensure that the appropriate blend (color) is showing. Keep a minimum space of 0.6 meter between the net and the vehicle. Never drape a screen over the vehicles (see Figure 8-41). Always use an erection set and anchor net system.

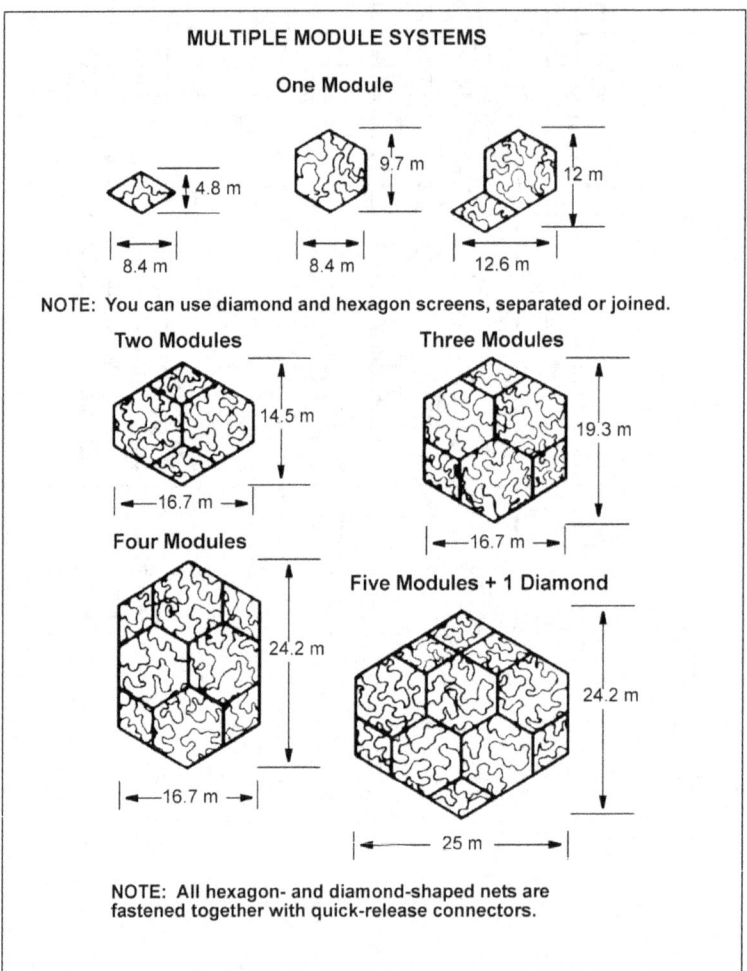

Figure 8-40. Lightweight camouflage screens

FM 5-34

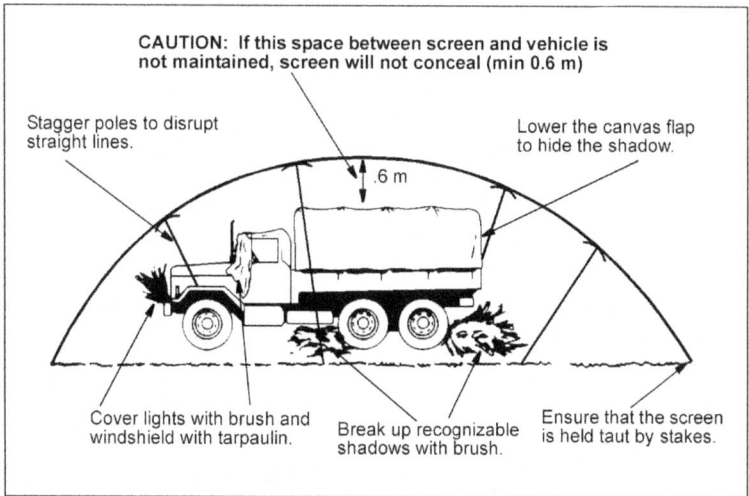

Figure 8-41. Placing net over vehicle

CHECKPOINT CONSTRUCTION

Checkpoints are established to control the movement of personnel and vehicles across a battlefield, preventing illegal actions or actions that aid the enemy. Checkpoints are either hasty (temporary, see Figure 8-42, page 8-56) or deliberate (permanent, see Figures 8-43 and 8-44, pages 8-56 and 8-57). They may also be used to—

- Ensure that classified routes carry only authorized traffic.
- Prevent a black-market transport of contraband.
- Prevent enemy sympathizers from supplying the enemy with food, medicine, ammunition, or other items of military use.

In a hasty checkpoint—

- Picket the wire at both ends to prevent run through.
- Ensure that the terrain requires vehicles and dismounted personnel to pass through the checkpoint.
- Position weapons and personnel to cover the entire checkpoint adequately.

In a one-way deliberate checkpoint—

- Picket the wire at both ends to prevent bull through.
- Replace shicane wire with any passive vehicle barrier such as jersey barriers or 55-gallon drums filled with earth.

Survivability 8-55

FM 5-34

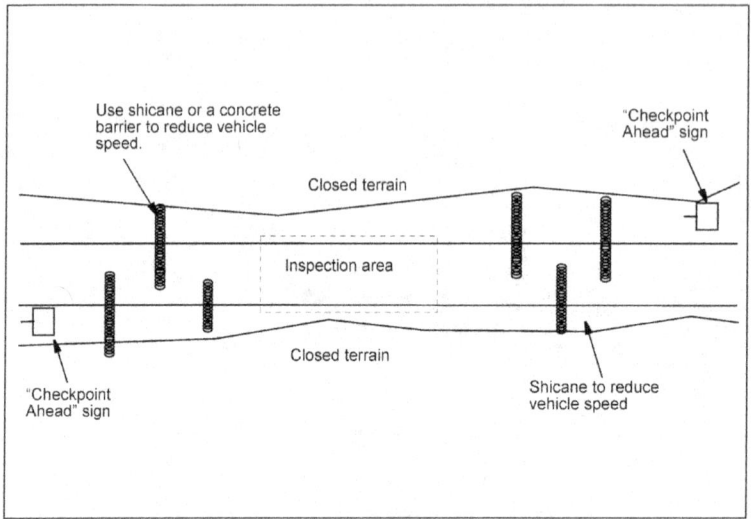

Figure 8-42. Typical hasty checkpoint

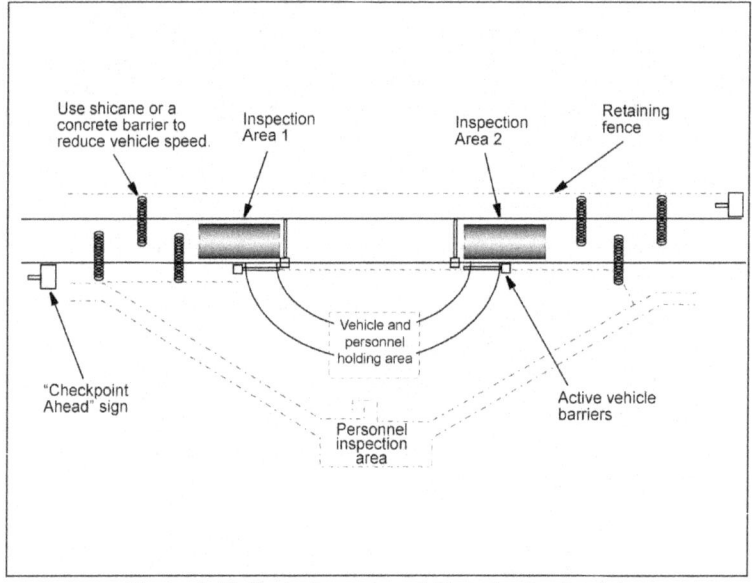

Figure 8-43. Typical one-way deliberate checkpoint

8-56 Survivability

FM 5-34

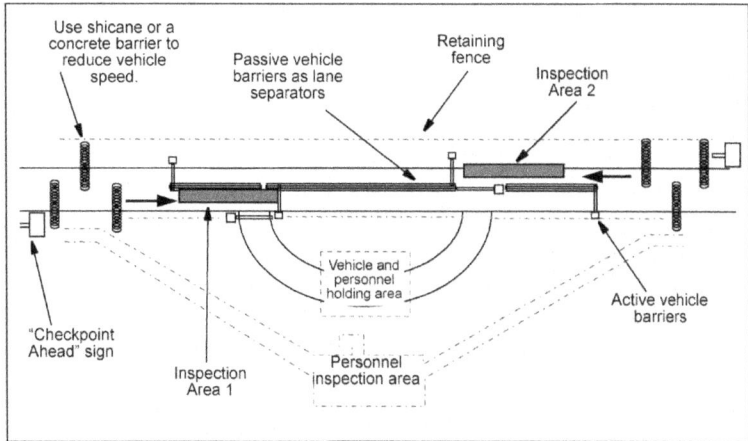

Figure 8-44. Typical two-way deliberate checkpoint

- Use a retaining fence as it prevents the vehicle's operator and personnel from escaping. The fence can be four-strand cattle fence, double lucas, or triple-standard concertina.

- Ensure that dismounted personnel are routed separately to a personnel-inspection area.

- Route vehicles which have questionable or illegal cargo that was discovered during inspection through the active vehicle barriers to a vehicle-holding area for further inspection and proper handling.

- Place weapons and personnel fighting positions so that they provide adequate coverage of an entire checkpoint.

In a two-way deliberate checkpoint—

- Use shicane wire or concrete barriers at both ends to prevent bull through. Refer to Figure 6-21, page 6-15, for concrete obstacle placement.

- Ensure that passive vehicle barriers, such as jersey barriers, used as lane separators provide protection to personnel from small-arms attack.

- Replace shicane wire with any passive vehicle barrier such as anchored 55-gallon drums filled with earth or steel post obstacles, if necessary.

- Use a retaining fence to prevent vehicle operators and personnel from escaping. The fence can be four-strand cattle fence, shicane, or triple-standard concertina.

Survivability 8-57

FM 5-34

- Route dismounted personnel separately to a personnel-inspection area.
- Route vehicles which have questionable or illegal cargo that was discovered during inspection through the active vehicle barriers to a vehicle-holding area for further inspection and proper handling.
- Place weapons and personnel fighting positions so they provide adequate coverage of an entire checkpoint.

TOWER CONSTRUCTION

Figure 8-45 and Figure 8-46, pages 8-60 and 8-61, show two different size guard towers.

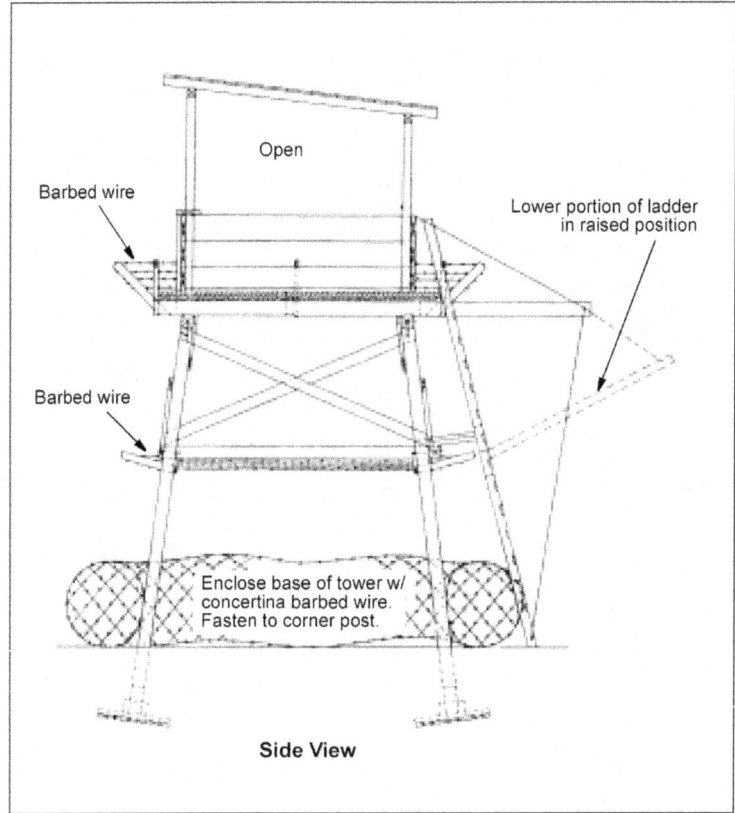

Figure 8-45. 11- x 11-foot guard tower

8-58 Survivability

FM 5-34

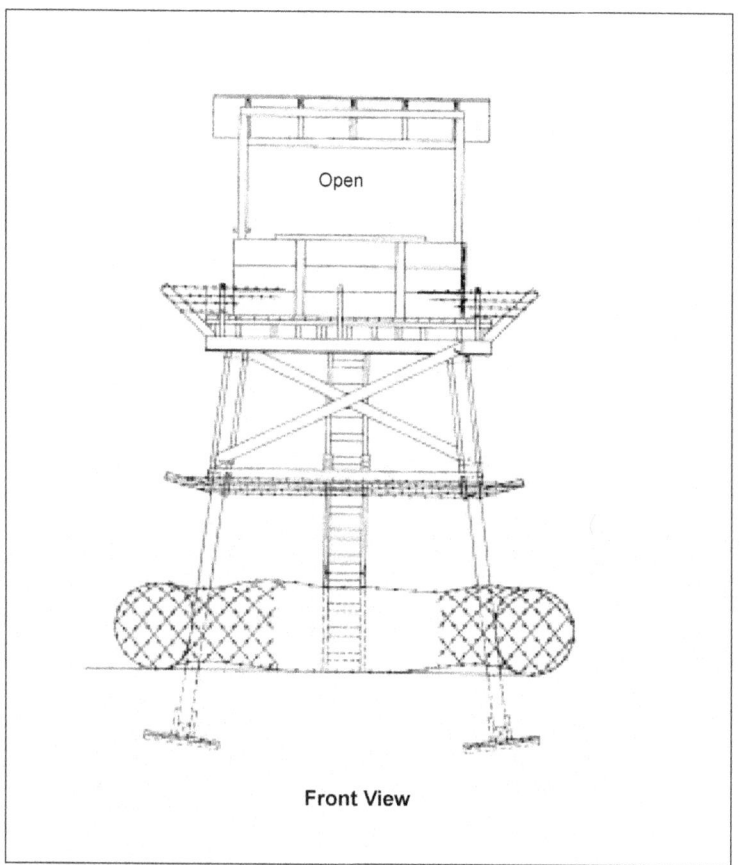

Figure 8-45. 11- x 11-foot guard tower (continued)

Survivability 8-59

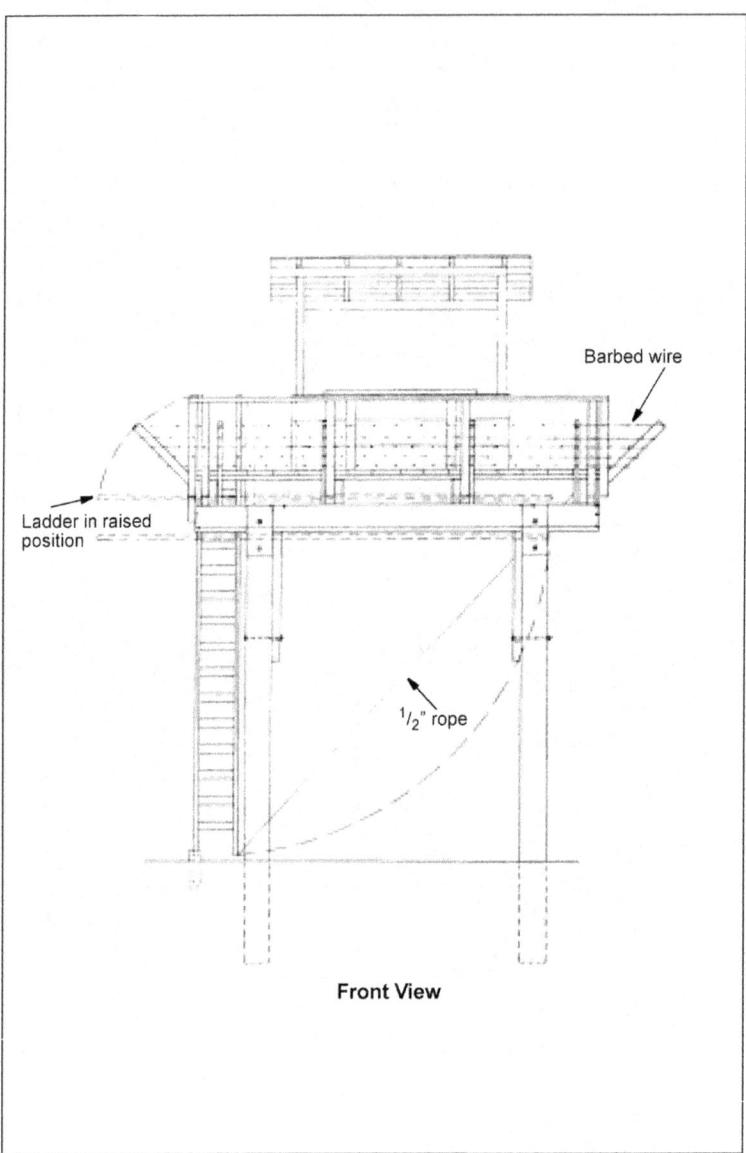

Figure 8-46. 12- x 12-foot guard tower

FM 5-34

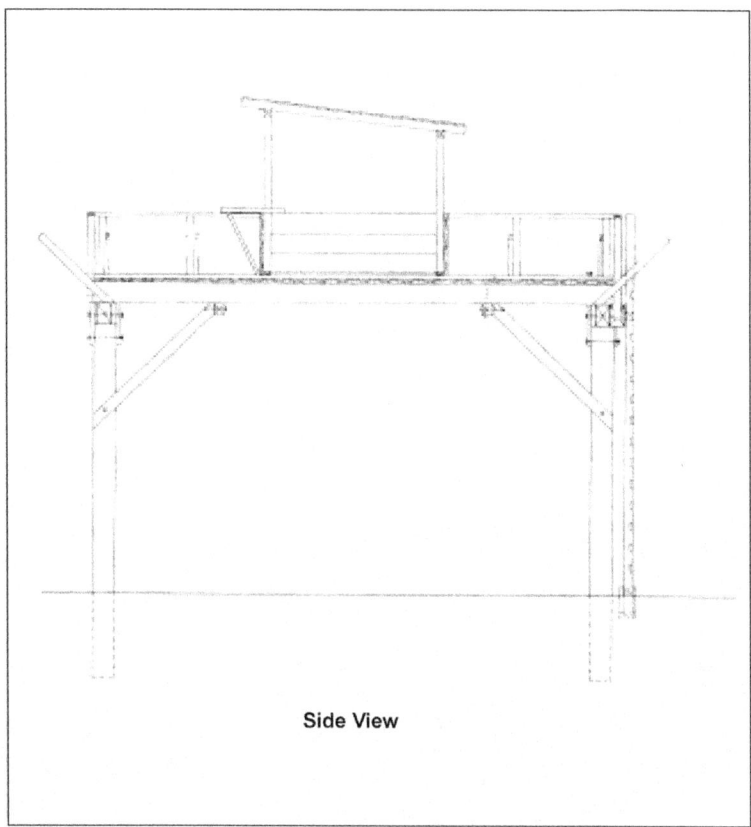

Figure 8-46. 12- x 12-foot guard tower (continued)

Survivability 8-61

Chapter 9

Demolitions and Modernized Demolition Initiators (MDI)

SECTION I. DEMOLITIONS

SAFETY CONSIDERATIONS

- Conduct risk-management operations, completing risk assessments and identifying all hazards and control measures.
- Do not attempt to conduct a demolitions mission if you are unsure of demolition procedures; review references or get assistance.
- Do not let inexperienced personnel handle explosives.
- Avoid dividing responsibility for demolition operations.
- Use the minimum number of personnel necessary to accomplish the demolition mission.
- Take your time when working with explosives; make your actions deliberate.
- Post guards to prevent access inside the danger radius.
- Maintain control of the blasting machine or initiation source.
- Use the minimum amount of explosives required to accomplish the mission while keeping sufficient explosives in reserve to handle any possible misfires.
- Maintain accurate accountability of all explosives and accessories. Always store blasting caps separately and at a safe distance from other explosives.
- Ensure that all personnel and equipment are accounted for before detonating a charge.
- Ensure that you give warnings before initiating demolitions; give the warning "Fire in the hole!" three times.
- Guard the firing points.
- Assign a competent safety officer for every demolition mission.
- Dual initiate all demolitions, regardless if they are single- or dual-primed.
- Avoid using deteriorated or damaged explosives.

- Do not dismantle or alter the contents of any explosive material.
- Avoid mixing live and inert (dummy) explosives.
- Assess the environmental impact of all demolition operations according to the environmental risk assessment (see Chapter 13).

Table 9-1 lists the minimum safe distance for personnel in the open when detonating explosives.

Table 9-1. Minimum safe distances for personnel in the open

Explosives (lb)	Safe Distance (m)	Explosives (lb)	Safe Distance (m)
27 or less	300	150	534
30	311	175	560
35	327	200	585
40	342	225	609
45	356	250	630
50	369	275	651
60	392	300	670
70	413	325	688
80	431	350	705
90	449	375	722
100	465	400	737
125	500	425	750
		500	800

NOTES:
1. For explosives over 500 pounds, use the following formula to calculate the safe distance:

Safe distance (meters) = $100 \times \sqrt[3]{\text{pounds of explosives}}$

2. The minimum safe distance for personnel in a missile-proof shelter is 91.4 m

Induced currents can prematurely detonate explosives. Table 9-2 lists the distances which transmitters with transmitter-induced currents can detonate explosives.

MISFIRES

Misfires occur for several reasons, most of which are preventable by using proper procedures. For MDI misfires, see Section II of this chapter.

NONELECTRIC-MISFIRE CLEARING PROCEDURES

- Delay investigating any detonation problem, after attempting to fire the demolition, for at least 30 minutes plus the time remaining on the secondary system. However, tactical conditions may require an investigation before the 30-minute limit.

9-2 Demolitions and Modernized Demolition Initiators (MDI)

Table 9-2. Minimum safe distance from transmitter antennas

Average or Peak Transmitter Power* (watts)	Minimum Distance to Transmitter (m)
0 to 29	30
3 to 49	50
5 to 99	110
10 to 249	160
25 to 499	230
50 to 999	305
1,000 to 2,999	480
3,000 to 4,999	610
5,000 to 19,999	915
20,000 to 49,999	1,530
50,000 to 100,000	3,050

*When the transmission is a pulsed or pulsed-continuous-wave type and its pulse width is less than 10 microseconds, the left-hand column indicates average power. For all other transmissions, including those with pulse widths greater than 10 microseconds, the left-hand column also indicates peak power.

NOTE: Do not conduct electric firing within 155 meters of energized power transmission lines. When conducting blasting operations at distances closer than 155 meters to electric power lines, use nonelectric firing systems or de-energize the power lines.

CAUTION

When transporting electric blasting caps near operating transmitters or in vehicles (including helicopters) that have operating transmitters, place the caps in a metal can. The can's cover must have a snug fit and lap over the can's body to a minimum depth of 1/2 inch (ammo can). Do not remove any caps from a container near any operating transmitter unless the hazard is deemed acceptable.

- Ensure that the soldier who placed the charges investigates and corrects any problems with the demolition.
- Do the following for above-ground misfires of charges primed with blasting caps: place a primed, 1-pound charge next to each misfired charge and detonate the new charge. Each misfired charge or charge separated from the firing circuit that contains a blasting cap requires a 1-pound charge for detonation. Do not touch the scattered charges that contain blasting caps; destroy them in-place. For the charges that are primed with detonating cord, do not investigate them until the charges have stopped burning. Wait 30-minutes if the charge is underground, reprime, and attempt to detonate the charge. You can collect scattered charges that do not contain blasting caps and detonate them together.

- Dig to within 1 foot of a buried charge; place a primed, 2-pound charge on top or to the side of the charge; and detonate the new charge.

ELECTRIC-MISFIRE CLEARING PROCEDURES

- Try to fire the charge a second time.
- Use a secondary firing system, when present.
- Check the wire connections, blasting machine, or power-source terminals.
- Disconnect the blasting machine or power source and test the blasting circuit. Check the continuity of the firing wire with a circuit tester.
- Use another blasting machine or power source and attempt to fire the charge again or change operators.
- Disconnect the blasting machine, when employing only one electrical initiation system, shunt the wires, and investigate immediately. When employing more than one electrical initiation system, wait 30 minutes before inspecting it. However, tactical conditions may require an investigation before the 30-minute limit.
- Inspect the entire circuit for wire breaks or short circuits.
- Do not attempt to remove or handle an electric blasting cap if you suspect it is the problem. Place a primed, 1-pound charge next to the misfired charge and detonate the new charge.

EXPLOSIVE CHARACTERISTICS

Table 9-3 shows the main characteristics and uses of military explosives.

WATERPROOFING

You must waterproof M1 dynamite that will be submerged in water for more than 24 hours. Seal it in plastic or dip it in pitch. Keep the composition 4 (C4) for underwater use in packages; this prevents erosion. The adhesive backing on all demolitions will not stick to surfaces when wet or submerged. Cratering charges will malfunction if the ammonium nitrate is exposed to moisture. Semipermanent waterproof sealant is available for use on the connections between time fuses or detonating cords and nonelectric blasting caps. Any demolitions should be fired as soon as possible.

Table 9-3. Military explosive characteristics

Explosive	Use	Detonation Velocity (fps)	Relative-Effectiveness Factor	Size, Weight, and Packaging
TNT	Breaching	23,000	1.00	1 lb: 48/box; 1/2 lb: 96/box; 1/4 lb: 192/box
Tetrytol	Breaching	23,000	1.20	Eight $2^1/_2$ lb/sack: 2 sacks/box
C4, M5A1, and M112	Cut and breach	26,000	1.34	M5A1: twenty-four $2^1/_2$-lb blk/box; M112: thirty $1^1/_4$-lb blk/box (1 blk 2"x 1" x 10" = 20 cu in, which may vary, depending on manufacturer data.)
M118 sheet explosive	Cutting	24,000	1.14	Four $^1/_2$-lb sheets/pack with 20 packs/box (one sheet is 3- x $^1/_4$- x 12-in)
M1 dynamite	Quarry/stump/ditching operations	20,000	0.092	One hundred $^1/_2$-lb sticks/box
Detonating cord	Priming	20,000 to 24,000		Three 1,000-ft rolls or eight 500-ft rolls/box
Crater charge	Craters	8,900	0.42	One 40-lb canister/box
M1A2, bangalore	Wire and breaching	25,600	1.17	Ten 5-ft sections/kit (176 lb)
M2A4 15-lb shaped charge	Hole cutting	25,600	1.17	Three 15-lb shaped charges/box
M3A1, 40-lb shaped charge	Hole cutting	25,600	1.17	One 40-lb shaped charge/box

PRIMING

You can prime explosives with detonating cord (see Figure 9-1, page 9-6), electrically or nonelectrically. Prime individual explosives exactly at the rear center of the charge, unless otherwise indicated.

Demolitions and Modernized Demolition Initiators (MDI) 9-5

FM 5-34

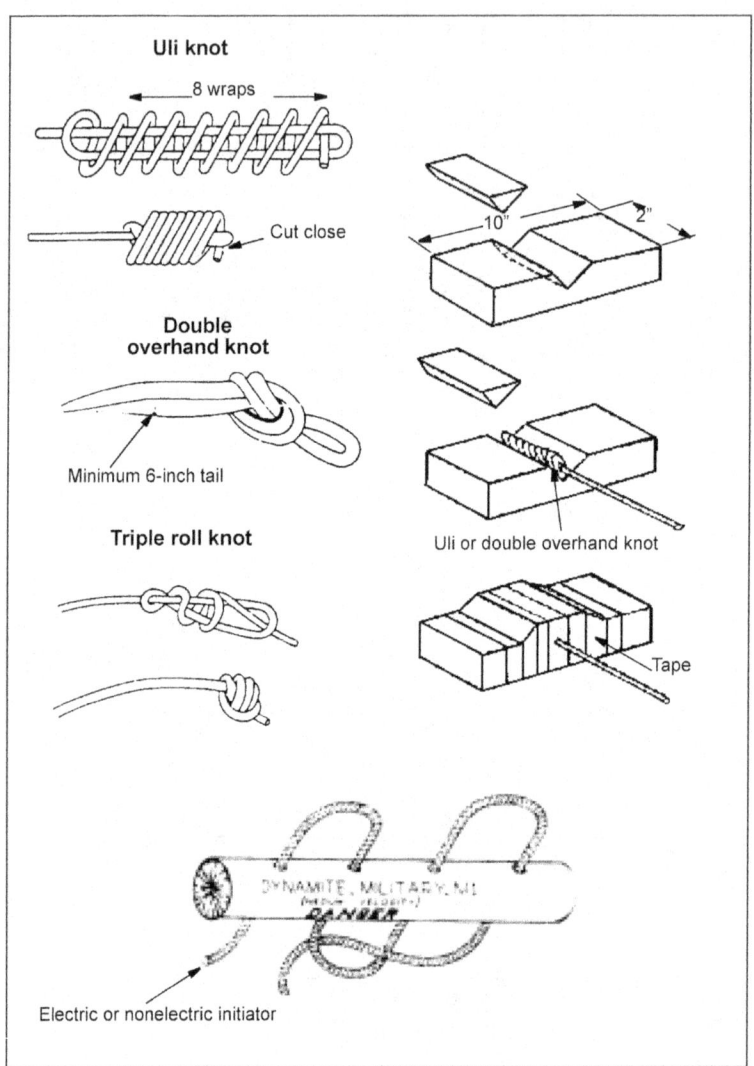

Figure 9-1. Priming with detonating cord

9-6 Demolitions and Modernized Demolition Initiators (MDI)

FM 5-34

FIRING SYSTEMS

Firing systems may be electric or nonelectric. A dual-firing system is two separate systems that may be initiated by dual electric, dual nonelectric, or a combination. See Figure 9-2 for details.

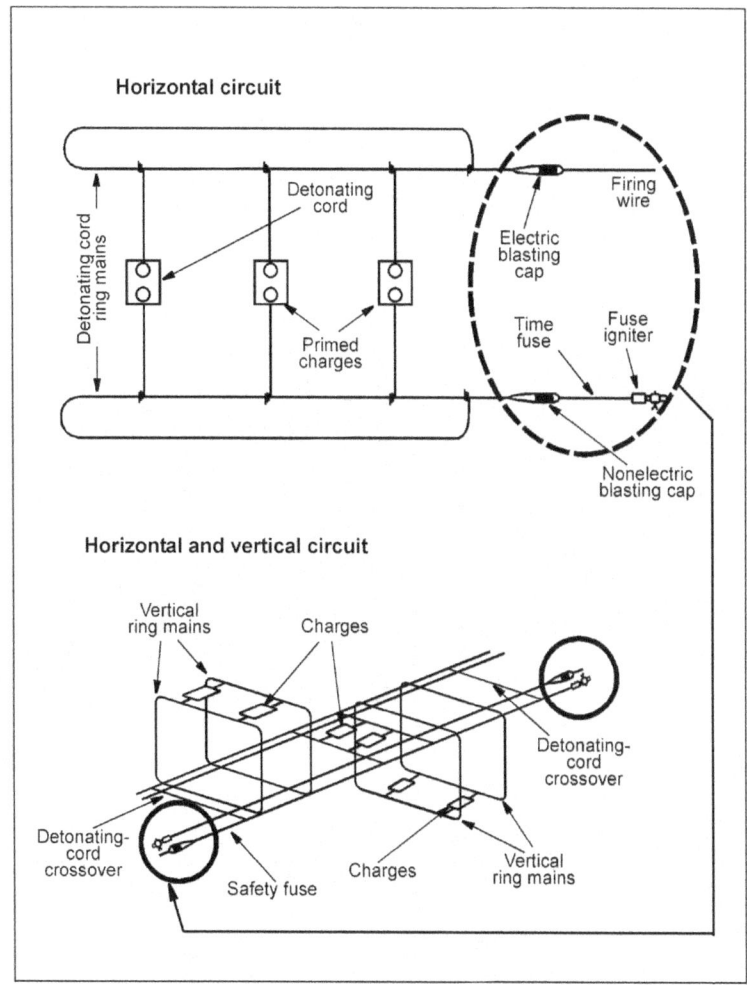

Figure 9-2. Combination dual-firing system

Demolitions and Modernized Demolition Initiators (MDI) 9-7

CHARGE CALCULATIONS

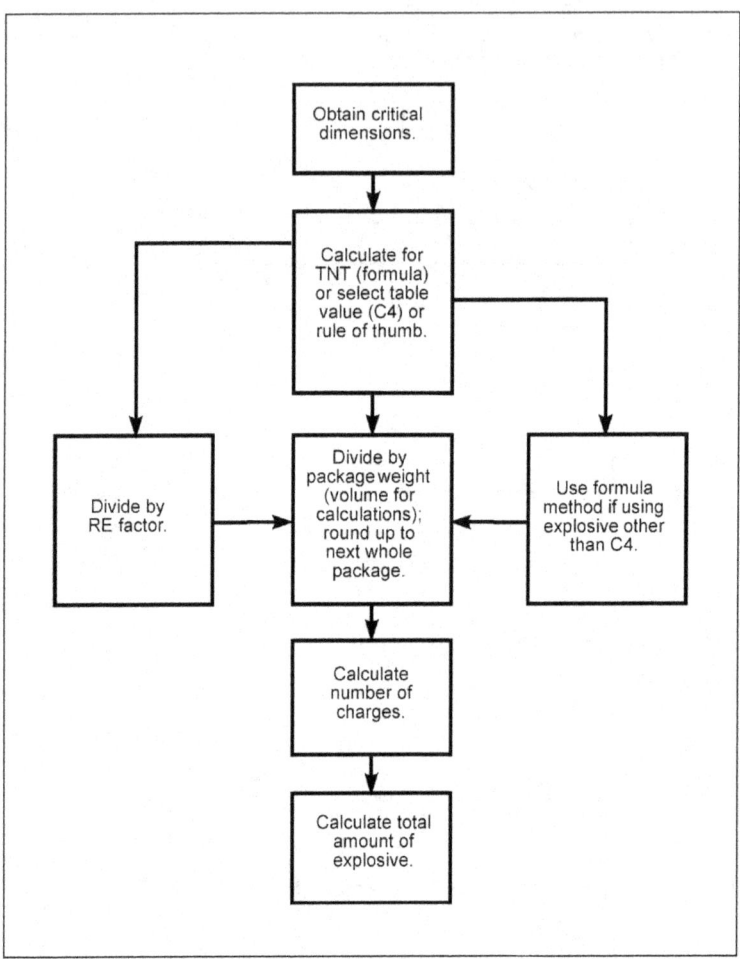

Figure 9-3. Calculation steps for explosives

STEEL-CUTTING CHARGES

Table 9-4 shows the formulas for steel-cutting charges; Table 9-5 lists the C4 requirements for rectangular steel sections.

9-8 Demolitions and Modernized Demolition Initiators (MDI)

Table 9-4. Steel-cutting formulas

Formula	Use
$P = {}^{3}/_{8}A$	Cut beams, columns, girders, steel plates, any structural steel section, bars that are 2 or more inches thick.
$P = D^2$	Cut high carbon or alloy steel (2 inches or less).
P = TNT, in pounds A = cross-sectional area of the steel member, in square inches D = thickness or diameter of section to be cut, in inches	

Table 9-5. C4 required to cut rectangular steel sections of given dimensions

Section Thickness (in)	Weight of C4 (in pounds) Required for Rectangular Steel Sections (height or width, in inches)												
	2	3	4	5	6	8	10	12	14	16	18	20	22	24
1/4	0.2	0.3	0.3	0.4	0.5	0.6	0.8	.09	1.0	1.2	1.3	1.5	1.6	1.8
3/8	0.3	0.4	0.5	0.6	0.7	0.9	1.1	1.6	1.5	1.8	2.0	2.1	2.4	2.6
1/2	0.3	0.5	0.6	0.8	0.9	1.2	1.5	1.8	2.1	2.3	2.6	2.9	3.2	3.4
5/8	0.4	0.6	0.8	0.9	1.1	1.5	1.8	2.2	2.5	2.9	3.2	3.5	3.9	4.3
3/4	0.5	0.7	0.9	1.1	1.3	1.8	2.1	2.6	3.0	3.4	3.8	4.3	4.7	5.1
7/8	0.6	0.8	1.1	1.3	1.5	2.1	2.5	3.0	3.5	4.0	4.5	5.0	5.5	5.9
1	0.6	0.9	1.2	1.5	1.8	2.3	2.9	3.4	4.0	4.5	5.1	5.6	6.2	6.8

Procedure (round UP to the nearest 1/10 pound when calculating charge sizes):
Measure each rectangular section of the total member separately.
Find the appropriate charge size for the rectangular section from the table. If the section dimension is not listed, use the next-larger dimension.
Add the individual charges for each section to get the total charge weight.

For rails (cut preferably at crossings, switches, or curves), cut at alternate rail splices for a distance of 500 feet. Use the following amounts of explosive:

- One-half pound for rails that are less than 5 inches high.
- One pound for rails that are 5 inches or higher.
- One pound for crossings and switches.
- Two pounds for frogs.

Use the following amounts of explosive for cables, chains, rods and bars:

- One pound for diameters up to 1 inch.
- Two pounds for diameters over 1 inch and up to 2 inches.

Demolitions and Modernized Demolition Initiators (MDI)

FM 5-34

- $P = (3/8)A$ or a suitable dimensional-type charge for diameters of 2 or more inches.
- One pound if you can bridge or fit a block of explosive snuggly between the links.

NOTE: Chain and cable rules are for those under tension. You must cut both sides of chain link.

Figure 9-4 shows emplacing a charge and a sample problem. Figure 9-5 lists information on special steel-cutting charges.

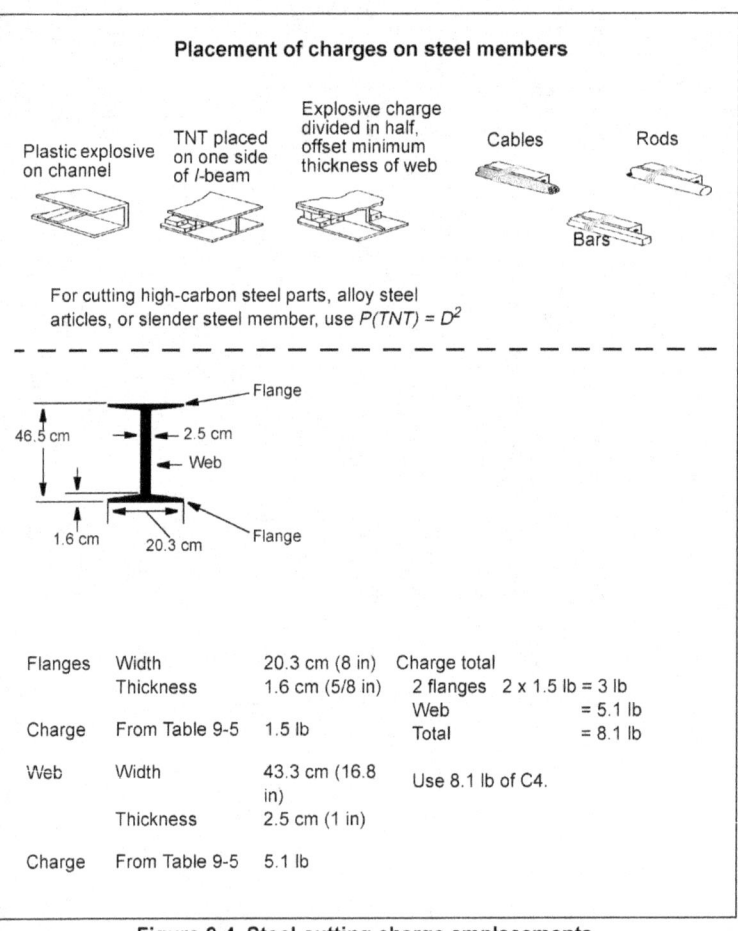

Figure 9-4. Steel-cutting charge emplacements

9-10 Demolitions and Modernized Demolition Initiators (MDI)

FM 5-34

Charge Type	Use and Dimensions	Remarks
Beams less than 2 inches thick: Offset flange charge so that one edge is opposite center of the C-shaped charges. Ribbon 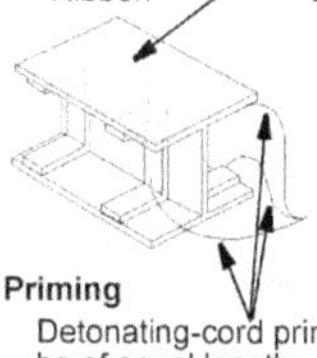 **Beams 2 inches thick or more:** Offset flange charge so that one edge is opposite an edge of the C-shaped charges. **Priming** Detonating-cord primers must be of equal length.	Cut flat steel up to 3" thick. (Plates, beams, columns) Depth: $1/2$ thickness of target Width: 3 times thickness of target Length: Same as length of cut desired	1/2" minimum charge thickness Cut explosive. DO NOT mold. Explosive target contact must exist over entire area. Darken ribbon charge lines (see FM 5-250, page 3-15).
Detonation at apex Base = $1/2$ circumference Saddle 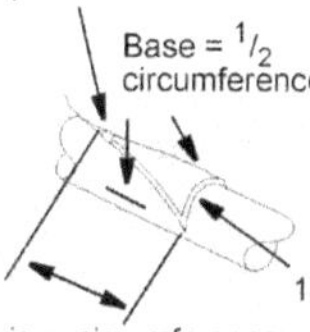 1" thick Long axis = circumference	Cut solid bars up to 8" thick (mild steel). See diagram for charge dimensions.	Explosive must be cut rather than molded. Difficult
Short axis = $1/2$ circumference and points of detonation Diamond 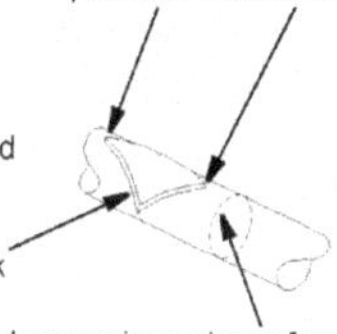 1" thick Long axis = circumference	Cut solid bars up to 8" thick (high carbon or steel alloy). See diagram for charge dimensions.	Detonating cord primers at apexes must be equal length.

Figure 9-5. Special steel-cutting charges

TIMBER-CUTTING CHARGES

Figure 9-6, page 9-12, shows charge placement, formulas, and amount of explosive for timber-cutting charges. Whenever possible, conduct a test shot to determine the exact amount of explosive required to get the desired effect. Use the values or formulas from Figure 9-6 for an initial test shot. After analyzing the initial result, increase or decrease the amount of explosive. When you do not need full removal, use the ring charges from Figure 9-6. See Figure 9-7, page 9-13, for stumping operations.

Demolitions and Modernized Demolition Initiators (MDI) 9-11

FM 5-34

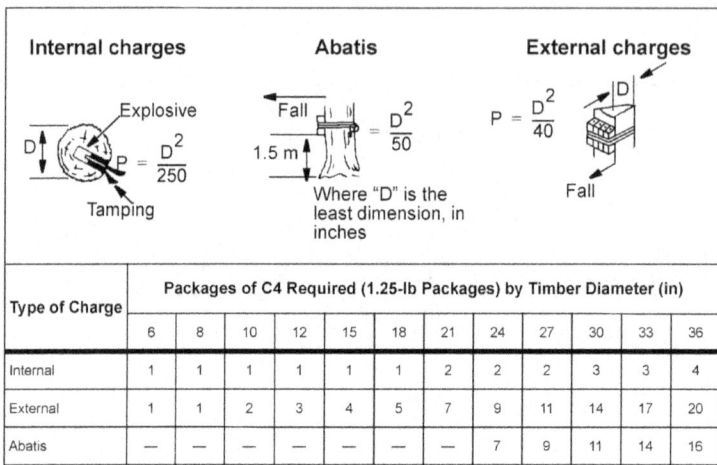

Type of Charge	Packages of C4 Required (1.25-lb Packages) by Timber Diameter (in)										
	6	8	10	12	15	18	21	24	27	30	33	36
Internal	1	1	1	1	1	1	2	2	2	3	3	4
External	1	1	2	3	4	5	7	9	11	14	17	20
Abatis	—	—	—	—	—	—	—	7	9	11	14	16

NOTES:
1. Packages required are rounded UP to the next whole package.
2. For external timber cutting, the charge should be twice as wide as it is high.
3. For internal charges, drill a 2-inch diameter to a depth of 2D/3 and use two holes drilled at right angles to each other without intersecting.

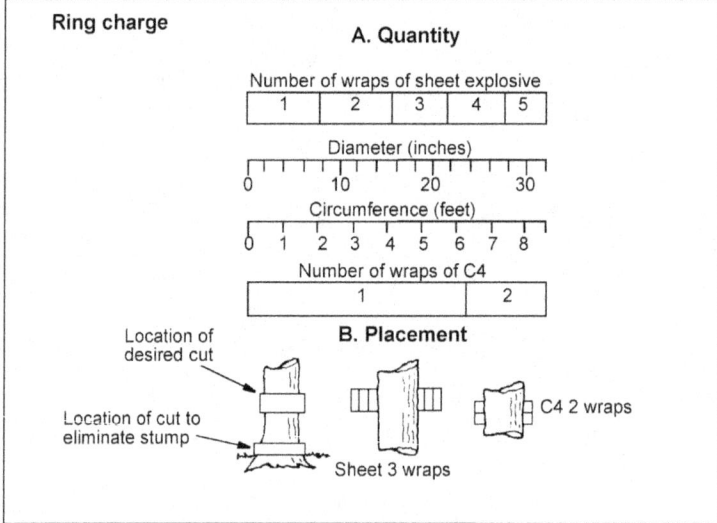

Figure 9-6. Timber-cutting charges

9-12 Demolitions and Modernized Demolition Initiators (MDI)

FM 5-34

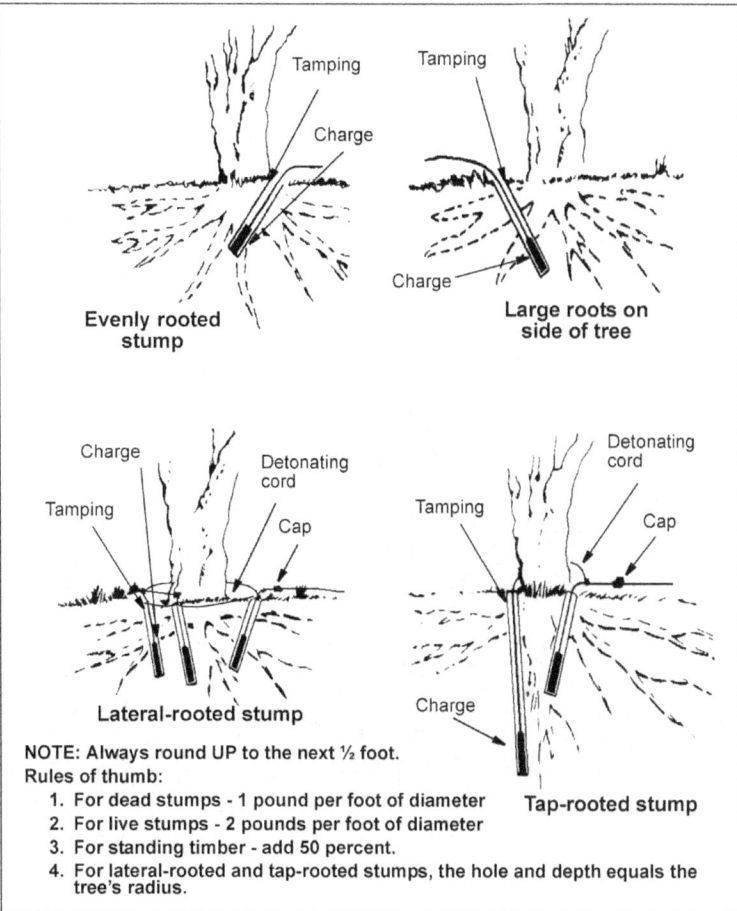

Figure 9-7. Stump-blasting charge placement

ABATIS

Figure 9-8, page 9-14, shows an abatis. Consider the following when constructing one:

- Make the depth a minimum of 75 meters.
- Use trees with a diameter of 60 centimeters and a height of 6 meters.
- Fell the trees at a 45° angle toward the enemy.

Demolitions and Modernized Demolition Initiators (MDI) 9-13

FM 5-34

- Use a test shot to determine the actual amount of demolition needed.
- Calculate the amount of trinitrotoluene (TNT) needed using the equation below:

$$P = D^2/50$$

where—

P = TNT required per tree, in pounds.

D = diameter or least dimension of dimensioned timber, in inches.

Figure 9-8. Abatis

BREACHING CHARGES

Figure 9-9 lists the quantity of explosive for reinforced concrete. Use the conversion factor from the figure to get the quantities for other materials. The breaching formula is below.

$$P = R^3 KC$$

where—

P = amount of TNT needed, in pounds

R = breaching radius, from Figure 9-10, page 9-16

K = material factor, from Table 9-6, page 9-17

C = tamping factor (see Figure 9-9).

9-14 Demolitions and Modernized Demolition Initiators (MDI)

FM 5-34

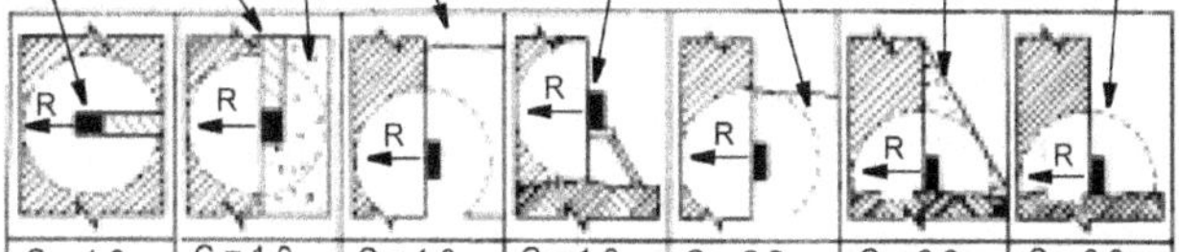

Reinforced-Concrete Thickness (ft)	Placement Methods and Tamping Factor (C) for Breaching Charges						
	Placed in center of mass C = 1.0	Tamped or stemmed Fill C = 1.0	Deep water C = 1.0	Elevated untamped C = 1.8	Shallow water C = 2.0	Earth tamping C = 2.0	Ground placed, untamped C = 3.6
	Packages of M112 (C4)						
2.0	1	5	5	9	10	10	17
2.5	2	9	9	17	18	18	33
3.0	2	13	13	24	26	26	47
3.5	4	21	21	37	41	41	74
4.0	5	31	31	56	62	62	111
4.5	7	44	44	79	88	88	157
5.0	9	48	48	85	95	95	170
5.5	12	63	63	113	126	126	226
6.0	13	82	82	147	163	163	293
6.5	17	104	104	186	207	207	372
7.0	21	111	111	200	222	222	399
7.5	26	137	137	245	273	273	490
8.0	31	166	166	298	331	331	595
Conversion Factor for Table (Material Factor [K]. Use with Table)							
Earth	Ordinary masonry, hardpan, shale, ordinary concrete, rock, good timber, and earth construction					Dense concrete first class masonry	
0.1	0.5					0.7	

Figure 9-9. Breaching-charge calculations

Demolitions and Modernized Demolition Initiators (MDI) 9-15

FM 5-34

To calculate breaching charges, do the following:
1. Determine the type of material in the object you plan to destroy. If in doubt, assume the material to be of the stronger type; for example, assume concrete to be reinforced unless you know differently.

2. Measure the thickness of the object.

3. Decide how you will place the charge against the object. Compare your method of placement with the diagrams at the top. If there is any question as to which column to use, always use the column that will give you the greater amount of C4.

4. Determine the amount of C4 that would be required if the object were made of reinforced concrete.

5. Determine the approriate conversion factor.

6. Multiply the number of pounds of C4 (see the above columns) by the conversion factor.

Example

You have a timber earth wall that is 2 meters (6.5 feet) thick with an explosive charge (without tamping) placed at its base. (If this wall was made of reinforced concrete, you would need 465 pounds of C4 to breach it.) The conversion factor is 0.5, so multiply the 465 pounds of C4 by 0.5. The result is that you will need 232.5 pounds of C4 to breach the timber earth wall.

Figure 9-9. Breaching-charge calculations (continued)

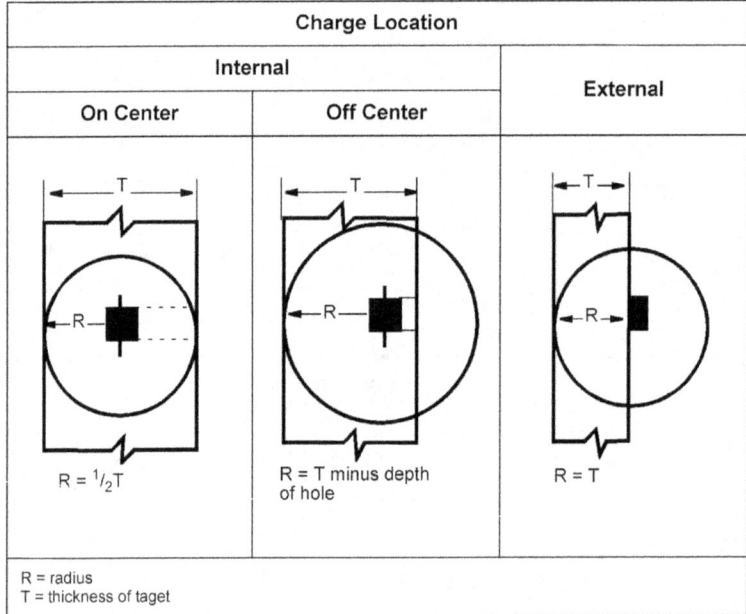

Figure 9-10. Breaching radius

9-16 Demolitions and Modernized Demolition Initiators (MDI)

Table 9-6. Values of K for breaching charges

Material	R (m)	K
Earth	All values	0.07
Poor masonry, shale, hardpan, good timber, and earth construction	Less than 1.5 1.5 or more	0.32 0.29
Good masonry, concrete block, rock	0.3 or less Over 0.3 to less than 0.9 0.9 to less than 1.5 1.5 to less than 2.1 2.1 or more	0.88 0.48 0.40 0.32 0.27
Dense concrete, first-class masonry	0.3 or less Over 0.3 to less than 0.9 0.9 to less than 1.5 1.5 to less than 2.1 2.1 or more	1.14 0.62 0.52 0.41 0.35
Reinforced concrete (concrete only; will not cut reinforcing steel)	0.3 or less Over 0.3 to less than 0.9 0.9 to less than 1.5 1.5 to less than 2.1 2.1 or more	1.76 0.96 0.80 0.63 0.54

Use the following formula and Table 9-7, page 9-18, to find the number of charges and thickness:

$$N = \frac{W}{2R}$$

where—

N = number of charges

W = pier, slab, or wall width, in feet

R = breaching radius, in feet

For R, round up to the next $1/2$ foot for external charge and to the next $1/4$ foot for internal charges.

For N, use the following rules:

- Use 1 charge if N is less than 1.25.
- Use 2 charges if N is 1.25 to 2.49.
- Round off to the next whole number if N is 2.5 or more.

For best results, place the charge as a flat, square shape with the flat side to the target. For breaching of hard surface pavements, use 1 pound of explosive for each 2 inches of surface.

FM 5-34

Table 9-7. Thickness of breaching charge

Amount of Explosive (lb)	Charge Thickness (in)
Less than 5	1
5 to less than 40	2
40 to less than 300	4
300 or more	8

NOTE: Thickness of breaching charge is approximate values when using TNT.

COUNTERFORCE CHARGES

Counterforce charges are pairs of opposing charges used to fracture small concrete or masonry blocks and columns. They are not effective against a thickness over 4 feet. Figure 9-11 shows a counterforce charge. Use the following formula to determine the amount of explosive for a counterforce charge:

$$P = 1.5 \times T$$

where—

P = amount of plastic explosive, in pounds

T = thickness, in feet (round fractional measurements UP to next higher 0.5 foot before multiplying). Divide the charge into two equal parts, place them opposite each other, and detonate simultaneously.

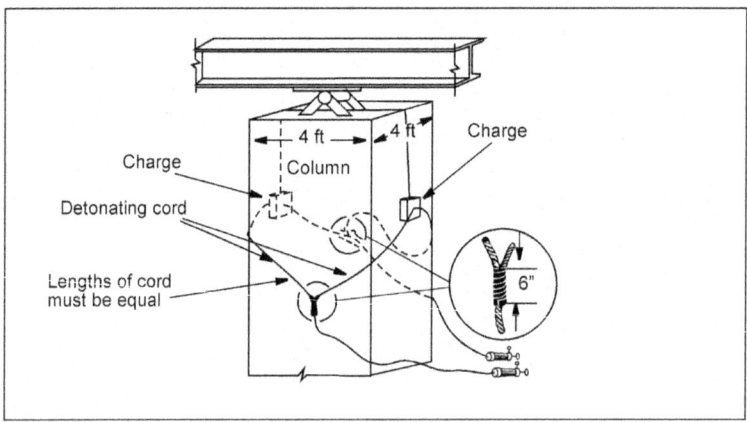

Figure 9-11. Counterforce charge

9-18 Demolitions and Modernized Demolition Initiators (MDI)

FM 5-34

BOULDER-BLASTING CHARGES

Figure 9-12 shows and explains a boulder-blasting charge.

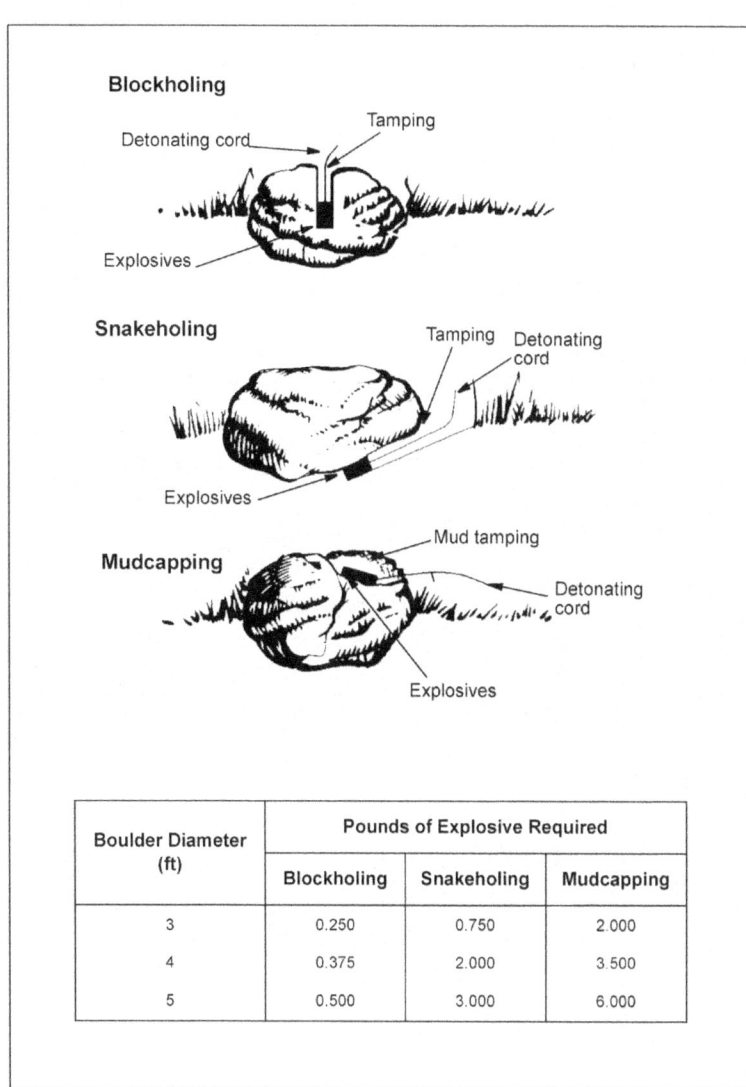

Figure 9-12. Boulder blasting

Demolitions and Modernized Demolition Initiators (MDI) 9-19

CRATERING CHARGES

The three types of craters are hasty, deliberate, and relieved-face (see Figures 9-13 through 9-15). Emplace craters by digging the holes by hand, mechanically, or with 15- or 40-pound shaped charges. Then load the holes with the required amount of explosive. (Place explosive on top of cratering charges to achieve 10 pounds of explosives per foot of hole.)

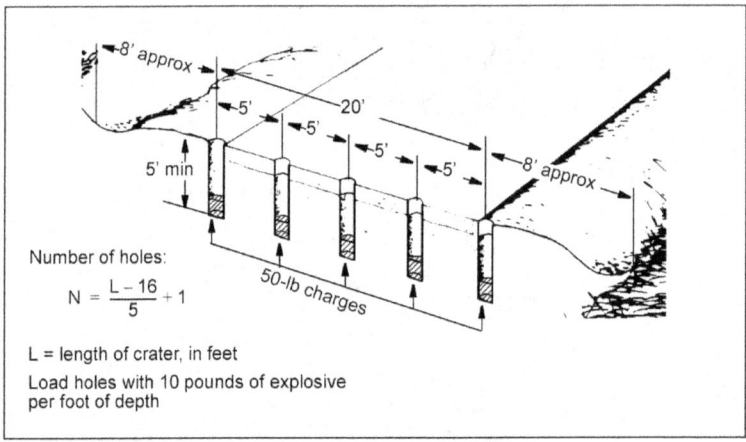

Figure 9-13. Hasty crater

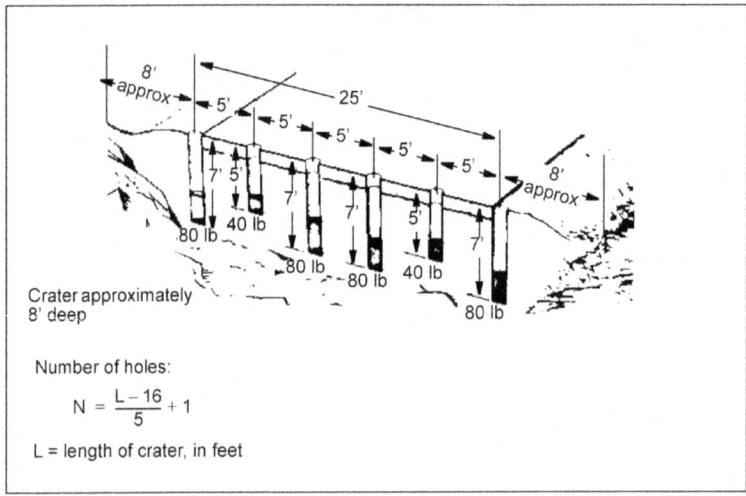

Figure 9-14. Deliberate crater

9-20 Demolitions and Modernized Demolition Initiators (MDI)

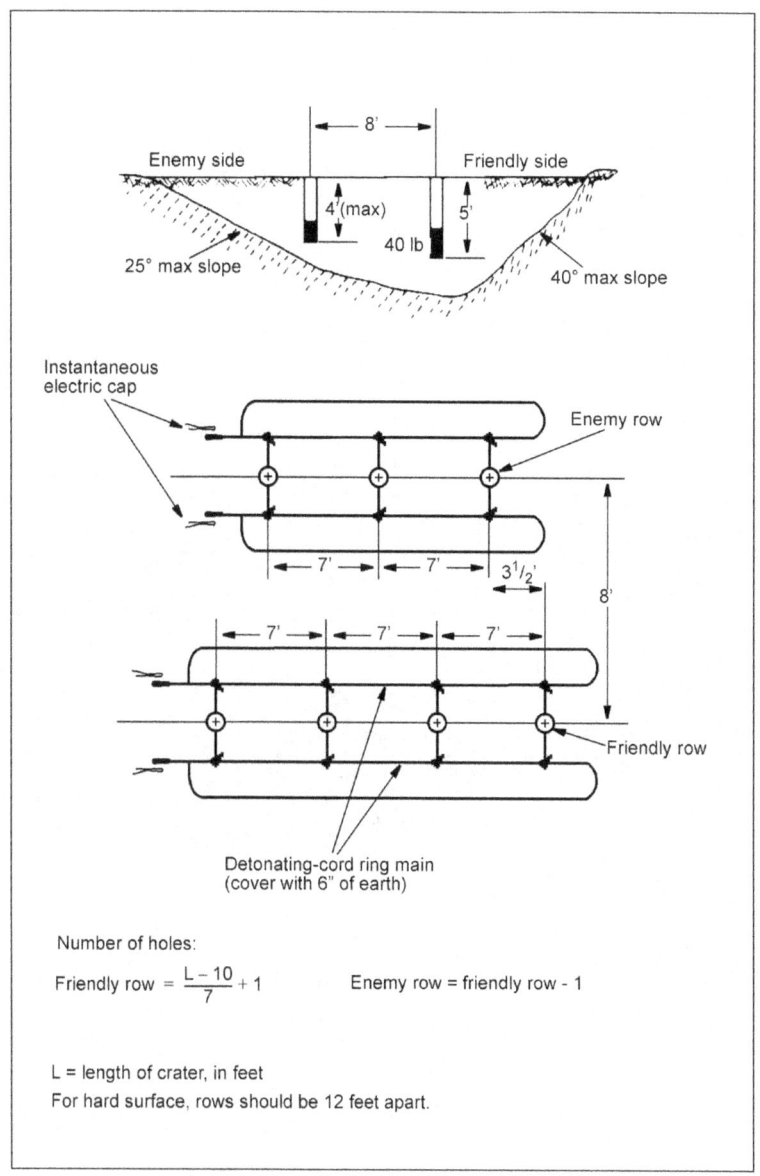

Figure 9-15. Relieved-face crater

FM 5-34

BREACHING PROCEDURES

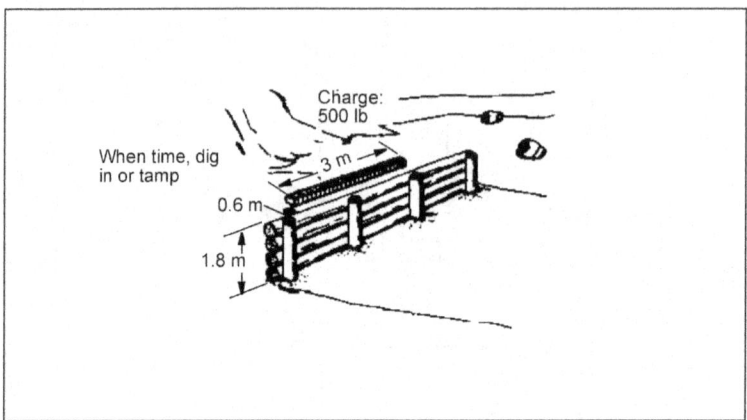

Figure 9-16. Backfilled log-wall breaching

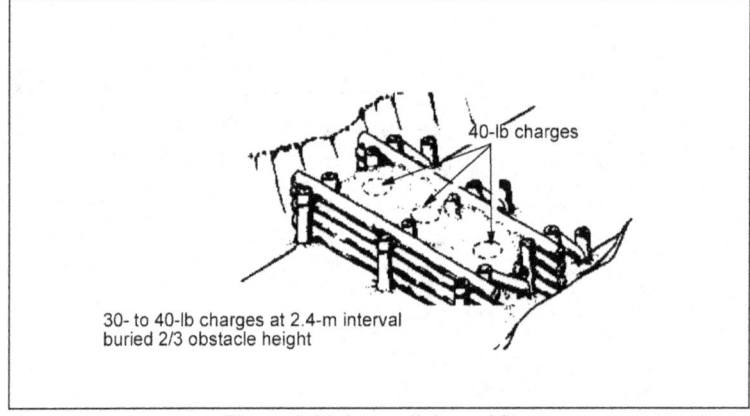

Figure 9-17. Log-crib breaching

9-22 Demolitions and Modernized Demolition Initiators (MDI)

FM 5-34

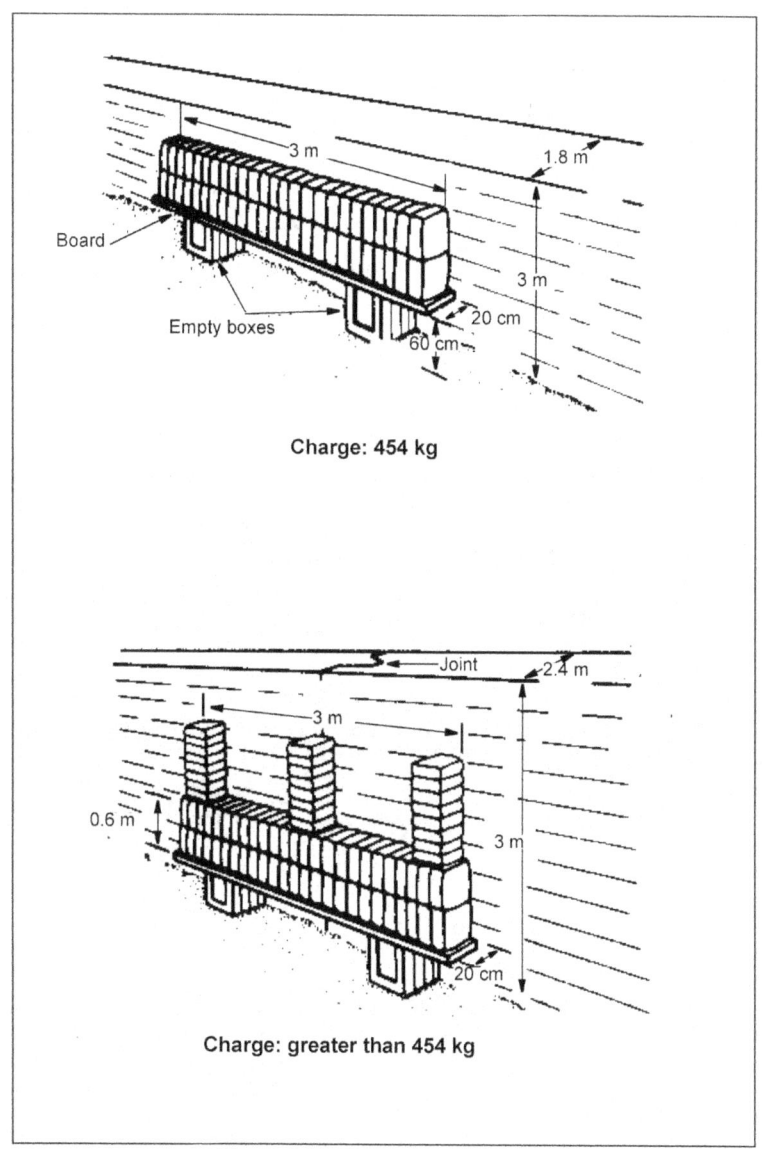

Figure 9-18. Placement of charges

Demolitions and Modernized Demolition Initiators (MDI) 9-23

FM 5-34

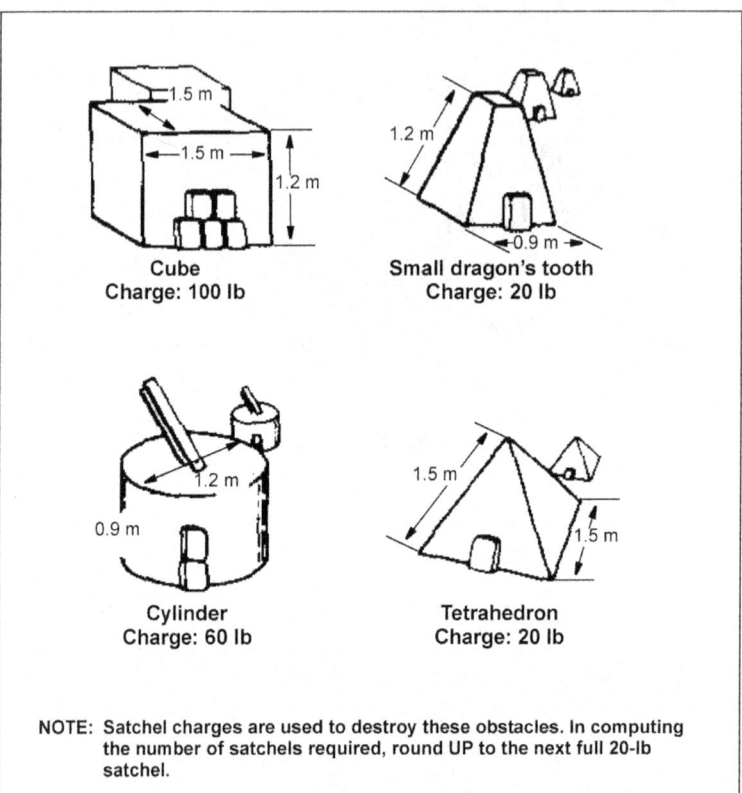

Figure 9-19. Explosive packs for destroying small concrete obstacles

9-24 Demolitions and Modernized Demolition Initiators (MDI)

BRIDGE DEMOLITIONS

When faced with unusual construction methods or materials (for example, hayricks, which are linear-shaped charges used by host NATO countries), the responsible engineer should adapt one of the recommended methods or recategorize the bridge as a miscellaneous bridge. Use Table 9-8 and Tables 9-9 and 9-10, pages 9-26 and 9-27, to determine the required clearance to prevent jamming. Use Figures 9-20 and 9-21, pages 9-28 through 9-38, for attack methods. The methods of attack shown are for the most common types of bridges; however, they are not all inclusive.

Table 9-8. Minimum E_R values for bottom attack (percent)

H/L	0.0100	0.0200	0.0300	0.0400	0.0500	0.0600	0.0700	0.0800	0.0900	0.1000
E_R/L	0.0002	0.0008	0.0020	0.0030	0.0050	0.0070	0.0100	0.0130	0.0160	0.0200
H/L	0.1100	0.1200	0.1300	0.1400	0.1500	0.1600	0.1700	0.1800	0.1900	0.2000
E_R/L	0.0240	0.0290	0.0340	0.0390	0.0440	0.0500	0.0570	0.0630	0.0700	0.0770

where—
H = beam, truss, and bow depth, in meters (includes the deck)
L = length of span for attack measured from end to end of the longitudinal memebers which support the deck, in meters
E_R = required end clearance, in meters

NOTES:
1. Go UP to the next higher value if the result H/L is not on the chart exactly as calculated. For example, if H/L = 0.076, use the column headed 0.08. Read down that column to determine E_R/L. In this case, E_R/L = 0.0130
2. Multiply the E_R/L value determined from the chart by L to get E_R.

FM 5-34

Table 9-9. Minimum L_c values for top attack (midspan)

H/L	Ratio of Section Removed to Span Length (L_s/L)															
	0.004	0.006	0.008	0.010	0.012	0.014	0.016	0.018	0.020	0.030	0.040	0.050	0.060	0.080	0.100
0.01	0.003	0.003	0.004	0.004	0.005	0.005	0.005	0.006	0.006	0.007	0.009	0.010	0.011	0.013	0.015
0.02	0.005	0.006	0.007	0.008	0.009	0.010	0.011	0.011	0.012	0.015	0.017	0.019	0.022	0.026	0.030
0.03	0.008	0.009	0.011	0.012	0.014	0.015	0.016	0.017	0.018	0.022	0.026	0.029	0.033	0.039	0.045
0.04	0.011	0.013	0.015	0.016	0.018	0.019	0.021	0.022	0.023	0.029	0.034	0.039	0.043	0.052	0.060
0.05	0.013	0.016	0.018	0.020	0.022	0.024	0.026	0.028	0.029	0.036	0.043	0.049	0.054	0.065	0.075
0.06	0.015	0.019	0.022	0.025	0.027	0.029	0.031	0.033	0.035	0.044	0.051	0.058	0.065	0.078	0.090
0.07	0.018	0.022	0.026	0.029	0.031	0.034	0.036	0.039	0.041	0.051	0.060	0.068	0.076	0.091	0.105
0.08	0.021	0.025	0.029	0.033	0.036	0.039	0.042	0.044	0.047	0.058	0.068	0.078	0.087	0.104	0.120
0.09	0.023	0.028	0.033	0.037	0.040	0.044	0.047	0.050	0.053	0.065	0.077	0.087	0.097	0.116	0.135
0.10	0.026	0.032	0.036	0.041	0.045	0.049	0.052	0.055	0.058	0.073	0.085	0.097	0.108	0.129	0.150
0.11	0.028	0.035	0.040	0.045	0.049	0.053	0.057	0.061	0.064	0.080	0.094	0.107	0.119	0.142	0.165
0.12	0.031	0.038	0.044	0.049	0.054	0.058	0.062	0.066	0.070	0.087	0.102	0.116	0.130	0.155	0.180
0.13	0.033	0.041	0.047	0.053	0.058	0.063	0.067	0.072	0.076	0.095	0.111	0.126	0.140	0.168	0.195
0.14	0.036	0.044	0.051	0.057	0.063	0.068	0.073	0.077	0.082	0.102	0.119	0.136	0.151	0.181	0.210
0.15	0.038	0.047	0.054	0.061	0.067	0.073	0.078	0.083	0.088	0.109	0.128	0.145	0.162	0.194	0.225
0.16	0.041	0.050	0.058	0.065	0.072	0.078	0.083	0.088	0.093	0.116	0.136	0.155	0.173	0.207	0.240
0.17	0.043	0.053	0.062	0.069	0.076	0.082	0.088	0.094	0.099	0.124	0.145	0.165	0.184	0.220	0.255
0.18	0.046	0.056	0.065	0.073	0.080	0.087	0.093	0.099	0.105	0.131	0.154	0.175	0.194	0.23	0.270
0.19	0.019	0.060	0.069	0.077	0.085	0.092	0.099	0.105	0.111	0.138	0.162	0.184	0.205	0.246	0.285
0.20	0.051	0.063	0.073	0.081	0.089	0.097	0.104	0.110	0.117	0.145	0.171	0.194	0.216	0.259	0.300

NOTE: If the results of L_s/L or H/L are not on the chart exactly as you calculate, round UP to the next higher value on the chart. For example, if H/L = 0.021, use 0.03; if L_s/L = 0.0142, use 0.016. Intersect the L_s/L and H/L values on the chart to get the value of L_c/L. Multiply the L_c/L value by L to get L_c.

9-26 Demolitions and Modernized Demolition Initiators (MDI)

Table 9-10. Minimum L_C values for arch and portal with pinned-footing bridge attacks

$\dfrac{H}{L}$	0.040	0.060	0.080	0.100	0.120	0.140	0.160	0.180	0.200
$\dfrac{L_C}{L}$	0.003	0.007	0.013	0.020	0.030	0.040	0.053	0.067	0.083
$\dfrac{H}{L}$	0.220	0.240	0.260	0.280	0.300	0.320	0.340	0.360	
$\dfrac{L_C}{L}$	0.100	0.130	0.150	0.170	0.200	0.230	0.270	0.300	

where—
H = rise for arch or portal bridges; measure the rise (meters) from the springing or bottom of the support leg to the deck or top of the arch, whichever is greater.
L = length of span for attack between the centerlines of the bearings, in meters
L_C = required length of the span removed, in meters

NOTE: If the result of H/L is not on the chart exactly as calculated, go UP to the next higher value on the chart. For example, if H/L = 0.089, use the column headed 0.100 to determine L_C/L. In the case, L_C/L = 0.020. Multiply the L_C/L value by L to get L_C; for example, 0.020 x L = L_C.

FM 5-34

Serial	Sub-category	Type	Attack Method	Remarks
a	b	c	d	e
1	Steel beam	Through bridge, Method I	Top attack: 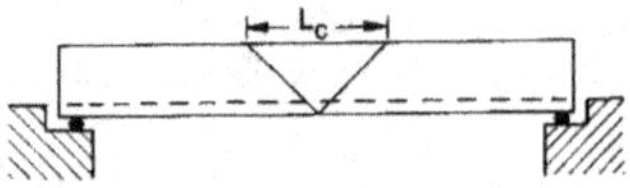 1. Cut at the midspan. 2. Cut beams, including bottom, flange in a "V." 3. Do not consider cutting the deck.	None
2		Through bridge, Method II	Bottom attack: E is greater than E_R 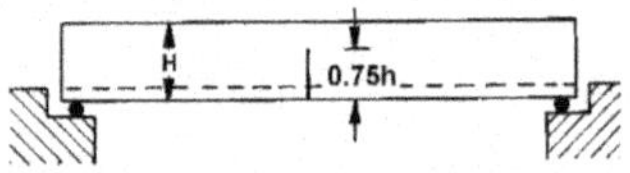 1. Cut at the midspan to 0.75h, as shown. 2. Cut the deck across the full bridge width.	None
3		Through bridge, Method III	Angled attack: 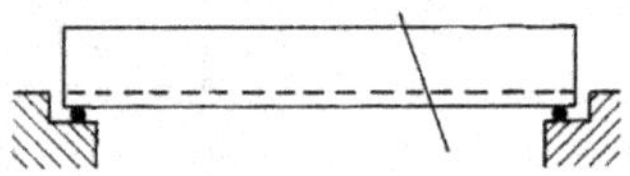 1. Cut between 1/3 span and the mid-span. 2. Cut the deck across the full bridge width.	End clearance is not a consideration.
4		Through bridge, Method IV	Bottom attack: E is less than E_R 1. Cut at the midspan to 0.75h. 2. Cut the deck across the full bridge width. 3. Attack one abutment or pier to create sufficient end clearance.	None
5		Through bridge, Method V	Top attack: 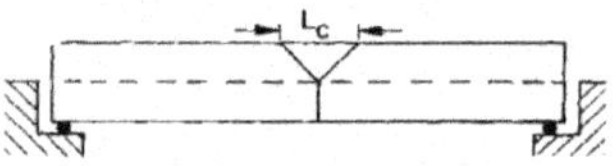 1. Cut at the midspan. 2. Cut the bridge as shown where the deck is located well above the beam bottom. 3. Do not consider cutting the deck.	None

Figure 9-20. Methods of attack on simply supported bridges

Serial	Sub-category	Type	Attack Method	Remarks
a	b	c	d	e
6	Steel beam	Deck bridge, top support	Angled attack:  1. Cut between 1/3-span and the midspan. 2. Cut the deck across the full bridge width.	1. Configuration is found in cantilever and suspended-span bridges. 2. End clearance is not a consideration.
7		Deck bridge, bottom support, Method I	Bottom attack: E is greater than E_R 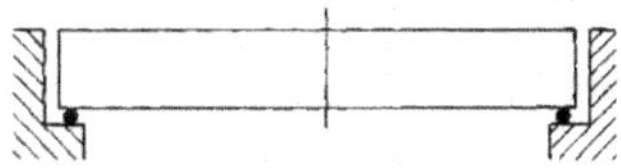 1. Cut at the midspan. 2. Do not consider cutting the deck.	None
8		Deck bridge, bottom support, Method II	Bottom attack: E is less than E_R 1. Cut at midspan. 2. Do not consider cutting deck. 3. Attack one abutment or pier to create sufficient end clearance.	None
9		Deck bridge, bottom support, Method III	Angled attack:  1. Cut between 1/3-span and the midspan. 2. Cut the deck across the full bridge width.	End clearance is not a consideration.

Figure 9-20. Methods of attack on simply supported bridges (continued)

Demolitions and Modernized Demolition Initiators (MDI) 9-29

FM 5-34

Serial	Sub-category	Type	Attack Method	Remarks
a	b	c	d	e
10	Steel truss	Through bridge, Method I	Top attack: 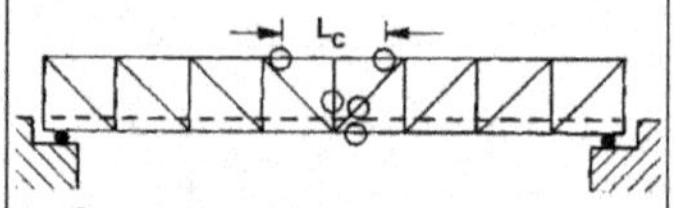 1. Cut at the midspan. 2. Cut the top chord twice, vertically (if necessary), and diagonals and bottom chord. 3. Remove the wind bracing over the midspan. 4. Do not consider cutting the deck.	None
11		Through bridge, Method II	Angled attack: 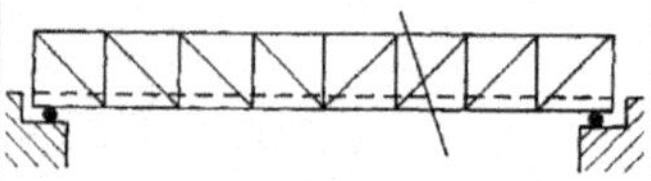 1. Cut between 1/3 span and the midspan. 2. Cut top chord, diagonals, and bottom chord in one bay only. 3. Cut the deck across the full bridge width.	None
12		Deck bridge, top support	Bottom attack: 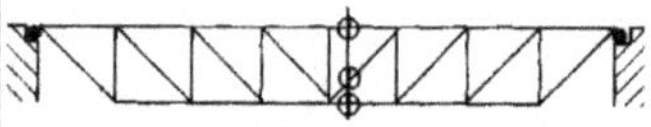 1. Cut between 1/3 span and the midspan. 2. Cut the top chord, diagonals, and bottom chord in one bay only. 3. Do not consider cutting the deck.	1. Configuration is found in cantilever and suspended-span bridges. 2. End clearance is not a consideration.
13		Deck bridge, bottom support, Method I	Bottom attack: E is greater than E_R 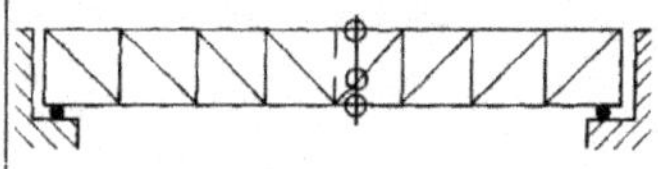 1. Cut at the midspan. 2. Cut top chord, diagonals, and bottom chord in one bay only. 3. Do not consider cutting the deck.	None

Figure 9-20. Methods of attack on simply supported bridges (continued)

9-30 Demolitions and Modernized Demolition Initiators (MDI)

Serial	Sub-category	Type	Attack Method	Remarks
a	b	c	d	e
14	Steel truss	Deck bridge, bottom support, Method II	Bottom attack: E is less than E_R 1. Cut at the midspan. 2. Cut top chord, diagonals, and bottom chord in one bay only. 3. Do not consider cutting the deck. 4. Attack one abutment or pier to create sufficient end clearance.	None
15		Deck bridge, bottom support, Method III	Angled attack: 1. Cut between 1/3 span and the midspan. 2. Cut the deck across the full bridge width.	End clearance is not a consideration.
16		Through bridge	Bottom attack: 1. Cut at the midspan. 2. Cut the deck across the full bridge width.	This method applies to slab bridges only.
17	Concrete	Deck bridge, top support	Top attack: Cut at the midspan with a concrete-stripping charge.	1. Configuration is found in cantilever and suspended-span bridges. 2. Remove concrete for L_c distance to full width and depth of beams.
18		Deck bridge, bottom support, Method I	Bottom attack: E is greater than E_R Cut at the midspan with hayricks.*	1. This method applies to slab bridges only. 2. Sufficient reinforcing bars are cut to cause bridge collapse.

*Hayricks are not in the US Army supply system.

Figure 9-20. Methods of attack on simply supported bridges (continued)

FM 5-34

Serial	Sub-category	Type	Attack Method	Remarks
a	b	c	d	e
19	Concrete	Deck bridge, bottom support, Method II	Bottom attack: E is less than E_R 1. Cut at the midspan with hayricks.* 2. Attack one abutment or pier to create sufficient end clearance.	This method applies to slab bridges only.
20		Deck bridge, bottom support, Method III	Top attack: E is less than E_R 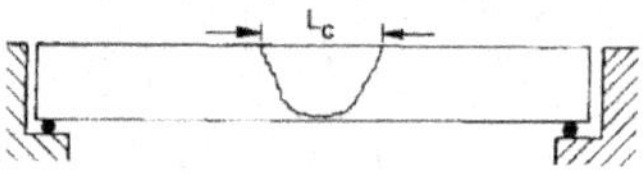 Cut at the midspan with a concrete-stripping charge.	Remove concrete for L_c distance to full width and depth of beams.
21	Bow-string	Normal	Top attack: 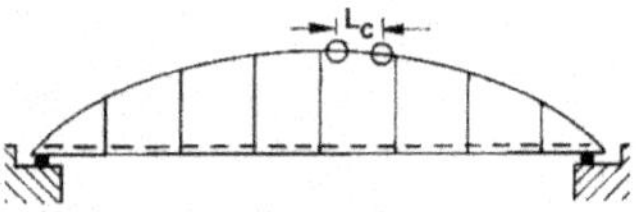 1. Cut at the midspan. 2. Cut the bow in two places. 3. Cut all hangers between the bow cuts. 4. Do not consider cutting the deck.	None
22		Reinforced beam or truss	Top attack, plus girders: 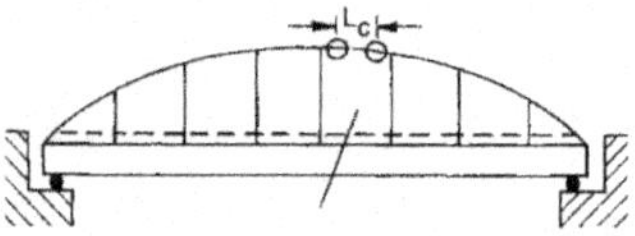 1. Cut the truss or beam with the appropriate method (Serials 1 through 15). 2. Cut the bow in two places, including the hangers.	None

* Hayricks are not in the US Army supply system.

Figure 9-20. Methods of attack on simply supported bridges (continued)

9-32 Demolitions and Modernized Demolition Initiators (MDI)

Serial	Sub-category	Type	Attack Method	Remarks
a	b	c	d	e
1	Concrete	Cantilever	Two cuts: 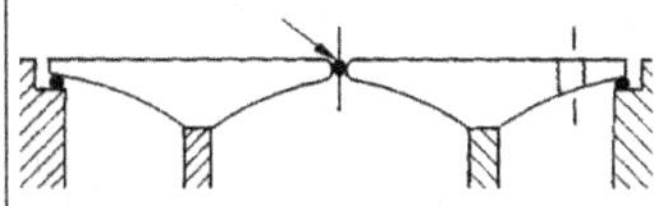 1. Cut the anchor span as closely to the pier as practical. 2. Cut the midspan shear joint.	1. Cutting the anchor span may require a two-stage attack. 2. Use a concrete-stripping charge for the first stage.
2		Cantilever and suspended span	One cut: 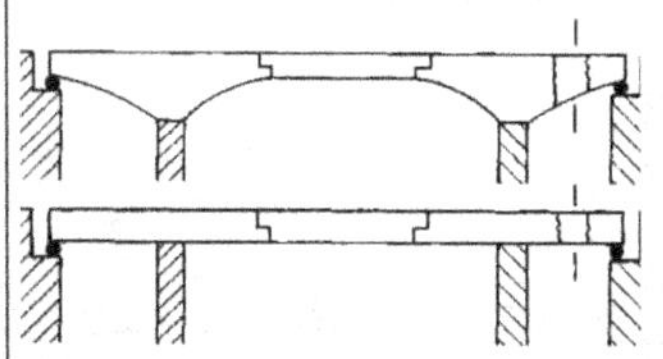 Cut the anchor as closely to the pier as practical.	1. Cutting the anchor span may require a two-stage attack. 2. Use a concrete-stripping charge for the first stage. 3. If demolition of the suspended span will create the desired obstacle, regard the span as simply supported and attack accordingly.
3		Beam or truss with short side span	One cut: 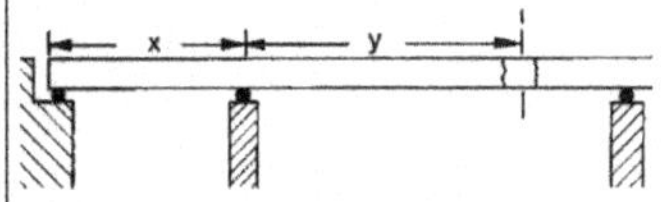 1. Cut interior span so y is greater than $1.25x$. 2. If necessary, cut other interior spans as in Serial.	1. Cutting longer spans may require a two-stage attack. 2. Use a concrete-stripping charge for the first stage.
4		Beam or truss without short side span	Two or more cuts: 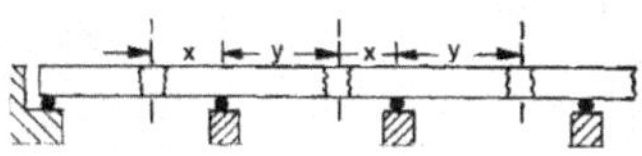 Cut the interior span so y is greater than $1.25x$.	1. Cutting these spans may require a two-stage attack. 2. Use a concrete-stripping charge for the first stage.

Figure 9-21. Methods of attack on continuous bridges

Demolitions and Modernized Demolition Initiators (MDI)

FM 5-34

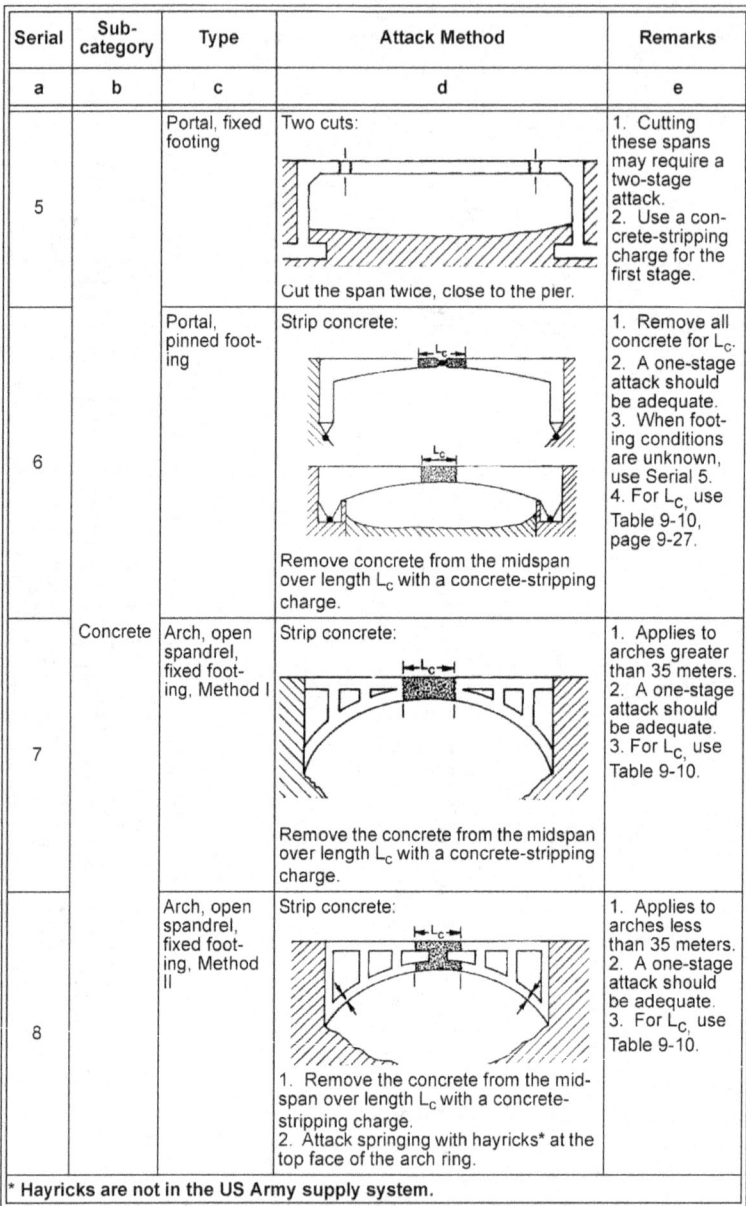

Serial	Sub-category	Type	Attack Method	Remarks
a	b	c	d	e
5		Portal, fixed footing	Two cuts: Cut the span twice, close to the pier.	1. Cutting these spans may require a two-stage attack. 2. Use a concrete-stripping charge for the first stage.
6	Concrete	Portal, pinned footing	Strip concrete: Remove concrete from the midspan over length L_c with a concrete-stripping charge.	1. Remove all concrete for L_c. 2. A one-stage attack should be adequate. 3. When footing conditions are unknown, use Serial 5. 4. For L_c, use Table 9-10, page 9-27.
7		Arch, open spandrel, fixed footing, Method I	Strip concrete: Remove the concrete from the midspan over length L_c with a concrete-stripping charge.	1. Applies to arches greater than 35 meters. 2. A one-stage attack should be adequate. 3. For L_c, use Table 9-10.
8		Arch, open spandrel, fixed footing, Method II	Strip concrete: 1. Remove the concrete from the midspan over length L_c with a concrete-stripping charge. 2. Attack springing with hayricks* at the top face of the arch ring.	1. Applies to arches less than 35 meters. 2. A one-stage attack should be adequate. 3. For L_c, use Table 9-10.

*Hayricks are not in the US Army supply system.

Figure 9-21. Methods of attack on continuous bridges (continued)

9-34 Demolitions and Modernized Demolition Initiators (MDI)

Serial	Sub-category	Type	Attack Method	Remarks
a	b	c	d	e
9		Arch, open spandrel, fixed footing, Method III	Four cuts: 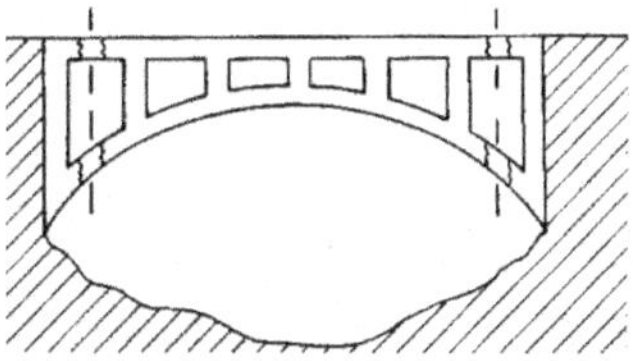	1. Alternative to Method II, applies to arches less than 35 meters. 2. Two-stage attack will probably be required. 3. Use a concrete-stripping charge for first stage. 4. For L_c, use Table 9-10, page 9-27.
10	Concrete	Arch, open spandrel, pinned footing	Strip concrete: 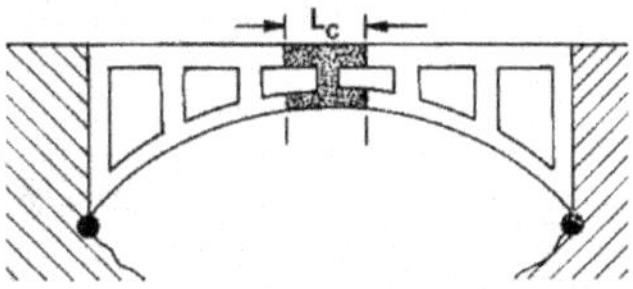 L_c Remove concrete from the midspan over length L_c with a concrete-stripping charge.	1. A one-stage attack should be adequate. 2. For L_c use Table 9-10.
11		Arch, solid spandrel, fixed footing, Method I	Strip concrete: 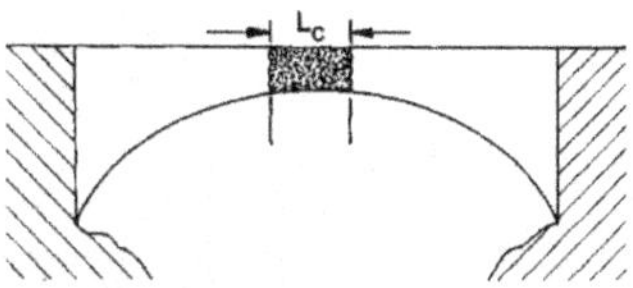 L_c Remove the concrete from the midspan over length L_c with a concrete-stripping charge.	1. This applies to arches of span greater than 35 meters only. 2. A one-stage attack should be adequate. 3. For L_c use Table 9-10.

Figure 9-21. Methods of attack on continuous bridges (continued)

FM 5-34

Serial	Sub-category	Type	Attack Method	Remarks
a	b	c	d	e

12	Concrete	Arch, solid spandrel, fixed footing, Method II	Strip concrete:
			
			Charges
			
			1. Remove concrete from the midspan over length L_c with a concrete-stripping charge.
			2. Attack both springing points with concrete-stripping charges:
			  a. Against bottom face of arch ring.
			  b. Against the top face (must remove the fill beneath the roadway to access the arch ring).
				1. Applies to arches less than 35 meters.
				2. A one-stage attack should be adequate.
				3. For L_c use Table 9-10, page 9-27.

13		Arch, solid sprandral, pinned footing	Strip concrete: Remove concrete from the midspan over length L_c with a concrete-stripping charge.	1. A one-stage attack should be adequate. 2. For L_c use Table 9-10.
14	Steel	Cantilever	Two cuts: Shear joints 1. Cut the anchor span as closely to the pier as practical. 2. Cut the midspan shear joints.	None
15		Cantilever and suspended span	One cut: Cut anchor span as closely to the pier as practical.	If demolition of the suspended span will create the desired obstacle, regard the span as simply supported and attack accordingly.

Figure 9-21. Methods of attack on continuous bridges (continued)

9-36 Demolitions and Modernized Demolition Initiators (MDI)

FM 5-34

Serial	Sub-category	Type	Attack Method	Remarks
a	b	c	d	e
16		Beam or truss with short side span	One cut: 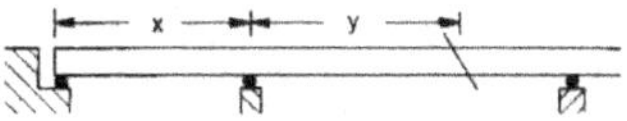 1. Cut interior span so y is greater than 1.25x. 2. If necessary, cut other interior spans as in Serial 17.	None
17		Beam or truss without short side span	Two or more cuts: 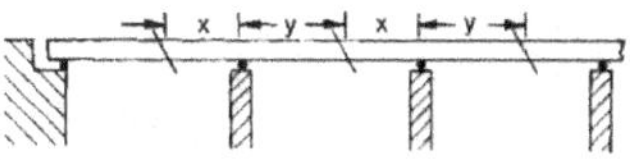 Cut spans so y is greater than 1.25x.	None
18	Steel	Portal, fixed footing	Two cuts: 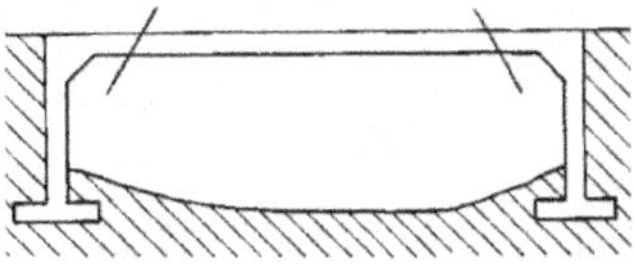 Cut the span twice, close to the piers.	None
19		Portal, pinned footing	Two Cuts: 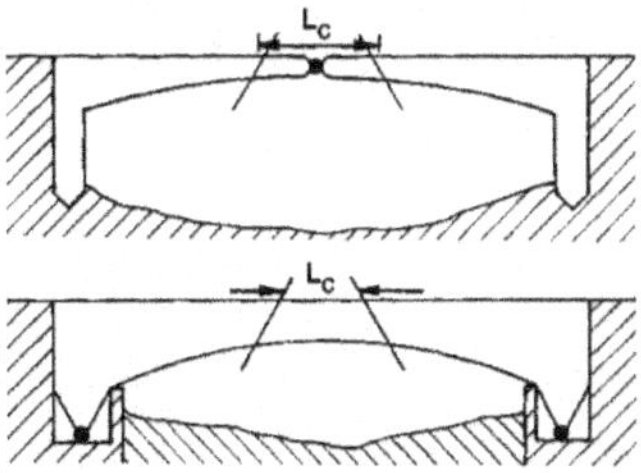 Remove section from midspan over length L_c.	For L_c use Table 9-10, page 9-27.

Figure 9-21. Methods of attack on continuous bridges (continued)

Demolitions and Modernized Demolition Initiators (MDI) 9-37

FM 5-34

Serial	Sub-category	Type	Attack Method	Remarks
a	b	c	d	e
20	Steel	Arch, open spandrel, fixed footing	Four cuts:	1. Angle cuts about 70 degrees. 2. For L_c use Table 9-10, page 9-27.
21		Arch, open spandrel, pinned footing	Two cuts: L_c Remove section from the midspan over length L_c.	For L_c use Table 9-10.
22	Masonry	Arch, Method I	Two cuts: 1. Cut at haunches. 2. Attack arch ring, spandrel walls, and parapet.	None
23		Arch, Method II	One cut: Crown Breach arch ring at the crown.	1. Use this method as an alternate to Method I, only when time is insufficient to allow attack at the haunches. 2. For L_c use Table 9-10.

* Hayricks are not in the US Army supply system.

Figure 9-21. Methods of attack on continuous bridges (continued)

9-38 Demolitions and Modernized Demolition Initiators (MDI)

ABUTMENT AND INTERMEDIATE-SUPPORT DEMOLITIONS

See Figures 9-22 and 9-23 and Figure 9-24, page 9-40. Single-abutment destruction should be on the friendly side.

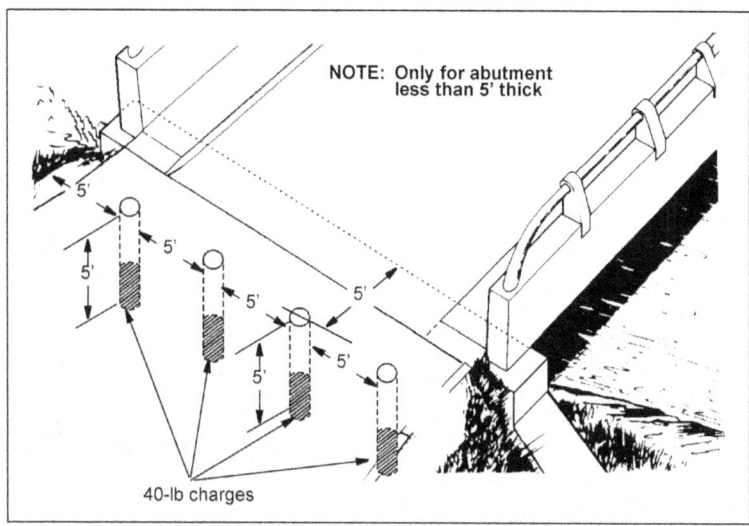

Figure 9-22. Placement of 5-5-5-40 charge (triple-nickle forty)

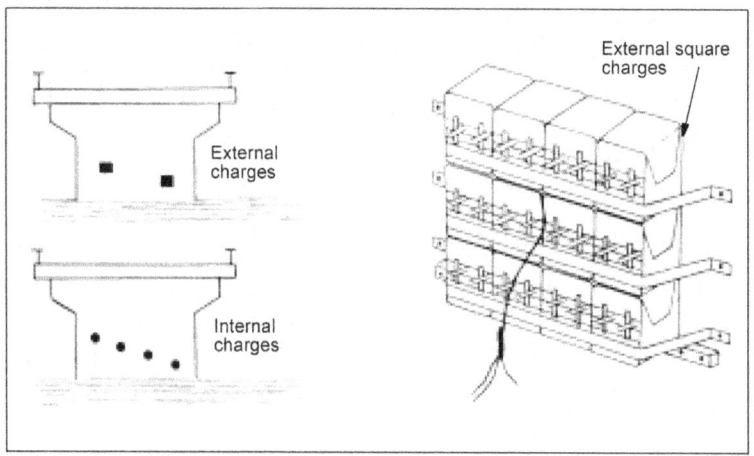

Figure 9-23. Pier demolition

Demolitions and Modernized Demolition Initiators (MDI) 9-39

FM 5-34

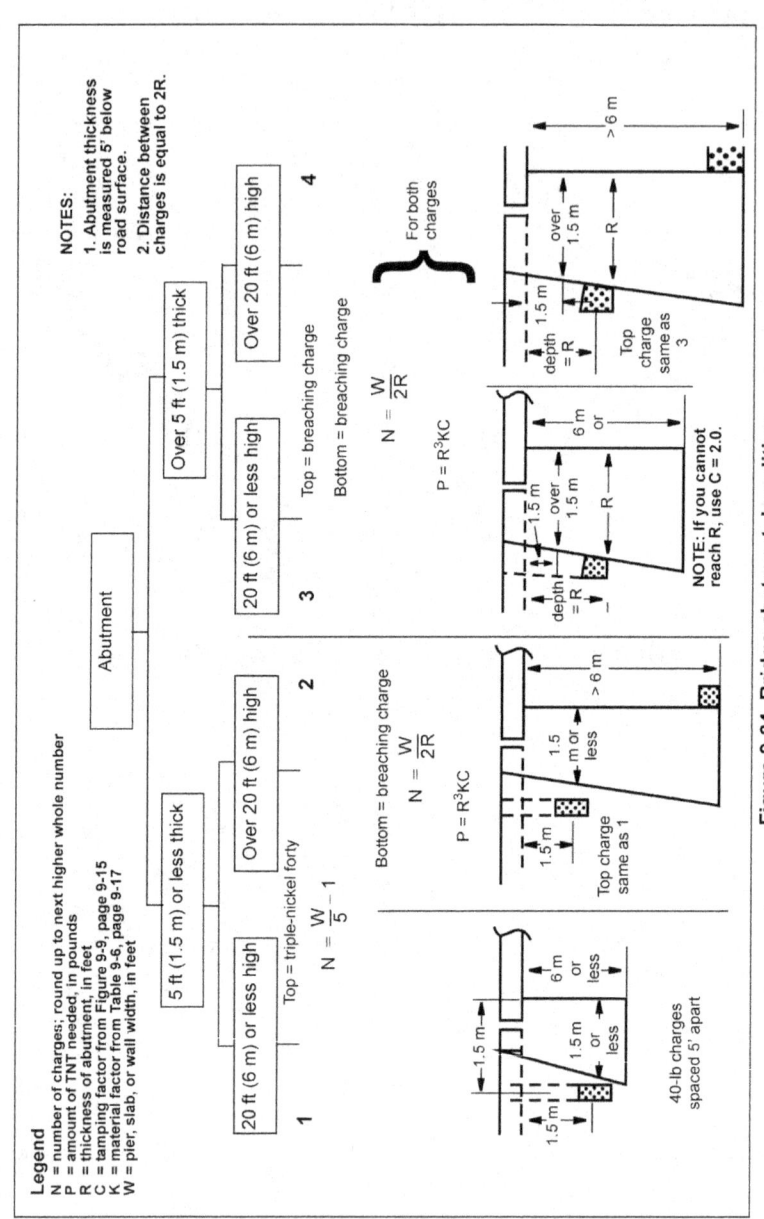

Figure 9-24. Bridge-abutment demolition

FM 5-34

DEMOLITION RECONNAISSANCE

Figure 9-25 shows a sample of DA Form 2203-R. To use this form with reconnaissance procedures, see Chapter 3. Refer to page 4 of DA Form 2203-R for instructions on completing the Demolition Reconnaissance Record.

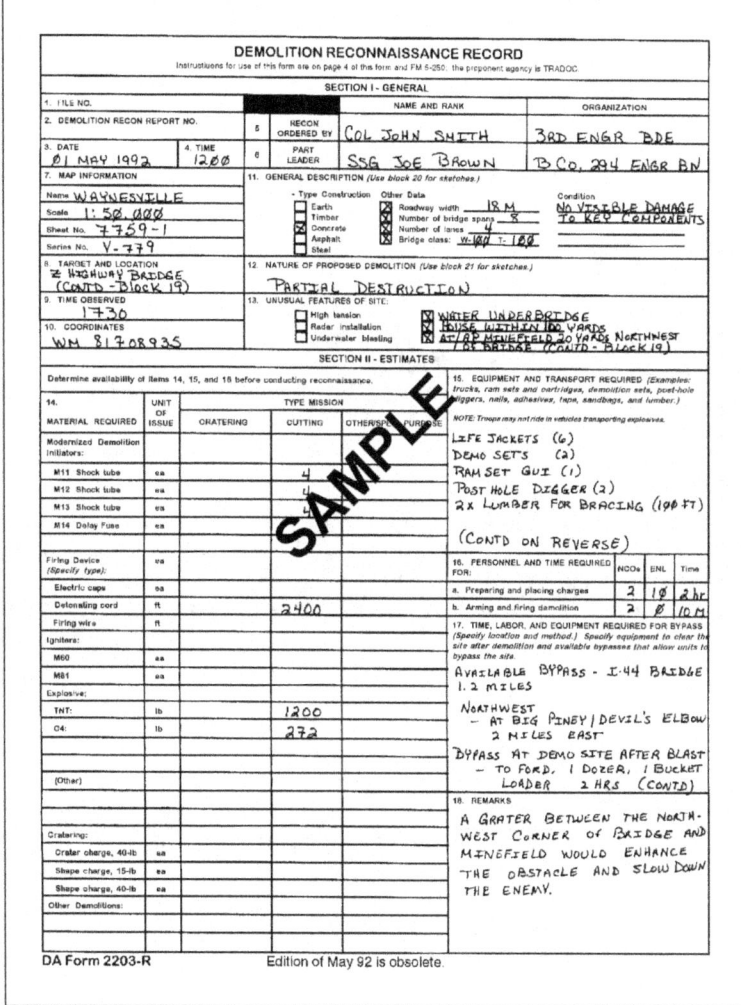

Figure 9-25. Sample, DA Form 2203-R

Demolitions and Modernized Demolition Initiators (MDI) 9-41

FM 5-34

DEMOLITION RECONNAISSANCE RECORD

Place additional comments in the appropriate blocks.

15. EQUIPMENT AND TRANSPORT REQUIRED *(Continued)*

HMMWV (1 - TRANSPORT OF CAPS)
2 1/2 -T CARGO (1 - TRANSPORT OF DEMO)
ROPE 3/4" (200')
SQUAD VEHICLE (1 - TRANSPORT OF TROOPS)

17. TIME, LABOR, AND EQUIPMENT REQUIRED FOR BYPASS *(Continued)*

TO BRIDGE AT SITE: 1 DOZER, 1 BUCKET LOADER, 3.5 HOURS TO CLEAR AND IMPROVE APPROACH. RIBBON BRIDGE ASSEMBLY - 10 INTERIOR BAYS, 2 RAMP BAYS. 30 MINUTES MGB - 1 COMPANY, 3 HOURS

18. REMARKS *(Continued)*

SAMPLE

19. ADDITIONAL COMMENTS *(Specify block)*

BLOCK 8 CONTD. BRIDGE IS OVER BIG PINEY RIVER, NEAR TOWN OF DEVIL'S ELBOW.

BLOCK 13 CONTD. ANTI-VEHICLE DITCH (WHEEL ONLY) ON SOUTHEAST SIDE OF BRIDGE.

PAGE 2, DA Form 2203-R

Figure 9-25. Sample, DA Form 2203-R (continued)

FM 5-34

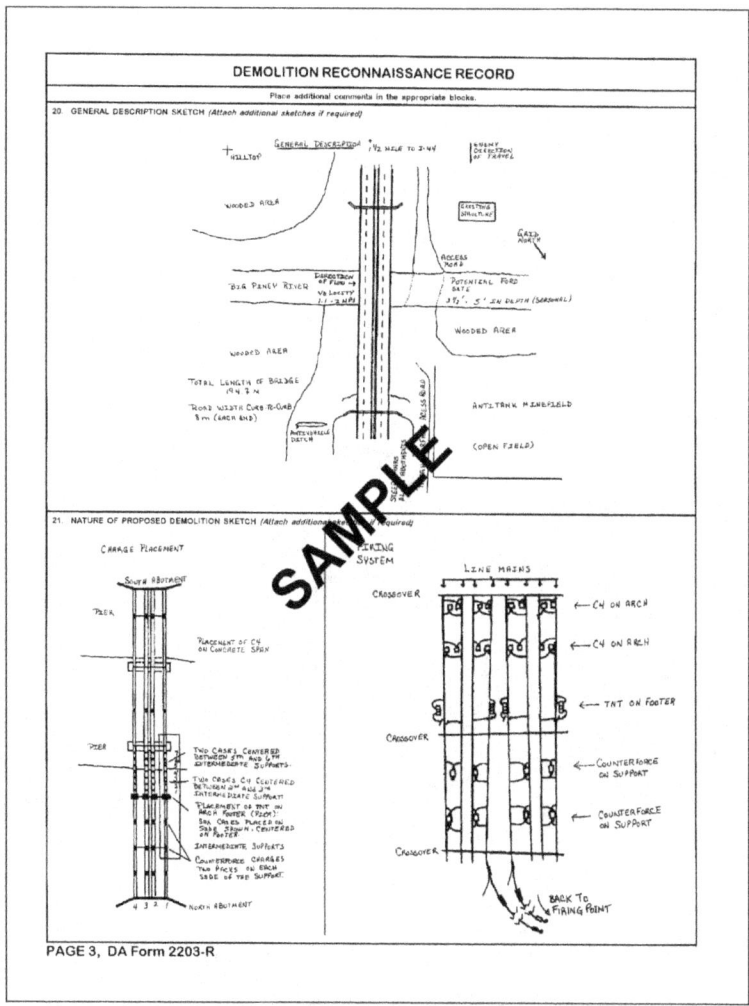

Figure 9-25. Sample, DA Form 2203-R (continued)

Demolitions and Modernized Demolition Initiators (MDI) 9-43

DEMOLITION RECONNAISSANCE RECORD

Instructions for Completing the DA Form 2203-R

Use the following instructions to complete DA Form 2203-R. This form may be locally reproduced on 8 1/2- by 11-inch paper.

1. *Block 1 (FILE NO.).* Leave blank unless a higher headquarters provides this number. Higher headquarters provides this number or enters it after you submit the form.

2. *Block 2 (DEMOLITION RECON REPORT NO.).* Leave blank unless a higher headquarters provides this number. Higher headquarters provides this number or enters it after you submit the form. Company SOP may specify the procedures for determining this number.

3. *Block 3 (DATE).* Enter the date the reconnaissance was performed.

4. *Block 4 (TIME).* Enter the time the reconnaissance party arrived at the target site (local or Zulu time).

5. *Block 5 (RECON ORDERED BY).* Enter the command authority authorizing the reconnaissance action.

6. *Block 6 (PARTY LEADER).* Enter the name of the NCOIC or OIC of the reconnaissance party who was physically at the site when the reconnaissance was performed.

7. *Block 7 (MAP NAME, SCALE, SHEET NO., and SERIES NO.).* Obtain this information from a map of the reconnaissance area and enter the information in this block.

8. *Block 8 (TARGET AND LOCATION).* Enter a brief description of the target and the distance and direction from an identifiable landmark (railroad bridge, crossroads, hilltop, and so forth). For example, "Target is 275 degrees, 300 meters from the railroad bridge, 2 miles east of Hanesville, on Route 2."

9. *Block 9 (TIME OBSERVED).* Enter the time you last saw the target as you departed the site.

10. *Block 10 (COORDINATES).* Enter the complete 8-digit map coordinates of the target.

11. *Block 11 (GENERAL DESCRIPTION (Use block 20 for sketches.)).* When applicable, include the type of construction, width of the roadway, number of lanes or tracks, type of pavement, number of spans, condition of spans or entire bridge, and bridge categorization and classification. For example, "Prestressed-concrete T-beam bridge, four simple spans supported by six concrete columns, two lanes; total bridge length is 140 feet; roadway width is 30 feet, overall bridge width is 38 feet, height is 16 feet; Class 80; very good condition."

12. *Block 12 (NATURE OF PROPOSED DEMOLITION (Use block 21 for sketches.)).* State the expected amount of destruction and the priority for placing charges, if feasible. Provide a sketch showing the number and type of charges to use (tamped or untamped), where the charges should be placed, and the type of firing system required.

13. *Block 13 (UNUSUAL FEATURES OF SITE).* Include any special features of the target or site that might affect the method of demolition (high-tension lines, radar installation, underwater blasting, and so forth). Give any details that may affect the safety of the target and the demolition work party.

14. *Block 14 (MATERIAL REQUIRED).* Indicate the types, quantities, caps, detonators, and so forth proposed for the demolition.

15. *Block 15 (EQUIPMENT AND TRANSPORT REQUIRED).* Specify the amount and type of transportation required (for example, two 5-ton dump trucks, one ram set with 50 cartridges, two post-hole diggers, two demolition sets, 10 po[illegible]d nails, twelve 8-foot 2 by 4s). Comments may be continued in block 18 on page 2 of the form.

16. *Block 16 (PERSONNEL AND TIME REQUIRED FOR).* Complete subsections a and b, indicating the number of personnel and amount of time necessary for placing the demolitions. The distance between the firing points and firing systems will be a consideration for determining the amount of time necessary to arm and fire the explosives.

17. *Block 17 (TIME, LABOR, AND EQUIPMENT REQUIRED FOR BYPASS (Specify location and method.)).* Specify the equipment necessary to clear the site after demolition and the available bypasses that allow units to bypass the site. Comments may be continued in Block 18 on page 2 of the form.

18. *Block 18 (REMARKS).* Include any appropriate remarks that are not covered in blocks 1 through 17.

19. *Block 19 (ADDITIONAL COMMENTS).* Use this block as a continuation for blocks 1 through 18. Identify the block being continued.

20. *Block 20 (GENERAL DESCRIPTION SKETCH).* The sketch should include–

- The avenues of approach to the target and possible bypasses in the vicinity of the target. Indicate route numbers and the direction of cities or towns.
- Rivers or streams including name, direction of flow, and velocity in meters per second.
- Terrain features, including observation points, cover and concealment, swampy areas, deep valleys, and so forth.
- A compass arrow indicating north (indicate grid or magnetic).
- Dimensions of the proposed target.
- Number and length of bridge spans.
- Height of the bridge from the ground or water.

21. *Block 21 (NATURE OF PROPOSED DEMOLITION SKETCH).* This sketch should include–

- Dimensions of members to be cut.
- Placement of charges.
- Charge calculations. Use either the formula or table method, but show your work.
- Priming of charges.
- Branch lines.
- Ring mains.
- Firing systems.
- Firing points.

SAMPLE

PAGE 4, DA Form 2203-R

Figure 9-25. Sample, DA Form 2203-R (continued)

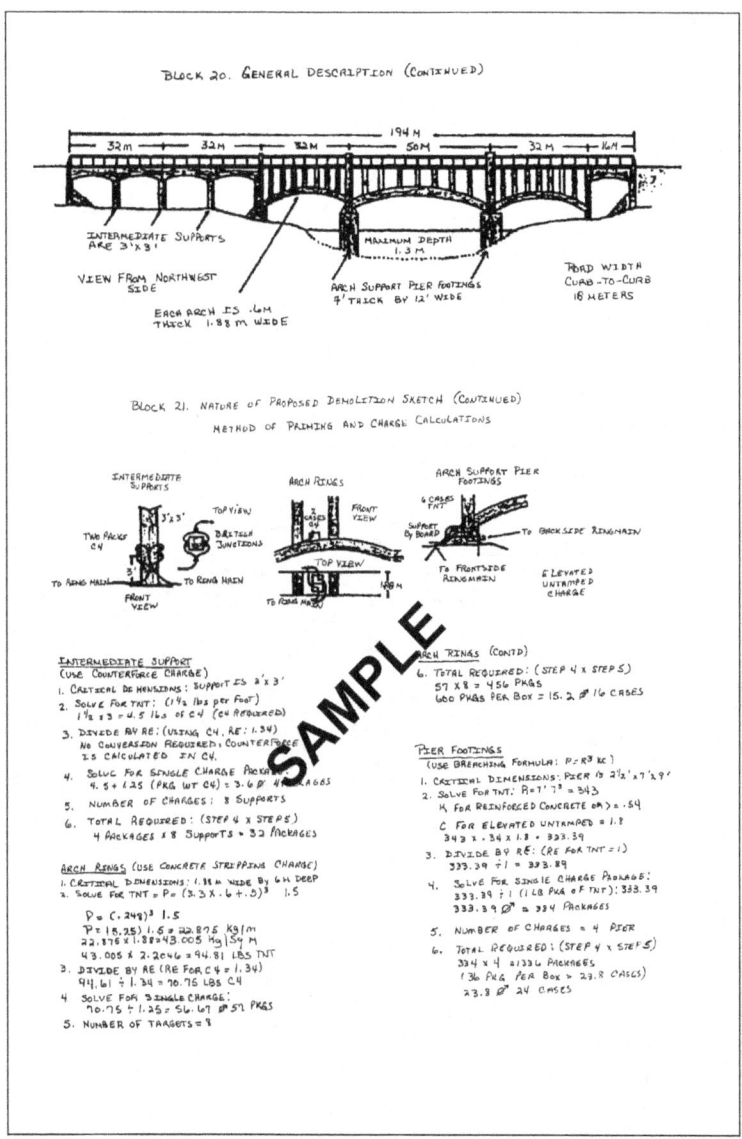

Figure 9-25. Sample, DA Form 2203-R (continued)

EQUIPMENT/AMMUNITION DESTRUCTION

AMMUNITION

You can destroy caches of weapons and ammunition to prevent enemy use. The general rule of thumb for destroying ammunition is to use 1 pound of explosive for every 1 pound of explosive contained in the ammunition. All ammunition must be removed from their shipping container and must touch each other to create a continuous line of explosive.

GUNS

To prepare a gun for demolition, first block the barrel just above the breach. For small-caliber guns that use combined projectile-propellant munitions, solidly tamp the first meter of the bore with earth. Table 9-11 details the charge size required for standard barrel sizes. Pack the explosive, preferably C4, into the breach, immediately behind the tamping. Place the plastic explosive in close contact with the chamber.

Table 9-11. Gun-destruction charge sizes

Barrel Size (mm)	Charge Size(lb)
76	10
105	18
120	23
155	38
203	66
Formula for determining amount of explosive: $$P = \frac{D^2}{636}$$ P = quantity of explosive (any HE), in pounds D = bore size of the barrel, in millimeters	
NOTE: If the actual barrel size is not listed, use the nearest gun size.	

ARMORED FIGHTING VEHICLES (AFVS)

You can destroy armored fighting vehicles using a 25-pound charge inside the hull. Make sure that all hatches, weapons slits, and openings are sealed and that the ammunition inside the hull detonates simultaneously. If it is not possible to enter the vehicle, place charges under the gun mantle, against the turret ring, and on the final drive (see Figure 9-26). If explosives are not available, destroy the AFV by using AT weapons or fire, or destroy the main gun with its own ammuntion. Insert and seat one round in the muzzle end and a second charge, complete with propellant charge (if required), in the breach end of the tube. Use a long lanyard and fire the gun from a safe distance. Make sure that the firing party is undercover before firing the gun.

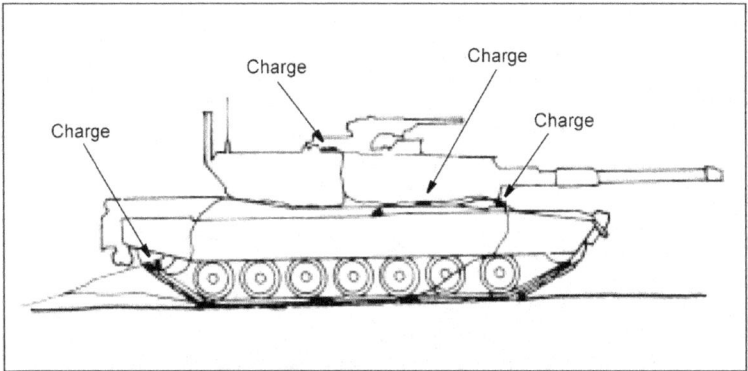

Figure 9-26. Placing charges on the AFV

WHEELED VEHICLES

Destroy wheeled vehicles by attacking the vital parts with explosives or even a sledge hammer. If you use explosives, place a 2-pound charge on the cylinder head, axles, and frame.

EXPEDIENT DEMOLITIONS

Expedient techniques are intended for use only by personnel experienced in demolitions and demolitions safety. Do not use expedient techniques to replace standard demolition methods.

CRATERING CHARGE

Cratering charges are used to supplement the 40-pound cratering charge or as an improvised cratering charge. To make a cratering charge—

- Use a mixture of ammonium-nitrate fertilizer (at least $33 \frac{1}{3}$ percent nitrogen) and liquid (diesel fuel, motor oil, or gasoline) at a ratio of 25 pounds of fertilizer to 1 quart of liquid. Mix the fertilizer with liquid and allow it to soak for an hour.

Demolitions and Modernized Demolition Initiators (MDI) 9-47

FM 5-34

- Pour half of the charge weight in a hole, place two 1-pound primed blocks of explosive, and then pour in the other half of the charge.
- Place the mixture inside of a sandbag or plastic bag or cardboard box, for transportation, and then place the entire package in the hole.

  NOTE: Bore holes should receive 10 pounds of explosives for every foot of depth and must be dual-primed.

SHAPED CHARGE

- Use a container, such as a can, jar, bottle or drinking glass, and remove both ends (see Figure 9-27). (Some containers come with built-in cavity liners, such as champagne or cognac bottles, with the stems removed.)
- Place the plastic explosive inside the container, and mold a cone in the base of the explosive. If possible, use a cone-shaped liner made from copper, aluminum foil, or glass. The optimum angle for the cone is 42 to 45 degrees, but cavity angles between 30 and 60 degrees will work.
- Ensure that the standoff or legs of the container are $1^1/_2$ times the cone's diameter.
- Detonate the charge from the top dead center of the charge.

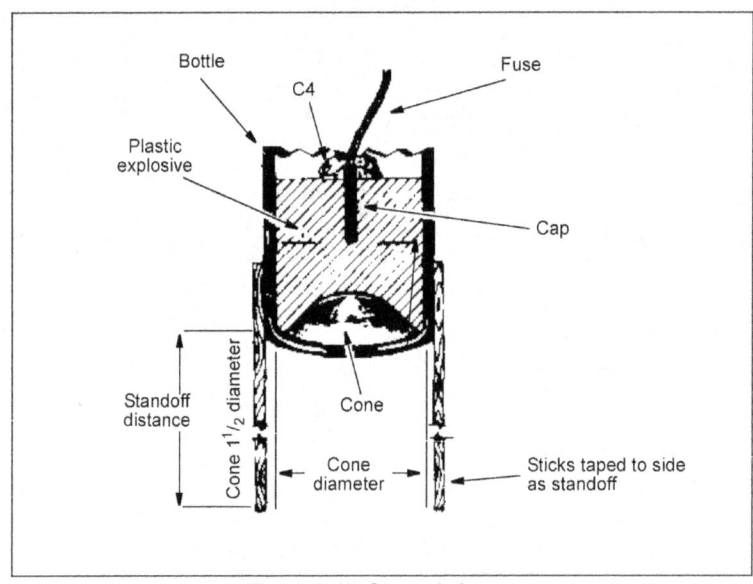

Figure 9-27. Shaped charge

9-48 **Demolitions and Modernized Demolition Initiators (MDI)**

PLATTER CHARGE

- Get a steel platter, preferably round, that weighs 2 to 5 pounds. Uniformly pack the explosive to the back of the platter.
- Make sure that the explosive weighs the same as the platter. You can tape the explosive to the platter.
- Prime the charge at the exact rear center. Cover the blasting cap with a small quantity of C4 if any part of the cap is exposed.
- Gut an M60 fuse igniter and tape it to the top of the charge as a sight. You can fabricate legs from sticks to help in aiming the platter. Make sure that the explosive is on the side of the platter opposite the target (see Figure 9-28).

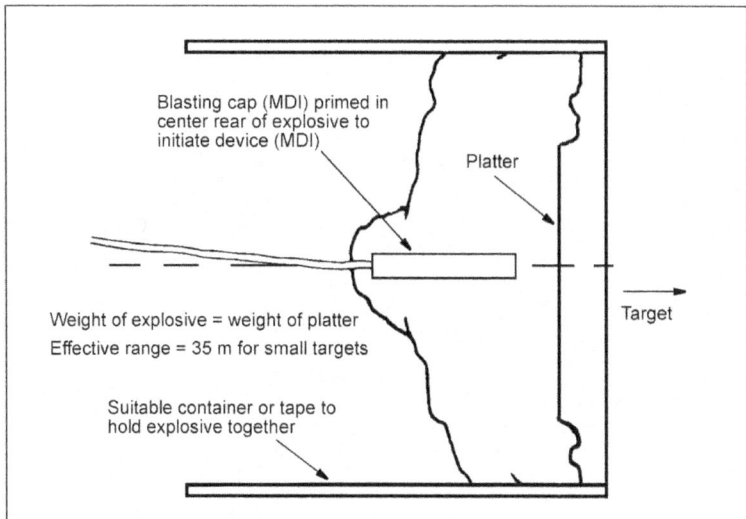

Figure 9-28. Platter charge

GRAPESHOT CHARGE

- For a grapeshot charge (see Figure 9-29, page 9-50), get a container, projectiles, buffer material, an explosive charge, and detonating cord.
- Make a large enough hole in the center bottom of the container to accept the detonating cord.
- Slip the detonating cord branch line through the hole and tie a double overhand knot.
- Place and tamp the C4 uniformly in the bottom of the container.

Demolitions and Modernized Demolition Initiators (MDI) 9-49

FM 5-34

- Place two inches of buffer material (leaves, dirt, cardboard) on top of the explosive. Place the projectiles (nails, bolts, rocks) on top of the buffer material and secure the opening of the container using tape or plastic wrap.

   NOTE: The United Nations Convention of Certain Conventional Weapons (CCW) mandates that all fragmentary munitions produce fragments that are visible by x-ray (metal, rock).

- Tie the detonating cord branch line to a line or ring main.

- Aim the charge at the center of the target from about 100 feet.

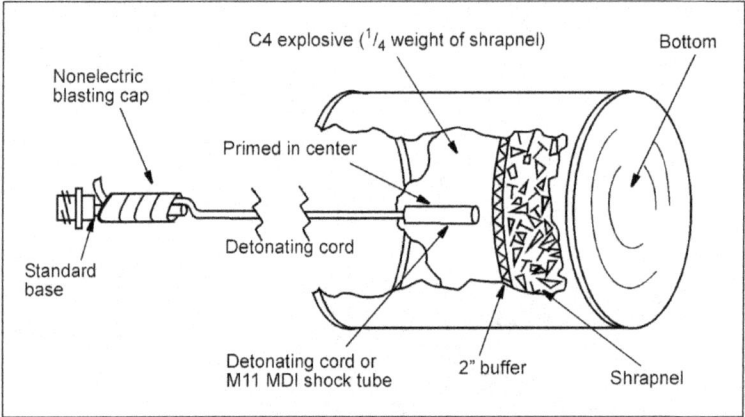

Figure 9-29. Grapeshot charge

Ammonium Nitrate Satchel Charge

This charge is a mixture of ammonium-nitrate fertilizer with melted wax instead of oil. The mixing ratio is four pounds of fertilizer to one pound of wax. To make this charge—

- Melt the wax in a container and stir in the ammonium-nitrate pellets, making sure that the wax is hot while mixing. Before the mixture hardens, add a $1/2$-pound block of explosive primed with detonating cord. Ensure that the primed charge is in the center of the mixture and that there is sufficient detonating cord available to attach initiation sets.

- Pour the mixture into a container. Add shrapnel material to the mixture if desired, or attach the shrapnel on the outside of the container to give a shrapnel effect. Detonate the charge by attaching intiation sets to the detonating cord coming from the satchel charge.

9-50 Demolitions and Modernized Demolition Initiators (MDI)

BANGALORE TORPEDO

- Separate the packaging material from C4 (M112), and place it in the concave portion of two U-shaped pickets which are not bent or damaged.
- Mold the C4 explosive, using a nonsparking tool, into the concave position that runs the entire length of the U-shaped pickets.
- Place a line of detonating cord, after tamping the C4, on top of the C4 of one of the pickets, and make a single overhand knot every 6 to 8 inches. Make sure the detonating cord runs several feet past the U-shaped picket length so that it can be tied into a firing system.
- Place the other U-shaped picket tamped with C4 onto the picket with the detonating cord previously set in. The C4 explosive from each picket will be touching, with the detonating cord in the middle.
- Secure the two U-shaped pickets together with tape or wire.

DETONATING-CORD WICK (BOREHOLE METHOD)

Use this method (see Figure 9-30, page 9-52) to enlarge boreholes in soil. You will get the best results in hard soil.

- Tape together several strands of detonating cord 5 to 6 feet long. Generally, one strand enlarges the diameter of the hole by about 1 inch. Tape or tie the strands together into a wick for optimum results.
- Make a hole by driving a steel rod about 2 inches in diameter into the ground to the depth required. According to the rule of thumb, a hole 10 inches in diameter requires 10 strands of detonating cord.
- Place the detonating-cord wick into the hole using an inserting rod or some other field expedient. The strands must extend the full length of the hole.
- Fire the cord either electrically or nonelectrically. An unlimited number of wicks can be fired at one time by connecting them with the detonating-cord ring main or line main. If you place successive charges in the holes, blow out excess gases and inspect the hole for excessive heat.

TIME FUSE

Soak length of clean string ($^1/_8$-inch diameter) in gasoline and hang to dry. After drying, store it in a tightly sealed container. Handle it as little as possible, and test it extensively before use.

GREGORY KNOT (BRANCH-LINE CONNECTION)

The Gregory knot (see Figure 9-31, page 9-52) is a detonating-cord knot tied at the end of a branch line to connect the branch line to a

Demolitions and Modernized Demolition Initiators (MDI)

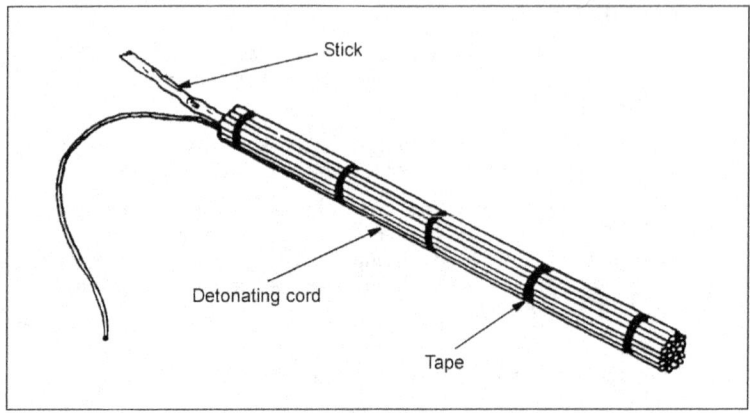

Figure 9-30. Detonating-cord wick

firing system. The Gregory knot saves time on a target when tied before arriving at the mission site. This knot does not take the place of the girth hitch with an extra turn or detonating-cord clips.

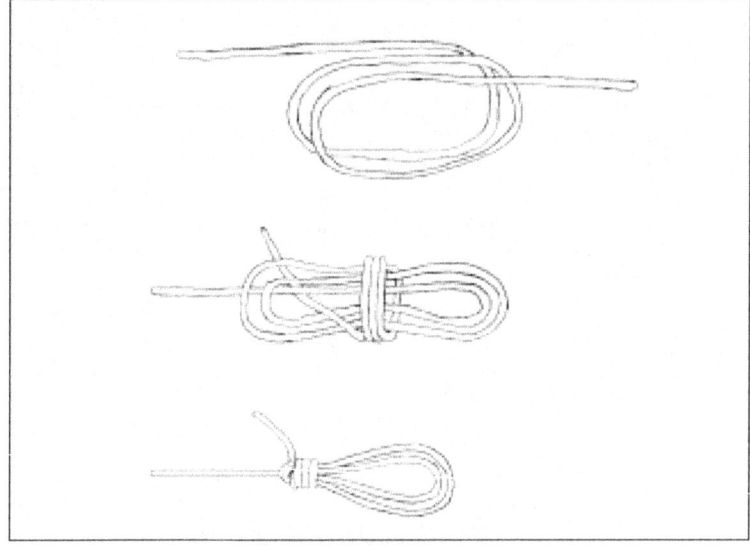

Figure 9-31. Gregory knot

SECTION II. MODERN DEMOLITION INITIATORS

MDI FIRING SYSTEMS

The MDI are a family of nonelectric blasting caps and associated items (see Table 9-12, page 9-54). The snap-together components simplify initiation systems and some types of explosive priming. The MDI was developed to effectively replace electric demolition systems. The MDI system removes the requirement to dual-initiate demolition systems except when there is a high probability of the system becoming cut.

Nonelectric priming with MDI is safer and more reliable than the current nonelectric priming methods. MDI blasting caps are factory-crimped to precut lengths of shock tube or time-blasting fuse. Because the caps are sealed units, they are moisture-resistant and will not misfire in damp conditions. A shock tube may be spliced using excess shock tube from an M12 or M13 or a precut splicing-tube splicing kit. Every splice in a shock tube reduces the reliability of the priming system. Prime military explosives with the MDI the same as with standard, nonelectric initiation systems. Use only high-strength MDI blasting caps (M11, M14, M15) to prime explosive charges. M12 and M13 relay-type blasting caps do not have sufficient power to detonate most explosives. You can use all MDI blasting caps to initiate a shock tube. Use only the M11, M14, or M15 blasting caps to initiate detonating cord or military explosives directly.

> **WARNING**
> Use care when cutting and splicing the shock tube. When cutting the shock tube, always tie and overhand knot in the left over shock tube.

With the introduction of MDI components, there will be two types of firing systems: a stand-alone firing system and a combination firing system. Both systems can be emplaced as single- or dual-firing systems. The choice of which system to use for a particular demolition mission is left to the experience of the engineer commander. However, the combination firing system is the preferred method for reserved demolition targets. See FM 5-250, Chapter 7, for detailed instructions on both systems.

Table 9-12. MDI components

Components	Description	Packaging
M11	High-strength, nonelectric blasting cap, factory crimped to a 30-ft length of shock tube—used to prime all standard military explosives, including det cord, or to initiate the shock tube of other MDI blasting caps. A red flag is attached 1 meter from the cap, and a yellow flag is attached 2 meters from the cap.	6/pkg, 10 pkg per box
M12	Low-strength, nonelectric blasting cap[1], factory crimped to a 500-ft length of shock tube—used as a transmission line from an initiator to another relay cap or to a high-strength, shock-tube blasting cap which initiates military explosives. Can actuate up to five shock tubes held by the connector.	8 spools/ cardboard box, 6 boxes/packing box
M13	Low-strength, nonelectric blasting cap[1], factory crimped to a 1,000-ft length of shock tube; used as a transmission line from an initiator to another relay cap or to a high-strength, shock-tube blasting cap which initiates military explosives. Can actuate up to five shock tubes held by the connector.	4 spools/ cardboard box, 6 boxes/packing box
M14	High-strength, nonelectric, delay blasting cap, factory crimped to a 7 1/2-ft length of time-blasting fuse—instead of the usual yellow band every 18 in, a marker band and the minimum burning time in minutes (from the band to the detonator) are marked on the fuse. Used to detonate all standard military explosives or initiate shock-tube blasting caps and detonating cord about 5 minutes after being ignited.	1/pkg, 60/ wooden box
M15	Nonelectric blasting cap, delay[2] —consists of two blasting caps, factory crimped at each end of a 70-ft length of shock tube. One blasting cap is low-strength to initiate another piece of shock tube, while the second is high-strength to initiate other explosives. A red flag is attached 1 meter from the high-strength blasting cap, and a yellow flag is attached 2 meters from the low-strength cap. Used to create staged detonations, as required for quarrying, ditching, and cratering operations.	30/box, 4 boxes/packing box
M9	Blasting cap and shock tube holder—clamping device used to hold the shock tube's branch lines secure to a high-strength blasting cap of the M11 or M14. Can hold up to five shock tubes and one blasting cap. Can also connect a MDI blasting cap to detonating cord.	
M81	Time-blasting fuse igniter with shock-tube capability—M81 and M60 fuse igniters are almost identical except the plug and screw end cap are colored black on the M81.[3]	5/paperboard box/ pkg, 6 pkg/ wooden box

NOTE: The M81 has a stronger primer than the M60
[1]Does not have enough output to initiate most military explosives.
[2]Blasting caps are slightly larger than standard military blasting caps and will not fit into standard cap wells.
[3]The M60 fuse igniter will not reliably initiate the shock tube.

9-54 Demolitions and Modernized Demolition Initiators (MDI)

STAND-ALONE SYSTEM

The stand-alone firing system is one in which the initiation sets and transmission and branch lines are constructed using only MDI components and the explosive charges are primed with MDI blasting caps. It is important to ensure that the firing system is balanced. All charges must have the same distance in shock-tube length from the firing point to the charge. Figure 9-32 shows the single-firing MDI system; Figure 9-33, page 9-56, shows the dual-firing MDI system; and Figure 9-34, page 9-56, shows a branch-line array.

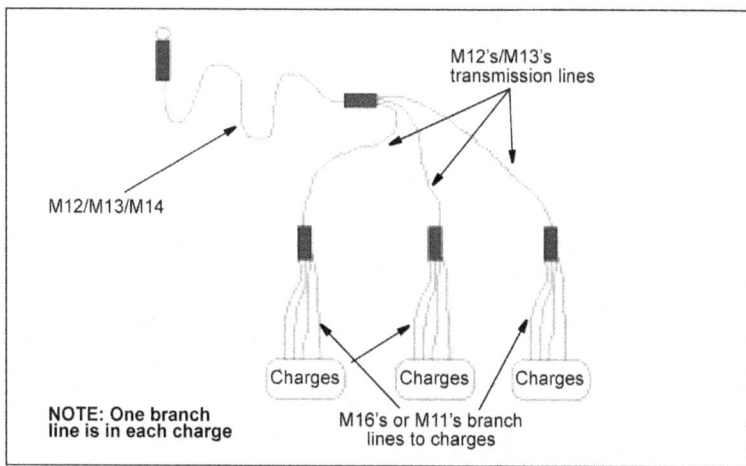

Figure 9-32. MDI single-firing system (single-primed)

The disadvantage of a single-firing system is that if the transmission line is cut, any charges down line from the cut will not detonate. If there is a possibility of the transmission lines being cut (for example, through artillery fires) a second firing system should be added as shown in Figure 9-33. Note that the charges in this case are now dual-primed, the transmission line is laid in the opposite direction of the first transmission line, and the system is a balanced system.

CAUTION
When making multi-shock-tube installations, take care to protect the shock tubes from the effects of nearby relay caps and charges. The shrapnel produced by a cap or charge could easily cause a (partial or complete) misfire. When there are many shock tubes involved in a shot, place them carefully away from the junction.

Demolitions and Modernized Demolition Initiators (MDI) 9-55

FM 5-34

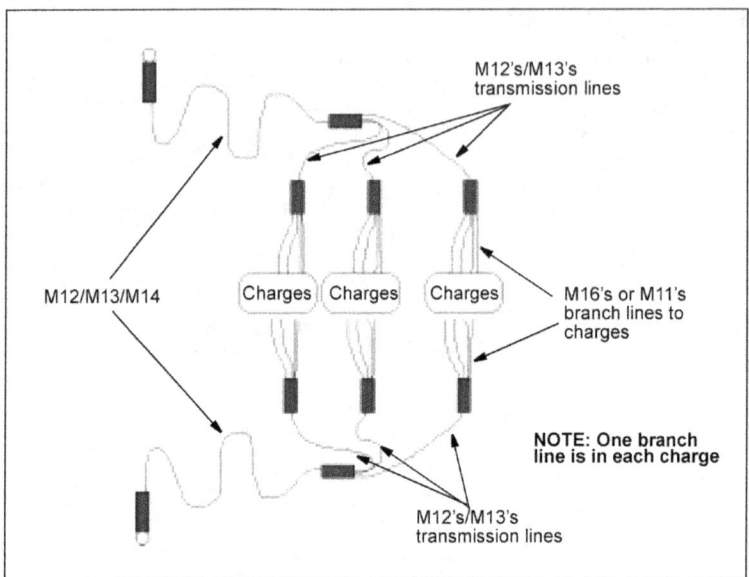

Figure 9-33. MDI dual-firing system (dual-primed)

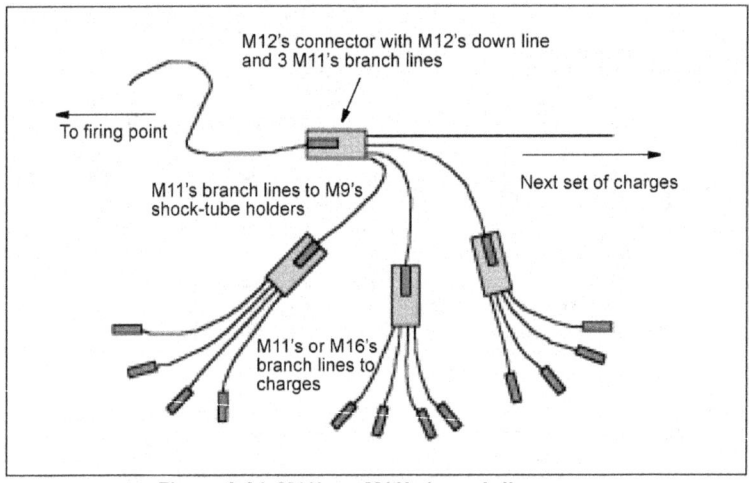

Figure 9-34. M11's or M16's branch-line array

Use the stand-alone MDI firing system for all types of demolition missions, including bridge demolitions. The MDI firing system can be used to initiate reserved demolition targets. However, under

9-56 Demolitions and Modernized Demolition Initiators (MDI)

current internationally agreed upon doctrine, charges cannot be primed with blasting caps until a change of readiness from state 1 (safe) to state 2 (armed) is ordered. Priming every charge with MDI blasting caps at this critical moment would take a considerable amount of time and be unacceptable to the maneuver commander. Priming charges with detonating cord is the preferred method on reserved demolition targets.

COMBINATION FIRING SYSTEM

A combination firing system is one which consists of the MDI initiation set; either a detonating-cord line or ring main; and branch lines that can be either MDI, detonating cord, or a mix of both. Figure 9-35 shows a combination firing system.

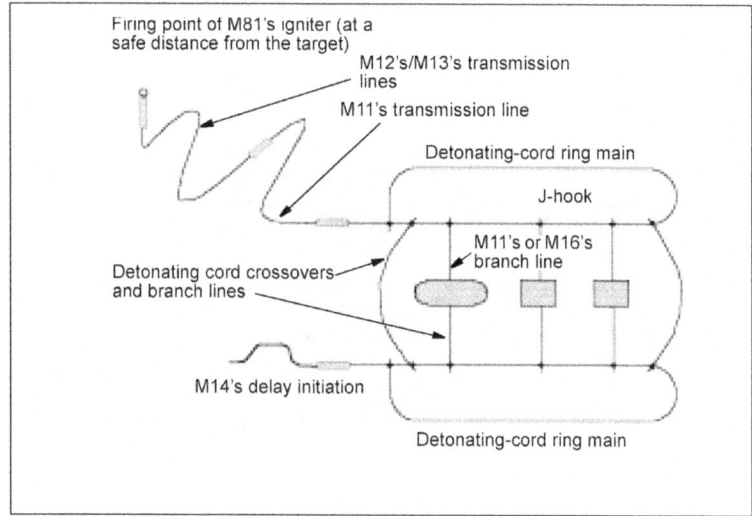

Figure 9-35. Combination (MDI and detonating cord) firing system (dual)

Use the combination (MDI and detonating cord) firing system for all types of demolition missions. It combines the advantages of MDI components with the simplicity and flexibility of detonating cord. The combination firing system is the preferred method for reserved demolition targets, underwater operations, and operations where subsurface-laid charges are used.

> **WARNING**
> Do not dispose of used shock tubes by burning them because of potentially toxic fumes given off from the burning plastic.

SPLICING THE SHOCK TUBE

The MDI are extremely reliable because all of the components are sealed. Unlike standard nonelectric priming components, they cannot be easily degraded by moisture. Cutting the shock tube makes the open ends vulnerable to moisture. Dampening the explosive film on the inside of the shock tube will stop a detonation from going beyond such a damp spot. Use care when cutting and splicing the shock tube. When cutting the shock tube, always tie an overhand knot in the leftover shock tube. Use splicing to repair a break in the shock tube of a transmission or branch line (caused, for example, by shrapnel from artillery fires) or to extend the shock tube of another MDI blasting cap, but only when necessary. This is done by using excess shock tube from an M12's or M13's shock-tube blasting cap when the entire length is not needed. Every splice in the shock tube reduces the reliability of the firing system. Keep the number of splices in a shock-tube line to as few as practicable. Unless splicing is absolutely necessary, use of a full, sealed MDI component is recommended. (Do not splice the shock tube while conducting water or diving demolition missions).

> **CAUTION**
> Taping two cut ends of the shock tube together does not make a reliable splice.

SAFETY PROCEDURES

When conducting training and missions with MDI, follow the general safety considerations for demolitions as given in Chapter 6 and AR 385-63.

Because MDI components are delivered from the factory precrimped, they are more reliable and safer to handle and use than the current standard military blasting caps. During testing of the MDI components, it was found that the blasting caps would always function correctly if the shock tube was properly initiated. Misfires only occurred when the—

- M81 fuse igniter was not properly connected to the shock tube before initiation.

- Shock tube was cut by shrapnel during the initiation process.
- Shock tube was incorrectly inserted into the holders on the M12 or M13 blasting caps or into the M9 holder.
- Shock tube was cut using crimpers.

> **WARNING**
> MDI is not authorized for below-ground or internal charges.

When transporting or storing MDI blasting caps, do not mix them with other explosives. Transporting blasting caps requires special consideration. The caps must be placed in a suitable container or in a separate vehicle.

MDI MISFIRE CLEARING PROCEDURES

- In most misfires of the shock-tube blasting caps, which are nonelectric, apply the standard rules.
- If the primer in the M81 does not fire (the most common problem), recock the M81 by pushing in on the pull rod to reset the firing pin, and then actuate the igniter again. If two or three retries result in a nonfiring, cut the shock tube, replace the igniter with a new one, and repeat the firing procedure.
- If the M81 fires and blows the shock tube out of its securing mechanism without it firing, cut about 3 feet from the end of the shock tube, replace with a new igniter, and repeat the firing procedure.
- If the M81 appears to have functioned properly but the charge did not fire, cut a 1-foot section from the shock tube starting 6 inches from the igniter. Hold the 1-foot piece of shock tube so one end is over your palm; gently blow through the other end. If a fine powder comes out from the shock tube, it has not fired. Install a new igniter on the freshly cut end of the priming shock tube and repeat the firing procedure. If no fine powder comes out from the shock tube or the shock tube was heard to fire or its flash was seen, wait for 30 minutes before moving downrange to check the components in the firing system.
- After waiting 30 minutes, proceed downrange and check all components in the firing system. The most likely cause of a misfire is the incorrect placement of the shock tube in the plastic connectors of the M12/13s or the M9 holder. If incorrect placement was the problem, replace the fired section and properly connect and refire the device.

FM 5-34

- If the first component of the firing train did not fail, check out each succeeding component until you find the one that failed. Replace the failed or fired relay components back to the initiating site and refire.
- If the final high-strength blasting cap seems to be the failed component, replace it if it is easily accessible. However, if it is used to prime an explosive charge, do not disturb it. Place a new, primed 1-pound explosive charge next to the misfired charge and detonate it when it is safe.

Chapter 10
Bridging

RIVER-CROSSING OPERATIONS

River-crossing operations can be hasty, deliberate, or retrograde. Deliberate crossings normally involve using assault-crossing equipment, rafts, and bridges (see Table 10-1).

Table 10-1. Assault-crossing equipment

Equipment	Transportation	Capabilities	Assembly/ Propulsion	Remarks/ Limitations
Pneumatic, 15-man assault boat	A 2 1/2-ton truck holds 20 deflated boats (250 lb per boat); inflated boat is an 8-man carry.	Carries either— 12 inf and 3 eng w/paddles, 12 inf and 2 eng w/OBM, or 3,375 lb of equipment	Inflation time is 5 to 10 min with pumps. Speed w/paddles is 1.5 MPS. Speed w/OBM is 4.6 MPS.	Max current velocity w/paddle is 1.5 MPS and w/ OBM, 3.5 MPS. Each boat has 3 pumps and 11 paddles. OBMs - on request
Pneumatic, 3-man recon boat	One man carries the boat by backpack; total weight is 37 lb.	Carries 3 soldiers with equipment or 600 lb of equipment	Inflation time is 5 min with pump. Speed with paddles is 1 MPS.	Max current velocity is 1.5 MPS. Each boat has 1 pump and 3 paddles. No provisions for OBMs
APC	A self-propelled, Class 13 vehicle	Carries 12 soldiers with equipment	Preparation time for swimming is 10 min. It has track propulsion in the water. Swim speed is 1.6 MPS. Fords up to 1.5 m	Max current velocity is 1.5 MPS. $D = \frac{C}{2} \times W$ D = drift (meters) C = current (MPS) W = river's width (meters)
BIFV	A self-propelled, Class 2A vehicle	Carries 10 soldiers with equipment	Preparation time for swimming is 18 min.	Max current velocity is 0.9 MPS. $D = \frac{C}{1.6} \times W$ D = drift (meters) C = current (MPS) W = river's width (meters)
AVLB	Bridge (weighs 15 tons) is carried on a launcher (modified M48A5 or M60A1 chassis); 20-ton crane transfers bridge to launcher in 20 to 30 minutes.	A Class 60 vehicle (one at a time can cross); measures 19.2 m and spans 18.3 m using prepared abutments or measures 17 m using unprepared abutments.	Launched in 2 to 5 min by buttoned-up 2-man crew Retrieved from either end; one soldier exposed, guide and connect Needs 9-m bearing for an unprepared abutment and 0.5 m for a prepared abutment	Scissors launch requires 10-m overhead clearance. Max launch slope— Uphill - 2.7 m Downhill - 2.7 m Sideslope - 0.3 m Fords 1.2 m

Bridging 10-1

BRIDGING/RAFTING

Boats

The standard boat in use today is the bridge erection boat - shallow draft (BEB-SD). However, the older 27-foot BEB is still in use. See Training Circular (TC) 5-210 for more information on this boat. Table 10-2 shows information about BEBs.

Table 10-2. BEBs

Equipment	Transportation	Capabilities	Assembly/ Propulsion	Remarks/ Limitations
BEB-SD	Carried by one 5-ton bridge truck w/ cradle or one medium-lift helicopter; boat weighs 8,800 lb	Carries a 3-man crew and either 12 soldiers with equipment or 4,400 lb of equipment	Launch time from the cradle is 5 minutes, and maximum speed is 25 knots.	Draft: normal opns is 22 inches; fully loaded is 26 inches; launch from cradle is 48 inches
BEB, 27 feet	Carried by one 5-ton bridge truck w/ cradle or one 2 1/2-ton truck w/pole trailer or one medium-lift helicopter, when procedures are certified	Carries a 3-man crew and either 9 soldiers with equipment or 3,000 lb of equipment	Launch time from the cradle is 5 minutes. Launch time from the 2 1/2-ton truck, when using a crane or wrecker, is 30 minutes. Maximum speed is 15 knots.	Draft is 40 inches

Improved Float Bridge (Ribbon)

A ribbon bridge's major components are the interior bay, which weighs 5,443 kilograms, and the ramp bay, which weighs 5,307 kilograms. For more information, see TM 5-5420-209-12. Table 10-3 lists allocations and Table 10-4 lists launch restrictions.

Table 10-3. Ribbon-bridge allocations (L-series TOE)

Allocation	Corps Ribbon Company
Number of bridge platoons	2
Number of interior bays	30
Number of ramp bays	12
Number of BEBs	14
Longest bridge that can be constructed (m)	215

Table 10-4. Launch restrictions

Restrictions	Free Launch	Controlled Launch	High-Bank Launch
Minimum depth of water required in centimeters (inches)	Ramp bay - 112 Interior bay - 92[1]	76 (30)[1]	76 (30)[1]
Bank's height restrictions in meters (feet)	0 to 1.5 (0 - 5)	0	1.5 to 8.5 (5 - 28)
Bank's slope restrictions	0 to 30 percent	0 to 20 percent	Level ground unless the front of the truck is restrained

NOTE: The launch is based on a 10 percent slope with the transporter backed into the water. The required water depth for a 30 percent slope with a 5-foot bank height is 183 centimeters. Interpolate between these values when needed.

[1] This is recommended water depth. Launch could technically be conducted in 43 centimeters (17 inches) of water.

Use the following formula to determine the number of ribbon interior bays you need to construct a ribbon bridge:

The number of interior bays =

$$\frac{gap\ (meters) - 14}{6.7}$$

or

$$\frac{gap\ (feet) - 45}{22}$$

NOTES:

1. Two ramp bays are required for all ribbon bridges.

2. During daylight hours, a ribbon bridge can be constructed at the rate of 200 meters (600 feet) per hour and during nighttime hours, at the rate of 133 meters (437 feet) per hour.

3. Two hundred vehicles per hour, with 30-meter spacing at 16 kilometers per hour, can cross the bridge.

FM 5-34

- Table 10-5 lists the bridge classification for wheeled and tracked vehicles.
- Normally, you anchor a ribbon bridge by tying BEBs to the downstream side of the bridge. Table 10-6 lists the number of boats you will need.
- Table 10-7, page 10-6, gives the load classifications of ribbon rafts based on the rafting site, type of rafting being conducted (longitudinal/conventional), and the current velocity.
- Table 10-8, page 10-7, gives planning factors for the number of round trips a raft can make based on river width, as well as the number of centerlines that can be supported.
- Table 10-9, page 10-7, gives unit rafting requirements.

FM 5-34

Table 10-5. Bridge classification

Crossing Type	Load Classification	Current Velocity (MPS)						
		0 to 0.9	1.2	1.5	1.75	2	2.5	2.7	3
Normal	Wheeled/tracked	96/75	96/75	96/70	96/70	82/70	65/60	45/45	30/30
Caution	Wheeled/tracked	105/85	105/85	100/80	100/80	96/80	75/65	50/50	35/35
Risk	Wheeled/tracked	110/100	110/95	105/90	105/90	100/90	82/75	65/65	40/40

Table 10-6. Boat requirements for anchoring a ribbon bridge

Current Velocity (MPS)	Number of Boats : Number of Bridge Bays (ratio)
0 to 2.0	1 : 6
2.0 to 2.6	1 : 3
2.7	1 : 2
Over 2.7	Bridge must be anchored using an overhead cable system.

Bridging 10-5

FM 5-34

Table 10-7. Ribbon-raft design

Raft Size	Assembly Time (increase by 50% at night)	Load Space (m)	Load Class	Current Velocity (MPS)							
				0 to 0.9	1.2	1.5	1.75	2	2.5	2.7	3
3 bays (2 ramps/ 1 interior)	8 min	6.7	L	45	45	45	40	40	35	30	25
			C	45	45	35	25	15	10	0	0
4 bays (2 ramps, 2 interiors)	12 min	13	L	70	70	70	60	60	60	55	45
			C	60	60	60	*55	*40	*30	*15	0
5 bays (2 ramps/ 3 interiors)	15 min	20.1	L	75	75	75	70	70	70	60	60
			C	75	70	70	*70	*60	*50	*25	0
6 bays (2 ramps/ 4 interiors)	20 min	26.8	L (W/T)	96/80	96/80	96/80	96/70	96/70	96/70	70/70	70/70
			C (W/T)	96/75	96/70	96/70	*70/70	*70/70	*55/55	*30/30	0

NOTES:
1. When determining raft classification, L refers to the longitudinal rafting and C refers to conventional rafting.
2. If the current's velocity in the loading/unloading area is greater than 1.5 MPS (5 fps), then conventional rafting must be used.
3. The roadway width of a ribbon raft is 4.1 meters (13 feet 5 inches).
4. The draft of a fully loaded ribbon raft is 61 centimeters (24 inches).
5. Vehicles should only be loaded on the interior bays.
6. Each raft requires a minimum of two BEBs for propulsion.
* Three BEBs are required for conventional rafting of 4, 5, 6 bay rafts in current velocities greater than 1.5 MPS (5 fps).

Longitudinal

Conventional

10-6 Bridging

Table 10-8. Planning factors for rafting operations, raft's centerline data

River Width (m)	Minutes per Round Trip	Maximum Number of Rafts per Centerline	Round Trips per Hour
75	7	1	8
100	8	1	7
125	9	1	6
150	10	2	6
188	11	2	5
225	12	2	5
263	14	3	4
300	16	3	3

NOTES:
1. This table provides approximate crossing times for a ribbon bridge and an M4T6 raft in current velocities of 0 to 1.5 MPS.
2. All round-trip times include the time required to load and unload the rafts.
3. Increase crossing times by 50 percent at night.
4. If the river width falls between two values, use the higher value.

Table 10-9. Unit rafting requirements

Units	Vehicles	Raft Trips Required		
		4 Bays	5 Bays	6 Bays
Armored battalion	161	119	101	86
Mechanized battalion	153	112	65	55
FA battalion	165	97	61	52
Engineer battalion	139	77	59	50
ACR	208	171	110	98

NOTE: Assume that current velocities are less than 0.9 MPS and that battalions/regiments are at 100 percent MTOE strength.

Long-Term Anchorage Systems

All heavy floating bridges require constructing a long-term anchorage, to include approach guys, an upstream (primary) anchorage, and a downstream (secondary) anchorage. See TC 5-210 for details.

Approach Guys

Attach approach guys to one end of the first floating support of all floating bridges. Secure them to the other end using deadmen, pickets, or natural holdfasts. Use a minimum of 1/2 inch improved plough steel (IPS) cable. When installed, the approach guys should form a 45-degree angle with the bridge.

Upstream Anchorage

An upstream anchorage system holds a bridge in position against a river's main current. This system should be based mainly on the current velocity and the bottom conditions. Table 10-10 contains information on designing an upstream anchorage system.

Table 10-10. Design of upstream (primary) anchorage systems

Current Velocity (MPS)	Bottom Conditions	
	Soft	Solid/Rocky
0 to 0.9	Kedge anchors every float upstream or shore guys every 6th float upstream	Shore guys every 6th float upstream
1.0 to 1.5	Combination of kedge anchors and shore guys	Overhead cable system
1.6 to 3.5	Overhead cable system	Overhead cable system

Downstream Anchorage

A downstream anchorage system protects floating bridges from reverse currents (tides) as well as from storms or severe winds which might change the direction of river flow. Table 10-11 lists information on the design of a downstream anchorage system.

Installation.

Table 10-12 contains information on how to install a long-term anchorage system.

Table 10-11. Design of downstream (secondary) anchorage systems

Reverse Current (MPS)	Bottom Conditions	
	Soft	Solid/Rocky
None expected	Kedge anchors every 3d float downstream or shore guys every 10th float downstream	Shore guys every 10th float downstream
0 to 0.9	Kedge anchors every float downstream or shore guys every 6th float downstream	Shore guys every 6th float downstream
1.0 to 1.5	Combination of kedge anchors and shore guys	Overhead cable system
1.6 to 3.5	Overhead cable system	Overhead cable system

Table 10-12. Procedures for installing long-term anchorage systems

System	Installation Procedures
Kedge anchor	1. Attach anchors to anchor lines, which must be a minimum of 1-inch manila rope. 2. Set or lay anchors. The horizontal distance from the anchor to the float must be at least 10 times the depth of the river. 3. Attach anchor lines to floats.
Shore guy	1. Attach shore guys to floats. Shore guys must be a minimum of 1/2-inch IPS cable and placed at a 45° angle with the bridge. 2. Ensure that shore guys are above the water; use floating supports if necessary. 3. Attach shore guys to deadman or holdfasts. 4. Ensure that the current's velocity does not exceed 0.9 MPS.
Combination	1. Emplace a kedge-anchor system. Attach anchor lines to every float. 2. Emplace a shore-guy system after installing the kedges. Attach shore guys to every sixth float. 3. Ensure that the current's velocity does not exceed 1.5 MPS.
Overhead cable	1. Design the system. 2. Construct Class 60 towers and install a deadman. 3. Install master cable. Check initial sag. 4. Attach every float to the master cable using bridle lines. 5. Ensure that the current's velocity does not exceed 3.5 mps.

Overhead-Cable Design Sequence

To design an overhead-cable anchorage system, you will need to calculate the information from Table 10-13 in the proper sequence. The numbers in the table correspond to the steps listed below.

Table 10-13. Data for overhead-design sequence

Cable Data	Tower Data	Deadman Data
1. The number of master cables 2. The diameter of master cable(s) (C_D) 3. The length of the master cable(s) (C_L) 4. The number of clips at each end of the cable 5. The spacing of cable clips 6. Initial sag (S)	7. Actual tower height (H) 8. Tower-waterline distance (A) 9. Tower-bridge offset (O_1)	10. Deadman face (D_f) and thickness (D_t) of largest timber 11. Deadman mean depth $1D_{Dmax}$) to determine actual mean depth (D_D) 12. Deadman length (D_L) 13. Deadman minimum thickness (D_L, D_t) 14. Tower-to-deadman distance (C) 15. Tower-to-deadman offset (O_2) 16. Bearing-plate dimensions (x, y, z) for each deadman

Use Figure 10-1 to determine where to measure for an overhead-cable anchorage system.

Cable Data

Step 1. Determine the size and number of master cables required. See Table 10-14, page 10-12, for float bridges.

Step 2. Determine the distance between the towers.

$$L = 1.1(G) + 100 \text{ feet}$$

where—

L = distance between towers, in feet

G = width of the wet gap, in feet

Step 3. Determine the length of the master cable.

$$C_L = L + 250 \text{ feet}$$

where—

C_L = length of the master cable, in feet

L = distance between towers, in feet

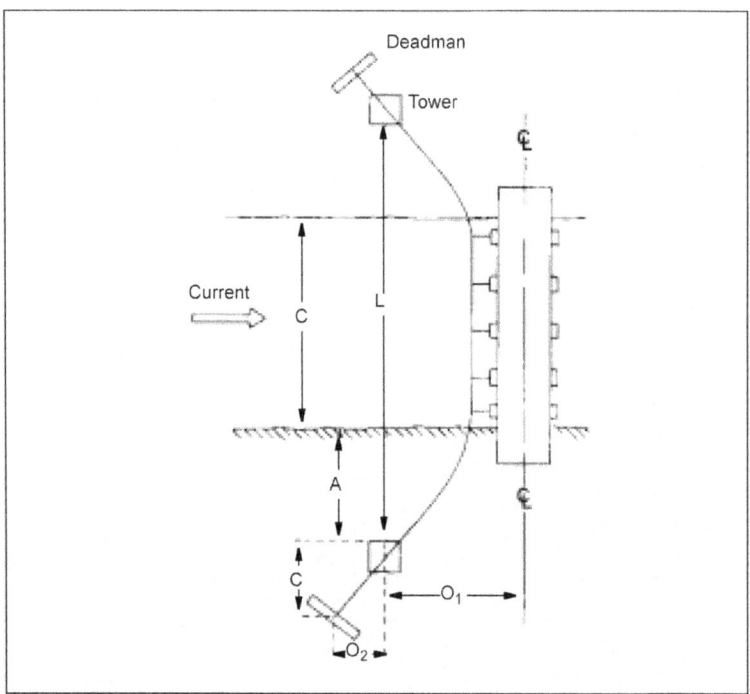

Figure 10-1. Measuring for an overhead-cable anchorage system

NOTE: The 250 feet (safety factor) is an approximation based on the most extreme circumstances (see Table 10-15, page 10-13).

Step 4. Determine the number of cable clips required to secure one end of the master cable.

$$Number\ of\ clips = (3C_D) + 1$$

where—

C_D = cable diameter, in inches

Step 5. Determine the spacing of the cable clips.

$$Clip\ spacing = 6C_D$$

where—

C_D = cable diameter, in inches

Bridging 10-11

FM 5-34

Table 10-14. Size and number of master cables (C_D) for float bridges

Wet-Gap Width (G) (ft)	Bridge-Assembly Type	1.5 MPS (5 fps)			2.1 MPS (7 fps)			2.7 MPS (9 fps)			3.4 MPS (11 fps)		
		Single	Dual	Triple	Single	Dual	Triple	Single	Dual	Triple	Single	Dual	Triple
200	Normal	0.500	0.375	0.375	0.625	0.500	0.500	0.750	0.625	0.500	0.875	0.750	0.625
	Reinforced	0.625	0.500	0.375	0.750	0.625	0.500	0.875	0.750	0.625	1.125	0.875	0.750
400	Normal	0.625	0.500	0.500	0.750	0.625	0.500	1.000	0.875	0.625	1.250	1.000	0.750
	Reinforced	0.750	0.625	0.500	1.000	0.750	0.625	1.250	1.000	0.750	1.500	1.250	0.875
600	Normal	0.750	0.625	0.500	1.000	0.750	0.625	1.250	1.000	0.750	1.500	1.250	0.875
	Reinforced	1.000	0.750	0.625	1.125	1.000	0.750	1.500	1.250	0.875	*	1.500	1.125
800	Normal	0.875	0.750	0.625	1.125	0.875	0.750	1.375	1.125	0.875	*	1.500	1.125
	Reinforced	1.125	0.875	0.750	1.375	1.125	0.875	*	1.375	1.000	*	*	1.250
1,000	Normal	1.000	0.875	0.750	1.250	1.000	0.875	1.500	1.375	1.000	*	*	1.250
	Reinforced	1.250	1.000	0.750	1.500	1.250	1.000	*	*	1.125	*	*	1.750
1,200	Normal	1.125	0.875	0.750	1.375	1.125	0.875	*	1.500	1.125	*	*	1.375
	Reinforced	1.375	1.125	0.875	*	1.375	1.000	*	*	1.250	*	*	*

NOTE: All values are based on IPS cable and a 2 percent initial sag.
* It is unsafe to construct this system.

Table 10-15. Weight and breaking strengths for common cables (cable capacity)

Cable Diameter	0.375	0.500	0.625	0.750	0.875	1.000	1.125	1.250	1.625	1.500
Weight (lb per Foot)	0.203	0.400	0.630	0.900	1.230	1.600	2.030	2.500	3.030	3.600
Type of Cable	Breaking strength (lb)									
IPS	10,000	17,000	26,200	37,400	50,800	66,000	83,000	102,000	123,000	145,000
MPS	11,000	18,800	28,800	41,200	56,000	73,000	92,000	113,000	136,000	161,000
Plough steel	12,600	21,600	33,200	47,400	64,400	84,000	106,000	130,000	157,000	185,000

NOTES:
1. The strength varies slightly with the strand construction and the number of strands.
2. The strength varies approximately with the square of the diameter of the cable; for example, a 3/4-inch cable is 4 times as strong as a 3/8-inch cable made of the same materials; see the equation below:

$$\left(\frac{3}{4}\right)^2 + \left(\frac{3}{8}\right)^2 = 4$$

Step 6. Determine the initial sag.

$$S = 0.02(L)$$

where—

S = sag, in feet

L = distance between towers, in feet

Tower Data

Step 7. Determine the tower height (H). Do this calculation for the near shore (NS) and far shore (FS) since bank heights may be different.

$$H_R = 3 feet + S - BH$$

where—

H_R = required tower height, in feet

S = sag, in feet

BH = bank height, in feet

After determining H_R, use Table 10-16, which lists actual tower heights, and select the smallest possible tower that is greater than or equal to H_R.

NOTE: If the NS and FS towers have different heights, calculate for each in steps 9 through 16.

Table 10-16. Tower heights

Number of Tower Sections	Tower Height (H)
Cap, base, and pivot unit	3 feet 8 1/4 inches
With 1 tower section	14 feet 6 1/4 inches
With 2 tower sections	25 feet 4 1/4 inches
With 3 tower sections	36 feet 2 1/4 inches
With 4 tower sections	47 feet 1/4 inch
With 5 tower sections	57 feet 10 1/4 inches
With 6 tower sections	68 feet 8 1/4 inches

Step 8. Determine the distance from each tower to the waterline. Calculate for the NS and FS.

$$A = \frac{L-G}{2}$$

where—

A = distance from each tower to the waterline, in feet

L = distance between towers, in feet

G = gap width, in feet

Step 9. Determine the offset from each tower to the ridge centerline. Calculate for the NS and FS.

$$O_1 = H + 50 \text{ feet (if BH is less than or equal to 15 feet)}$$

or

$$O_1 = H + BH + 35 \text{ inches (if BH is greater than 15 feet)}$$

where—

O_1 = offset from tower to ridge centerline, in feet

H = actual tower height, in feet

BH = bank height, in feet

Deadman Data

Step 10. Identify the deadman dimensions. Select a deadman from the available timbers and logs. Generally, you would select the timber with the largest timber face/log diameter.

$$D_f \underline{\hspace{2cm}}$$
$$D_t \underline{\hspace{2cm}}$$

where—

D_f = largest face of deadman, in feet

D_t = deadman thickness, in feet

Step 11. Determine the mean depth of a deadman. Make sure that at least 1 foot of undisturbed soil is between the bottom of the deadman and the groundwater level (GWL). Calculate for the deepest a deadman can be; calculate for NS and FS.

$$D_{Dmax} = GWL - 1\,ft - \frac{D_f}{2}$$

where—

D_{Dmax} = maximum depth of deadman, in feet

GWL = groundwater-level depth, in feet

D_f = deadman face, in feet

Compare D_{Dmax} to the minimum depth (3 feet) and maximum depth (7 feet) to determine the actual mean depth of a deadman (D_D).

Step 12. Determine the length of a deadman. Calculate for NS and FS.

$$D_L = \frac{CC}{HP \times D_f} + 1$$

where—

D_L = deadman length, in feet

CC = capacity of anchorage cable, in lb/1,000 (see Table 10-17)

HP = required holding power (HP), in lb/sq ft (see Table 10-18)

D_f = deadman face, in feet (for log deadman, use log's diameter)

Table 10-17. Anchorage-cable capacities

Cable Type	Cable Size (inches) (C_D)								
	3/8	1/2	5/8	3/4	7/8	1	1 1/8	1 1/4	1 3/8	1 1/2
IPS	12.6	21.6	33.2	47.4	64.4	84	106	130	157	185
PS	11	18.8	28.8	41.2	56	73	92	113	136	161
MPS	10	17	26.2	37.4	50.8	66	83	102	123	145

Table 10-18. Required HP (lb/sq ft)

Deadman Depth (ft) (D_D)	Tower to Deadman Slope (ratio)			
	1:1 (45°)	1:2 (26.5°)	1:3 (18.5°)	1:4 (14°)
3	0.95	1.3	1.45	1.5
4	1.75	2.2	2.60	2.7
5	2.80	3.6	4.00	4.1
6	3.80	5.1	5.80	6.0
7	5.10	7.0	8.00	8.4

NOTES:
1. For hardpan or rock, multiply HP by 5.
2. For fine-grained soils with high moisture content, multiply HP by 1/2.
3. For this table, assume loamy soil.

Step 13. Check the minimum thickness of deadman for timber and logs. Calculate for NS and FS.

$\dfrac{D_l}{D_t}$ for timber, must be less than or equal to 0

10-16 Bridging

$$\frac{D_L}{d} \text{ for logs, must be less than or equal to 5}$$

where—

D_L = deadman length, in feet

D_t = deadman thickness, in feet

d = log diameter, in feet

Step 14. Determine the tower-to-deadman distance. Calculate for NS and FS.

$$C = \frac{H + D_D}{slope}$$

where—

C = distance from the tower to the deadman, in feet

H = actual tower height, in feet

D_D = mean depth of deadman, in feet

slope = tower-to-deadman slope

Step 15. Determine the tower-to-deadman offset. Calculate for NS and FS.

$$O_2 = C(O_2 ft)$$

where—

O_2 = tower-to-deadman offset, in feet

C = tower-to-deadman distance, in feet

O_2 ft = a factor determined from Table 10-19.

Table 10-19. O_2 ft factor

Assembly Type	Current Velocity				
	3 fps	5 fps	7 fps	9 fps	11 fps
Normal	0.09	0.11	0.14	0.17	0.19
Reinforced	0.11	0.14	0.17	0.19	0.23

Step 16. Design a bearing plate for each deadman. Given the deadman face (D_f) or log diameter (d) and the size of the master cable (C_D), use Table 10-20, page 10-18, to determine the length (y), thickness (x), and face (z) of the deadman bearing plate. The values in Table 10-20 and Table 10-21, page 10-19, are based on the use of IPS cable.

Table 10-20. Flat bearing-plate dimensions

Deadman Face (D_f)		Cable Size (C_D) (inches)								
		3/8	1/2	5/8	3/4	7/8	1	1 1/8	1 1/4	1 1/2
8	x	7/16	7/8	1 1/4						
	y	4	8	11						
	z	6	6	6						
10	x	7/16	11/16	1	1 3/8					
	y	4	6	9	12					
	z	8	8	8	8					
12	x	7/16	9/16	13/16	1 1/8	1 7/16				
	y	4	5	7	10	13				
	z	10	10	10	10	10				
14	x	7/16	7/16	11/16	7/8	1 1/4	1 9/16	2		
	y	4	4	6	8	11	14	18		
	z	12	12	12	12	12	12	12		
16	x	7/16	7/16	9/16	13/16	1 1/8	1 3/8	1 11/16	2 1/8	
	y	4	4	5	7	10	12	15	19	
	z	14	14	14	14	14	14	14	14	
18	x	7/16	7/16	7/16	11/16	7/8	1 1/4	1 9/16	1 13/16	
	y	4	4	4	6	8	11	14	16	
	z	16	16	16	16	16	16	16	16	
20	x	7/16	7/16	7/16	11/16	7/8	1 1/8	1 3/8	1 11/16	
	y	4	4	4	6	8	10	12	15	
	z	18	18	18	18	18	18	18	18	
24	x	7/16	7/16	7/16	9/16	11/16	7/8	1 1/8	1 3/8	1 7/8
	y	4	4	4	5	6	8	10	12	17
	z	22	22	22	22	22	22	22	22	22

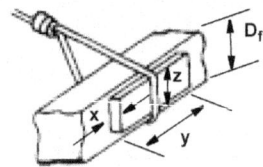

10-18 Bridging

FM 5-34

Table 10-21. L-shaped flat bearing-plate dimensions

Deadman Face (D_f)		Cable Size (C_D) (inches)								
		3/8	1/2	5/8	3/4	7/8	1	1 1/8	1 1/4	1 1/2
6	x	1/8	3/16							
	y	4	7							
8	x	1/8	1/8	3/16						
	y	3	5	8						
10	x	1/8	1/8	1/8	1/4					
	y	2	4	7	10					
12	x	1/8	1/8	1/8	1/8	1/4				
	y	2	4	6	8	11				
14	x	1/8	1/8	1/8	1/8	1/4	5/16			
	y	2	3	5	7	9	12	15		
16	x	1/8	1/8	1/8	1/8	1/8	3/16	1/4	3/8	
	y	2	2	4	6	8	11	14	17	
18	x	1/8	1/8	1/8	1/8	1/8	1/8	3/16	1/4	
	y	2	2	4	6	7	10	12	15	
20	x	1/8	1/8	1/8	1/8	1/8	1/8	1/8	3/16	3/8
	y	2	2	3	5	7	9	11	13	19
24	x	1/8	1/8	1/8	1/8	1/8	1/8	1/8	1/8	1/4
	y	2	2	3	4	6	8	9	11	16
30	x	1/8	1/8	1/8	1/8	1/8	1/8	1/8	1/8	1/8
	y	2	2	3	4	5	6	7	9	13
36	x	1/8	1/8	1/8	1/8	1/8	1/8	1/8	1/8	1/8
	y	2	2	2	3	4	5	6	8	10

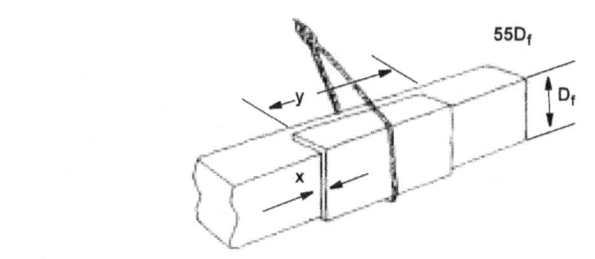

Bridging 10-19

MEDIUM GIRDER BRIDGE (MGB)

See TM 5-5420-212-12 for information on component descriptions, construction, palletizing, and maintenance procedures for the MGB and TM 5-5420-212-12-1 for information on the link-reinforcement set (LRS). The following list of abbreviations are used in the text, figures, and tables on the MGB:

Abbreviation	Meaning
A	edge of the gap, far bank (FB)
A'	edge of the gap, near bank (NB)
AA	anchor assembly
AA(L)	long link of the anchor assembly
AA(S)	short link of the anchor assembly
AF	antiflutter tackle
AR	angle of repose which is marked on-site with A and A'
AR gap	distance from A' to A
BES	bridge-erection set
boom marker	carrying bar (painted orange) which marks the position of the next booming/ launching point
BP	building pedestal (SS only), baseplate (SS and DS)
BSP	bank-seat beam
C	distance of the water below the line joining the FRB and the final position of the far end of the bridge as marked with the F peg (F) at distance of end taper panel from the FRB, for maximum deflection (W) from the FRB (negative), fine for up to 2 end of bridge (E)+12; need a CRB for 13 through 22 bays.
CG marker	carrying bar (painted blue) which marks the center of gravity of the bridge during construction
CRB	capsill roller beam: must be used for 2E +13 through 2E +22 bays DS bridges w/ or w/o LRS
D	deflection of bridge during launch in relation to line joining FRB and F pegs
DS	double-story bridge construction
DU	deck unit
E	end of the bridge
F	final position of the far end of the bridge as marked with the F peg
FRB	front roller beam
G	distance between the O peg and the baseline
H	far-bank height at F peg, relative to the baseline
Ht	height
L	bridge length
LLN	light launching nose
LNCG	launching-nose cross girder

Abbreviation	Meaning
LNH	launching nose, heavy
LR	landing roller: used by itself for 4 through 8 bays, SS; used in the LRP for all other bridge lengths.
LRD	long ramp-deck pallet; the last pallet to be used on a bridge site should be loaded on the push vehicle to maintain a proper counterweight.
LRP	landing-roller pedestal (MK 1 for 2E + 1 through 2E + 12 bays DS, MK2 for 2E + 13 through 2E + 22 bays DS w/ or w/o LRS)
LRS	link-reinforcing set
LT	light tackle
LZ	landing zone
MLC	military load classification
N	nose-tip height above the baseline
*N1	launching-nose, heavy, one-story high
**N2	launching-nose, heavy, two-stories high
O	distance R from the RB (SS), FRB (DS), and CRB (DS w/ or w/o LRS) as marked with the O peg
PT	post-tensioning assembly
R	maximum distance to the rear of the bridge during construction (excluding the push bar and vehicle)
RB	roller beam
RRB	rear roller beam
SS	single-story bridge construction
T	height of the home bank end of the bridge in relation to the baseline
V	for delaunching purposes, the distance from the FRB or CRB to the LRP for DS bridges requiring a launching nose
W	distance of the end-taper panel from the FRB, for maximum deflection
WL	waterline
1LL	one long link
1SL	one short link
*6N1, 7N1, and 8N1	SS nose construction: the first number shows the number of heavy nose sections used, N1= single nose.
**6N1 + 3N2	DS nose construction: 6N1, same as above; 3N2 means three heavy nose sections used in the second story; N2 means nose, DS
2+3+ or 8 through 10	number of bays to add: 2 + 3 + means add bays 2 and 3; 8 through 10 means add bays 8 through 10.
Boom to	movement of the bridge until the panel point given is over the RB (for SS) or the RRB (for DS)
Launch to	movement of the bridge until the panel point given is over the RB, FRB, or CRB.
3D, 8D, 20D, 27D +6C, and 37D+6C	counterweight codes giving the number of deck units and curbs required
(4pO), (2p4), and (Bp3)	examples of the way that the center of gravity is shown

MGB DESIGN—SS, 4 THROUGH 12 BAYS

Step 1. Measure the AR gap: A to $A' =$ _____ meters.

Use Table 10-22, columns a and b, for MLCs 30 through 70 and Table 10-23, columns a and b, for MLCs 16 through 24.

Table 10-22. Dimensions for SS bridges, 4 through 8 bays

a AR Gap Ranges (m)	b MLC	c Number of Bays	d Bridge Length (m)	e R Distance (m)	f Nose Construction
3.7 to 7.1	70*	4	7.9	5.8	LLN only
5.6 to 9.0	70*	5	9.8	6.7	LLN only
7.3 to 10.8	40	6	11.6	7.6	LLN only
9.1 to 12.6	30	7	13.4	9.5	LLN only
11.0 to 14.4	30	8	15.2	11.3	LLN only
*See step 7.					

Table 10-23. Dimensions for SS bridges, 9 through 12 bays

a AR Gap Ranges (m)	b MLC	c Number of Bays	d Bridge Length (m)	e R Distance (m)	f Nose Construction
12.9 to 16.3	24	9	17.1	10.4	5N1
14.6 to 18.1	20	10	18.9	12.2	5N1
16.5 to 19.9	16	11	20.7	12.2	6N1
18.3 to 21.7	16	12	22.6	14.0	6N1

Step 2. Select the number of bays from column c = _____.

Step 3. Select the bridge length from column d = _____ meters.

Step 4. Select the R distance from column e = _____ meters.

Step 5. Select the nose construction from column f = _____ (see Figure 10-2 and Figure 10-3, page 10-24).

Step 6. See Figures 10-4 and 10-5, pages 10-24 and 10-25, for key construction points, dimensions, and elevations for a push launch and a jack launch.

C1, FM 5-34

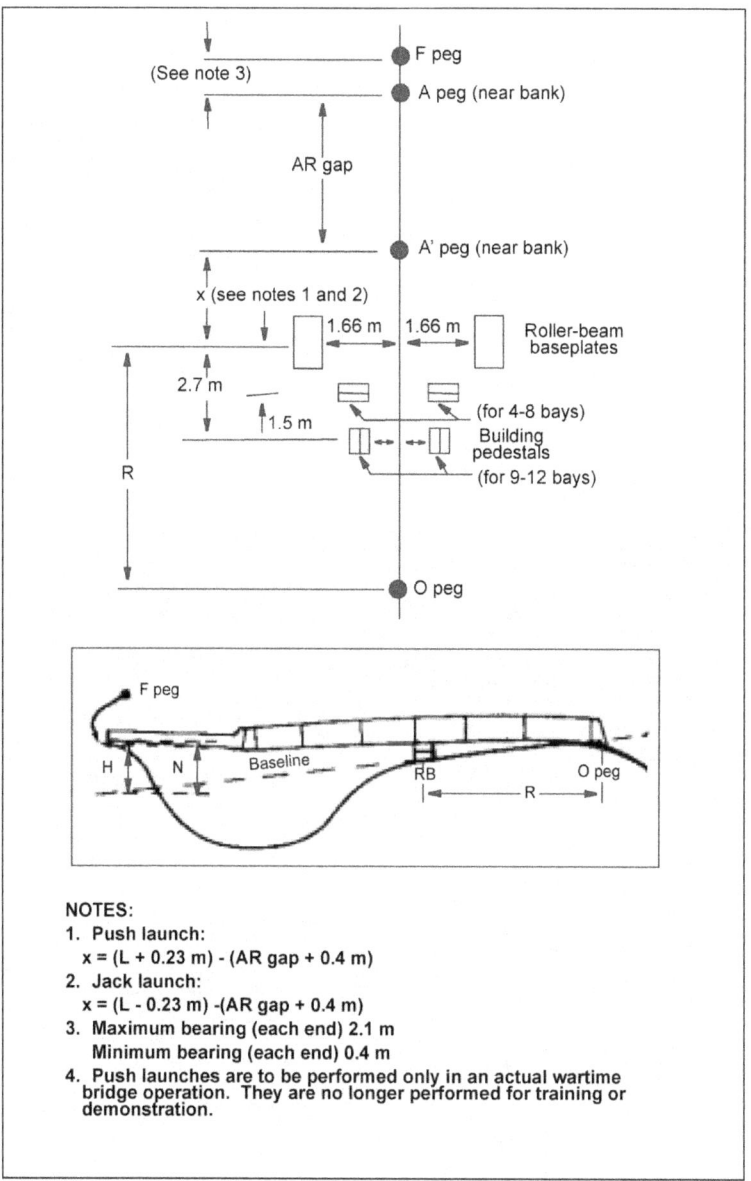

NOTES:
1. Push launch:
   x = (L + 0.23 m) - (AR gap + 0.4 m)
2. Jack launch:
   x = (L - 0.23 m) - (AR gap + 0.4 m)
3. Maximum bearing (each end) 2.1 m
   Minimum bearing (each end) 0.4 m
4. Push launches are to be performed only in an actual wartime bridge operation. They are no longer performed for training or demonstration.

Figure 10-2. SS MGB site layout (4 through 12 bays)

Bridging 10-23

FM 5-34

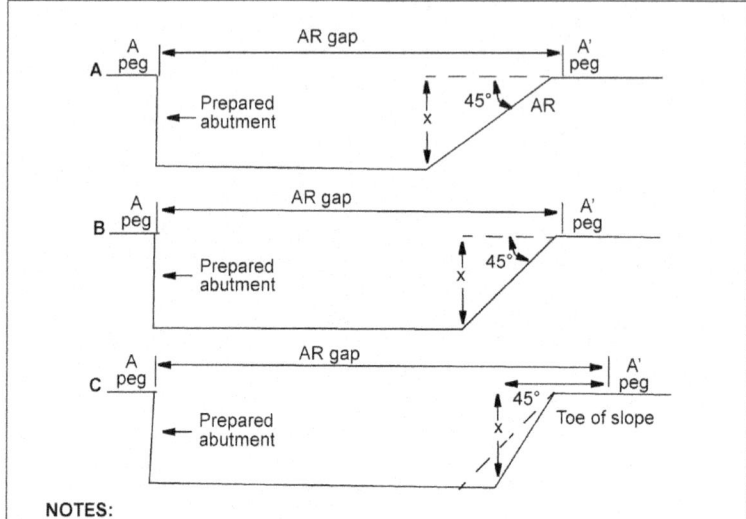

NOTES:
1. If the actual slope of the bank does not exceed 45° from the horizontal, place A, A' peg as shown in A or B.
2. If the actual slope of the bank does exceed 45° from the horizontal, place A, A' peg a distance equal to the height of the bank, which is measured from the toe of slope (see example C above, distance x).
3. The gaps above are shown with one prepared and one unprepared abutment. The actual sites may be any combination of the examples shown.

Figure 10-3. Measuring AR gap

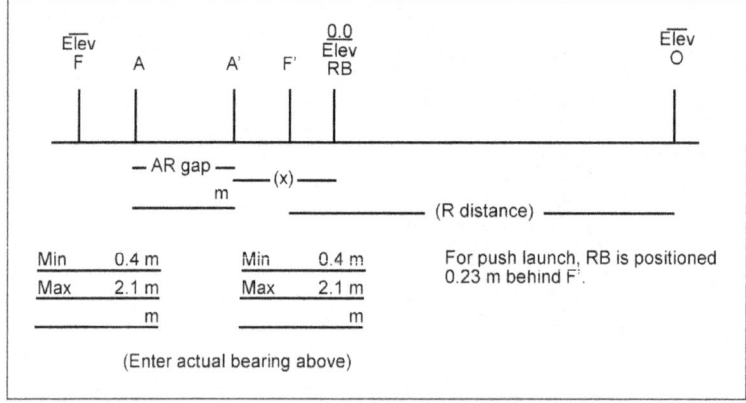

Figure 10-4. Construction elements for a push launch

10-24 Bridging

FM 5-34

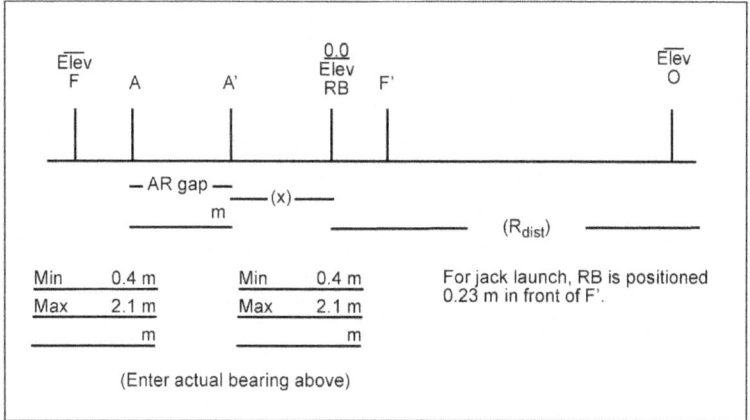

Figure 10-5. Construction elements for a jack launch

Step 7. Check the slope. For longitudinal slope, ensure that the difference in elevation between the F' (use elevation of the RB) and F peg does not exceed 1/10th of the total bridge length. If it does, you are going to have to crib up, undertake a major construction project, or find another site. For a transverse/cross slope, ensure that the transverse slope on both banks does not exceed 1/10th of the total bridge width (4 meters) for an MLC of 61 and over.

Step 8. Calculate H (for later comparison against N) (see Figure 10-6).

For a push launch—

$$H = elev\, F + \frac{elev\, O\, (L + 0.23\ meter)}{R_{dist}} = \underline{\qquad}$$

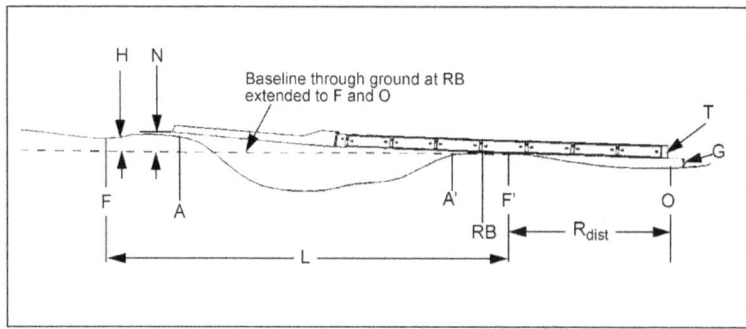

Figure 10-6. Calculating H to compare against N

Bridging 10-25

For a jack launch—

$$H = elev\ F + \frac{elev\ O\ (L - 0.23\ meter)}{R_{dist}} =$$

Step 9. Determine an RB setup and packing, if required. For 4 through 8 bays, choose an RB setup and/or packing, if required, to give $N>H$ (see Table 10-24). For 9 through 12 bays, choose an RB setup, an LNCG setting, and/or packing, if required (see Table 10-25). Use the guidelines below:

- You can get an extra 75 millimeters of clearance by lifting the nose to take out the pin sag (4 through 8 bays). If you estimate the levels, you do not have to consider this during the design, but you must compensate for any errors when calculating the value of H.

- You can get an extra 0.6 meters of clearance by lifting the nose to take out the pin sag (for bridges 9 through 12 bays).

- Placing extra packing under the RB will increase the vertical-lift interval N by three times the thickness of the packing (for example, if the packing is 75-millimeters thick, N would increase by 225 millimeters).

- Table 10-25 incorporates an allowance to ensure that the nose clears the LR when it is placed 230 millimeters in front of the F peg.

- The height of the RB on the baseplate is only 0.43 meter. The height of the baseplate and deck unit is 0.69 meter.

Table 10-24. RB setup and packing

Number of Bays	g N (m) BP	h N (m) BP + DU
4	1.30	1.98
5	1.14	1.91
6	1.07	1.83
7	0.76	1.14
8	0.38	1.07
NOTE: Read right from number of bays until N>H. Then read up to select the RB setup and/or packing, if required.		

Step 10. Calculate the number of truck and trailers needed to haul bridge components (see Table 10-26).

Truck_____ Trailer_____

Table 10-25. RB setup and packing (LNCG setting)

Number of Bays	SS Bridges, 9 Through 12 Bays					
	LNCG Setting #4 Black		LNCG Setting #2 Black		LNCG Setting #1 Black	
	g BP (m)	h BP + DU (m)	g BP (m)	h BP + DU (m)	g BP (m)	h BP + DU (m)
9	-0.76	0.15	0.61	1.37	1.83	2.59
10	-0.99	-0.38	0.38	0.9	1.60	2.21
11	-1.37	-0.84	0.15	0.71	1.83	2.67
12	-2.13	-1.37	-0.46	0.31	1.07	1.83

NOTE: Use an LNCG setting so that N>H and T>G.

Table 10-26. SS pallet loads

Pallet Type	Number of Bays			
	4 to 5	6 to 7	8 to 9	10 to 12
Erection	1	1	1	1
Bridge	1	2	3	4
Total	2	3	4	5

NOTE: Additional vehicles are required to transport personnel. Erection pallets may only be partial, depending on the bridge being built.

Step 11. Determine the manpower and time required to accomplish the mission (see Table 10-27).

Table 10-27. Manpower and time requirements

a Manpower and Time	b 4 to 5 Bays	c 6 to 8 Bays	d 9 to 12 Bays
Working party	1 + 8	1 + 16	1 + 16
Time during daylight (hours)	1/2	3/4	1
Time during darkness (hours)	3/4	1	1 1/4

PROCEDURE:
Add 20 percent for unskilled personnel.
Add 30 percent for inclement weather.
Add 30 percent for adverse site conditions.
Calculate all timings consecutively.
Exclude any work on approaches.

Step 12. Identify the final bridge design.
- Number of bays____
- LNCG setting____
- Setup of roller beam and packing____
- Bearings: NB____FB____
- Total loads: Trucks____Trailers____
- Work party____
- Total time____: Day____Night____

MGB DESIGN—DS, 2E + 1 THROUGH 12 BAYS

Step 1. Measure the AR gap: A to A' = _____meters; see Table 10-28, column a.

Step 2. Select the number of bays from column c = 2E+_____.

Step 3. Select the bridge length from column d = _____meters.

Step 4. Select the R distance from column e = _____meters.

Step 5. Select the nose construction from column f = _____.

Table 10-28. Dimensions for DS, 2E + 1 through 12 bays

a Gap Ranges (m)	b MLC*	c Number of Bays	d Bridge Length (m)	e R Distance (m)	f Nose Construction
6.4 to 9.0	70 (T)	1	11.0	10.0	2N1
8.2 to 10.8	70 (T)	2	12.8	11.9	3N1
10.0 to 12.6	70 (T)	3	14.6	12.2	3N1
11.9 to 14.5	70 (T)	4	16.5	13.1	3N1
13.7 to 16.3	70 (T)	5	18.3	14.9	4N1
15.5 to 18.4	70 (T)	6	20.1	14.9	4N1
17.3 to 19.9	70 (T)	7	21.9	15.8	4N1
19.2 to 21.8	70 (T)	8	23.8	16.8	5N1
21.0 to 23.6	70 (T)	9	25.6	17.7	5N1
22.8 to 25.4	70 (T)	10	27.4	19.5	5N1
24.7 to 27.3	70 (T)	11	29.3	20.4	6N1
26.5 to 29.1	70 (T)	12	31.1	21.6	6N1

*See step 7. (T) = Tracked

Step 6. Calculate the FB and NB bearings as follows (assume the FB bearing to be 0.6 meter)(see Figure 10-7):

$$NB = L - AR\ gap - FB\ bearing$$

$$X = NB\ bearing - 0.5\ meter$$

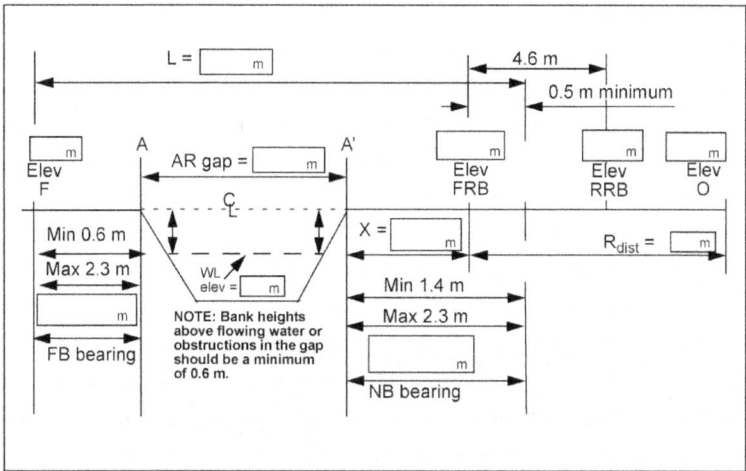

Figure 10-7. Constructing NB and FB bearings (DS, 2E + 1 through 12 bays)

Step 7. Check the slope. For longitudinal slope, ensure that the difference in elevation between the F' (use elevation of the FRB) and F peg does not exceed 1/10 of the total bridge length. If it does, you will either crib it up, undertake a major construction project, or find another site. For a transverse/cross slope, ensure that the transverse slope on both banks does not exceed 1/10th of the total bridge width (4 meters), regardless of the MLC.

Step 8. Calculate H, G, and C (for later comparison against N, T, and D) (see Figure 10-8, page 10-30).

$$H = elev\ F + \frac{elev\ RRB\ (L - 0.5\ meter)}{4.6\ meters} =$$

$$G = elev\ O - \frac{elev\ RRB(R_{dist})}{4.6\ meters} =$$

$$C = elev\ WL - \frac{elev\ F(W_{dist})}{(L - 0.5\ meter)} =$$

FM 5-34

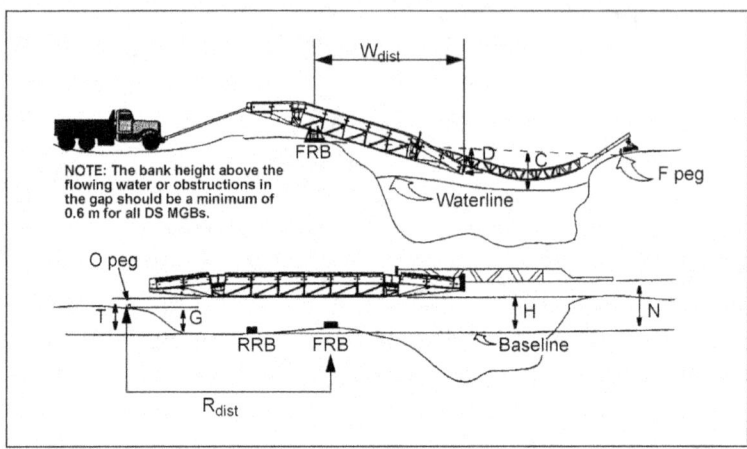

Figure 10-8. Calculating *H, G,* and *C* to compare against *N, T,* and *D*

Step 9. Identify the LNCG using Table 10-29. If the LNCG is allowed, see columns h, i, and j.

Table 10-29. Rule 1 for LNCG, 2E + 1 through 12 bays

2E+ Number of Bays	D: Given LNCG Settings With FRB and RRB in Low Position			
	g W Distance	h #6 Yellow	i #4 Yellow	j #2 Yellow
1 to 6	-	-	-	-
7	13.1	0.70	0.31	-0.09
8	15.0	0.67	0.25	-0.20
9	16.5	0.64	0.21	-0.30
10	17.6	0.60	0.12	-0.40
11	18.5	0.50	0.04	-0.43
12	19.2	0.46	-0.06	-0.58
PROCEDURE: • Go to Table 10-30 if *D* is > *C*. • Go to Table 10-30 if *D* is not > *C* and water is not flowing. • Choose another site if *D* is not > *C* and the current velocity is 5 mph. • Do not adjust the LCNG setting if *D* is not >*C* and immersion in water is < 0.3 meter and the current velocity is < 5 MPS. Go to Table 10-31. • Go to Table 10-31, if both bank heights are ≥ 0.6 meter (depth) from centerline. • Choose an LNCG setting that ensures that the value of *D* (from FRB/F line to the point of maximum deflection) is greater than the depth of *C* (from FRB/F to WL).				

Step 10. Choose the LNCG setting so N>H and T>G; use Table 10-30 (rule 2).

Step 11. Identify *N* and *T* using Table 10-31.

Step 12. Identify *N* using Table 10-32, page 10-32, rule 4A.

10-30 Bridging

Table 10-30. Rule 2 for LNCG, 2E + 1 through 12 bays

2E+ Number of Bays	N: Given LNCG Settings With FRB and RRB in Low Position—DS Bridges			
	k #6 Yellow	l #4 Yellow	m #2 Yellow	n Tail Lift
1	1.02	1.48	2.04	0.55
2	0.89	1.53	2.30	0.55
3	0.86	1.50	2.28	0.55
4	0.81	1.45	2.23	0.55
5	0.70	1.52	2.51	0.52
6	0.65	1.48	2.47	0.52
7	0.53	1.36	2.36	0.52
8	0.49	1.48	2.69	0.46
9	0.33	1.35	2.55	0.46
10	0.25	1.28	2.49	0.46
11	0.16	1.23	2.63	0.40
12	-0.20	1.02	2.47	0.40

PROCEDURE:
- Use an LCNG setting so that $N>H$ and $T>G$.
- Choose an LNCG setting so that $N>H$; see columns k, l, and m.
- Choose the highest value available if none of the choices meet the criterion. Read right from number of bays, then up to determine the LNCG number.
- Record N and T: $N =$ _____ $T =$ _____ (column n).
- Record LNCG setting:___ (setting cannot be lower than setting from rule 1).
- Go to Table 10-31 if $N \leq H$.
- Go to Step 14 if $N > H$.

Table 10-31. Rule 3 for N and T, 2E + 1 through 12 bays

2E+ Number of Bays	o N	p T
1 to 4	$N = N$ (Table 10-30) + 0.69 m	1.24
5 to 7	$N = N$ (Table 10-30) + 0.69 m	1.21
8 to 10	$N = N$ (Table 10-30) + 0.69 m	1.15
11 to 12	$N = N$ (Table 10-30) + 0.69 m	1.09

PROCEDURE:
- Raise the FRB and RRB by 0.69 meter. Both roller beams are now in the highest position.
- Determine N:_____ $N = N$ (Table 10-30) + 0.69 meter.
- Check to see if $T>G$: $T =$ _____
- Go to step 14 if $N>H$ and $T>G$; the design is correct.
- Go to Table 10-32, page 10-32, (rule 4A) if $N \leq H$.
- Go to Table 10-32 (rule 4B) if $T \leq G$.

FM 5-34

Table 10-32. Rules 4A and 4B for N and T, 2E + 1 through 12 bays

Rule 4A—Lowering RRB to Increase N		Rule 4B—Lowering FRB to Increase T
2E+ Number of Bays	q N FRB in High	r T RRB in High
1 to 4	1.75 (1.24 - G)	0.2 (N [Table 10-31] - H)
5 to 7	1.75 (1.21 - G)	0.2 (N [Table 10-31] - H)
8 to 10	1.75 (1.15 - G)	0.2 (N [Table 10-31] - H)
11 to 12	1.75 (1.09 - G)	0.2 (N [Table 10-31] - H)

PROCEDURE:
• Determine N:_____ N = N (Table 10-31, page 10-31) + answer to column q.
• Go to step 14 if N>H.
• Consider another site or additional packing if N<H.

Step 13. Identify T using Table 10-32 (rule 4B). Lower the FRB to the low position; the RRB will remain in the high position.

$T = T$ (Table 10-31) + answer from Table 10-32 column r. = _____

If $T>G$, go to step 14.

Step 14. Determine the required truck and trailer loads using Table 10-33.

Truck_____Trailer_____

Table 10-33. DS pallet loads, 1 through 12 bays

Pallet Type	Number of Bays			
	1 to 3	4 to 6	7 to 9	10 to 12
Erection	1	1	1	1
Bridge	4	5	6	7
Total	5	6	7	8

NOTE: Additional vehicles are required to transport personnel. Erection pallets may only be partial, depending on the bridge being built.

10-32 Bridging

Step 15. Determine the manpower and time required to accomplish the mission (see Table 10-34).

Table 10-34. Manpower and time requirements, 1 through 12 bays

a Manpower and Time	b 1 to 4 Bays	c 5 to 8 Bays	d 9 to 12 Bays
Working party	1 + 24	1 + 24	1 + 24
Time during daylight (hours)	3/4	1	1 1/2
Time during darkness (hours)	1 1/4	1 1/2	2
PROCEDURE: •Add 20 percent for unskilled personnel. •Add 30 percent for inclement weather. •Add 30 percent for adverse site conditions. •Calculate all timings consecutively.			

Step 16. Identify the final bridge design.

• 2E + ____ bays

• LNCG setting ____

• FRB setting ____

• RRB setting ____

• Bearings: NB ____ FB ____

• Total loads ____ : Trucks ____ Trailers ____

• Total time ____ : Day ____ Night ____

MGB Design—DS, 2E + 13 Through 22 Bays (w/o LRS)

Step 1. Measure AR gap: A to A' = ____ meters; see Table 10-35 (page 10-34), column a.

Step 2. Select the number of bays from column c = 2E+ ____ .

Step 3. Select bridge length from column d = ____ meters.

Step 4. Select R distance from column e = ____ meters.

Step 5. Select nose construction from column f = ____ .

Step 6. Calculate the FB and NB bearings as follows (assume the FB bearing to be 0.6 meter)(see Figure 10-9, page 10-34):

$$NB = L - AR\ gap - FB\ bearing$$

$$X = NB\ bearing - 0.5\ meter$$

Table 10-35. DS dimensions, 2E + 13 through 22 bays w/o LRS

a Gap Ranges (m)	b MLC*	c Number of Bays	d Bridge Length (m)	e R Distance (m)	f Nose Construction
28.3 to 30.9	60(T)	13	32.9	27.7	6N1
30.2 to 32.8	50	14	34.8	29.0	7N1
32.0 to 34.6	40	15	36.6	29.0	7N1
33.8 to 36.4	40	16	38.4	29.6	7N1
35.6 to 38.2	30	17	40.2	29.6	8N1
37.5 to 40.1	30	18	42.1	29.6	8N1
39.3 to 41.9	24	19	43.9	35.1	6N1 + 3N2
41.1 to 43.7	24	20	45.7	38.7	6N1 + 3N2
43.0 to 45.3	20	21	47.6	38.7	6N1 + 3N2
44.8 to 47.4	16	22	49.4	40.5	6N1 + 3N2

*See step 7. (T) = Tracked

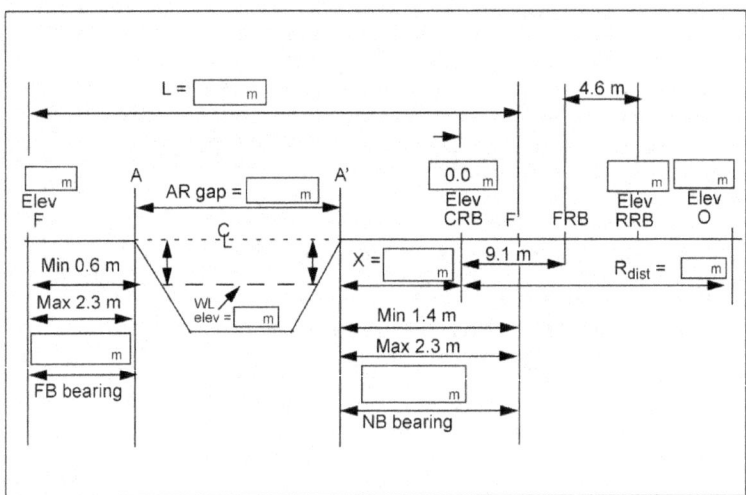

Figure 10-9. Constructing NB and FB bearings, 2E + 13 through 22 bays w/o LRS

Step 7. Check the slope. For longitudinal slope, ensure that the difference in elevation between the F' peg (use the elevation of the CRB) and the F peg does not exceed 1/10th of the total bridge length. If it does, you will either crib it up, undertake a major construction project, or find another bridge site. For a transverse/cross slope, ensure that the transverse slope on both of the banks does not exceed 1/10th of the total bridge width (4 meters) on both banks, regardless of the MLC.

Step 8. Calculate H and G (for later comparison against N and T) (see Figure 10-10):

$$H = elev\ F + \frac{elev\ RRB\ (1.\ -\ 0.5\ meter)}{13.7\ meters} = \underline{\qquad}$$

$$G = elev\ O - \frac{elev\ RRB\ (R\ distance)}{13.7\ meters} = \underline{\qquad}$$

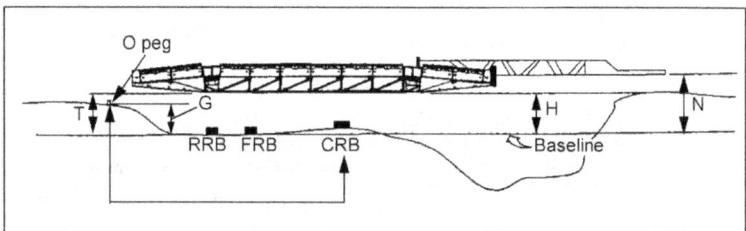

Figure 10-10. Calculating *H* and *G* to compare against *N* and *T* (2E + 13 through 22 bays w/o LRS)

Step 9. Identify the LNCG using Table 10-36, page 10-36, so that N>H and T>G. The CRB and RRB are in the low position.

Step 10. Identify N and T using Table 10-37, page 10-36. Raise the CRB and RRB by 0.25 meter. The CRB and RRB are in the high position.

Step 11. Identify N using Table 10-38, page 10-37, (rule 3A). Lower the RRB; the CRB is in the high position.

$$N = N\ (Rule\ 2) + answer\ to\ column\ m = \underline{\qquad}$$

If $N>H$, go to step 13; the design is correct.

Step 12. Identify T using Table 10-38 (rule 3B). Lower the CRB; the RRB is in the high position.

$$T = T\ (rule\ 2) + answer\ to\ column\ n = \underline{\qquad}$$

If $T>G$, go to step 13; the design is correct.

Table 10-36. Rule 1 for LNCG, 2E + 13 through 12 bays w/o LRS

2E+ Number of Bays	D: Given LNCG Settings With the FRB and RRB in Low Position			
	g #6 Yellow	h #4 Yellow	i #2 Yellow	j Tail Lift
13	-0.07	1.49	2.68	0.40
14	-0.38	1.00	2.65	0.37
15	-0.49	0.90	2.55	0.34
16	-0.61	0.79	2.43	0.30
17	-0.15	0.75	2.69	0.27
18	-1.33	0.54	2.54	0.21
19	-2.04	-0.19	1.72	0.21
20	-1.93	-0.31	1.61	0.21
21	-2.65	-0.52	1.17	0.18
22	-2.58	-0.68	1.04	0.15

- PROCEDURE:
- Choose an LNCG setting so that $N>H$ and $T>G$. If none of the choices meet the criterion, choose the highest setting available.
- Determine the LNCG setting: _____. See columns g, h, and i.
- Record N and T: $N=$_____ $T=$_____
- Check to see if $T>G$. See column j.
- Go to Table 10-37 if $N \leq H$ and/or $T \leq G$. Go to step 13 if $N>H$, and $T>G$.
- Choose another site or prepare to dig out under the NB end of the bridge before launching if $T<G$.

Table 10-37. Rule 2, identifying N, 2E + 13 through 22 bays w/o LRS

2E+ Number of Bays	N: Raise CRB and RRB by 0.25 meter	
	k	l
13	2.93	0.65
14	2.90	0.62
15	2.80	0.59
16	2.68	0.55
17	2.94	0.52
18	2.79	0.49
19	1.97	0.46
20	1.86	0.46
21	1.42	0.43
22	1.29	0.40

PROCEDURE:
- Raise the CRB and RRB by 0.25 meter. Both are in the high position.
- Check to see if $N>H$. See column k.
- Record N_____
- Check to see if $T>G$. See column l.
- Record T:_____
- Go to step 13 if $N>H$ and $T>G$; the design is correct.
- Go to Table 10-38 (rule 3A) if $N \leq H$ or to rule 3B if $T \leq G$.

Table 10-38. Rule 3A and 3B for N and T, 2E + 13 through 22 bays w/o LRS

Rule 3A —Lowering RRB to Increase N		Rule 3B —Lowering CRB to Increase T
2E+ Number of Bays	m N	n T
13	1.9 (0.82 - G)	0.2 (2.93 - H)
14	1.9 (0.79 - G)	0.2 (2.90 - H)
15	1.9 (0.76 - G)	0.2 (2.80 - H)
16	1.9 (0.72 - G)	0.2 (2.68 - H)
17	1.9 (0.69 - G)	0.2 (2.94 - H)
18	1.9 (0.66 - G)	0.2 (2.79 - H)
19	1.9 (0.63 - G)	0.2 (1.97 - H)
20	1.9 (0.63 - G)	0.2 (1.86 - H)
21	1.9 (0.60 - G)	0.2 (1.42 - H)
22	1.9 (0.57 - G)	0.2 (1.29 - H)

PROCEDURE:
• Determine N:_____; N = N (rule 2) + answer from column m, this table.
• Go to step 13 if N>H; the design is correct.
• Determine T:_____; T = T(rule 2) + answer from column n, this table.
• Go to step 13 if T>G; the design is correct.

Step 13. Determine the required truck and trailer loads using Table 10-39.

Truck_____Trailer_____

Table 10-39. DS pallet loads, 13 through 22 bays w/o LRS

Pallet Type	Number of Bays			
	13 to 15	16 to 18	19 to 21	22
Erection set	1	1	1	1
Bridge	8	9	10	11
Total	9	10	11	12
NOTE: Additional vehicles are required to transport personnel.				

Bridging 10-37

FM 5-34

Step 14. Determine the manpower and time required to accomplish the mission (see Table 10-40).

Table 10-40. Manpower and time requirements, 13 through 22 bays w/o LRS

a Manpower and Time	b 13 Bays	c 14 to 18 Bays	d 19 to 22 Bays
Working party	1 + 24	1 + 24	1 + 24
Time during daylight (hours)	1 1/2	1 3/4	2
Time during darkness (hours)	2	2 3/4	3
PROCEDURE: • Add 20 percent for unskilled personnel. • Add 30 percent for inclement weather. • Add 30 percent for adverse site conditions. • Calculate all timings consecutively.			

Step 15. Identify the final bridge design.

- 2E + ____ bays
- LNCG setting ____
- CRB setting ____
- RRB setting ____
- Bearings: NB ____ FB ____
- Total loads ____ : Trucks ____ Trailers ____
- Work party ____
- Total time ____ : Day ____ Night ____

MGB DESIGN—DS, 2E + 13 THROUGH 22 BAYS (W/LRS)

Step 1. Measure the AR gap: A to A' = ____ meters; see Table 10-41, column a.

Step 2. Select the number of bays from column c = 2E+ ____ .

Step 3. Select the bridge length from column d = ____ meters.

Step 4. Select the R distance from column e = ____ meters.

Step 5. Select the nose construction from column f = ____ .

Step 6. See Figure 10-11. Calculate the FB and NB bearings using the following equation; Figure 10-12, page 10-40; and Table 10-42, page 10-40 (assume the FB bearing to be 0.5 meter):

$$NB\ bearing = L - AR\ gap - FB\ bearing$$

$$X = NB\ bearing - 0.5\ meter$$

C1, FM 5-34

Table 10-41. Dimensions for DS, 2E + 13 through 22 bays w/LRS

a Gap Ranges (m)	b MLC*	c Number of Bays	d Bridge Length (m)	e R Distance (m)	f Nose Construction
28.3 to 30.9	70 (T)	13	32.9	27.7	7N1
30.2 to 32.8	70 (T)	14	34.8	29.0	7N1
32.0 to 34.1	70 (T)	15	36.6	29.0	7N1
33.8 to 36.4	70 (T)	16	38.4	29.6	8N1
35.6 to 38.2	70 (T)	17	40.2	29.6	8N1
37.5 to 40.1	70 (T)	18	42.1	29.6	6N1 + 3N2
39.3 to 41.9	70 (T)	19	43.9	35.1	6N1 + 3N2
41.1 to 43.7	70 (T)	20	45.7	38.7	6N1 + 3N2
43.0 to 45.1	70 (T)	21	47.6	38.7	6N1 + 3N2
44.8 to 46.2	70 (T)	22	49.4	40.5	6N1 + 3N2
*See step 7. (T) = Tracked					

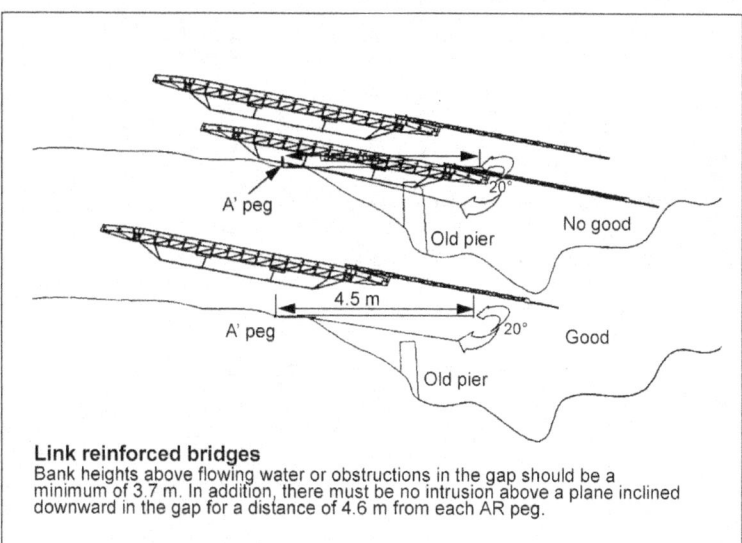

Link reinforced bridges
Bank heights above flowing water or obstructions in the gap should be a minimum of 3.7 m. In addition, there must be no intrusion above a plane inclined downward in the gap for a distance of 4.6 m from each AR peg.

Figure 10-11. Gap-rule obstructions

Bridging 10-39

FM 5-34

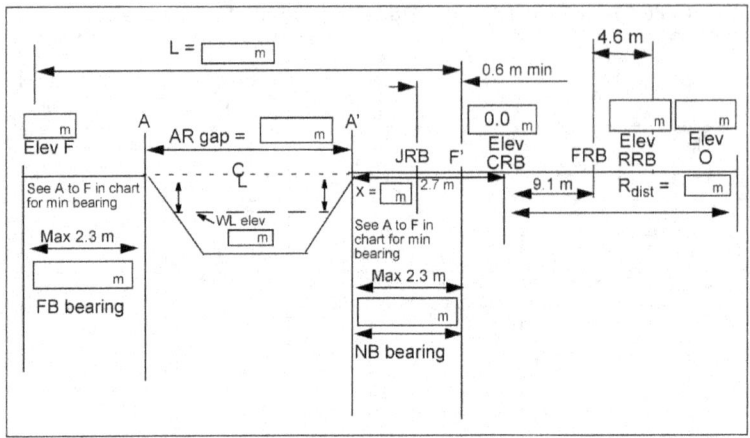

Figure 10-12. Constructing NB and FB bearings (DS, 2E + 13 through 22 bays w/ LRS)

Table 10-42. Minimum distances

Number of Bays	A to F (m)	A' to F' (m)
13, 14, 16 through 20	0.6	1.4
15	1.1	1.4
21	1.14	1.4
22	1.6	1.4

Step 7. Check the slope. For the longitudinal slope, ensure that the difference in elevation between the F (use elevation of CRB) and F' pegs does not exceed 1/20th of the total bridge length. If it does, you will either crib it up, undertake a major construction project, or find another bridge site. For a transverse/cross slope, ensure that the transverse slope on both banks does not exceed 1/20th of the total bridge width (4 meters).

Step 8. Calculate H and G (for later comparison against N and T) (see Figure 10-13):

$$H = elev\ F - \frac{elev\ RRB\ (L - 0.5\ meter)}{13.7\ meters}$$

$$G = elev\ O - \frac{elev\ RRB\ (R\ distance)}{13.7\ meters}$$

Step 9. Identify the LNCG setting by using Table 10-43

10-40 Bridging

Step 7. Check the slope. For longitudinal slope, ensure that the difference in elevation between the F' peg (use the elevation of the CRB) and the F peg does not exceed 1/10th of the total bridge length. If it does, you will either crib it up, undertake a major construction project, or find another bridge site. For a transverse/cross slope, ensure that the transverse slope on both of the banks does not exceed 1/10th of the total bridge width (4 meters) on both banks, regardless of the MLC.

Step 8. Calculate H and G (for later comparison against N and T) (see Figure 10-10):

$$H = elev\ F + \frac{elev\ RRB\ (1. - 0.5\ meter)}{13.7\ meters} = \underline{\qquad}$$

$$G = elev\ O - \frac{elev\ RRB\ (R\ distance)}{13.7\ meters} = \underline{\qquad}$$

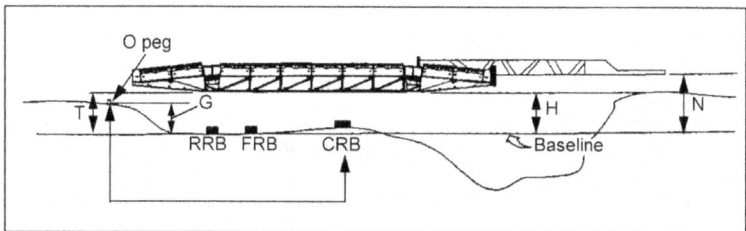

Figure 10-10. Calculating *H* and *G* to compare against *N* and *T* (2E + 13 through 22 bays w/o LRS)

Step 9. Identify the LNCG using Table 10-36, page 10-36, so that N>H and T>G. The CRB and RRB are in the low position.

Step 10. Identify N and T using Table 10-37, page 10-36. Raise the CRB and RRB by 0.25 meter. The CRB and RRB are in the high position.

Step 11. Identify N using Table 10-38, page 10-37, (rule 3A). Lower the RRB; the CRB is in the high position.

$$N = N\ (Rule\ 2) + answer\ to\ column\ m = \underline{\qquad}$$

If $N>H$, go to step 13; the design is correct.

Step 12. Identify T using Table 10-38 (rule 3B). Lower the CRB; the RRB is in the high position.

$$T = T\ (rule\ 2) + answer\ to\ column\ n = \underline{\qquad}$$

If $T>G$, go to step 13; the design is correct.

FM 5-34

Table 10-36. Rule 1 for LNCG, 2E + 13 through 12 bays w/o LRS

2E+ Number of Bays	D: Given LNCG Settings With the FRB and RRB in Low Position			
	g #6 Yellow	h #4 Yellow	i #2 Yellow	j Tail Lift
13	-0.07	1.49	2.68	0.40
14	-0.38	1.00	2.65	0.37
15	-0.49	0.90	2.55	0.34
16	-0.61	0.79	2.43	0.30
17	-0.15	0.75	2.69	0.27
18	-1.33	0.54	2.54	0.21
19	-2.04	-0.19	1.72	0.21
20	-1.93	-0.31	1.61	0.21
21	-2.65	-0.52	1.17	0.18
22	-2.58	-0.68	1.04	0.15

- PROCEDURE:
- Choose an LNCG setting so that $N>H$ and $T>G$. If none of the choices meet the criterion, choose the highest setting available.
- Determine the LNCG setting: _____. See columns g, h, and i.
- Record N and T. $N=$_____ $T=$_____
- Check to see if $T>G$. See column j.
- Go to Table 10-37 if $N \leq H$ and/or $T \leq G$. Go to step 13 if $N>H$, and $T>G$.
- Choose another site or prepare to dig out under the NB end of the bridge before launching if $T<G$.

Table 10-37. Rule 2, identifying N, 2E + 13 through 22 bays w/o LRS

2E+ Number of Bays	N: Raise CRB and RRB by 0.25 meter	
	k	l
13	2.93	0.65
14	2.90	0.62
15	2.80	0.59
16	2.68	0.55
17	2.94	0.52
18	2.79	0.49
19	1.97	0.46
20	1.86	0.46
21	1.42	0.43
22	1.29	0.40

PROCEDURE:
- Raise the CRB and RRB by 0.25 meter. Both are in the high position.
- Check to see if $N>H$. See column k.
- Record N_____
- Check to see if $T>G$. See column l.
- Record T:_____
- Go to step 13 if $N>H$ and $T>G$; the design is correct.
- Go to Table 10-38 (rule 3A) if $N \leq H$ or to rule 3B if $T \leq G$.

10-36 Bridging

Table 10-38. Rule 3A and 3B for *N* and *T*, 2E + 13 through 22 bays w/o LRS

Rule 3A —Lowering RRB to Increase *N*		Rule 3B —Lowering CRB to Increase *T*
2E+ Number of Bays	m N	n T
13	1.9 (0.82 - *G*)	0.2 (2.93 - *H*)
14	1.9 (0.79 - *G*)	0.2 (2.90 - *H*)
15	1.9 (0.76 - *G*)	0.2 (2.80 - *H*)
16	1.9 (0.72 - *G*)	0.2 (2.68 - *H*)
17	1.9 (0.69 - *G*)	0.2 (2.94 - *H*)
18	1.9 (0.66 - *G*)	0.2 (2.79 - *H*)
19	1.9 (0.63 - *G*)	0.2 (1.97 - *H*)
20	1.9 (0.63 - *G*)	0.2 (1.86 - *H*)
21	1.9 (0.60 - *G*)	0.2 (1.42 - *H*)
22	1.9 (0.57 - *G*)	0.2 (1.29 - *H*)

PROCEDURE:
• Determine *N*:_____; *N* = *N* (rule 2) + answer from column m, this table.
• Go to step 13 if *N*>*H*; the design is correct.
• Determine *T*:_____; *T* = *T*(rule 2) + answer from column n, this table.
• Go to step 13 if *T*>*G*; the design is correct.

Step 13. Determine the required truck and trailer loads using Table 10-39.

Truck_____ Trailer_____

Table 10-39. DS pallet loads, 13 through 22 bays w/o LRS

Pallet Type	Number of Bays			
	13 to 15	16 to 18	19 to 21	22
Erection set	1	1	1	1
Bridge	8	9	10	11
Total	9	10	11	12
NOTE: Additional vehicles are required to transport personnel.				

Step 14. Determine the manpower and time required to accomplish the mission (see Table 10-40).

Table 10-40. Manpower and time requirements, 13 through 22 bays w/o LRS

a Manpower and Time	b 13 Bays	c 14 to 18 Bays	d 19 to 22 Bays
Working party	1 + 24	1 + 24	1 + 24
Time during daylight (hours)	1 1/2	1 3/4	2
Time during darkness (hours)	2	2 3/4	3
PROCEDURE: •Add 20 percent for unskilled personnel. •Add 30 percent for inclement weather. •Add 30 percent for adverse site conditions. •Calculate all timings consecutively.			

Step 15. Identify the final bridge design.

- 2E + ____ bays
- LNCG setting ____
- CRB setting ____
- RRB setting ____
- Bearings: NB ____ FB ____
- Total loads ____ : Trucks ____ Trailers ____
- Work party ____
- Total time ____ : Day ____ Night ____

MGB Design—DS, 2E + 13 Through 22 Bays (w/LRS)

Step 1. Measure the AR gap: A to A' = ____ meters; see Table 10-41, column a.

Step 2. Select the number of bays from column c = 2E+ ____.

Step 3. Select the bridge length from column d = ____ meters.

Step 4. Select the R distance from column e = ____ meters.

Step 5. Select the nose construction from column f = ____.

Step 6. See Figure 10-11. Calculate the FB and NB bearings using the following equation; Figure 10-12, page 10-40; and Table 10-42, page 10-40 (assume the FB bearing to be 0.5 meter):

$$NB\ bearing = L - AR\ gap - FB\ bearing$$

$$X = NB\ bearing - 0.5\ meter$$

C1, FM 5-34

Table 10-41. Dimensions for DS, 2E + 13 through 22 bays w/LRS

a Gap Ranges (m)	b MLC*	c Number of Bays	d Bridge Length (m)	e R Distance (m)	f Nose Construction
28.3 to 30.9	70 (T)	13	32.9	27.7	7N1
30.2 to 32.8	70 (T)	14	34.8	29.0	7N1
32.0 to 34.1	70 (T)	15	36.6	29.0	7N1
33.8 to 36.4	70 (T)	16	38.4	29.6	8N1
35.6 to 38.2	70 (T)	17	40.2	29.6	8N1
37.5 to 40.1	70 (T)	18	42.1	29.6	6N1 + 3N2
39.3 to 41.9	70 (T)	19	43.9	35.1	6N1 + 3N2
41.1 to 43.7	70 (T)	20	45.7	38.7	6N1 + 3N2
43.0 to 45.1	70 (T)	21	47.6	38.7	6N1 + 3N2
44.8 to 46.2	70 (T)	22	49.4	40.5	6N1 + 3N2
*See step 7. (T) = Tracked					

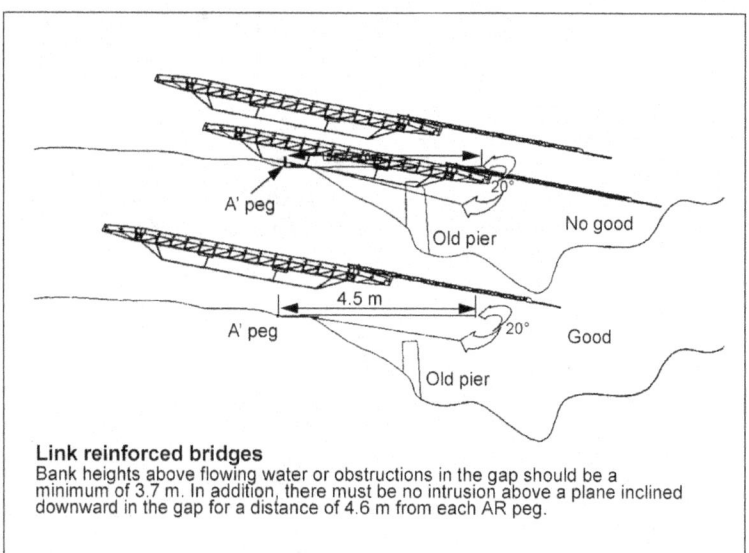

Link reinforced bridges
Bank heights above flowing water or obstructions in the gap should be a minimum of 3.7 m. In addition, there must be no intrusion above a plane inclined downward in the gap for a distance of 4.6 m from each AR peg.

Figure 10-11. Gap-rule obstructions

Bridging 10-39

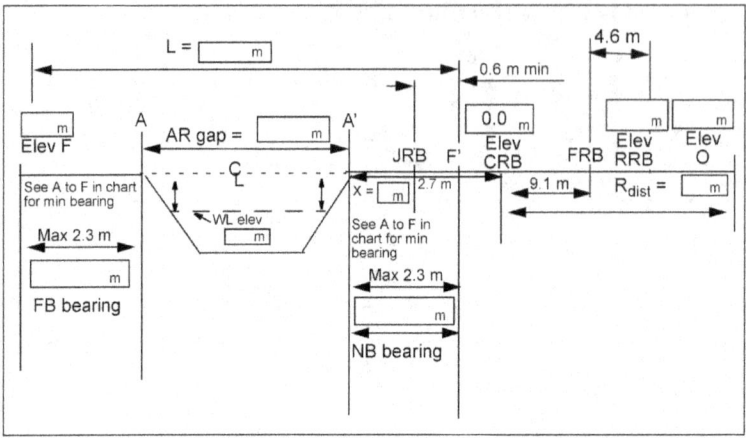

Figure 10-12. Constructing NB and FB bearings (DS, 2E + 13 through 22 bays w/ LRS)

Table 10-42. Minimum distances

Number of Bays	A to F (m)	A' to F' (m)
13, 14, 16 through 20	0.6	1.4
15	1.1	1.4
21	1.14	1.4
22	1.6	1.4

Step 7. Check the slope. For the longitudinal slope, ensure that the difference in elevation between the F (use elevation of CRB) and F' pegs does not exceed 1/20th of the total bridge length. If it does, you will either crib it up, undertake a major construction project, or find another bridge site. For a transverse/cross slope, ensure that the transverse slope on both banks does not exceed 1/20th of the total bridge width (4 meters).

Step 8. Calculate H and G (for later comparison against N and T) (see Figure 10-13):

$$H = elev\ F - \frac{elev\ RRB\ (L - 0.5\ meter)}{13.7\ meters} - \_\_\_\_$$

$$G = elev\ O - \frac{elev\ RRB\ (R\ distance)}{13.7\ meters} - \_\_\_\_$$

Step 9. Identify the LNCG setting by using Table 10-43.

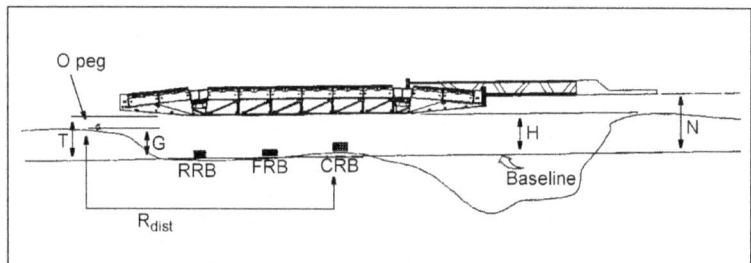

Figure 10-13. Calculating H and G to compare against N and T (DS, 2E + 13 through 22 bays w/LRS)

Step 9. Identify the LNCG setting by using Table 10-43.

Table 10-43. Rule 1 for LNCG, 2E + 13 through 22 bays w/ LRS

2E+ Number of Bays	N: Nose Lift, Given LNCG Setting With CRB and RRB in High Position			
	g #6 Yellow	h #4 Yellow	i #2 Yellow	j Tail Lift
13	0.48	1.87	3.52	0.40
14	0.31	1.72	3.35	0.37
15	0.25	1.64	3.29	0.34
16	-0.62	1.27	3.25	0.30
17	-0.77	1.12	3.10	0.27
18	-1.06	0.80	2.71	0.21
19	-1.46	0.40	2.32	0.21
20	-1.75	0.11	2.03	0.21
21	-2.08	0.05	1.75	0.18
22	-2.44	-0.31	1.40	0.15

PROCEDURE:
- Choose an LNCG setting so that the value of $N>H$.
- Determine the LNCG setting_____. See columns g, h, and i. See column j for tail lift.
- Record $N=$_____ and $T=$_____
- Go to Table 10-44, page 10-42, if $N \leq H$.
- Go to step 11 if $N > H$.
- Choose another site or prepare to dig out under the NB end of the bridge before launching if $T \leq G$.

FM 5-34

Step 10. Identify N using Table 10-44. Place the RRB in the low position. The CRB is in the high position.

Table 10-44. Rule 2, identifying N, 2E + 13 through 22 bays w/ LRS

Lowering RRB to Increase N (RRB is in low position and CRB is in high position.)	
2E+ Number of Bays	(k)
13	1.9 (0.82 - G)
14	1.9 (0.79 - G)
15	1.9 (0.76 - G)
16	1.9 (0.72 - G)
17	1.9 (0.69 - G)
18	1.9 (0.66 - G)
19 to 20	1.9 (0.63 - G)
21	1.9 (0.60 - G)
22	1.9 (0.57 - G)
PROCEDURE: • Determine N:_____. N = (Table 10-43, page 10-41) + answer to column k. • Go to step 11 if $N>H$; design is correct. • Consider another site or crib up if $N<H$.	

Step 11. Determine the required truck and trailer loads using Table 10-45.

Truck_____Trailer_____

Table 10-45. DS pallet loads, 2E + 13 through 22 bays with LRS

Pallet Type	Number of Bays			
	13 to 15	16 to 18	19 to 21	22
Erection set	1	1	1	1
Bridge	8	9	10	11
Link	2	2	2	2
Total	11	12	13	14
Note: Additional vehicles are required to transport personnel.				

10-42 Bridging

Step 12. Determine the manpower and time required to accomplish the mission (see Table 10-46).

Table 10-46. Manpower and time requirements, 2E + 13 through 22 bays w/ LRS

a Manpower and Time	b 13 Bays	c 14 to 18 Bays	d 19 to 22 Bays
Working party	2 + 32	2 + 32	2 + 32
Time during daylight (hours)	2	2 3/4	3
Time during darkness (hours)	3	4	4 1/2
PROCEDURE: •Add 20 percent for unskilled personnel. •Add 30 percent for inclement weather. •Add 30 percent for adverse site conditions. •Calculate all timings consecutively.			

Step 13. Identify the final bridge design.

- 2E + ____ bays
- LNCG setting ____
- CRB setting ____
- RRB setting ____
- Bearings: NB ____ FB ____
- Total load ____ : Trucks ____ Trailers ____
- Work party ____
- Total time ____ : Day ____ Night ____

BAILEY BRIDGE, TYPE M-2

TRUSS

The Bailey bridge trusses are formed from 10-foot panels and may be constructed in any configuration (see Table 10-47).

Table 10-47. Truss/story configuration

Truss	Story	Nomenclature	Abbreviation
Single	Single	Single single	SS
Double	Single	Double single	DS
Triple	Single	Triple single	TS
Double	Double	Double double	DD
Triple	Double	Triple double	TD
Double	Triple	Double triple	DT
Triple	Triple	Triple triple	TT

SITE RECONNAISSANCE

A site reconnaissance must be conducted. The construction area must provide enough space for equipment layout (see Figure 10-14). Figure 10-15 shows a roller layout for a triple-truss or multistory bridge.

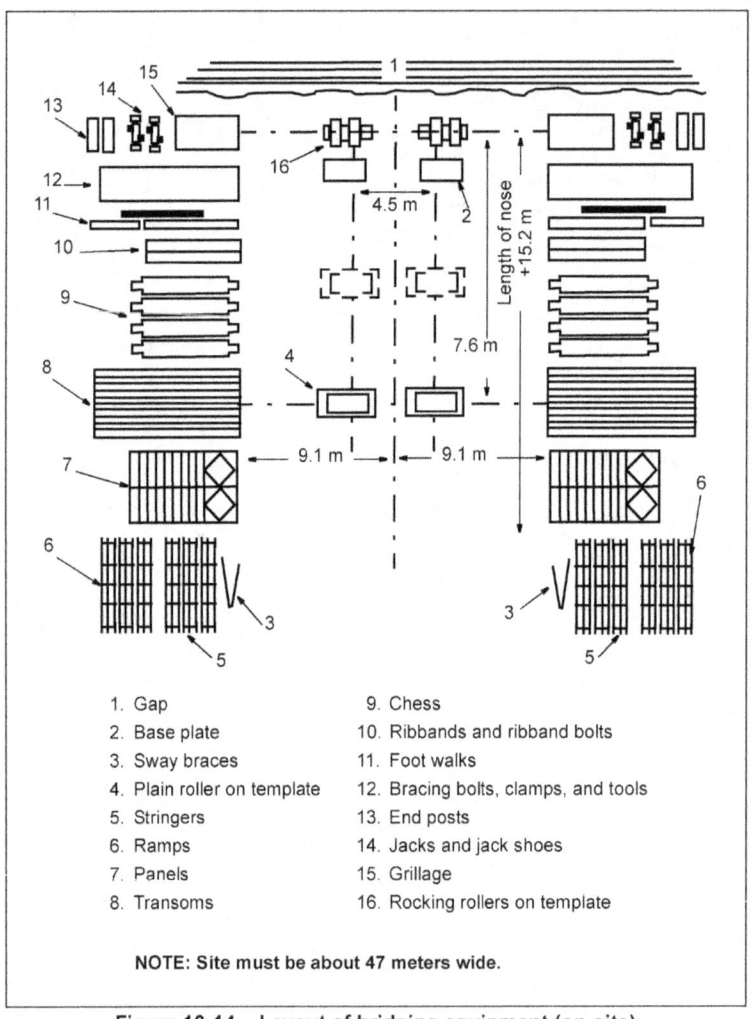

Figure 10-14. Layout of bridging equipment (on-site)

FM 5-34

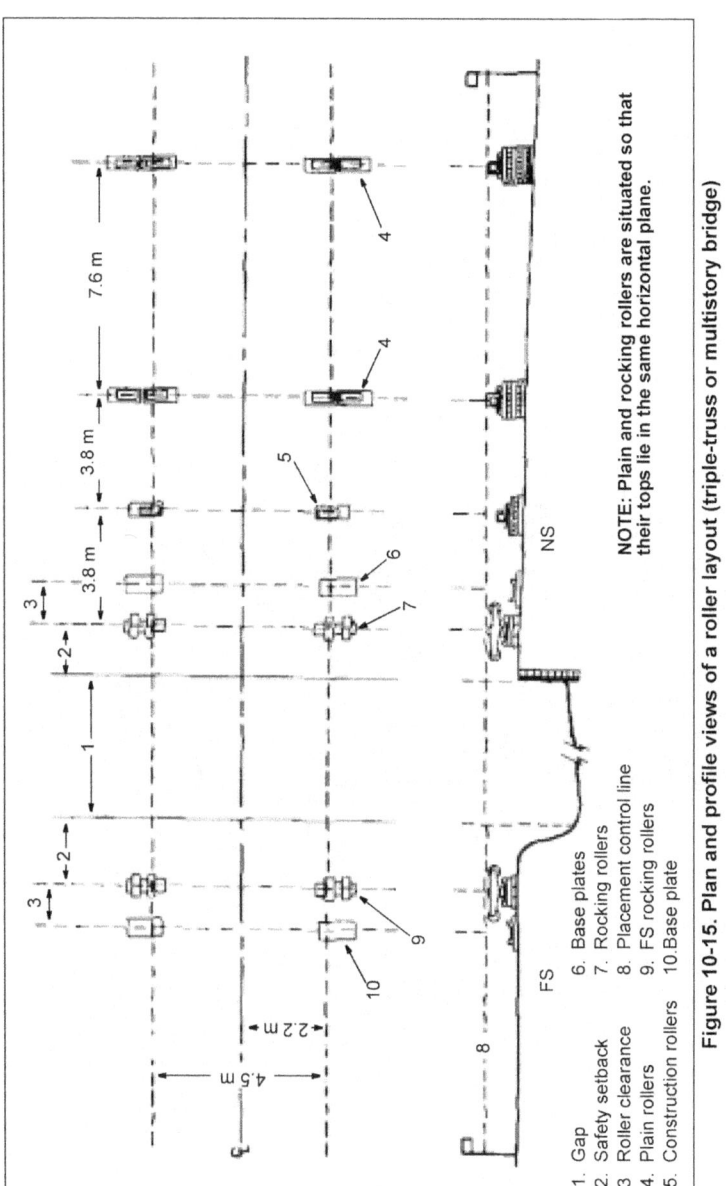

Figure 10-15. Plan and profile views of a roller layout (triple-truss or multistory bridge)

Bridging 10-45

FM 5-34

BRIDGE DESIGN

Figure 10-16 shows an example of a site profile. Use the steps below Figure 10-16 to design a bridge.

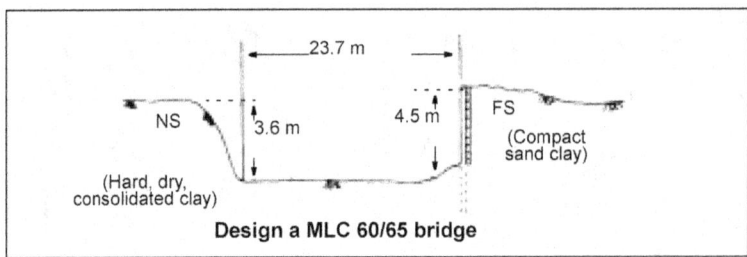

Figure 10-16. Site profile example

Step 1. Gap as measured during reconnaissance: 79 feet

Step 2. Safety setback:

• Prepared abutment = constant of 3.5 feet   FS: 3.5 feet

• Unprepared abutment = 1.5 x bank height   NS: 1.5 x 12 = 18

Step 3. Initial roller clearance: always a constant of 2.5 feet   NS and FS: 2.5 feet

Step 4. Initial bridge length: Add steps 1, 2, and 3. If the value is not a multiple of 10 feet, round up to the next 10-foot value.   Initial length: 103 feet
Rounded up: 110 feet

Step 5. Initial truss/story type (see Table 10-48, page 10-51):   DD

Step 6. Initial bridge class (see Table 10-48): The classification must meet or exceed the requirements designated in the mission statement. The truss/story type selected is always based on a normal crossing unless the tactical commander directs otherwise.   65/70

Step 7. Selection of grillage:

• Safe soil bearing (see Table 10-49, page 10-53)   NS = 5 tons/ft^2
FS = 3 tons/ft^2

• Safe soil pressure (see Table 10-50, page 10-54). If the soil-bearing capacity values are not in Table 10-50, round down to the closest listed. Use these values for the grillage.   NS = 3.5 tons/ft^2
FS = 2.5 tons/ft^2

• Grillage required.   NS and FS = Type 1

Step 8. Adjusted bridge length: distance required for new roller clearance (see Table 10-51, page 10-55).   NS = 4.5 feet; FS = 4.5 feet

- Add steps 1, 2, and the above value from Table 10-51.   109.5 feet

- Round up to next highest 10-foot value if the value is not a multiple of 10 feet.   110 feet

NOTE: Compare the values in steps 4 and 8. If they are different, you must redesign the bridge using steps 9 through 12 below. If the values are the same, go to step 13.

Step 9. Final truss/story type (see Table 10-48, page 10-51):   Try 1   Try 2

Step 10. Final bridge classification (see Table 10-48):   Try 1   Try 2

- Classification must meet or exceed the requirements designated in the mission statement.

- The truss/story type selected is always based on a normal crossing unless the tactical commander designates otherwise.

Step 11. Final grillage section: Try 1   Try 2

- Safe soil bearing (see Table 10-49, page 10-53).   NS

- Safe soil pressure (see Table 10-50, page 10-54). If the soil-bearing-capacity values from above are not in Table 10-50, round down to the closest one listed. Use these values for grillage required.   FS   NS   FS

- Grillage required.

   NS   Type   Type
   FS   Type   Type

Step 12. Final bridge length: distance required for new roller clearance (see Table 10-51).   NS   FS

- Add steps 1 and 2 to figures from Table 10-51.

- Round up to the next highest 10-foot value if the above value is not a multiple of 10 feet.   =   =   =   =

FM 5-34

For Try 1—compare the rounded-up values in steps 8 and 12. If they are the same, go to step 13. If they are different, compare the rounded-up values in steps 4 and 12. If these values are the same, you can make a judgment call. Repeating the design sequence in the Try 2 column, using the bridge length from step 12, Try 1, will put you in an endless cycle unless you can reduce the final bridge length. Regardless, you will have to either overdesign the final bridge (Try 1 column) or choose a higher number of grillages than selected in step 7. This choice could reduce the roller clearance on one or both banks so that the required bridge length/final truss/story may be at the minimum to do the job. You could choose a higher number of grillages than allowed (step 11); however, be careful not to exceed the BP and RRT capacities listed in FM 5-277, Tables 4-2 and 4-3. After making a decision, go to step 13.

For Try 2 and higher—compare the current rounded-up value in step 12 to the previous Try's value. If they are the same, go to step 13. If they are different, repeat the design sequence until the rounded-up values of the current Try and previous Try match. Then go to step 13.

Step 13. Slope check: The maximum allowable bank-height difference is 1 to 30. Therefore, the maximum allowable bank-height difference equals the $\frac{\text{final bridge length}}{30}$ if— $\quad \frac{110}{30} + 3.7 \text{ feet} > 3 \text{ feet}$

- The above value > actual bank-height difference; the slope is acceptable.   GO/NO GO (circle one)

- The above value < actual bank height difference. Choose another site or crib up/excavate the FS or NS until the bridge slope is within acceptable limits.

Step 14. Final bridge requirements: Length

Truss/story type

Class

Step 15. Launching nose composition: Use Figures 10-17 through 10-24, pages 10-55 through 10-62, depending on the truss/story type.   7 bays, single truss

Step 16. Placing the launching nose links:

- The sag (use the same figure as in step 15).   22 inches

- The safety sag (constant of 6 inches).

- The lift required (add above two results).   22 + 6 = 28 inches

- The position of launching nose link (see Figure 10-23, page 10-61).   30 feet from nose tip

Step 17. Rocking-roller requirement: See Table 10-52, page 10-63.   NS = 4   FS = 2

Step 18. Plain-roller requirement:

- The SS and DS bridges only have two rollers per row; all others have four rollers per row. Use Table 10-53, page 10-63, to determine the number of rows and then multiply.   12

- Add two more plain rollers to allow for construction-roller requirement.   2

- Add the above two results.   12 + 2 = 14

Step 19. Jack requirements: See Table 10-54, page 10-63.   4

NOTE: Only one end of the bridge will be jacked down at a time.

Step 20. Ramp requirements:

- Slope requirements (check one).
    — Final bridge class ≤ 50 = 1 to 10. ( )
    — Final bridge class > 50 = 1 to 20. (x)

- Support for end ramp (check one).
    — Final bridge class ≤ 67 = 2 chess. ( )
    — Final bridge class > 67 = 4 chess. (x)

- Midspan ramp supports (check one).
    — Final bridge class ≤ 44 = not needed. ( )
    — Final bridge class > 44 = needed. (x)

- Pedestal supports (check one).
    — Not needed. ( )
    — Needed. (x)

NOTE: See FM 5-277 for criteria and drawings. Ramp lengths must be estimated from the site sketch.

- Support for end transom (check one).
    — Final bridge class ≤ 39 = not needed. ( )
    — Final bridge class > 39 = needed. (x)

Step 21. Personnel required: See Table 10-55, page 10-64. (Note the differences between manpower and crane construction.)   Total 5/66

FM 5-34

Step 22. Assembly time: See Table 10-56, page 10-65.   5 hours

NOTE: This time allows for ideal bridge construction conditions and does not allow for site preparation or roller layout.

FM 5-34

Table 10-48. Classification of Bailey bridge

Construction Type	Rating	30	40	50	60	70	80	90	100	110	120	130	140	150	160	170	180	190	200	210
SS	N	30/30	24	24	20	20	16	12												
SS	C	42/37	36/34	33/31	30/29	24	20	16	12											
SS	R	47/42	40/38	36/35	33/32	30/30	24	19	14											
DS	N					60/60	50/55	40/45	30/30	20	16									
DS	C			83/76	77/73	68/69	60/60	50/50	37/39	30/32	23	18	8							
DS	R			88/84	85/79	78/75	66/64	55/55	42/44	34/36	27/30	21	17							
TS	N						85/80	65/65	50/55	35/40	30/35	20	16	12	8					
TS	C						95/90	74/75	57/60	47/49	38/41	31/33	24	18	15	10				
TS	R						100*/90*	82/82	64/66	52/54	43/45	35/38	29/31	22	17	13	8			
DD	N								80/80	65/70	45/55	35/45	30/35	24	16	12	8			
DD	C								86/90	72/76	57/61	47/50	39/42	32/35	25	19	15			
DD	R								96/90	80/83	64/68	53/56	44/48	36/40	30/33	24	18			

NOTES:
1. N = Normal   C = Caution   R = Risk
2. Top number is wheeled-vehicle load class; bottom number is tracked-vehicle class.
3. * indicates limited by roadway width.

Bridging 10-51

FM 5-34

Table 10-48. Classification of Bailey bridge (continued)

Construction Type	Rating	\multicolumn{19}{c}{Span (feet)}
		30	40	50	60	70	80	90	100	110	120	130	140	150	160	170	180	190	200	210
TD	N									90/90*	75/80	55/60	45/55	35/45	30/35	20	16	12		
TD	C									100*/90*	83/90*	65/72	57/62	47/51	37/41	31/34	24	18		
TD	R									100*/90*	91/90*	74/80	64/70	54/58	45/48	37/40	29/32	22	20	16
DT	N											70/80	70/70	60/60	55/55	45/50	35/45	30/35		
DT	C											80/90*	80/90*	77/85	69/78	57/64	48/58	39/43	32/36	25
DT	R											90/90*	88/90*	85/90*	80/89	64/74	55/60	46/51	38/43	31/35
TT	N														80/75	70/70	55/60	45/55	35/40	24
TT	C														100/90*	80/90*	66/75	59/66	48/52	38/43
TT	R														100*/90*	90/90*	77/87	68/77	55/62	46/51

NOTES:
1. N = Normal   C = Caution   R = Risk
2. Top number is wheeled-vehicle load class; bottom number is tracked-vehicle class.
3. * indicates limited by roadway width.

Table 10-49. Safe bearing capacity for various soils

Soil Description	Bearing Values (tons per sq ft)
Hardpan overlying rock	12
Very compact sandy gravel	10
Loose gravel and sandy gravel, compact sand and gravelly sand; very compact sand, inorganic silt soils	6
Hard, dry consolidated clay	5
Loose, coarse-to-medium sand; medium-compact fine sand	4
Compact sand clay	3
Loose, fine sand; medium-compact sand, inorganic silt soils	2
Firm or stiff clay	1.5
Loose saturated-sand clay soils; medium-soft clay	1

FM 5-34

Table 10-50. Safe soil pressures

Construction Type	Safe Soil Pressure (ton per sq ft)	Span (feet)																		
		30	40	50	60	70	80	90	100	110	120	130	140	150	160	170	180	190	200	210
SS	0.5	6,7	5,6,7	5,6,7																
	1.0	4	3	3																
	2.0	1																		
	2.5																			
	3.5																			
DS	0.5			6	6	6,7	6,7	6,7	6,7	6,7	6,7	6,7	6,7	6,7	6,7					
	1.0			6,7	5,6,7	4	4	4	4	4	4	4	4	4	4					
	2.0			4	3	1	1	1	1	1	1	1	1	1	1					
	2.5			1	1	1	1	1	1	1	1	1	1	1	1					
	3.5			1			1	1	1	1	1	1	1	1	1					
TS	0.5				6		6,7	6,7	6,7	6,7	6,7	6,7	6,7	6,7	6,7	6,7	6,7			
	1.0				5,6,7		4	5,6,7	4	4	4	4	4	4	4	4,6,7	4,6,7			
	2.0				3		3	3		1	1	1	2	2	2	2	2,2			
	2.5						1	1	1	1	1	1	1	1	1	1	1,1			
	3.5						1	1	1	1	1	1	1	1	1	1	1			
DD	0.5							6	6	6	6	6	6	6	6,7	6,7	6,7	6,7		
	1.0							6,7	6,4	5,6,7	5,6,7	4	4,6,7	4,6,7	4,6,7	4,6,7	4,6,7	4,6,7		
	2.0							3	3	4	3	2	2	2	2,2	2,2	2	2		
	2.5							1	1	1	1	1	1	1	1	1	1	1		
	3.5																			
TD	0.5									6,7	6,7	6,7	6,7	6,7	6,7	6,7	6,7	6,7	6,7	
	1.0									4,6,7	4,6,7	4,6	4,6,7	4,6,7	4,6,7	4,6,7	4,6,7	4,6,7	4,6,7	
	2.0									4,6,7	3	2	2	2	2	2	2	2	2	
	2.5									1	1	1	2	2	2	2	2	2	2	
	3.5																			
DT	0.5											6	6,7	6,7	6,7		6,7	6,7	6,7	6,7
	1.0											6,7	4,6,7	4,6,7	6,7		6,7	6,7	6,7	6,7
	2.0											6,7	4,6,7	4,6,7	2		2	4,6,7	4,6,7	4,6,7
	2.5											2	2	2			2	2	2	2
	3.5																			
TT	0.5															6,7	6,7	6,7	6,7	6,7
	1.0															6,7	6,7	6,7	6,7	6,7
	2.0															6,7	6,7	6,7	6,7	6,7
	2.5															6,7	6,7	6,7	6,7	6,7
	3.5															2	2	2	2	2

10-54 Bridging

FM 5-34

Table 10-51. Roller clearance and grillage height

Grillage Type	Overall Height (in)	Base-Plate Height (in)	Roller Clearance (ft)
1	6	6	4.5
2	15	6	4.5
3	11	11	3.5
4	17	11	4.5
5	16	16	3.5
6	26	20	3.5
7	13	13	3.5

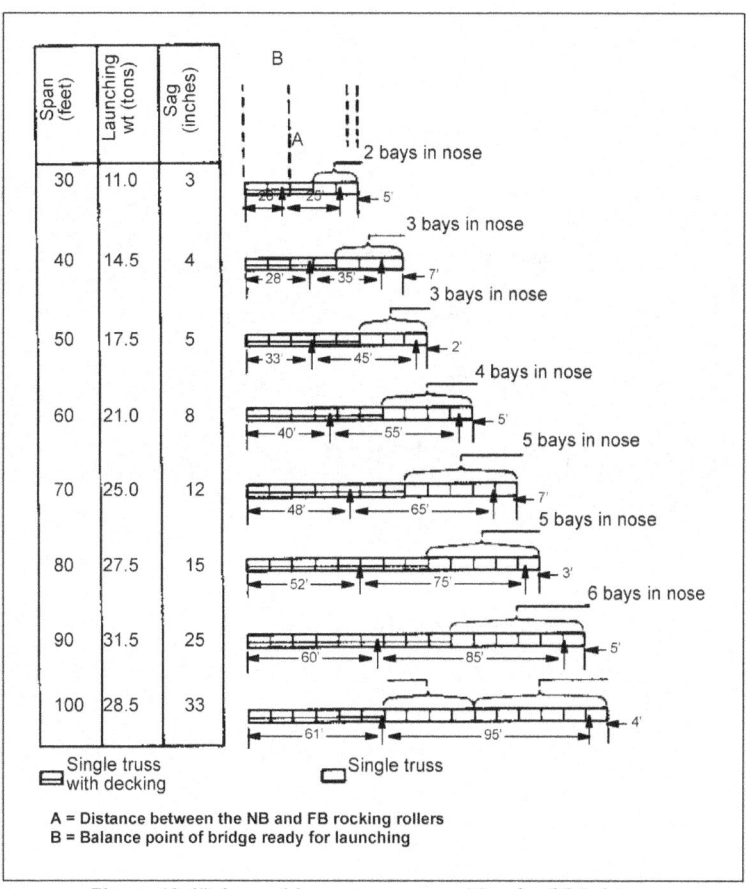

Figure 10-17. Launching-nose composition for SS bridges

Bridging 10-55

FM 5-34

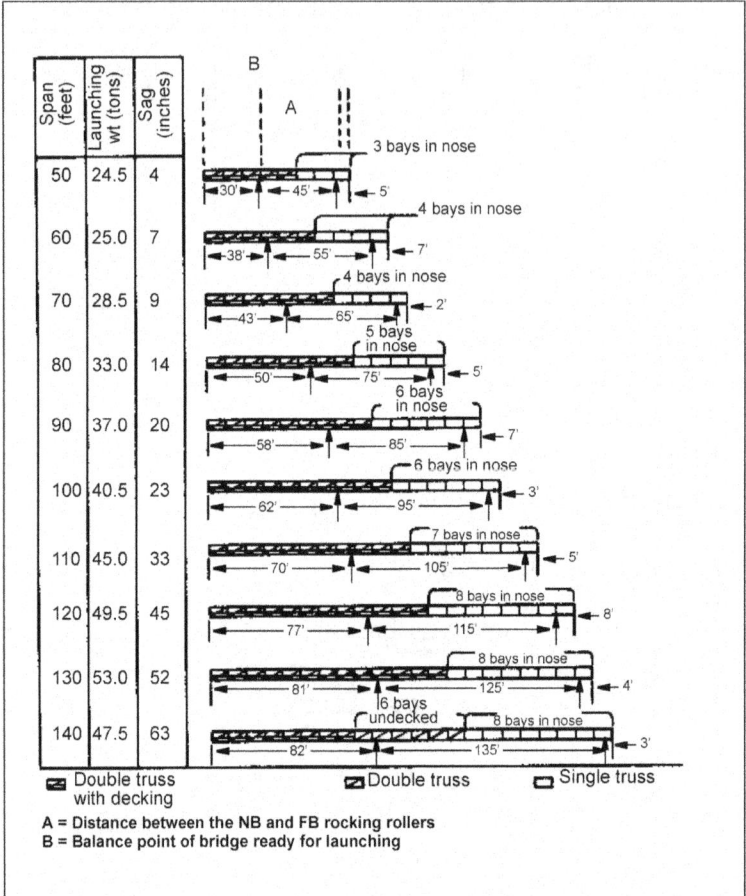

Figure 10-18. Launching-nose composition for DS bridges

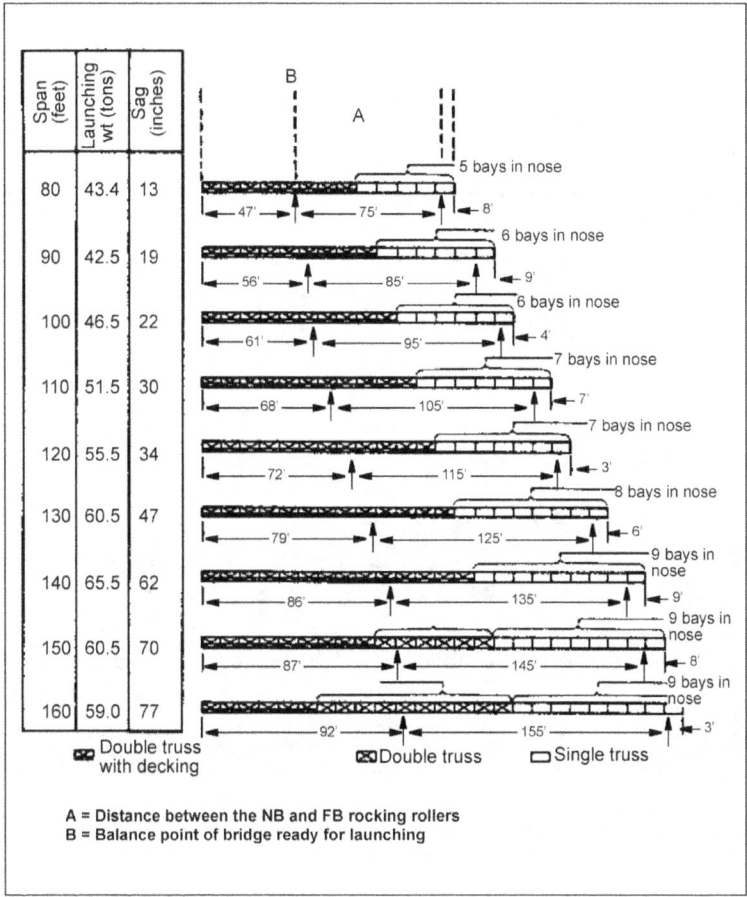

Figure 10-19. Launching-nose composition for TS bridges

FM 5-34

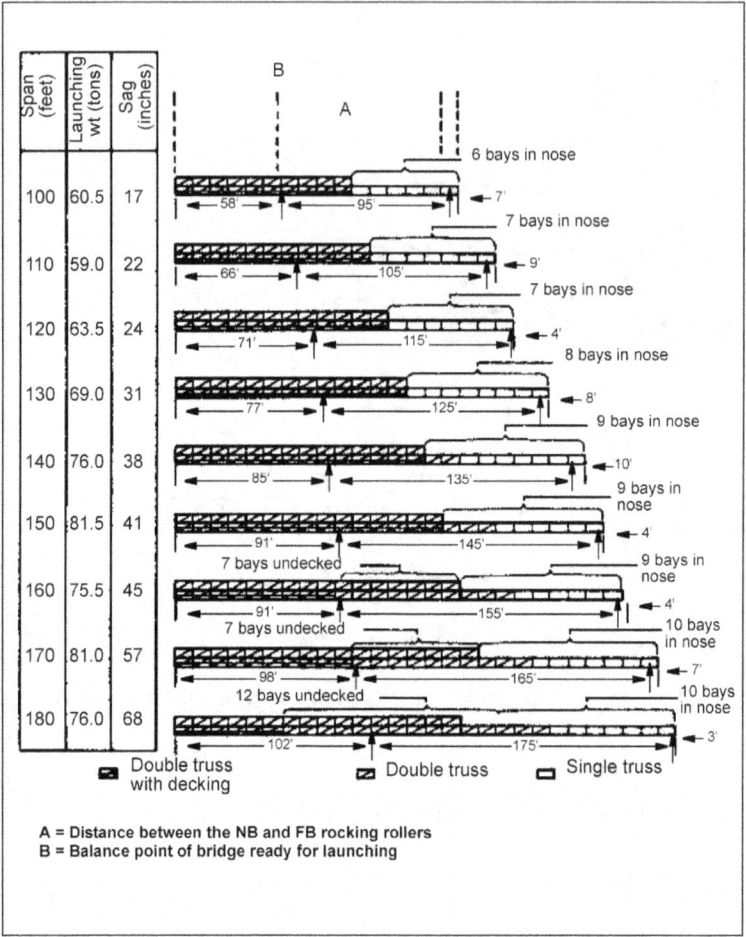

Figure 10-20. Launching-nose composition for DD bridges

FM 5-34

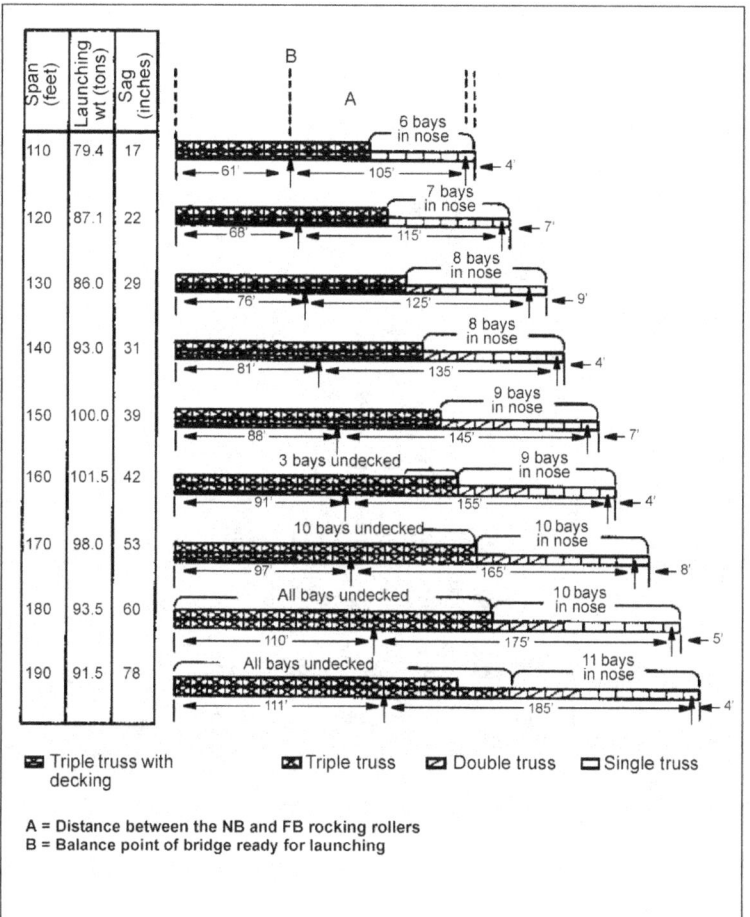

Figure 10-21. Launching-nose composition for TD bridges

Bridging 10-59

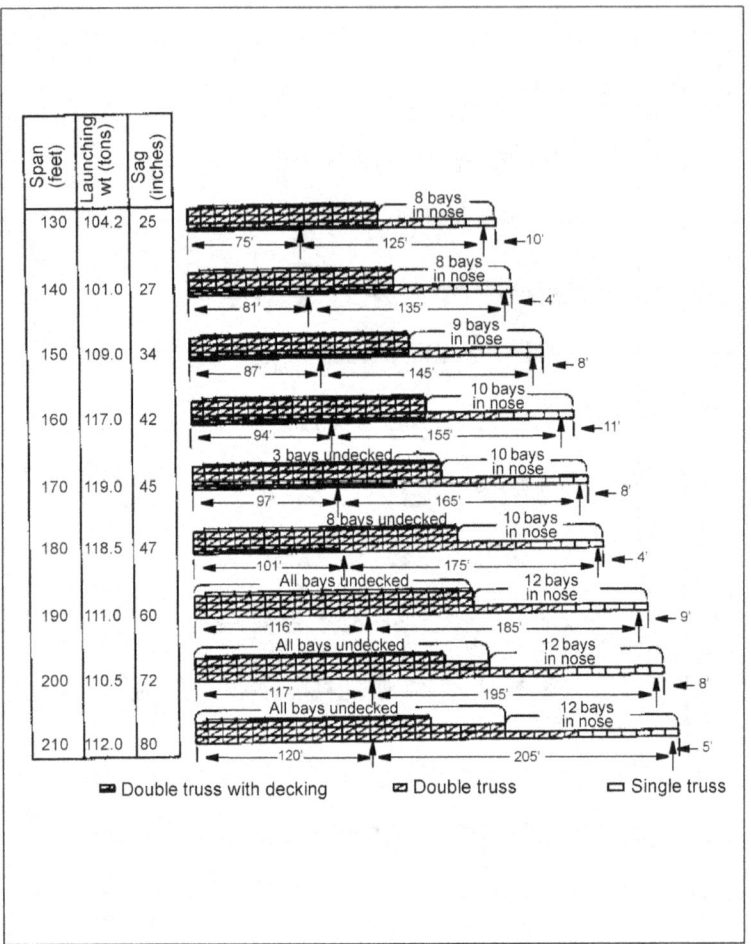

Figure 10-22. Launching-nose composition for DT bridges

10-60 Bridging

FM 5-34

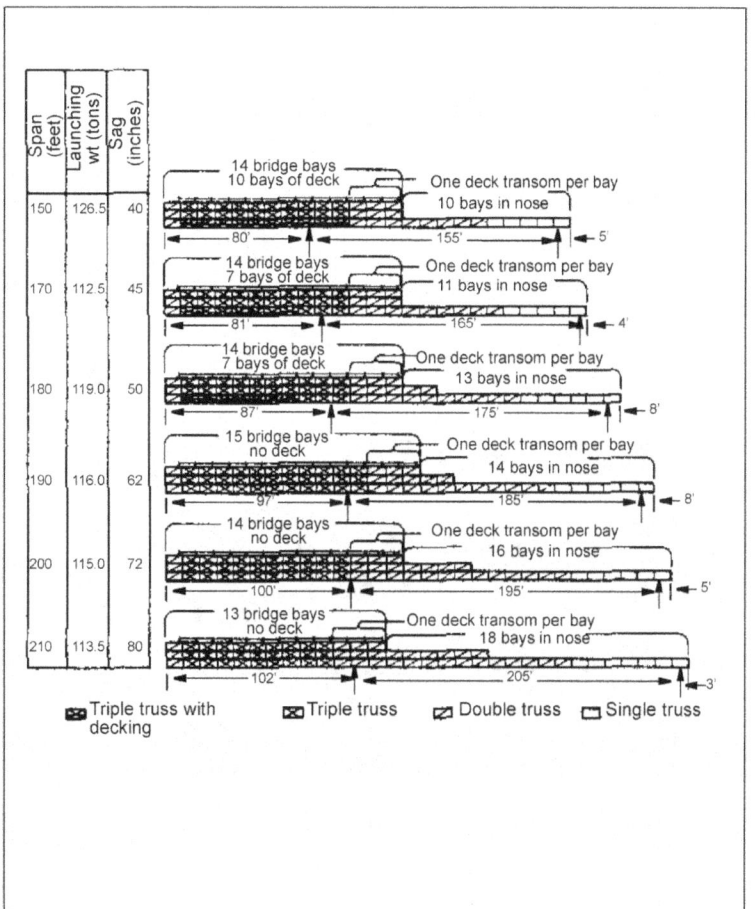

Figure 10-23. Launching-nose composition for TT bridges

Bridging 10-61

FM 5-34

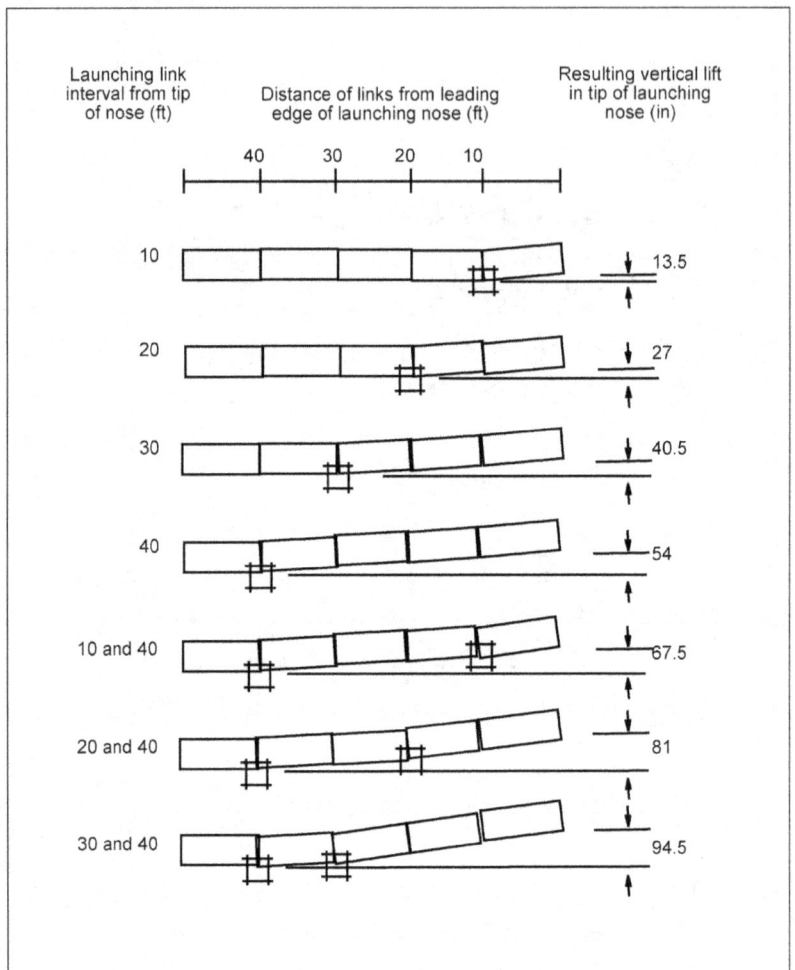

Figure 10-24. Upturned skeleton launching nose

10-62 Bridging

Table 10-52. Rocking-roller requirements

Construction Type	Span (ft)	NB	FB
SS	30 to 100	2	2
DS	50 to 80 90 to 100 110 to 140	2 2 4	2 2 2
TS	80 to 160	4	2
DD	100 to 130 140 to 180	4 4	2 4
TD	110 to 120 130 to 190	4 4	2 4
DT	130 to 210	4	4
TT	160 to 210	4	4

Table 10-53. Plain-roller requirements

Span (ft)	Construction Type						
	SS	DS	TS	DD	TD	DT	TT
30 to 50	1	1					
60 to 80	2	2	2				
90	3	2	2				
100	3	3	2	2			
110 to 120		3	3	3	3		
130		3	3	3	3	3	
140		3	4	4	3	3	
150			4	4	4	4	
160			4	4	4	4	3
170				4	4	4	3
180				4	5	4	4
190					5	5	4
200 to 210						5	4

Table 10-54. Jack requirements

Construction Type	Span (ft)	Number of Jacks, Each End
SS	30 to 100	2
DS	50 to 140	4
TS	80 to 140 150 to 160	4 6
DD	100 to 120 130 to 180	4 6
TD	110 to 140 150 to 190	6 8
DT	130 140 to 180 190 to 210	6 8 10
TT	160 to 170 180 to 210	10 12

FM 5-34

Table 10-55. Organization of an assembly party

Detail	Construction Type								
	Using Manpower Only							Using One Crane*	
	SS	DS	TS	DD	TD	DT	TT	DT	TT
	NCO/EM	NCO/EM	NCO/EM	NCO/EM	NCO/EM	NCO/EM	NCO/EM	NCO/EM	NCO/EM
Crane— • Truck driver • Crane operator • Hook man								/3** /1 /1 /1	/3** /1 /1 /1
Panel— • Carrying • Pin	1/14** /12 /2	1/14** /12 /2	2/28** /24 /4	2/32** /28 /4	3/50** /44 /6	3/50** /44 /6	3/68** /60 /8	3/30** /24 /6	3/30** /24 /6
Transom— • Carrying • Clamp	1/9** /8 /1	1/10** /8 /2	1/10** /8 /2	1/10** /8 /2	1/10 /8 /2	2/28 /24 /4	2/28** /24 /4	2/20** /16 /4	2/20** /16 /4
Bracing— • Sway brace • Raker • Bracing frame • Chord bolt • Tie plate • Overhead support	1/4** /2 /2	1/6** /2 /2	1/8** /2 /2 /2	1/12** /2 /2 /4 /4	1/20** /2 /4 /8 /4	1/32** /6 /2 /8 /10 /6	1/40** /6 /2 /8 /14 /4 /6	1/32** /6 /2 /10 /10 /4	1/38** /6 /2 /8 /14 /4 /4
Decking— • Stringer • Chess and ribband	1/12** /8 /4	1/12** /8 /4	1/12** /8 /4	1/12** /8 /4	1/12** /8 /4	1/12** /8 /4	1/12** /8 /4	1/12** /8 /4	1/12** /8 /4
Total	4/39	4/42	5/58	5/66	6/92	7/122	7/148	7/97	7/103

*Normally a crane is not used for an SS or a DS assembly.
**Represents the total number of noncommissioned officers (NCOs) and enlisted members (EMs) for the particular detail.

10-64 Bridging

Table 10-56. Estimated assembly times

Span (ft)	Construction Type								
	SS	DS	TS	DD	TD	DT	TT	DT	TT
	Using Manpower Only (hours)							Using 1 Crane (hours)	
40	1 1/2								
60	1 3/4	2							
80	2	2 1/2	3						
100	2 1/4	3	3 1/2	4 1/2					
120		3 1/2	4	5	6 3/4				
140		3 3/4	4 1/2	5 3/4	7 1/2	11 3/4		10 1/2	
160			5	6 1/4	8 1/2	13 1/4	19	11 3/4	16 1/4
180				7	9 1/2	14 3/4	21 1/4	13 1/4	18 1/4
200						16 1/4	24	14 1/2	20 1/2

NOTES:
1. Add 30 minutes to 4 hours for roller layout, depending on local conditions.
2. Add 30 percent for untrained troops, poor weather, terrain conditions, or enemy activity.
3. Add 50 percent for blackout conditions.
4. Consider the rough rule of thumb for adverse conditions: TOTAL time is one bridge bay per hour.

ENGINEER MULTIROLE BRIDGE COMPANY

The engineer multirole bridge company has the capabilities to perform fixed bridging with the MGB and float bridging with the ribbon bridge (see Figure 10-25, page 10-66). The company has four MGB sets with sufficient components for assembling various spans and load classes of single- and double-story bridges. Under normal conditions, the sets provide four 31.4-meter or two 49.7-meter Class 60 bridges with reinforcement kits. The company has about 213 meters of Class 75 (tank) and Class 96 (wheel) float bridge or six rafts of Class 75 (tank) and Class 96 (wheel), based on a 0 to 3 feet per second velocity. The company's assignment is to a corps. It is normally task-organized to a corps engineer battalion or combat engineer group to support bridging operations.

FM 5-34

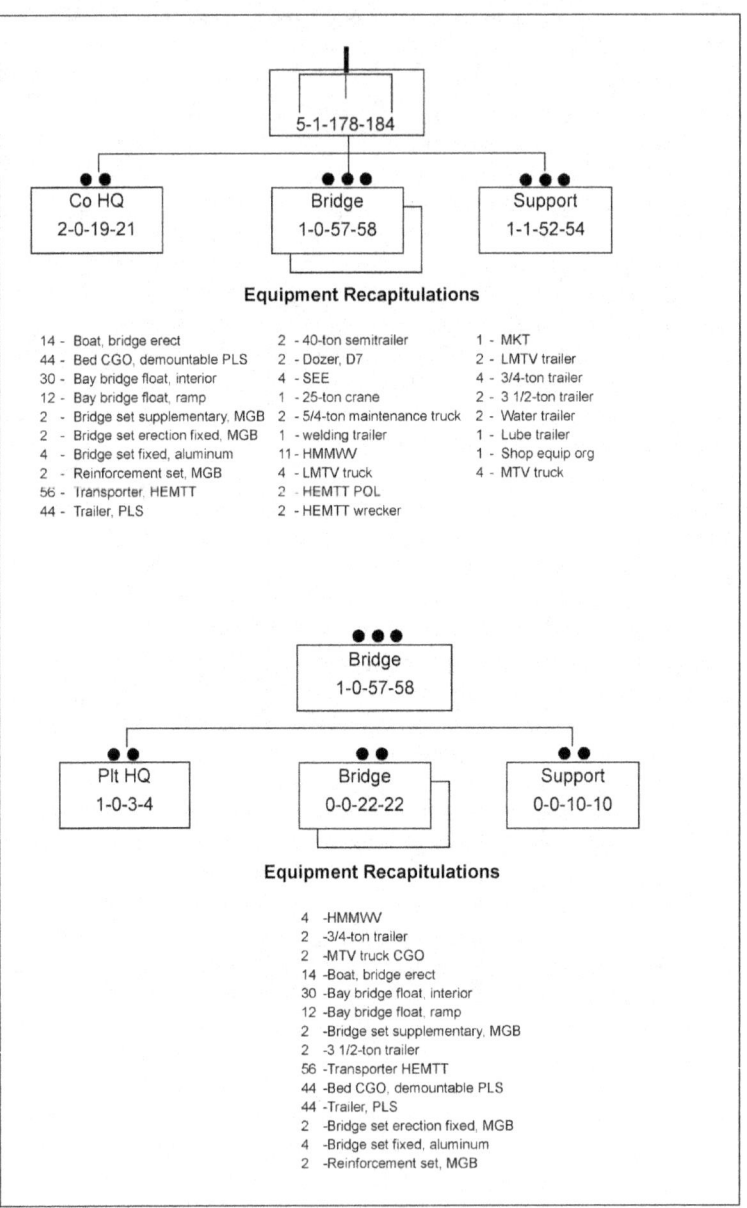

Figure 10-25. Engineer multirole bridge company

10-66 Bridging

FM 5-34

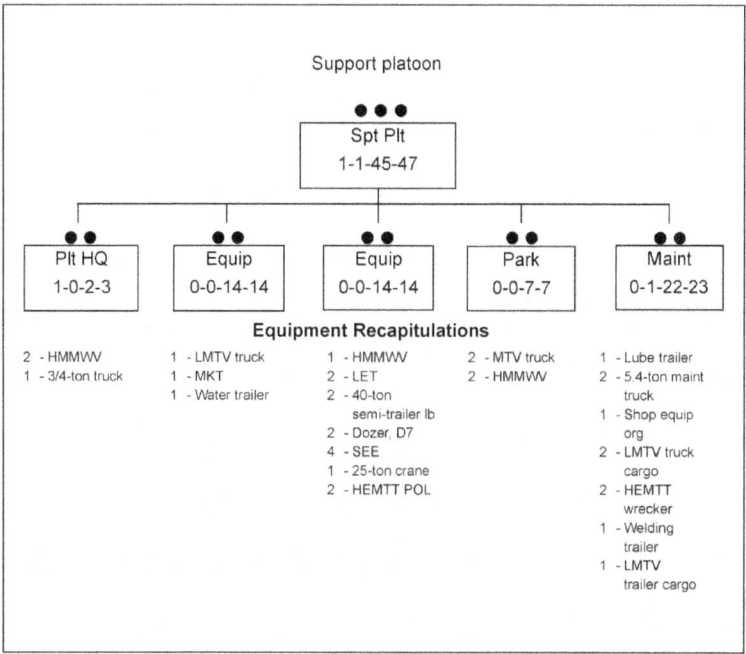

Figure 10-25. Engineer multirole bridge company (continued)

Bridging 10-67

Chapter 11

Roads and Airfields

SOILS AND GEOLOGY

CHARACTERISTICS

Table 11-1 shows the characteristics of specific soils. Figure 11-1, page 11-2, outlines the procedure for field identification of soils.

NOTE: This procedure will give a very hasty classification of soils. Do not use it to support permanent or semipermanent construction projects.

Table 11-1. Soil characteristics

Symbol	Description	Drainage Characteristics	Airfield Index (frost susceptibility)	Value as a Subgrade	Value as a Subbase	Value as a Base	Compaction Equipment
G	Gravels and sands Gravels with little or no fines	Excellent	None to very slight	Good to excellent	Good to excellent	Fair to good	Crawler tractor Rubber-tire roller Steel-wheel roller
GM	Silty gravels Gravel-sand silt mixture	Fair to practically impervious	Slight to medium	Good	Fair to good	Not suitable	Rubber-tire roller Sheepsfoot roller
GC	Clayey gravels Gravel Sand-clay mixtures	Poor to practically impervious	Slight to medium	Good	Fair	Not suitable	Rubber-tire roller
S	Sands and gravels Sands with little or no fines	Excellent	None to very slight	Fair to good	Fair to good	Not suitable	Crawler tractor Rubber-tire roller
SM	Silty-sands Sand-silt mixtures	Fair to practically impervious	Slight to medium	Fair to good	Poor to fair	Not suitable	Rubber-tire roller Sheepsfoot roller
SC	Clayey sands Sand-clay mixtures	Poor to practically impervious	Slight to high	Poor to fair	Poor	Not suitable	Rubber-tire roller Sheepsfoot roller
M	Inorganic silts and very fine sand Rock flour Clayey silts with slight plasticity	Fair to poor	Medium to high	Poor to fair	Not suitable	Not suitable	Rubber-tire roller Sheepsfoot roller
CL	Inorganic clays, low to medium plasticity Gravely or sandy clays	Practically impervious	Medium to high	Poor to fair	Not suitable	Not suitable	Rubber-tire-roller Sheepsfoot roller
CH	Inorganic clays of high plasticity	Practically impervious	Medium	Poor to fair	Not suitable	Not suitable	Sheepsfoot roller
O	Mineral grains containing highly organic matter	Poor to practically impervious	Medium to high	Poor to very poor	Not suitable	Not suitable	Rubber-tire roller Sheepsfoot roller
PT	Peat and other highly decomposed vegetable matter	Fair to poor	Slight	Not suitable	Not suitable	Not suitable	Compaction not practical

FM 5-34

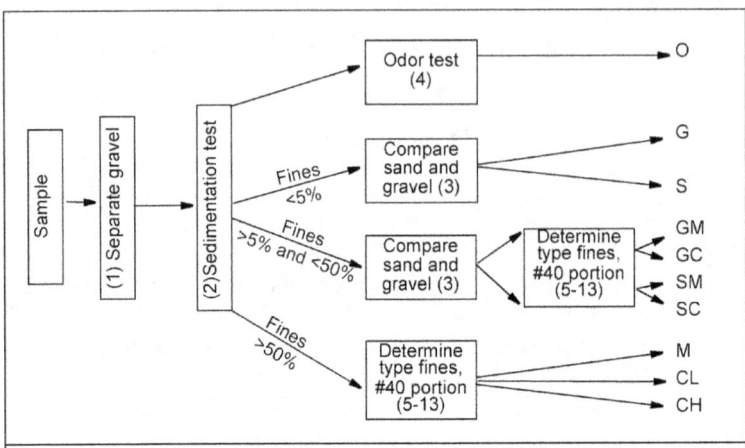

Procedure:
1. Separate the gravel.
   a. Remove all particles from the sample that are larger than 1/4-inch diameter (use a #4 sieve, if available).
   b. Estimate the percent of gravel (*G*) by volume.
2. Conduct a sedimentation test, using either method below, to determine the percent of sand (*S*):
   a. Mason-jar methodæ
      (1) Put about 1 inch of sample in a glass jar.
      (2) Place a line on the outside of the jar, using a grease pencil, to indicate the sample's depth.
      (3) Fill the jar with 5 or 6 inches of clear water. Leave 1 inch of air at the top.
      (4) Shake the mixture vigorously for 3 to 4 minutes.
      (5) Allow the sample to settle for 30 seconds.
      (6) Compare the sediment line to grease-pencil line, estimating the percent settled.
      (7) Determine the percent of *S* and fines (*F*): *S* + *F* = 100 - percent of *G*
      (8) Determine the percent of *S*:
         $S = \frac{(\text{percent of settled material})}{100} \times \text{percent of } S + F$
   b. Canteen-cup methodæ
      (1) Place a sample (less gravel) in a canteen cup and mark the level.
      (2) Fill the jar with water and shake vigorously.
      (3) Allow the mixture to settle for 30 seconds.
      (4) Pour off the water.
      (5) Repeat steps 2 and 4 until the water that pours off is clear.
      (6) Dry the soil left in the cup (*S*).
      (7) Estimate the percent of *S* by comparing its level to the first mark you made.
         S = 100 - percent of *G*

Figure 11-1. Field identification of soils

11-2 Roads and Airfields

FM 5-34

3. Compare the *G*, *S*, and *F*.
   a. Percent of *G* from 1b
   b. Percent of *S* from 2a(8)
   c. Percent of the *F* = 100 - 3a - 3b
4. Conduct an odor test by heating the sample with a match or an open flame. If the odor becomes musty or foul smelling, that is a strong indication of the presence of organic material.
5. Conduct a dry-strength test (use a #40 sieve).
   a. Form a moist pat 2 inches in diameter by 1/2 inch thick.
   b. Dry the pat (low heat).
   c. Place the dried pat between your thumb and index finger only and attempt to break it.
   d. Determine the contents using the breakage factors below:
      - Easy, contents is silt *(M)*.
      - Difficult, contents is low compressible clay *(CL)*.
      - Impossible, contents is high compressible clay *(CH)*.
6. Conduct a powder test.
   a. Scrape a portion of the broken pat with your thumbnail. Try to flake off any particles.
   b. Determine the contents using the following:
      - If the pat powders or flakes, the contents is *M*.
      - If the pat does not powder or flake, the contents is clay *(C)*.
7. Conduct a feel test.
   a. Rub a portion of dry soil over a sensitive portion of your skin (inside of wrist).
   b. Determine the contents based on the following:
      - If your skin feels harsh or is irritated, the contents is *M*.
      - If your skin feels smooth and floury, the contents is *C*.
8. Conduct a shine test.
   a. Use a knife blade or your thumbnail to draw a smooth surface over a pat of slightly moist soil.
   b. Determine the contents based on the following:
      - If the surface becomes shiny and lighter in texture, the contents is *C*.
      - If the surface is dull or granular, the contents is *M* or *S*.
9. Conduct a thread test.
   a. Form a ball of moist soil (marble size).
   b. Attempt to roll it into a 1/8 inch-diameter thread (wooden match size).
   c. Determine the contents based on the following:
      - If you can obtain the thread easily, the contents is *C*.
      - If cannot obtain the thread, the contents is *M*.
10. Conduct a ribbon test.
    a. Form a cylinder of moist soil, about cigar shape and size.
    b. Flatten it over your index finger with your thumb; try to form a ribbon that is 8 to 9 inches long, 1/8 to 1/4 inch thick, and 1 inch wide.
    c. Determine the contents based on the following:
       - If the ribbon is 8 inches or larger, the contents is *CH*.
       - If the ribbon is 3 to 8 inches long, the contents is *CL*.
       - If the ribbon is 0 to 3 inches long, the contents is *M*.

Figure 11-1. Field identification of soils (continued)

Roads and Airfields 11-3

11. Conduct a grit or bite test.
    a. Place a pinch of sample between your teeth and bite the sample.
    b. Determine the contents based on the following:
        - If the sample feels gritty, the contents is *M*.
        - If the sample feels floury, the contents is *C*.
12. Conduct a wet-shaking test.
    a. Place a pat of very moist soil (not sticky) in your palm. Close your hand.
    b. Shake your hand vigorously. Strike the pat with your other hand.
    c. Determine the contents based on the following:
        - If water rises to the surface quickly (positive reaction), the contents is *M*.
        - If no water rises (negative reaction), the contents is *C*.
13. Conduct a hand-washing test.
    a. Rub your hands with a portion of the sample.
    b. Wash your hands.
    c. Determine the contents based on the following:
        - If you can wash your hands easily, the contents is *M*.
        - If it is difficult to wash your hands, the contents is *C*.

NOTE: When classifying soils in the field, if you notice any unusual odors or characteristics to the soil (oil or petroleum) or any unknown substances, notify your chain of command immediately.

Figure 11-1. Field identification of soils (continued)

MOISTURE CONTENT

To determine whether or not soil is at or near optimum moisture content (OMC), mold a golf-ball-size sample of soil with your hands. Squeeze the ball between your thumb and fore finger. If the ball shatters into several fragments of rather uniform size, the soil is near or at the OMC. If the soil is difficult to roll into a ball or it crumbles under very little pressure, the soil is below the OMC.

STABILIZATION

Table 11-2. Recommended initial stabilizing agent (percent of weight)

Soil Type	Hydrated Lime	Portland Cement	Quicklime
GC, GM-GC	2 to 4		2 to 3
CL	5 to 10		3 to 8
CH	3 to 8		3 to 6
GW, SW		3 to 5	
GP, GM, SM		5 to 8	
GC, SC		5 to 9	
SP		7 to 11	
CL, ML		8 to 13	
CH		9 to 15	
MH, OH		10 to 16	

Table 11-3. Soil conversion factors

Soil Type	Initial Soil Condition	Converted to—		
		In Place	Loose	Compacted
Sand	In place		1.11	0.95
	Loose	0.90		0.86
	Compacted	1.05	1.17	
Loam	In place		1.25	0.90
	Loose	.080		0.72
	Compacted	1.11	1.39	
Clay	In place		1.43	0.90
	Loose	0.70		0.63
	Compacted	1.11	1.59	
Rock (blasted)	In place		1.50	1.30
	Loose	0.67		0.87
	Compacted	0.77	1.15	

ENGINEERING PROPERTIES OF ROCKS

Listed below are the engineering properties of rocks (Figure 11-2, page 11-6, also shows rock descriptions). Table 11-4, page 11-7, lists the characteristics of various rocks, in terms of their properties, and their relative uses as an aggregate, base course, or subbase.

- Toughness—mechanical strength, resistance to crushing or breaking. Estimate this property by trying to break a rock with a hammer or by measuring a rock's resistance to penetration using impact drills.

- Hardness—resistance to scratching or abrasion. Estimate this property by trying to scratch the rock with a steel knife or nail. Soft material will scratch readily; hard materials are difficult or impossible to scratch.

- Density—weight per unit volume. Estimate this property by *hefting* a rock sample and comparing two samples that are close in equal volume.

- Durability—resistance to slaking or disintegration due to alternating cycles of wetting and drying or freezing and thawing. Estimate this property by observing the effects of weathering on natural exposures of the rock.

- Chemical stability—resistance to reaction with alkali materials in portland cements. Several rock types contain impure forms of silica (opal and chalcedony) that react with alkalis in cement to form a gel which absorbs water and expands to crack or disintegrate the hardened concrete. Estimate this potential alkali-aggregate reaction (in the field only) by identifying the rock and comparing it to known reactive types or by investigating structures in which the aggregate has previously been used.

- Crushed shape—form a rock takes on after crushing; bulky angular fragments provide the best aggregate for construction. Estimate this property by breaking a sample of the rock into smaller pieces.

- Surface character—bonding characteristics; excessively smooth, slick, nonabsorbent aggregate surfaces bond poorly with cement. Excessively rough, jagged, or absorbent surfaces are undesirable because they resist compacting and placement and require excessive cementing material. Visually inspect the rock surface and feel the surface texture.

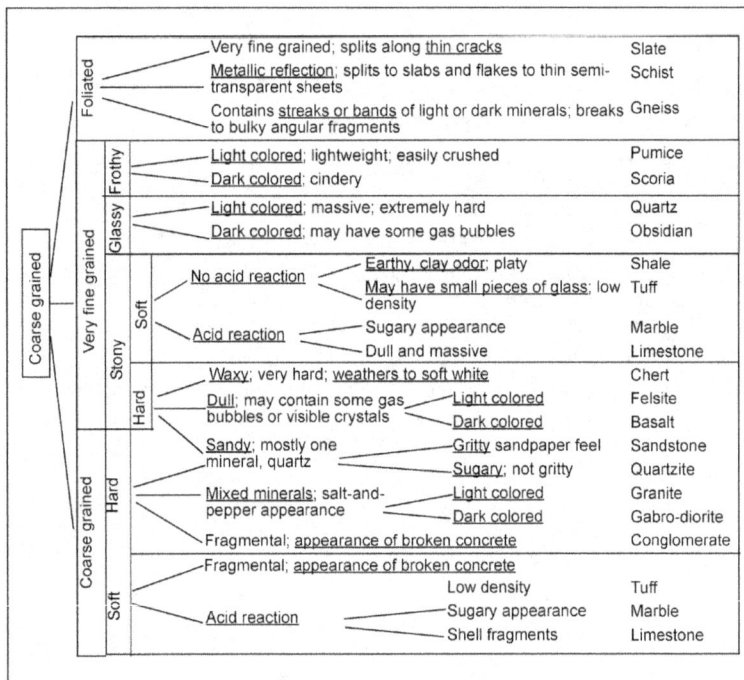

Figure 11-2. Rock identification

11-6 Roads and Airfields

FM 5-34

Table 11-4. Rock characteristics

Rock Type	Toughness	Hardness	Durability	Chemical Stability	Surface Character	Crushed Shape	Use as Aggregates			Use as Base Course or Subbase
							Concrete	Asphalt		
Granite	Good	Good	Good	Excellent	Fair to good	Good	Fair to good	Fair to good**	Good
Syenite							Good	Good	
Gabbro diorite	Excellent	Excellent	Excellent	Excellent	Excellent	Good	Excellent	Excellent	Excellent
Diabase basalt	Excellent	Excellent	Excellent	Excellent	Excellent	Fair	Excellent	Excellent	Excellent
Felsite	Excellent	Good	Good	Questionable	Fair	Fair	Poor*	Fair	Fair to good
Conglomerate breccia	Poor	Poor	Poor	Variable	Good	Fair	Poor	Poor	Poor
Sandstone	Variable	Variable	Variable	Good	Good	Good	Poor to fair	Poor to fair	Fair to good
Shale	Poor	Poor	Poor	Questionable	Fair to good	Poor	Poor	Poor	Poor
Limestone Dolomite	Good	Good	Fair to good	Good	Good	Good	Fair to good	Good	Good
Chert	Good	Excellent	Poor	Poor	Fair	Poor	Poor*	Poor**	Poor to fair
Gneiss	Good	Good	Good	Excellent	Good	Good to fair	Good	Good	Good
Schist	Good	Good	Fair	Excellent	Poor to fair	Poor to fair	Poor to fair	Poor to fair	Poor-fair
Slate	Good	Good	Fair to good	Excellent	Good	Poor	Poor	Poor	Poor
Quartzite	Excellent	Excellent	Excellent	Excellent	Good to fair	Fair	Good	Fair to good**	Fair to good
Marble	Good	Fair	Good	Good	Good	Good	Fair	Fair	Fair

*Reacts (alkali-aggregate)
**Use antistripping agents.

Roads and Airfields 11-7

FM 5-34

DRAINAGE

The most common drainage structures are open ditches and culverts.

RUNOFF ESTIMATE

You can estimate the volume of water that an open channel carries using the hasty method. Determine the water volume using the following formula:

$$A_w = \frac{(W_1 + W_2)H}{2}$$

where—

A_w = cross-sectional area of storm-water runoff, in square feet

W_1 = channel width at the watermark, in feet

W_2 = channel width at the bottom, in feet

H = water height, from the bottom of the stream to the watermark, in feet

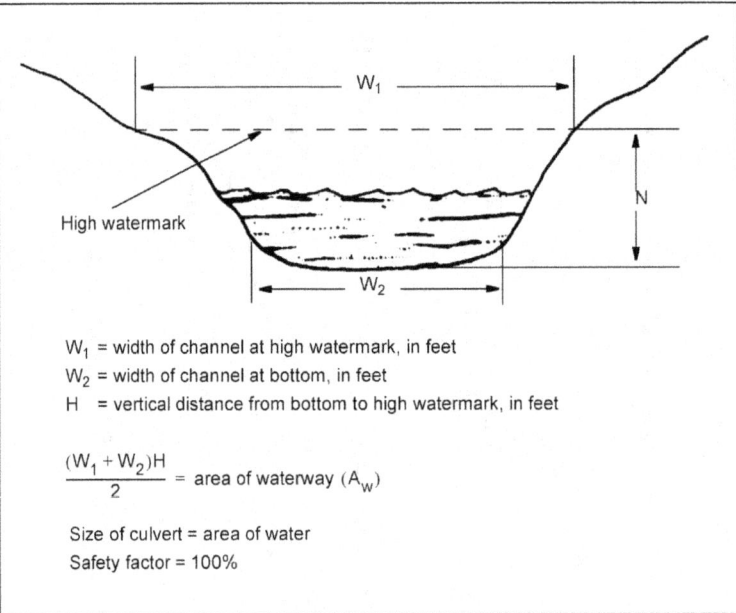

Figure 11-3. Cross-sectional area of water

11-8 Roads and Airfields

Estimate the quantity of runoff water, using A_w (previous formula) and the velocity of the water in the channel, with the following formula:

$$Q = A_w V$$

where—

Q = quantity of the runoff water, in cubic feet per second (cfs)

A_w = cross-section area of the runoff, in square feet

V = velocity of the water, in feet per second (fps)

For information on a more deliberate design method, rational method, see FM 5-430-00-1.

CULVERTS

To find the required diameter of pipe, use Q (previous formula), the desired slope for the culvert (between 0.5 and 4 percent), and Table 11-5, page 11-10. If the diameter of the available pipe is different from what you calculate, you may need more than one pipe. Use the following formula to determine the number of pipes you need:

$$NP = \frac{Q}{Q_P}$$

where—

NP = number of pipes required

Q = quantity of water, in cfs

Q_P = quantity of water a pipe can handle, in cfs

Design

Use the A_w to compute the culvert-design area (A_{des}), which is $2A_w$. See Figure 11-4, page 11-11, to determine the maximum allowable culvert diameter, fill, and cover. Round down to the next available culvert diameter. Determine the number of pipes using the following formula:

$$N = \frac{A_{des}}{\text{pipe area}}$$

where—

N = number of pipes

A_{des} = design area of the culvert, in square feet

Start with the largest available culvert that meets the maximum diameter requirement. Then go to smaller diameters until you find the most economical solution.

Table 11-5. Determining pipe diameter in relation to Q_p

S (%)	V (fps)	Pipe Diameter (in)							
		12	18	24	30	36	42	48	60	72
		Quantity of water in pipe (Q_P) (cfs)							
0.4		1.30	3.90	8.40	15.00	25.00	37.00	53.00	96.00	160.00
0.5	2.00	1.45	4.35	9.20	16.50	27.50	41.00	58.50	108.00	175.00
0.6		1.60	4.80	10.00	18.00	30.00	45.00	64.00	120.00	190.00
0.7		1.70	5.10	11.00	19.50	32.00	47.50	68.00	125.00	200.00
0.8		1.80	5.40	12.00	21.00	34.00	50.00	72.00	130.00	210.00
0.9	3.00	1.90	5.65	12.50	21.50	35.00	52.00	74.50	135.00	215.00
1.0		2.00	5.90	13.00	22.00	36.00	54.00	77.00	140.00	220.00
1.1		2.10	6.15	13.00	23.00	37.00	55.50	78.50	145.00	225.00
1.2		2.20	6.40	13.00	24.00	38.00	57.00	80.00	150.00	230.00
1.3		2.25	6.50	13.00	24.50	38.50	58.00	81.00	150.00	230.00
1.4		2.30	6.60	14.00	25.00	39.00	59.00	82.00	150.00	230.00
1.5		2.35	6.70	14.00	25.00	39.00	59.00	82.00		
1.6		2.40	6.80	14.00	25.00	40.00	59.00	83.00		
1.7	4.00	2.45	6.90	14.00	25.00	40.00				
1.8		2.50	7.00	14.00	26.00	40.00				
1.9		2.50	7.00	14.50	26.00					
2.0		2.50	7.00	15.00	26.00					
2.1		2.55	7.05	15.00						
2.2		2.60	7.10	15.00						
	5.00		6.00	7.00	8.00				9.00	10.00	11.00

NOTES:
1. The last value in each column is the maximum flow.
2. To use the chart, find the slope percentage and move right (horizontally) until you intersect the column for the pipe diameter you selected. Find the velocity by following the shaded or nonshaded area to the heavy bordered area and reading the value. For example—
A 36-inch pipe at 1.3 percent slope has a Q_P of 38.5 and a V of 4.
A 72-inch pipe at 0.9 percent slope has a Q_P of 215 and a V of 3.

FM 5-34

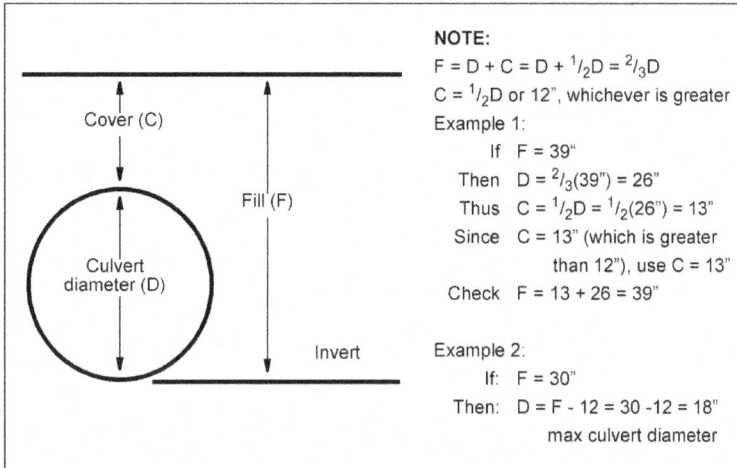

Figure 11-4. Minimum fill and cover

Length

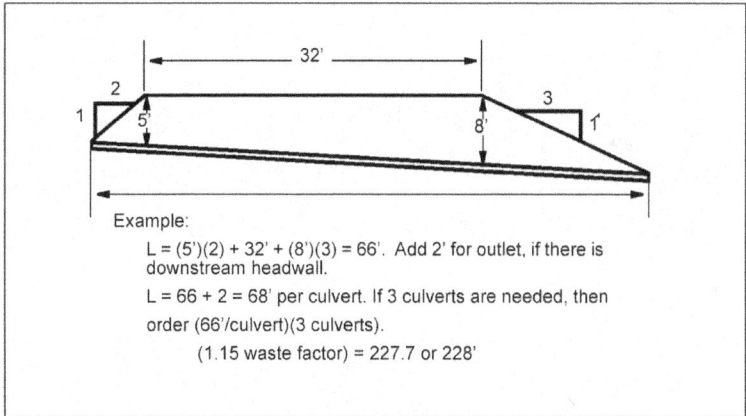

Figure 11-5. Culvert-length determination

Installation

Use the following criteria, when possible: Figure 11-6, page 11-12; and Tables 11-6 and 11-7, page 11-13, to install a culvert:

• Place the inlet elevation at or below the ditch bottom.

Roads and Airfields 11-11

FM 5-34

- Extend the culvert 2 feet minimum downstream beyond the fill slopes, if there is no downstream headwall.
- Use a minimum bedding of 1/10 diameter of the culvert.
- Space multiple culverts a minimum of half the diameter of the culvert.
- Ensure that the slope is a minimum of 0.5 percent; the desired slope is 2 to 4 percent.
- Use a headwall upstream.
- Use riprap downstream to control erosion.

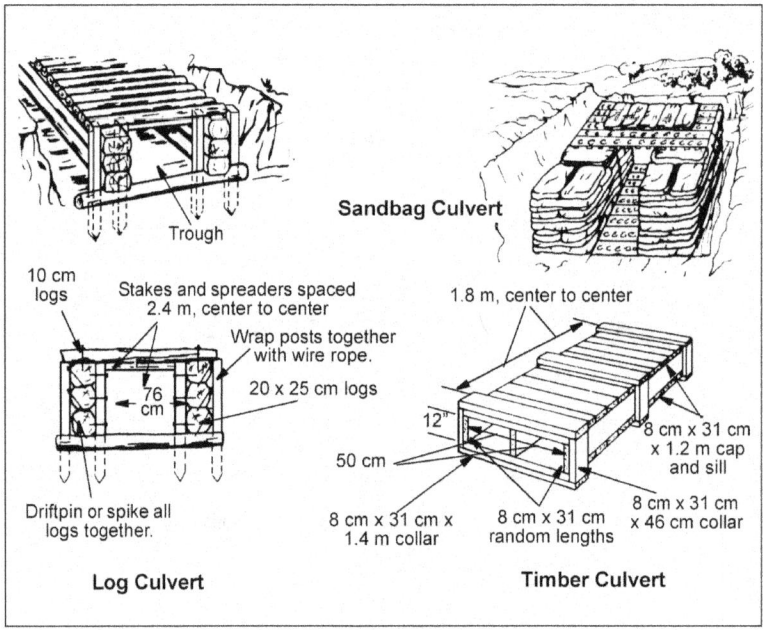

Figure 11-6. Expedient-culvert examples

11-12 Roads and Airfields

Table 11-6. Recommended gauges for nestable corrugated pipe

Diameter (in)	Waterway Area (sq ft)	Fill						
		Up to 8 Feet	Up to 16 Feet	20 Feet	25 Feet	30 Feet	35 Feet	40 Feet
8	0.35	16	16	16	16	16	16	16
10	0.55	16	16	16	16	16	16	16
12*	0.79	16	16	16	16	16	16	16
15	1.23	16	16	16	16	16	16	16
18*	1.77	16	16	16	16	16	16	16
21	2.41	16	16	16	16	16	16	16
24*	3.14	16	16	16	16	14	14	14
30*	4.91	14	14	14	14	14	12	12
36*	7.07	14	14	14	12	12	12	10
42*	9.62	14	14	12	12	10	10	8
48*	12.57	12	12	12	10	8	8	8
54	15.90	12	12	10	8	8	8	8
60*	19.64	12	10	8	8	8	8	8
66	23.76	**	10	8	8	8	8	
72*	28.27	**	10	8	8	8	Must be designed for these fill heights and others above 40 feet.	
78	33.18	**	8	8	8			
84	38.49	**	8	8	8			

NOTE: Culverts below the heavy line should be strutted during installation.
*Indicates corrugated-metal pipe sizes normally found in the theater of operations (TO)
**Indicates insufficient cover

Table 11-7. Strut spacing using 4- by-4 inch timbers with compression caps

Diameter (in)	Strut Length (in)	Strut Spacing (ft)		
		With fill of 5 to 10 ft	With fill of 10 to 20 ft	With fill of 20 to 30 ft
48	37.5	6*	6	6
54	43.6	6*	6	6
60	49.8	6*	6	6
66	56.0	6*	6	6
72	62.2	6*	6	5

*Ensures sufficient cover

OPEN-DITCH DESIGN

- Determine the area of water A_w (see formula in Runoff Estimate, page 11-8).

FM 5-34

- Select a site slope ratio based on soil stability (see Table 11-8), equipment capacity, and safety.

Table 11-8. Recommended requirements for slope ratios in cuts and fills - homogeneous soils

USCS Classification	Slopes Not Subject to Saturation		Slopes Subject to Saturation	
	Maximum Height of Earth Face	Maximum Slope Ratio	Maximum Height of Earth Face	Maximum Slope Ratio
GW, GP, GMd SW, SP, SMd	Not critical	$1\frac{1}{2}:1$	Not critical	2:1
GMu, GC SMu, SC ML, MH CL, CH	Less than 50 feet	2:1	Less than 50 feet	3:1
OL, OH, Pt	Generally not suitable for construction			

NOTES:
1. Recommended slopes are valid only in homogeneous soils that have either an in-place or compacted density equaling or exceeding 95% of CE 55 maximum dry density. For nonhomogeneous soils, or soils at lower densities, a deliberate slope stability analysis is required.
2. Back slopes cut into loose soil will seek to maintain a near-vertical cleavage. Do not apply loading above this cut face. Expect sloughing to occur.

- Determine the cutting depth according to Figure 11-7.

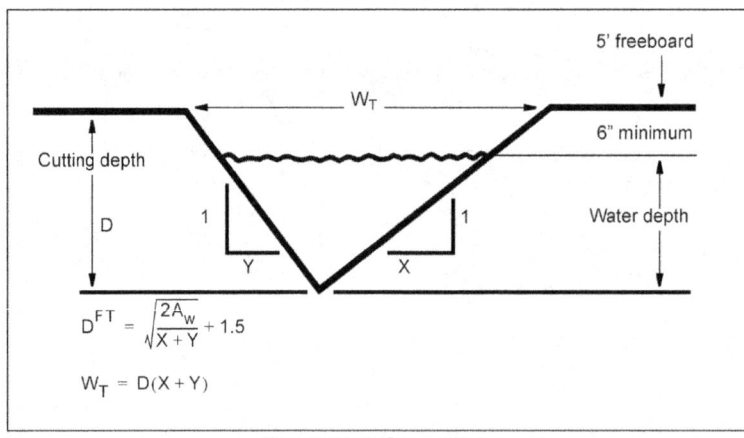

$$D^{FT} = \sqrt{\frac{2A_W}{X+Y}} + 1.5$$

$$W_T = D(X+Y)$$

Figure 11-7. Open ditch

11-14 Roads and Airfields

EXPEDIENT AIRFIELD SURFACES

- Calculate the requirements to prepare subgrade, lay membrane, and lay matting using Table 11-9 and Table 4-7, page 4-27.

Table 11-9. Mat characteristics

	M8A1	M8	M18b	M19	AM2
Bundle Volume (cu ft) Placing area (sq ft) Weight (lb)	24.7 269 2,036	22.7 269 1,960	74 432 2,400	86.7 534 2,484	62 288 1,980
Number of panels (full/half)	13/2	13/2	16/4	32/0	11/2
Panel Dimension (ft) Weight (lb) Placing area (sq ft)	1.6 x 11.8 144 19.2	1.0 x 11.8 140 19.2	2 x 12 120 24	4 x 4.1 68 16.7	2 x 12 140 24

- Start placing matting from one corner of the runway with the male hinges parallel with and toward the centerline. Lay the first strip along the edge of the roadway. Stagger the second strip so that the connectors from the first strip are at the center of the second strip's panels. You must fully insert the connecting bars (Figure 11-8, page 11-16).

Minimum Operating Strip (MOS)

The main focus in airfield repair is the MOS, which is 15 by 1,525 meters for fighter aircraft and 26 by 2,134 meters for cargo.

Work Priority

Use the procedure below and Figure 11-9, page 11-16, to construct MOSs.

- Establish the first MOS (15 by 1,525 meters).
- Use minimal effort to build 7.6-meter-wide access routes.
- Establish a second MOS (15 by 1,525 meters).
- Build more 7.6-meter-wide access routes.
- Lengthen the first MOS to 2,134 meters.
- Lengthen the second MOS to 2,134 meters.
- Widen the first MOS to 27.4 meters.
- Widen the second MOS to 27.4 meters.

FM 5-34

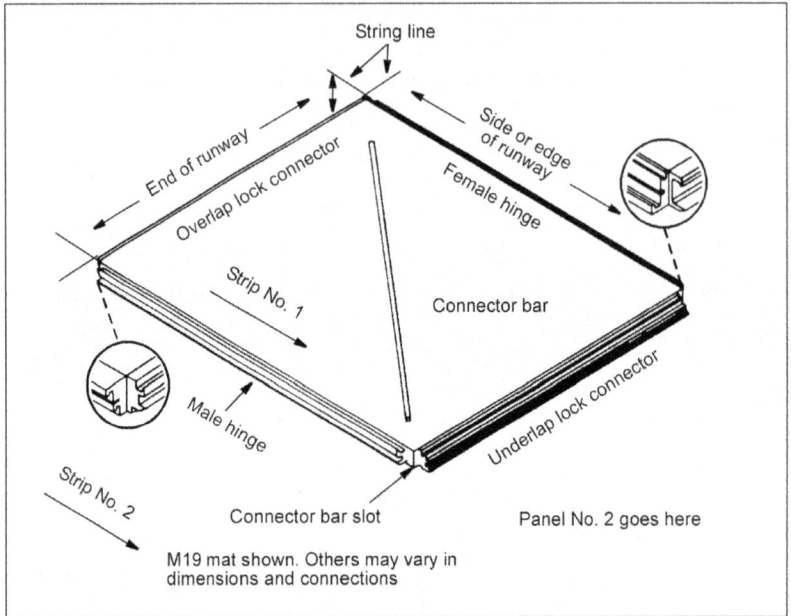

Figure 11-8. Typical mat and connectors

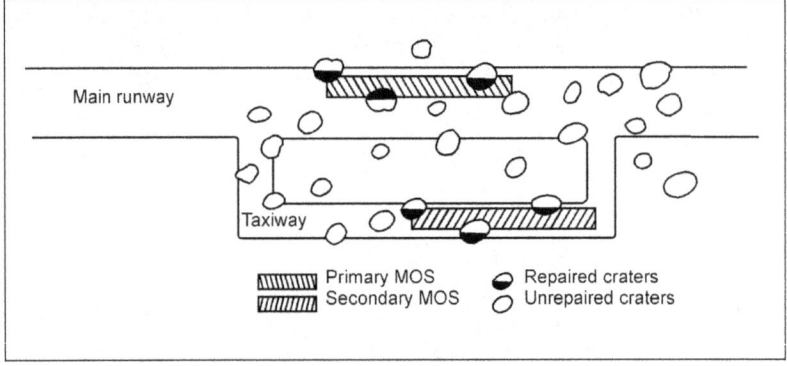

Figure 11-9. Constructing an MOS

11-16 Roads and Airfields

MEMBRANE AND MAT REPAIR

Membranes

Repair any tears in membranes by cutting an *X* and lifting the four flaps back. Place a new piece of membrane under the torn area to extend at least 30 centimeters beyond the torn area. Apply an adhesive to the top of the new membrane and the bottom of the old membrane. Allow the adhesive to become tacky. Fold the flaps back into position, and let the adhesive set for at least 15 minutes. Roll the patched area with a wheeled roller or vehicle.

Mats

M8A1

Unlock the end connector bars from the damaged panel; remove the locking lugs. Move the panel laterally until the hooks are centered on the slots. Pry the hooks out of the slots, and move the panel to clear the overlapping ends. Remove the damaged panel. Remove the locking lugs from the new panel, and orient it to the same position as the damaged panel. Reverse the removal procedures to install the new panel.

AM2

- Slide-out method. Slide out the entire run where damage to the panel is located. Remove the end connector bars. Replace the damaged panel. Push a new run in until it is 5 to 10 centimeters from the next panel. Continue the procedure until you have replaced all the panels. Push the run to its original position.

- Cutting method. Cut the damaged panel and remove the pieces (see Figure 11-10, page 11-18). Replace the damaged panels with special repair panel and accessories, if they are available (see Figure 11-11, page 11-19).

M19

Replace a single mat by first cutting it using a circular saw (see Figure 11-12, page 11-20). Use a pry-bar to lift the cut pieces. Unbolt the edges of the damaged panel and replace it (see Figure 11-13, page 11-20).

To repair large areas, create a pyramid (see Figure 11-14, page 11-21). Remove the maintenance access adapter, and start removing the panels from the outside in until you reach the damaged area. Replace the damaged area and panels.

FM 5-34

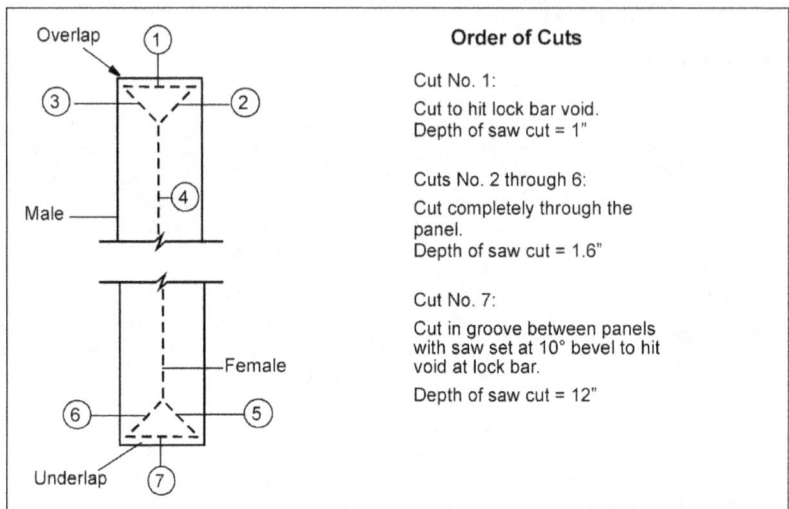

Figure 11-10. Cutting an AM2 mat

11-18 Roads and Airfields

FM 5-34

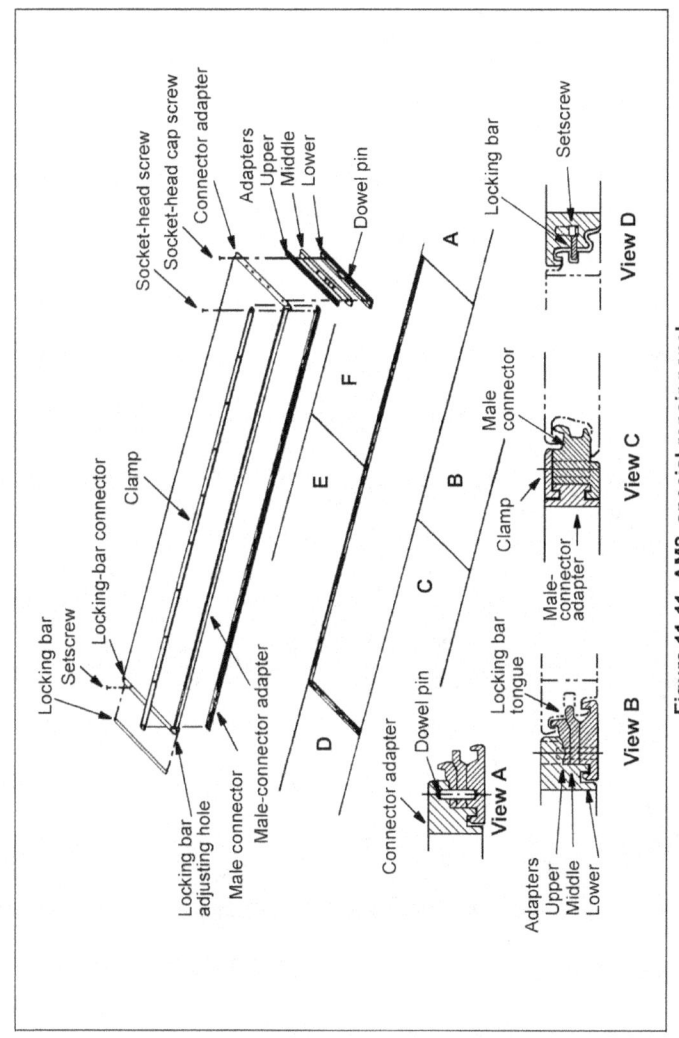

Figure 11-11. AM2, special repair panel

Roads and Airfields 11-19

FM 5-34

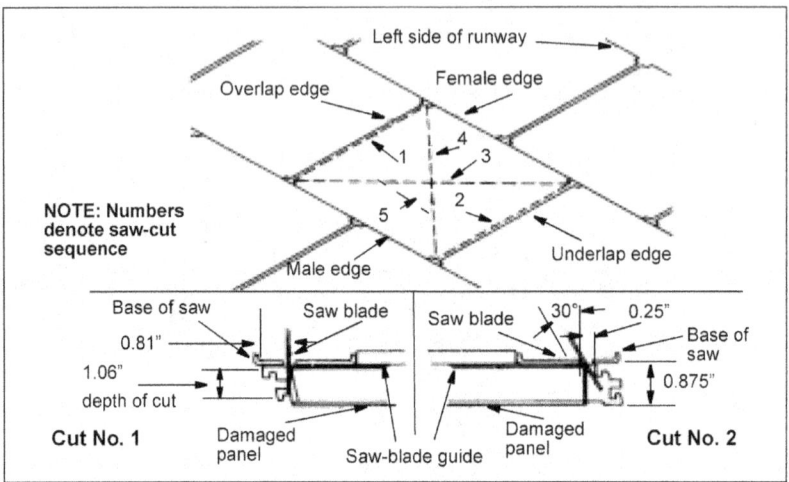

Figure 11-12. Cutting an M19 mat

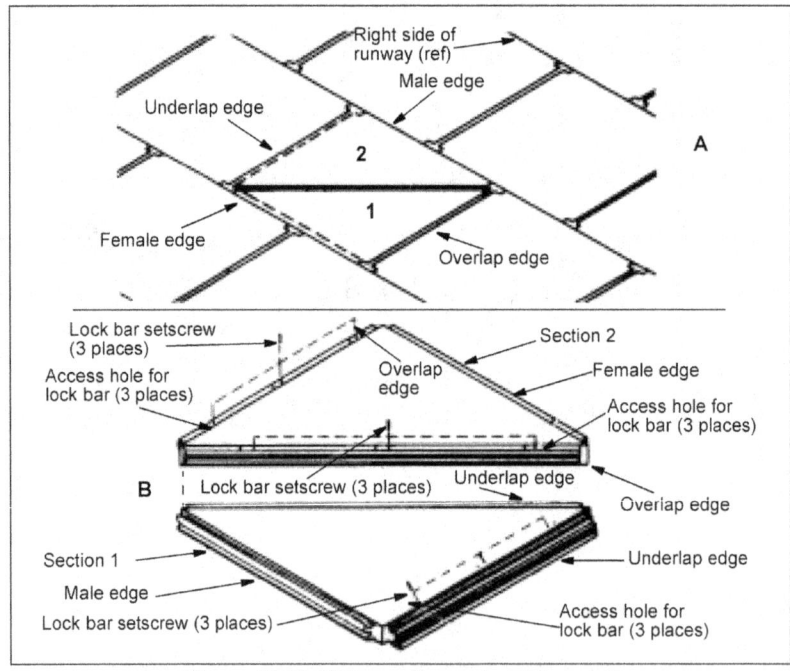

Figure 11-13. M19, repair-panel replacement

11-20 Roads and Airfields

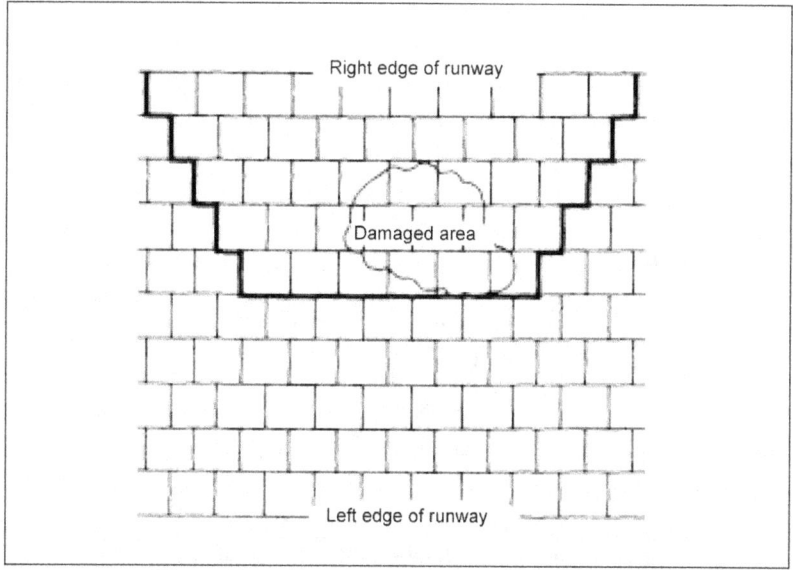

Figure 11-14. M19, repairing large damaged areas

OTHER REPAIRS

Figure 11-15 and Figures 11-16 and 11-17, page 11-23, show different emergency repair methods.

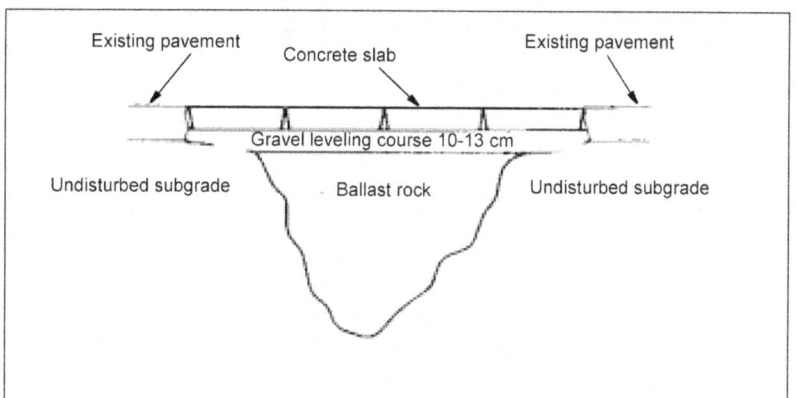

Figure 11-15. Precast concrete-slab-crater repair

FM 5-34

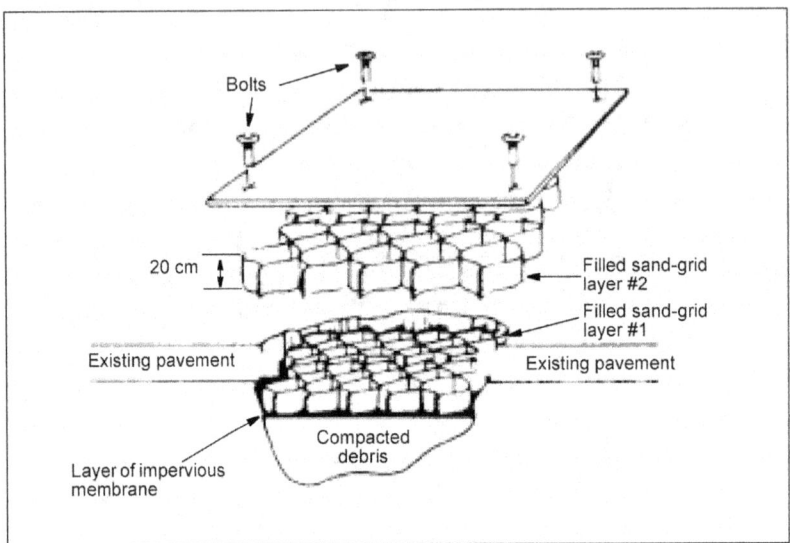

Figure 11-16. Sand-grid repair method

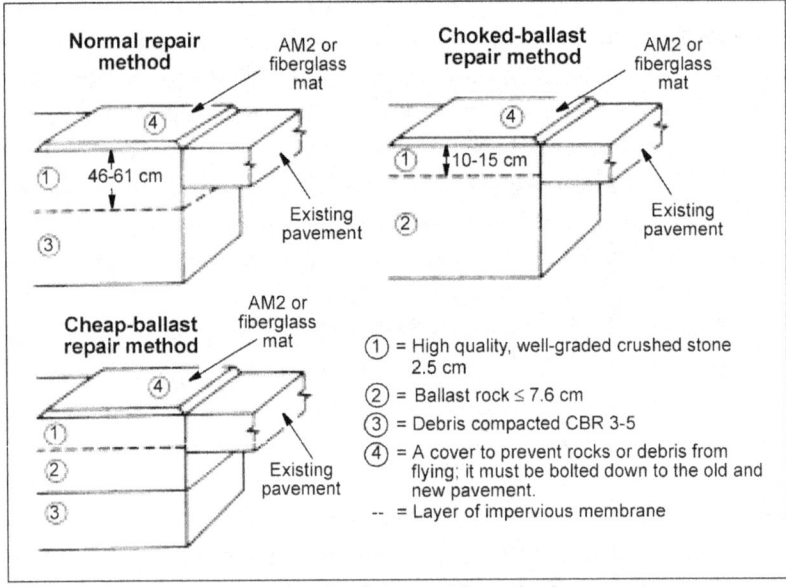

Figure 11-17. Other repair methods

11-22 Roads and Airfields

ROAD DESIGN

ELEMENTS OF A HORIZONTAL CURVE

The following are elements of a simple, horizontal curve (see Figure 11-18):

- PC is the point of tangent departure.
- PT is where the curve ends or joins tangent B.
- PI is the intersecting point of two tangents.
- T is the tangent distance from PI to PC or from PI to PT.
- R is the radius of the circle from PC to PT.
- L is the length of the curve.
- I (angle of intersection) is the exterior angle at PI formed by tangents between A and B.
- E (external distance) is the distance from PI to the midpoint of the curve.
- C (long chord) is the straight-line distance from PC to PT.
- M (middle ordinate) is the distance from the midpoint of the curve to the midpoint of the long chord.

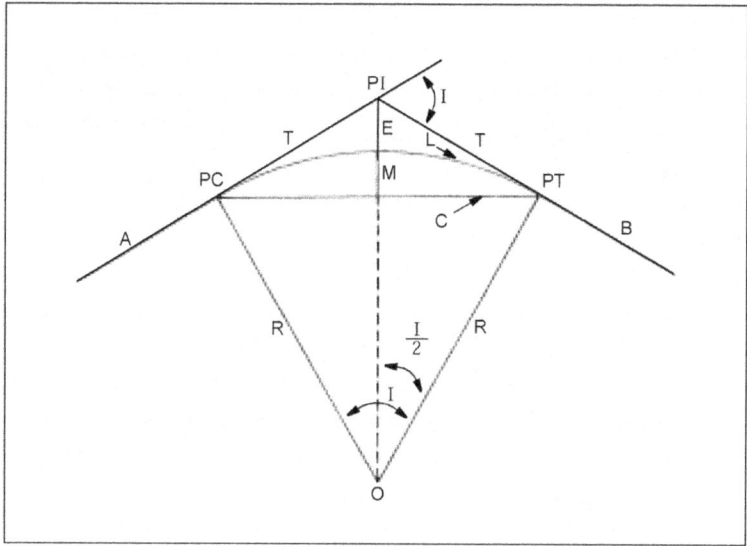

Figure 11-18. Elements of a simple, horizontal curve

DEGREE OF CURVATURE (D)

The connecting curve between two tangents may be short and sharp or long and gentle, depending on the properties of the circle chosen. Sharpness is defined by the radius of the circle. The common reference term for defining curve sharpness is D, which is established as a whole or half degree and may be stated in terms of either the arc or the chord.

Arc Definition

D is that angle which subtends a 100-foot arc along the curve (see Figure 11-19). This definition is used by state highway departments and the Corps of Engineers in road design.

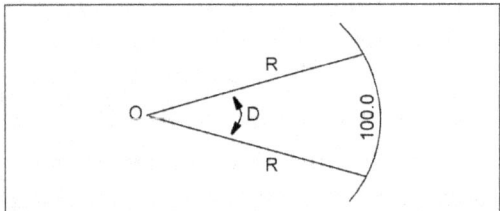

Figure 11-19. Arc definition for degree of curvature

Chord Definition

D is the angle which subtends a 100-foot chord on the curve (see Figure 11-20). This definition results in a slightly larger angle than the arc method, and it is used by the railroad industry and the Corps of Engineers in railroad design.

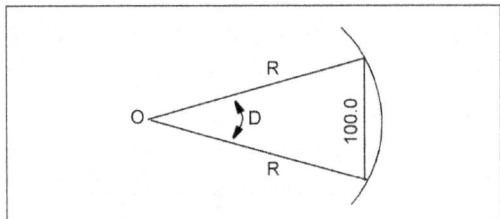

Figure 11-20. Chord definition for degree of curvature

The difference between the arc and chord definitions is very slight and nearly insignificant (frequently well below 1 percent) for TO construction. However, because the arc definition is the most widely used procedure in road design, only its definition will be used throughout the rest of the chapter.

EQUATIONS FOR THE SIMPLE, HORIZONTAL-CURVE DESIGN

The two methods commonly used to solve horizontal-curve problems are the 1-degree-curve method (see Figure 11-21) and the trigonometric method. Both methods may be used with the same degree of accuracy. The 1-degree-curve method requires the functions of a 1-degree-curve table shown in Appendix F of FM 5-430-00-1/AFPAM 32-8013, Vol 1.

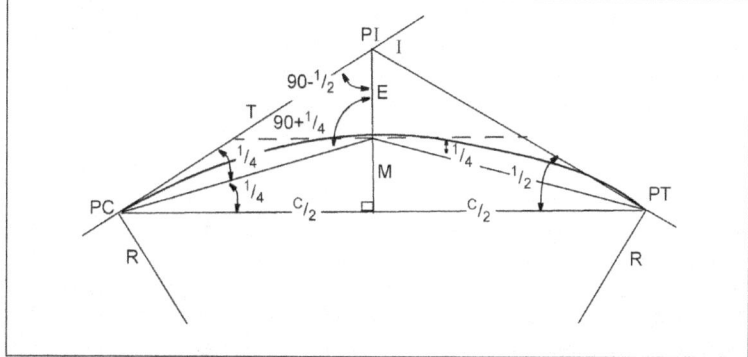

Figure 11-21. Derivation of external distance

Radius of Curvature

$$R = \frac{5,729.58}{D}$$

Tangent Distance (T)

Use the following equation if Appendix F of FM 5-430-00-1/AFPAM 32-8013, Vol 1 is not available:

$$T = R \tan \frac{I}{2}$$

or the following equation if Appendix F of FM 5-430-00-1/AFPAM 32-8013, Vol 1 is available:

$$T = \frac{T_{1°} I}{D} \quad \text{(arc definition)}$$

Roads and Airfields 11-25

External Distance (E)

Using the 1-degree curve method (see Figure 11-21, page 11-25), find E as follows if Appendix F of FM 5-430-00-1/AFPAM 32-8013, Vol 1 is available:

$$E = \frac{E_{1°}}{D} \text{ (arc definition)}$$

or the following equation if Appendix F of FM 5-430-00-1/AFPAM 32-8013, Vol 1 is not available:

$$E = R\left(\tan\frac{I}{2}\right)\left(\tan\frac{I}{4}\right)$$

Long Chord (C)

$$C = 2R\left(\sin\frac{I}{2}\right) = 2T\left(\cos\frac{I}{2}\right)$$

Middle Ordinate (M)

Using the 1-degree-curve method (see Figure 11-21), find M as follows if Appendix F of FM 5-430-00-1/AFPAM 32-8013, Vol 1 is available:

$$M = \frac{M_{1°}}{D} \text{ (arc definition)}$$

or the following equation if Appendix F of FM 5-430-00-1/AFPAM 32-8013, Vol 1 is not available:

$$M = R\left(1 - \cos\frac{I}{2}\right) = \frac{1}{2}C\left(\tan\frac{I}{4}\right)$$

Length of Curve (L)

$$L = \frac{I}{D} \times 100$$

The central angle subtended by the entire horizontal curve has sides that are radii to the PC and PT. Both of these radii are perpendicular to the tangents that form the intersection angle I. The quadrilateral formed by the four points of PI (180° - I), PC (90°), O (I), and PT (90°) must total 360 [(180-I) + 90 + I + 90 = 360]. Thus, the central angle is equal to the angle of intersection I.

DESIGNING HORIZONTAL CURVES

The engineer designing horizontal curves must know two facts about the curve from the preliminary survey: the location and station of the PI and the angle between intersecting tangent lines (I). The curves can be designed after this information is obtained. The engineer can use either the 1-degree-curve method or the trigonometric method. The following steps show the design of horizontal curves using the 1-degree-curve method:

Step 1. Find D using one of the following methods:

- If the curve is unrestricted—

$$D = \frac{5,729.58}{R}$$

where—

D = degree of curvature

R = the radius of the curve

- If the curve is restricted by the tangent distance—

$$D = \frac{T_{1°}}{T_{(restricted)}}$$

where—

D = degree of curvature

$T_{1°}$ = tangent distance for a 1-degree curve (found in Appendix F, FM 5-430-00-1/AFPAM 32-8013, Vol 1, based on the angle of intersection)

$T_{(restricted)}$ = restricted tangent distance for a horizontal curve

- If the curve is restricted by the external distance—

$$D = \frac{E_{1°}}{E_{(restricted)}}$$

where—

$E_{1°}$ = external distance for a 1-degree curve (found in Appendix F, FM 5-430-00-1/AFPAM 32-8013, Vol 1, based on the angle of intersection)

$E_{(restricted)}$ = restricted external distance for a horizontal curve

- If both tangent and external distance restrictions exist, choose the larger of the two Ds that result from the above equations.

Step 2. Round up D to the next half degree when possible.

Step 3. Determine T.

$$T = \frac{T_{1°}}{D}$$

Step 4. Find the stationing of PC.

$$PC = PI - T$$

Step 5. Calculate the length of the curve.

$$L = \left(\frac{I}{D}\right) 100$$

Step 6. Find the stationing of PT.

$$PT = PC + L$$

Horizontal-Curve Design Examples

This section describes the horizontal-curve design procedures for three common situations:

- No terrain restriction which limits T or E.
- Terrain restriction of the tangent distance.
- Terrain restriction of the external distance.

Example:

D with no terrain restriction—Figure 11-22 shows the following computations:

Given: $I = 50°$, PI at 14 + 28

Find the station and location of PC and PT for a Class C road.

Solution:

D of 6° is selected as a flat, gentle curve. D = 6° is far below the maximum allowable of D = 14.5° for Class C roads and is slightly sharper than the maximum allowable of D = 5.5° for Class B roads.

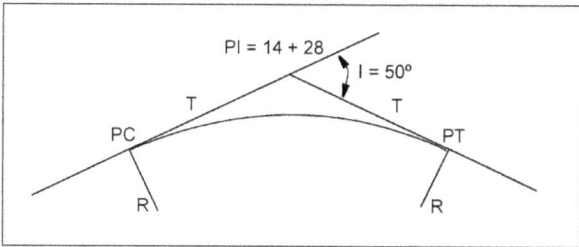

Figure 11-22. Horizontal curve with no sharpness restriction

Use D = 6°

$$R = \frac{5,729.58}{D} = \frac{5,729.58}{6} = 954.93'$$

$$T = R\tan\left(\frac{I}{2}\right) = 954.93'\left(\tan\frac{50°}{2}\right) = 445.29'$$

$$L = \frac{I}{D} \times 100 = \frac{50}{6} \times 100 = 833.33'$$

$$PC = PI - T = (14 + 28) - (445.29') = (9 + 82.71)$$

$$PT = PC + L = (9 + 82.71) + (833.33') = (18 + 16.04)$$

The PT is one point on the centerline, but it has two station values: PT "back" and PT "ahead." The PT station found above is the PT "back." This station is the overall distance from the beginning of the project at station 0+00 to the PT, measured along the centerline of the road.

PT "ahead" is determined from the equation PT = PI + T. The PT "ahead" station is needed to verify if adequate distance exists between the PT and the PC of the next curve. The distance between adjacent curves is the difference between the PT "ahead" station and the next PC station.

Roads and Airfields 11-29

Chapter 12

Rigging

ROPE

This chapter deals with information on rope, knots and attachments, chains and hooks, slings, and picket holdfasts. Most of the information is in table or figure format. For more detailed information on rigging, see FM 5-125.

Table 12-1. Properties of manila and sisal rope

Nominal Diameter (in)	Circumference (in)	Approximate Weight (lb per ft)	Number 1 Manila		Sisal	
			Breaking Strength (lb)	Safe Load (lb) FS = 4	Breaking Strength (lb)	Safe Load (lb) FS = 4
1/4	3/4	0.020	600	150	480	120
3/8	1 1/8	0.040	1,350	325	1,080	260
1/2	1 1/2	0.075	2,650	660	2,120	520
5/8	2	0.133	4,400	1,100	3,520	880
3/4	2 1/4	0.167	5,400	1,350	4,320	1,080
7/8	2 3/4	0.186	7,700	1,920	6,160	1,540
1	3	0.270	9,000	2,250	7,200	1,800
1 1/8	3 1/2	0.360	12,000	3,000	9,600	2,400
1 1/4	3 3/4	0.418	13,500	3,380	10,800	2,700
1 1/2	4 1/2	0.600	18,500	4,620	14,800	3,700
1 3/4	5 1/2	0.895	26,500	6,625	21,200	5,300
2	6	1.080	31,000	7,750	24,800	6,200
2 1/2	7 1/2	1.350	46,500	11,620	37,200	9,300
3	9	2.420	64,000	16,000	51,200	12,800

NOTES:
1. Breaking strengths and safe loads are for new rope that is used under favorable conditions. As rope ages or deteriorates, reduce safe loads progressively to one-half of the values given.
2. You can compute the safe-working load using an FS 4, but when the condition of the rope is doubtful, divide the computed further load by 2.

Table 12-2. Breaking strength of 6 by 19 standard wire rope

Nominal Diameter (in)	Approximate Weight (lb per ft)	Iron	Breaking Strength, in Tons*			
			Traction Steel	Plow Steel	Improved Plow Steel	Extra Improved Plow Steel
1/4	0.10	1.4	2.6	2.39	2.74	
3/8	0.23	2.1	4.0	5.31	6.10	7.55
1/2	0.40	3.6	6.8	9.35	10.70	13.30
5/8	0.63	5.5	10.4	14.50	16.70	20.60
3/4	0.90	7.9	14.8	20.70	23.80	29.40
7/8	1.23	10.6	20.2	28.00	32.20	39.80
1	1.60	13.7	26.0	36.40	4.18	51.70
1 1/8	2.03	17.2	32.7	45.70	52.60	65.00
1 1/4	2.50	21.0	40.6	56.20	64.60	79.90
1 1/2	3.60	29.7	56.6	80.00	92.00	114.00
1 3/4				108.00	124.00	153.00
2				139.00	160.00	198.00

*The maximum allowable working load is the breaking strength divided by the appropriate FS (Table 12-3).

Table 12-3. Wire-rope FS

Type of Service	Minimum FS
Track cables	3.2
Guys	3.5
Miscellaneous hoisting equipment	5.0
Haulage ropes	6.0
Derricks	6.0
Small electric and air hoists	7.0
Slings	8.0

FM 5-34

KNOTS AND ATTACHMENTS

Table 12-4. Knots

Group	Knot	Purpose
Knots at the end of a rope	Overhand	For preventing the end of a rope from untwisting, forming a knob at the end of a rope, or serving as a part of another knot
	Figure eight	For forming a larger knot at the end of a rope than would be formed by an overhand knot
	Wall	For preventing the end of a rope from untwisting when an enlarged end is not objectionable
Knots for joining two ropes	Square	For tying two ropes of equal size together so they will not slip
	Single sheet bend	For tying together two ropes of unequal size or tying a rope to an eye
	Double sheet bend	For tying together two ropes of equal or unequal size requiring more holding power than that of the single sheet bend, tying wet ropes together, or tying a rope to an eye
	Carrick bend	For heavy loads and for joining large hawsers of heavy rope
Knots for making loops	Bowline	For lowering men and material
	Double bowline	For slinging a man
	Running bowline	For tying a handline around an object you cannot safely reach, such as the end of a limb
	Bowline on a bight	For the same purpose as a boatswain's chair; use it when— • You need more strength than a single bowline will give. • You need to form a loop at some point in a rope other than at the end. • You do not have access to the end of a rope.
	Spanish bowline	For use in rescue work or for giving a twofold grip for lifting a pipe or other round object in a sling
	French bowline	For use as a sling to lift an injured man or when working alone and you need your hands free
	Spier	For use when you need a fixed loop, a non-slip knot, and a quick release
	Cat's paw	For fastening an endless sling to a hook, or making it at the end of a rope to fasten the rope to a hook
Knots for tightening a rope	Figure eight with an extra turn	For tightening a rope
	Butterfly	For pulling taught a high line, handline, tread rope for foot bridges or similar installations
	Baker bowline	For the same purpose as the butterfly knot and for lashing cargo

Rigging 12-3

FM 5-34

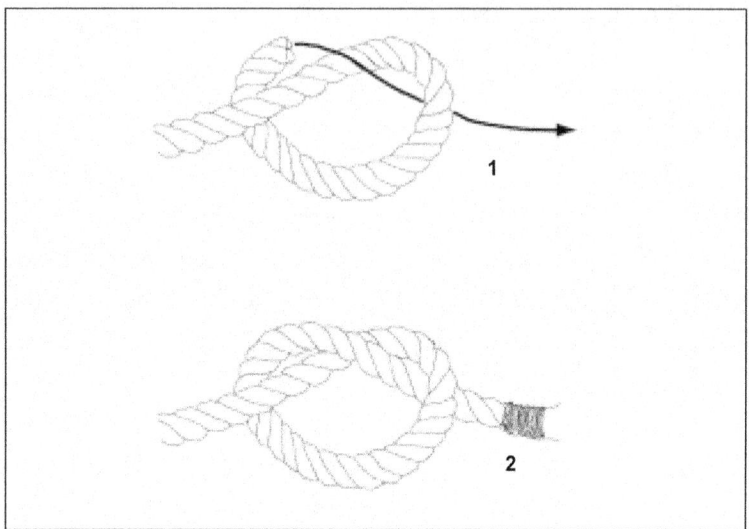

Figure 12-1. Overhand knot

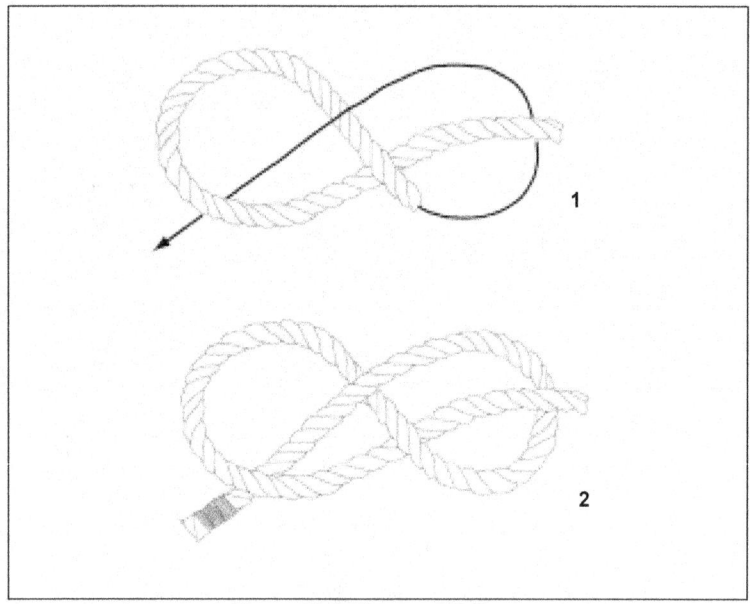

Figure 12-2. Figure-eight knot

12-4 Rigging

FM 5-34

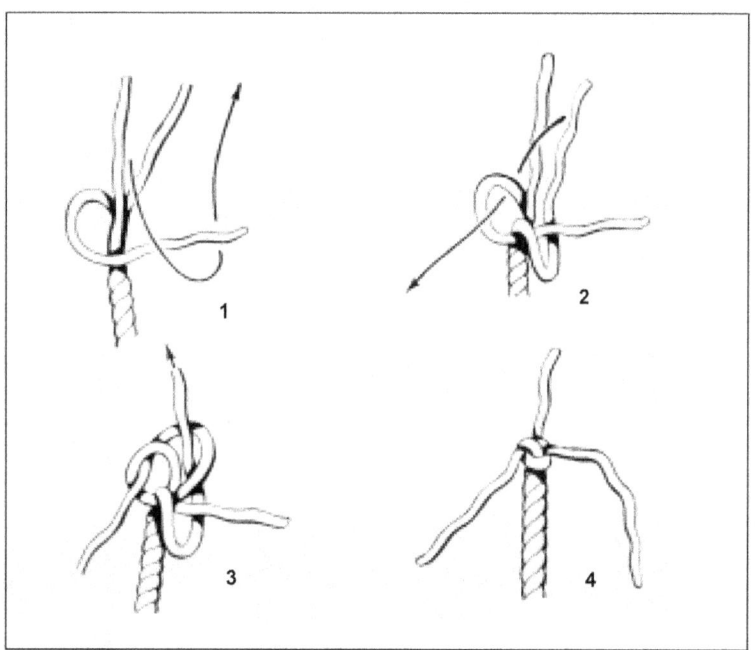

Figure 12-3. Wall knot

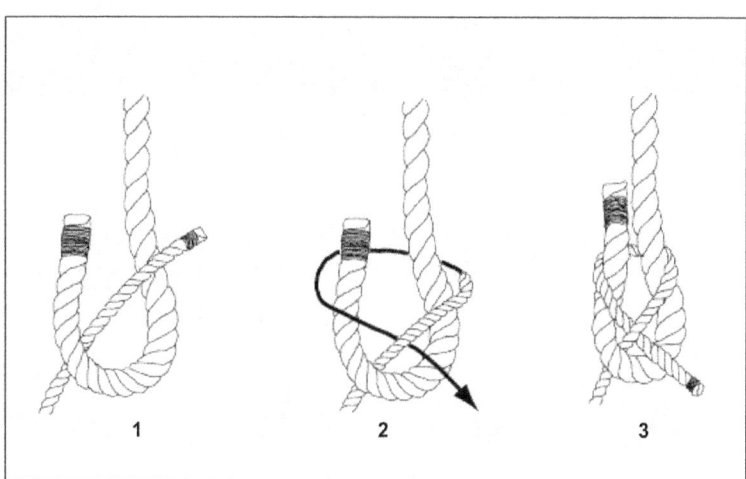

Figure 12-4. Single sheet bend

Rigging 12-5

FM 5-34

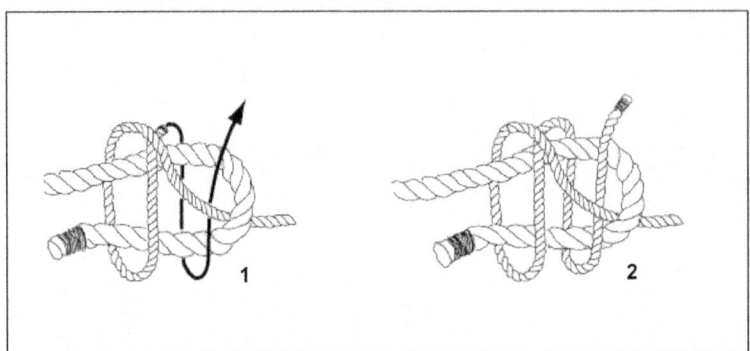

Figure 12-5. Double sheet bend

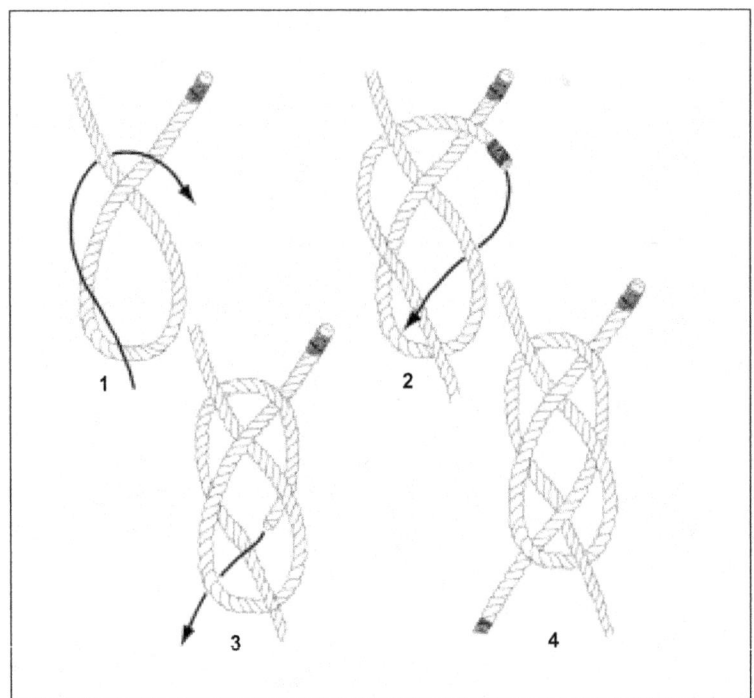

Figure 12-6. Carrick bend

12-6 Rigging

FM 5-34

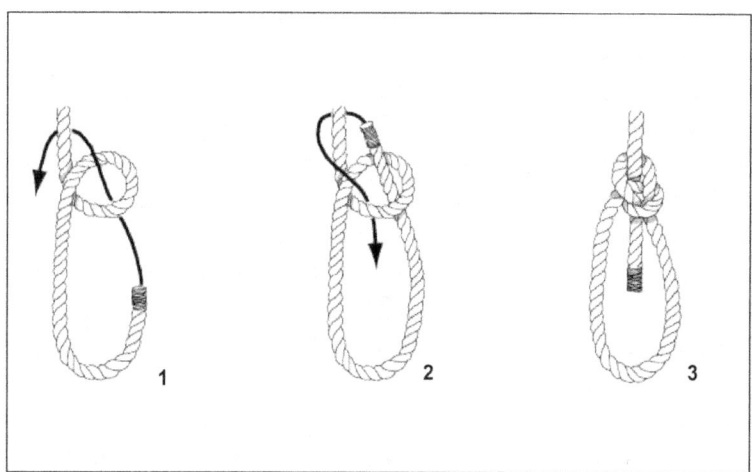

Figure 12-7. Bowline

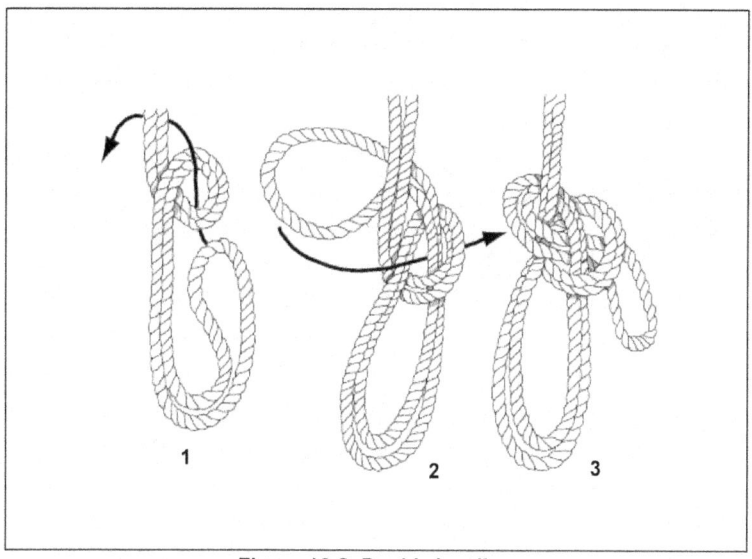

Figure 12-8. Double bowline

Rigging 12-7

FM 5-34

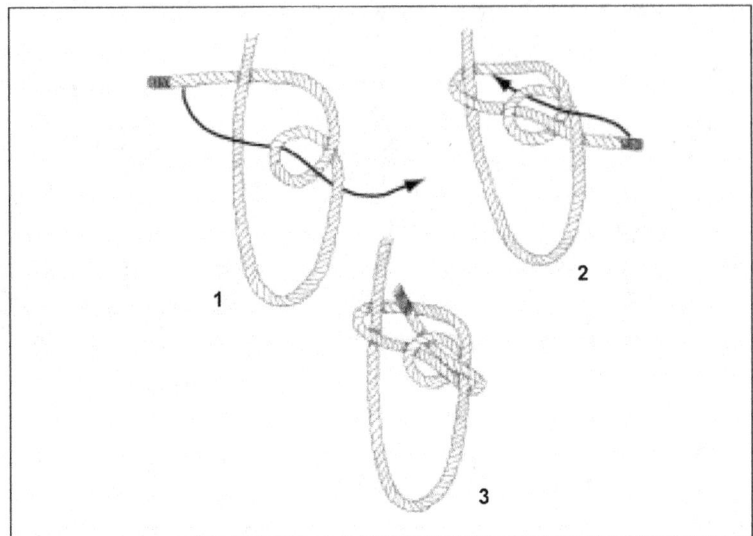

Figure 12-9. Running bowline

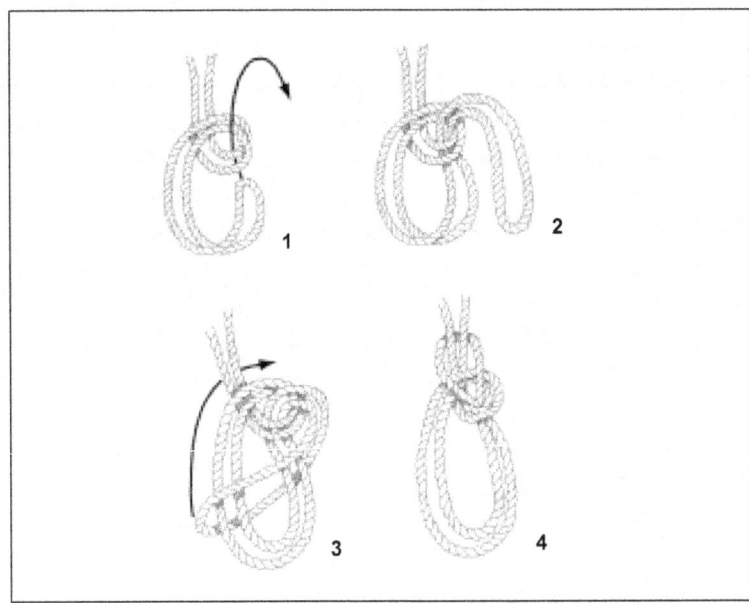

Figure 12-10. Bowline on a bight

12-8 Rigging

FM 5-34

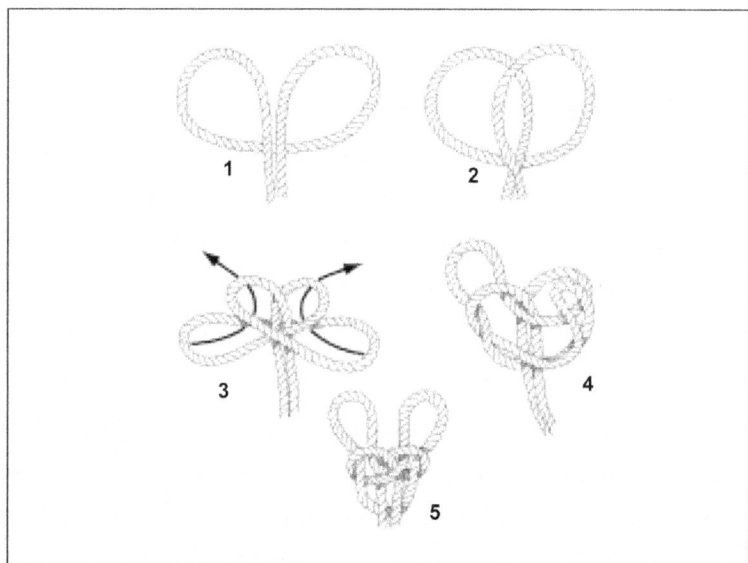

Figure 12-11. Spanish bowline

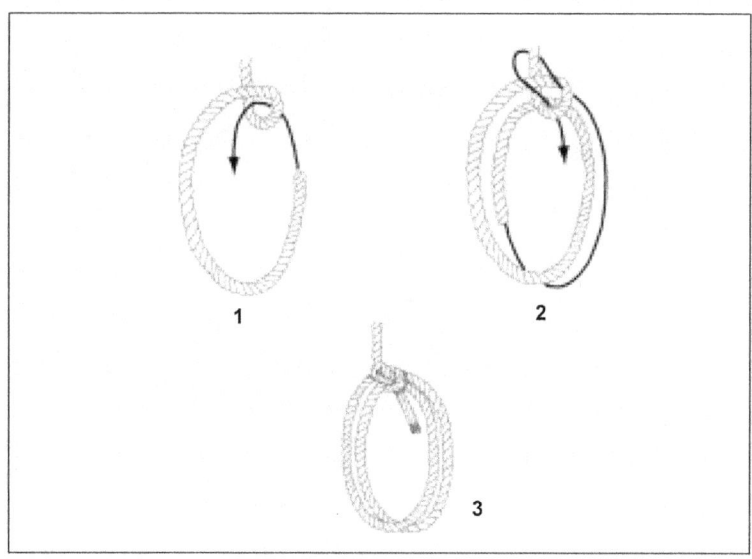

Figure 12-12. French bowline

Rigging 12-9

FM 5-34

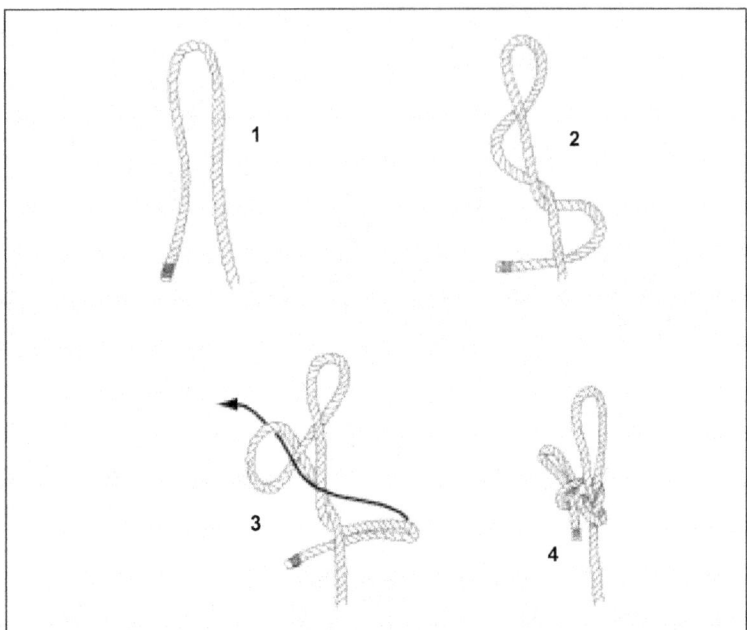

Figure 12-13. Speir knot

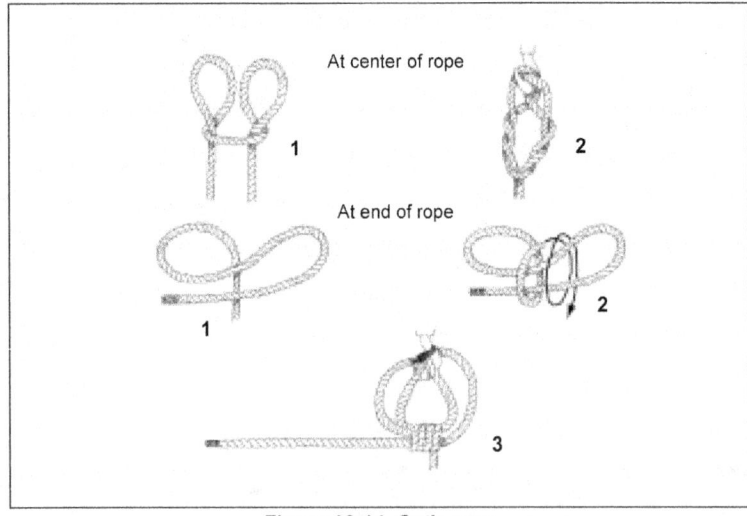

Figure 12-14. Cat's-paw

12-10 Rigging

FM 5-34

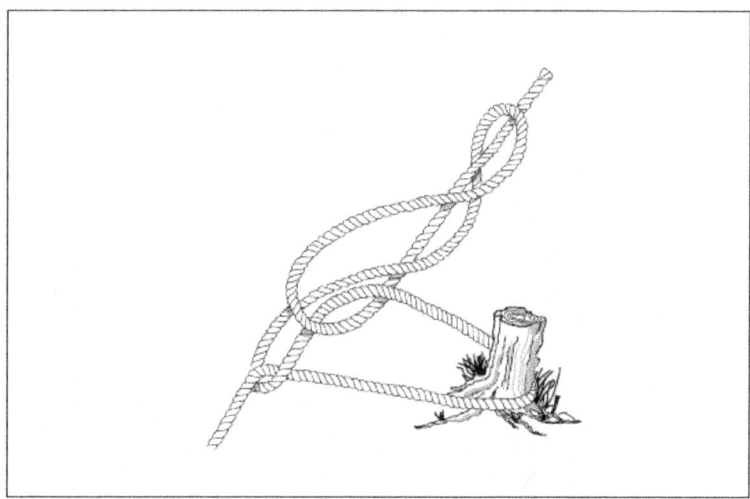

Figure 12-15. Figure eight with an extra turn

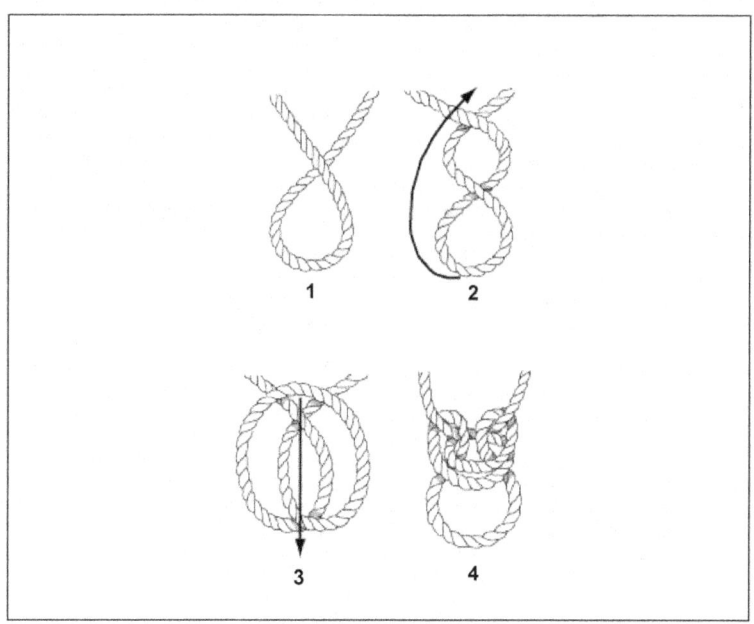

Figure 12-16. Butterfly knot

Rigging 12-11

FM 5-34

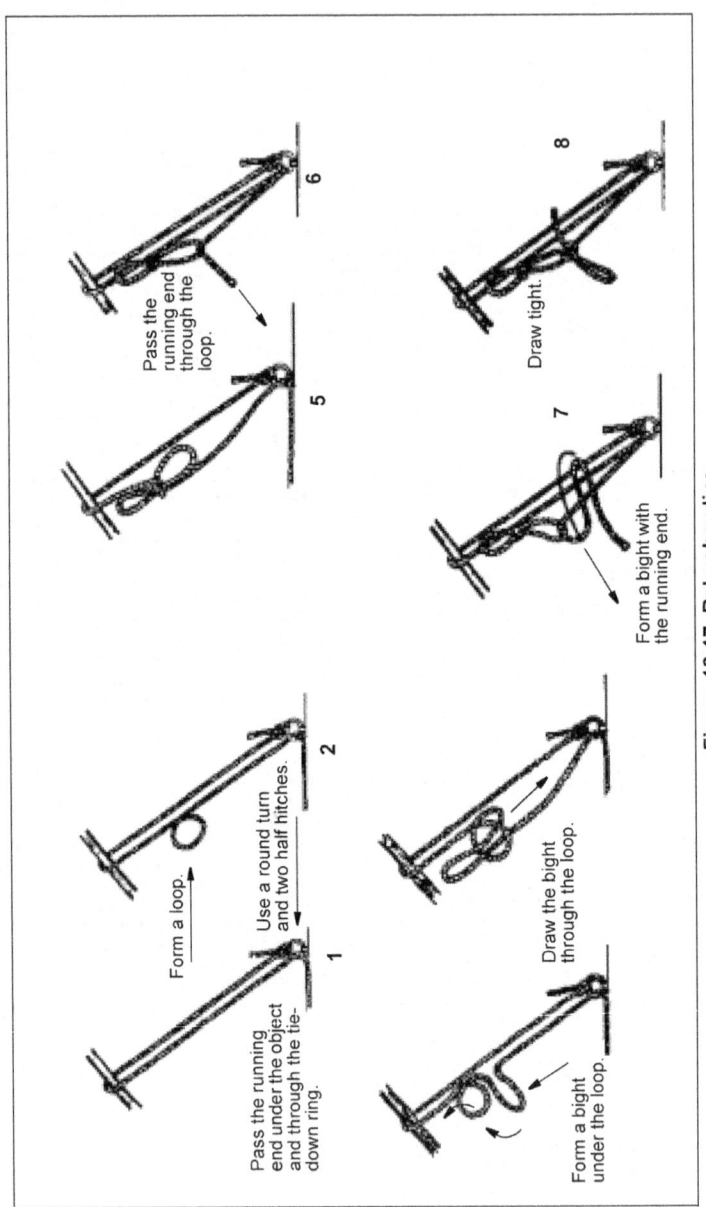

Figure 12-17. Baker bowline

12-12 Rigging

FM 5-34

Figure 12-18. Wire-rope clips

Table 12-5. Assembling wire-rope eye-loop connections

Wire-Rope Diameter		Nominal Size of Clips (in)	Number of Clips	Spacing of Clips		Torque to be Applied to Nuts of Clips	
(in)	(mm)			(in)	(mm)	(ft–lb)	(m–kg x 0.1382)
5/16	7.95	3/8	3	2	50	25	3.5
3/8	9.52	3/8	3	2 1/4	57	25	3.5
7/16	11.11	1/2	4	2 3/4	70	40	5.5
1/2	12.70	1/2	4	3	76	40	5.5
5/8	15.85	5/8	4	3 3/4	95	65	9.0
3/4	19.05	3/4	4	4 1/2	114	100	14.0
7/8	22.22	1	5	5 1/4	133	165	23.0
1	25.40	1	5	6	152	165	23.0
1 1/4	31.75	1 1/4	5	7 1/2	190	250	35.0
1 3/8	34.92	1 1/2	6	8 1/4	210	375	52.0
1 1/2	38.10	1 1/2	6	9	230	375	52.0
1 3/4	44.45	1 3/4	6	10 1/2	267	560	78.0

Note: Clip spacing should be six times the diameter of the wire rope. To assemble an end-to-end connection, increase the number of clips indicated above by two. Use the proper torque indicated above on all clips. Reverse the U-bolts at the center of connection so that they are on the dead end (reduced load) of each wire rope.

Rigging 12-13

FM 5-34

ROPE BRIDGES

ONE-ROPE BRIDGE

Construct a one-rope bridge using a 36 1/2-meter rope; however do not bridge obstacles that exceed 20 meters with that rope length. Anchor the rope with an anchor knot (round turn with two half hitches) on the far side of the obstacle, and tie it off at the near end with a tightening system.

You can build a one-rope bridge in many ways, depending on the tactical situation and area you are to cross. (For example, if you cross a gorge above a treeline, you may have to emplace artificial anchors.) Regardless, all one-rope bridges require similar elements for you to emplace safely:

- Two suitable anchors.
- Good loading and unloading platforms.
- One rope about 1-meter high for loading and unloading.
- A tightening system.
- A rope tight enough for ease of crossing.

The technique you use will determine on which side you place the tightening system and whether you use an anchor knot or a retrievable bowline (see Figure 12-19).

Figure 12-19. One-rope bridge

12-14 Rigging

FM 5-34

TWO-ROPE BRIDGE

Construct a two-rope bridge (see Figure 12-20) the same as a one-rope bridge except use two ropes. Space the ropes about 1 $^1/_2$ meters apart at the anchor points. The two-rope bridge is ideal for a platoon-size element. This bridge, however, does requires more time and equipment to construct than the one-rope bridge:

- Two climbing ropes.
- Two snaplinks.
- Seven soldiers for construction.
- One sling rope and two snaplinks for those using the bridge.

NOTE: **Construct the top rope using any transport-tightening-system technique.**

Figure 12-20. Two-rope bridge

Rigging 12-15

FM 5-34

CHAINS AND HOOKS

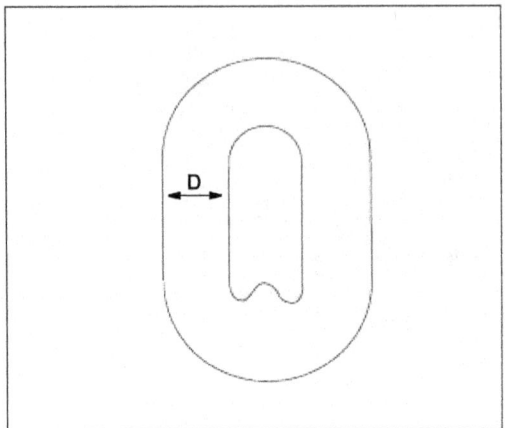

Figure 12-21. Link thickness

Table 12-6. Properties of chains (FS 6)

Size*	Approximate Weight per Linear Foot (lb)	SWC (lb)			
		Common Iron	High-Grade Iron	Soft Steel	Special Steel
1/4	0.8	512	563	619	1,240
3/8	1.7	1,350	1,490	1,650	3,200
1/2	2.5	2,250	2,480	2,630	5,250
5/8	4.3	3,470	3,810	4,230	7,600
3/4	5.8	5,070	5,580	6,000	10,500
7/8	8.0	7,000	7,700	8,250	14,330
1	10.7	9,300	10,230	10,600	18,200
1 1/8	12.5	9,871	10,858	11,944	21,500
1 1/4	16.0	12,186	13,304	14,634	26,300
1 3/8	18.3	14,717	16,188	17,807	32,051
*Size listed is the diameter, in inches, of one side of a link.					

12-16 Rigging

FM 5-34

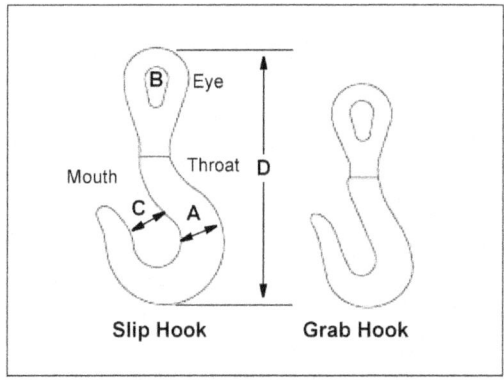

Figure 12-22. Types of hooks

Table 12-7. Safe loads on hooks

Diameter of Metal A* (in)	Inside Diameter of Eye B* (in)	Width of Opening C* (in)	Length of Hook D* (in)	SWC of Hooks (lb)
11/16	1/8	1 1/16	4 15/16	1,200
3/4	1	1 1/3	5 13/32	1,400
7/8	1 1/8	1 1/4	6 1/4	2,400
1	1 1/4	1 3/8	6 7/8	3,400
1 1/8	1 3/8	1 1/2	7 5/8	4,200
1 1/4	1 1/2	1 11/16	8 19/32	5,000
1 3/8	1 5/8	1 7/8	9 1/2	6,000
1 1/2	1 3/4	2 1/16	10 11/32	8,000
1 5/8	2	2 1/4	11 21/32	9,400
1 7/8	2 3/8	2 1/2	13 9/32	11,000
2 1/4	2 3/4	3	14 13/16	13,600
2 5/8	3 1/8	3 3/8	16 1/2	17,000
3	3 1/2	4	19 3/4	24,000
*See Figure 12-22.				

Rigging 12-17

FM 5-34

Table 12-8. SWCs for manila-rope slings (standard, three-strand, splice in each end)

Size		Single Sling	Double Sling			Quadruple Sling		
Circumference (in)	Diameter (in)	Vertical Lift (lb)	60° Angle (lb)	45° Angle (lb)	30° Angle (lb)	60° Angle (lb)	45° Angle (lb)	30° Angle (lb)
3/4	1/4	108	187	153	108	374	306	216
1 1/8	3/8	241	418	341	241	836	683	482
1 1/2	1/2	475	822	672	475	1,645	1,345	950
2	5/8	791	1,370	1,119	791	2,740	2,238	1,585
2 1/4	3/4	970	1,680	1,375	970	3,360	2,750	1,940
2 3/4	7/8	1,382	2,395	1,945	1,382	4,790	3,890	2,764
3	1	1,620	2,805	2,290	1,620	5,610	4,580	3,240
3 1/2	1 1/8	2,160	3,740	3,060	2,163	7,480	6,120	4,320
3 3/4	1 1/4	2,430	4,205	3,437	2,430	8,410	6,875	4,860
4 1/2	1 1/2	3,330	5,770	4,715	3,330	11,540	9,430	6,660
5 1/2	1 3/4	4,770	8,250	6,750	4,770	16,500	13,500	9,540
6	2	5,580	9,670	7,900	5,580	19,340	15,800	11,160
7 1/2	2 1/2	8,366	14,500	11,850	8,366	29,000	23,700	16,732
9	3	11,520	19,950	16,300	11,520	39,900	32,600	23,040

SLINGS

12-18 Rigging

FM 5-34

Table 12-9. SWCs for chain slings (new wrought-iron chains)

Link Stock Diameter (in)	Single Sling	Double Sling			Quadruple Sling		
	Vertical Lift (lb)	60° Angle (lb)	45° Angle (lb)	30° Angle (lb)	60° Angle (lb)	45° Angle (lb)	30° Angle (lb)
3/8	2,510	4,350	3,555	2,510	8,700	7,110	5,020
7/16	3,220	5,575	4,560	3,220	11,150	9,120	6,440
1/2	4,180	7,250	5,915	4,180	14,500	11,830	8,360
9/16	5,420	9,375	7,670	5,420	18,750	15,340	10,840
5/8	6,460	11,175	9,150	6,460	22,350	18,300	12,920
3/4	9,160	15,850	12,950	9,160	31,700	25,900	18,320
7/8	13,020	22,550	18,410	13,020	45,100	36,820	26,000
1	17,300	29,900	24,450	17,300	59,800	48,900	34,600
1 1/8	21,550	37,350	30,550	21,550	74,700	61,100	43,100
1 1/4	26,600	46,050	37,600	26,600	92,100	75,200	53,200
1 3/8	32,200	55,750	45,600	32,200	111,500	91,200	64,400
1 1/2	38,300	66,400	54,250	38,300	132,800	108,500	76,600
1 5/8	44,600	77,200	63,050	44,600	154,400	126,100	89,200
1 3/4	51,300	88,750	72,500	51,300	177,500	145,000	102,600
1 7/8	58,700	101,500	83,000	58,700	203,000	166,000	117,400
2	66,200	114,500	93,500	58,700	229,000	187,000	132,400

Rigging 12-19

Table 12-10. SWCs for wire-rope slings (new IPS wire rope)

Link Stock Diameter (in)	Single Sling Vertical Lift (lb)	Double Sling 60° Angle (lb)	Double Sling 45° Angle (lb)	Double Sling 30° Angle (lb)	Quadruple Sling 60° Angle (lb)	Quadruple Sling 45° Angle (lb)	Quadruple Sling 30° Angle (lb)
1/4	1,096	1,899	1,552	1,096	3,798	3,105	2,192
5/16	1,690	2,925	2,390	1,690	5,850	4,780	3,380
3/8	2,460	4,260	3,485	2,460	8,520	6,970	4,920
7/16	3,560	6,170	5,040	3,560	12,340	10,080	7,120
1/2	4,320	7,475	6,105	4,320	14,950	12,210	8,640
9/16	5,460	9,450	7,725	5,460	18,900	15,450	10,920
5/8	6,650	11,500	9,400	6,650	23,000	18,800	13,300
3/4	9,480	16,400	13,400	9,480	32,800	26,800	18,960
7/8	12,900	22,350	18,250	12,900	44,700	36,500	25,800
1	16,800	29,100	23,750	16,800	58,200	47,500	33,600
1 1/8	21,200	36,700	30,000	21,200	73,400	60,000	42,400
1 1/4	26,000	45,000	36,800	26,000	90,000	73,600	52,000
1 3/8	32,000	55,400	45,250	32,000	110,800	90,500	64,000
1 1/2	37,000	64,000	52,340	37,000	128,000	104,700	74,000
1 5/8	41,800	72,400	59,200	41,800	144,800	118,400	83,600
1 3/4	49,800	86,250	70,500	49,800	172,500	141,000	99,600
2	62,300	107,600	88,050	62,300	215,200	176,100	124,600
2 1/4	82,900	143,500	117,400	82,900	287,000	234,800	165,800
2 1/2	101,800	176,250	144,000	101,800	352,500	288,000	203,600
2 3/4	122,500	212,000	173,500	122,500	424,000	347,000	245,000

PICKET HOLDFASTS

You can drive a single picket, steel or wood, into the ground as an anchor. The holding power depends on the following:

- Diameter and kind of material you use.
- Type of soil.
- Depth and angle in which you drive the picket.
- Angle of the guy line in relation to the ground.

Table 12-11 lists the holding capacities of the various types of wooden picket holdfasts. Figure 12-23 shows the various picket holdfasts. Figure 12-24, page 12-22, shows how to prepare a picket holdfast.

FM 5-34

Table 12-11. Holding power of wooden picket holdfasts in loamy soil

Holdfasts	Pounds
Single picket	700
1-1 picket	1,400
1-1-1 picket	1,800
2-1 picket	2,000
3-2-1 picket	4,000
NOTE: Wet earth factors— Clay and gravel mixtures - 0.9 Riven clay and sand - 0.5	

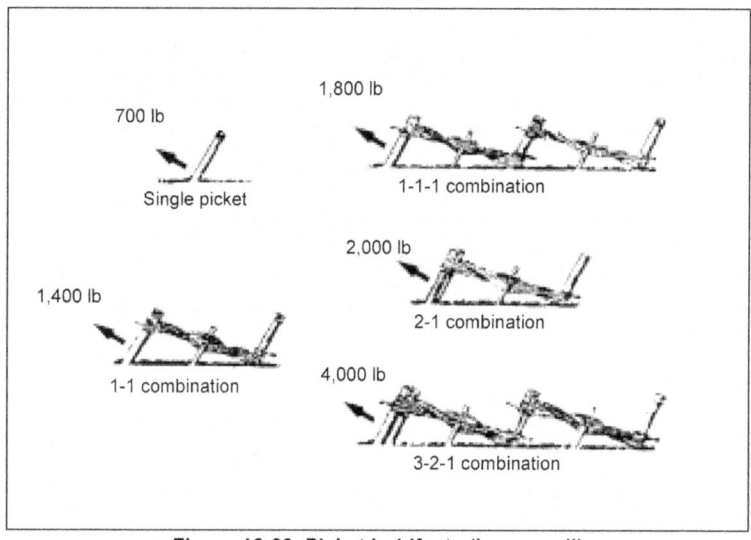

Figure 12-23. Picket holdfasts (loamy soil)

Rigging 12-21

FM 5-34

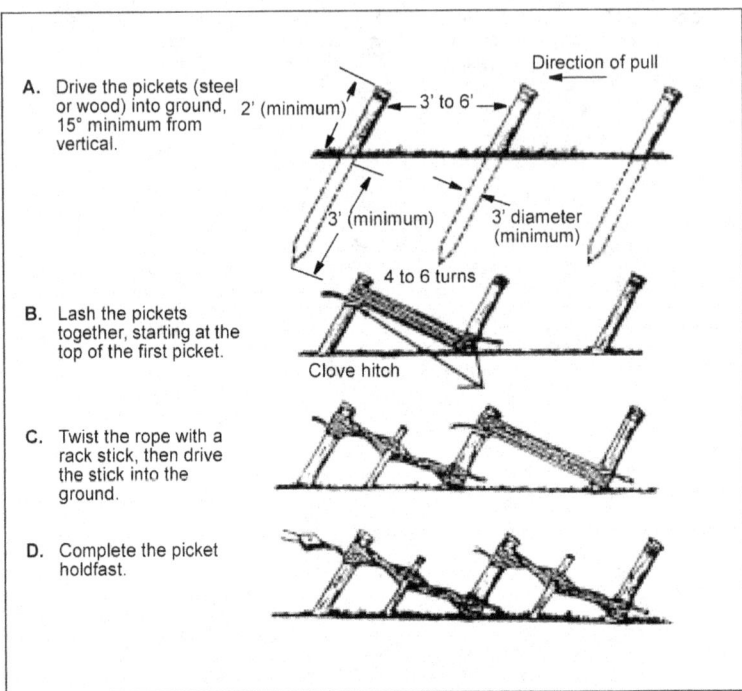

A. Drive the pickets (steel or wood) into ground, 15° minimum from vertical.

B. Lash the pickets together, starting at the top of the first picket.

C. Twist the rope with a rack stick, then drive the stick into the ground.

D. Complete the picket holdfast.

Figure 12-24. Preparing a picket holdfast

12-22 Rigging

Chapter 13
Environmental-Risk Management

Each day, commanders make decisions affecting the environment. These decisions affect resources entrusted to the Army. These decisions also have serious environmental and legal consequences for decision makers. The military's inherent responsibility to the nation is to protect and preserve its resources—a responsibility that resides at all levels. Risk management is an effective process to assist in preserving these resources. Unit leaders identify actions that may negatively impact the environment and take appropriate steps to prevent or mitigate damage.

PURPOSE

This chapter shows how to use the risk-management process of assessing and managing. It concentrates specifically on environmental-related risk; however, these risks would be incorporated into a company's overall risk-management plan. When assessing hazardous risks in operations, a commander and his staff must look at two types of risk:

- Tactical risk–is concerned with hazards that exist because of the presence of either an enemy or an adversary. It applies to all levels of war and across the spectrum of operations.

- Accidental risk–includes all operational-risk considerations other than tactical risk. It includes risk to friendly forces and the risk posed to civilians by an operation, as well as the impact of operations on the environment. Accidental risk can include activities associated with hazards concerning friendly personnel, civilians, equipment readiness, and environmental conditions.

Tactical and accident risks may be diametrically opposed. A commander may choose to accept a high level of environmental-related accident risk to reduce the overall tactical risk. For example, he may decide to destroy an enemy's petroleum storage area to reduce his overall tactical risk. Figure 13-1, page 13-2, shows the relationship of environmental hazards to the total risk-management process.

LEGAL AND REGULATORY RESPONSIBILITIES

Risk management does not convey authority to deliberately disobey local, state, national, or host nation (HN) laws and regulations. It does not justify ignoring regulatory restrictions and applicable standards, nor does it justify bypassing risk controls required by law.

FM 5-34

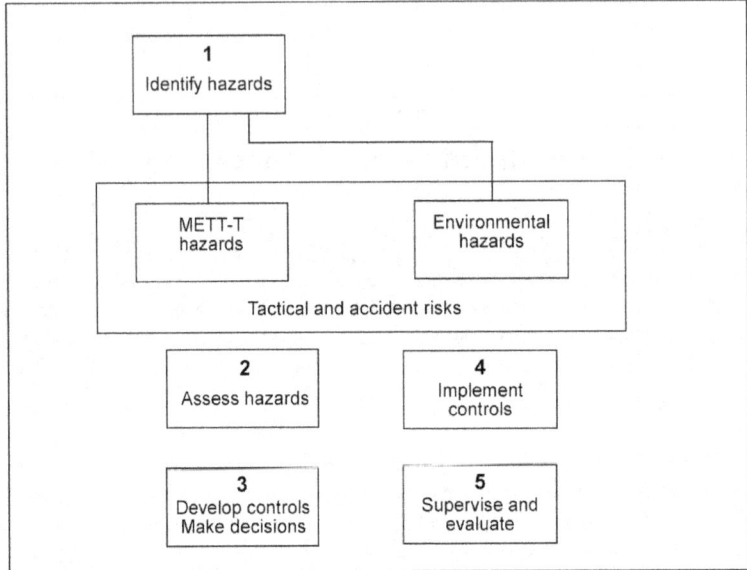

Figure 13-1. Environmental hazard relationship to the risk-management process

Examples include the provisions applicable to the transportation of hazardous material (HM) and hazardous waste (HW), the life safety and fire-protection codes, or the storage of classified material and physical security.

RISK-MANAGEMENT PRINCIPLES

A commander uses the three risk-management principles described in FM 100-14 to assist him in making environmental-risk decisions:

- Integrate risk management into mission planning, preparation, and execution.
- Make risk decisions at the appropriate level in the chain of command.
- Accept no unnecessary risk.

ENVIRONMENTAL BENEFITS OF RISK MANAGEMENT

Risk management assists a commander in complying with environmental regulatory and legal requirements and operating within the higher commander's intent. Risk management provides a commander a tool to do the following:

13-2 Environmental-Risk Management

- Identify applicable environmental standards, laws, and rules of engagement (ROE) that affect a mission.
- Identify alternate courses of action (COAs) or alternate standards that meet the intent of the law and the operational requirements.
- Identify feasible and effective control measures where specific standards do not exist.
- Ensure better use of limited resources, such as training areas and ranges.
- Ensure the health and welfare of soldiers and other affected personnel.
- Minimize or eliminate damage to natural and cultural resources.

THE RISK-MANAGEMENT PROCESS

Risk management is the process of identifying, assessing, and controlling risk that arises from operational factors and balancing risk with mission benefits. This description integrates risk management into the military decision-making process (MDMP). FM 100-14 outlines the risk-management process and provides the framework for making risk management a routine part of planning, preparing, and executing operational missions and everyday tasks. Assessing environmental-related risks is part of the total risk-management process. The five steps in the risk-management process are as follows:

Step 1. Identify the environmental hazards.

Step 2. Assess the environmental hazards to determine the risk.

Step 3. Develop the controls and make risk decisions.

Step 4. Implement the controls.

Step 5. Supervise and evaluate.

Knowledge of environmental factors is key to planning and decision-making. With this knowledge, a commander quantifies risks, detects problem areas, reduces risk of injury or death, reduces property damage, and ensures compliance with environmental laws and regulations. A unit commander should conduct risk assessments using the risk-management work sheet before conducting any training, operations, or logistical activities. Figure 13-2, page 13-4, shows this work sheet with all the blocks filled in. Blocks A through E contain general information. Steps 1 through 4 in the following paragraphs explain how to fill in blocks F through J.

FM 5-34

A. Mission or Task 586th Engineer Company (AFB)		B. Date/Time Group Begin: 010600RJun XX End: 061200RJun XX			C. Date Prepared: 22 May XX
D. Prepared By: (Rank, Last Name, Duty Position) 1LT Elizabeth Young, XO					
E. Task:	F. Identify Hazards:	G. Assess Hazards:	H. Develop Controls:	I. Determine Residual Risk:	J. Implement Controls: ("How To")
Conduct convoy operations to Camp Yukon	Vehicle accidents and breakdowns causing fuel and HM spills	Moderate (M)	1. Train all drivers on proper actions to take during a spill: protect themselves, stop the flow, notify chain of command, and confine the spill. 2. Provide vehicle-spill equipment	Low (L)	TACSOP, para 8(a), OPORD train all drivers before the exercise. Supply NCO will order and issue vehicle-spill equipment. Platoon leaders will brief soldiers before the convoy (ARTEP 5-145-32-MTP 05-2-1030).
	Spills during refueling stops	Moderate (M)	1. Train all fuel handlers on proper refueling procedures. 2. Provide spill equipment. 3. Ensure that only fuel handlers dispense fuel. 4. Locate refueling sites away from bodies of water and wetland areas.	Low (L)	TACSOP, para 11(a), OPORD support platoon leader will check status of spill equipment and brief all soldiers, before the convoy, on refueling procedures (FM 10-71, FM 20-400, ARTEP 5-145-32-MTP 05-2-1024).
	Maneuver damage from off-road movement	Moderate (M)	1. Brief all drivers to stay on primary and secondary roads. 2. Identify all sensitive areas and habitats along the route. 3. Conduct prior-route recon.	Low (L)	TACSOP, para 9(a), OPORD provide all drivers with strip map marking route and sensitive areas; leaders account for all vehicles at halts (ARTEP 5-145-32-MTP 05-2-1030).
K. Determine the overall Mission/Task risk level after controls are implemented (circle one): LOW (L) MODERATE (M) HIGH (H) EXTREMELY HIGH (E)					

Figure 13-2. Sample risk-management work sheet, all blocks filled in

STEP 1. IDENTIFY THE ENVIRONMENTAL HAZARDS

A commander and his staff identify environmental hazards during mission analysis (see Figure 13-2, column F). FM 100-14 defines a hazard as any actual or potential condition that can cause injury, illness, or death of personnel; damage to or loss of equipment or property; or mission degradation. Environmental hazards include all activities that may pollute, create negative noise-related effects, degrade archeological/cultural resources, or negatively affect

13-4 Environmental-Risk Management

threatened or endangered species habitat. Figure 13-3 lists common environmental hazards identified by environmental media areas.

Element	Hazard
Air	Equipment exhaust Convoy dust Range fires Open-air burning Pyrotechnics/smoke pots/smoke grenades Part-washer emissions Paint emissions Air-conditioner/refrigeration CFCs HM/HW release
Archeological/cultural	Maneuvering in sensitive areas Digging in sensitive areas Disturbing or removing artifacts Demolition/munitions effects HM/HW spills Sonic booms/prop wash
Noise	Low-flying aircraft (helicopters) Demolition/munitions effects Nighttime operations Operations near post/camp boundaries and civilian populace Vehicle convoys/maneuvers Large-scale exercises
Threatened/ endangered species	Maneuvering in sensitive areas Demolition/munitions effects, especially during breeding seasons Disturbing individual species or their habitats HM/HW spills or releases Poor field sanitation Improper cutting of vegetation Damage to coral reefs
Soil (terrain)	Over use of maneuver areas Demolition/munitions effects Range fires Poor field sanitation Poor maneuver-damage control Erosion Troop construction effects Refueling operations HM/HW spills Maneuvering in ecologically sensitive areas such as wetlands and tundra
Water	Refueling operations near water sources HM/HW spills Erosion and unchecked drainage Amphibious/water-crossing operations Troop construction effects Poor field sanitation Washing vehicles at unapproved sites

Figure 13-3. Common environmental hazards

STEP 2. ASSESS THE ENVIRONMENTAL HAZARDS TO DETERMINE THE RISK

Risk assessment is a three-stage process used to determine the risk of potential harm to the environment. A commander considers two factors, probability and severity. Probability is how often an environmental hazard is likely to occur. Severity is the effect that a hazard will have on the environment. Probability and severity are estimates that require an individual's judgment and a working knowledge of the risk-management process and its terminology. Figure 13-4 defines the five degrees of probability for a hazard; Figure 13-5 defines the four degrees of severity.

Frequent (A) Occurs very often, continuously experienced	
Single item	Occurs very often in service life; expected to occur several times over duration of a specific mission or operation; always occurs
Fleet or inventory of items	Occurs continuously during a specific mission or operation or over a service life
Individual soldier	Occurs very often in career; expected to occur several times during mission or operation; always occurs
All soldiers exposed	Occurs continuously during a specific mission or operation
Likely (B) Occurs several times	
Single item	Occurs several times in service life; expected to occur during a specific mission or operation
Fleet or inventory of items	Occurs at a high rate but experienced intermittently (regular intervals, generally often)
Individual soldier	Occurs several times in career; expected to occur during a specific mission or operation
All soldiers exposed	Occurs at a high rate but experienced intermittently
Occasional (C) Occurs sporadically	
Single item	Occurs sometime in service life; may occur about as often as not during a specific mission or operation
Fleet or inventory of items	Occurs several times in service life
Individual soldier	Occurs sometime in career; may occur during a specific mission or operation but not often
All soldiers exposed	Occurs sporadically (irregularly, sparsely, or sometimes)
Seldom (D) Remotely possible; could occur at sometime	
Single item	Occurs in service life but only remotely possible; not expected to occur during a specific mission or operation
Fleet or inventory of items	Occurs as isolated incidents; possible to occur sometime in service life but rarely; usually does not occur
Individual soldier	Occurs as isolated incident during a career; remotely possible but not expected to occur during a specific mission or operation
All soldiers exposed	Occurs rarely within exposed population as isolated incidents
Unlikely (E) Can assume will not occur, but not impossible	
Single item	Occurrence not impossible; but may assume will almost never occur in service life; may assume will not occur during a specific mission or operation
Fleet or inventory of items	Occurs very rarely (almost never or improbable); incidents may occur over service life
Individual soldier	Occurrence not impossible but may assume will not occur in career or during a specific mission or operation
All soldiers exposed	Occurs very rarely but not impossible

Figure 13-4. Hazard probability

Stage 1

A commander assesses the probability of each hazard. For each hazard he identified (see Figure 13-2, page 13-4), he would make the following determinations:

- Based on experience and the information in Figure 13-4 he determines that a vehicle accident or breakdown causing a fuel and/or HM spill would seldom happen.

13-6 Environmental-Risk Management

- Based on his judgment and the information in Figure 13-4, he determines that spills during refueling stops can occasionally be expected.

- Based on his working knowledge and the information in Figure 13-4, he determines that maneuver damage from off-road movement could happen frequently.

Stage 2

A commander assesses the severity of each hazard he identified. Definitions for the degrees of severity are not absolutes; they are more conditional and are mission, enemy, terrain, troops, and time available (METT-T) related. A commander must use his experience, judgment, lessons learned, and subject-matter experts to help determine the degrees of severity. Figure 13-5 defines the four degrees of severity.

Catastrophic (I)	Loss of ability to accomplish the mission or mission failure, death or permanent total disability (accident risk), loss of major or mission-critical system or equipment, major property (facility) damage, severe environmental damage, mission-critical security failure, unacceptable collateral damage
Critical (II)	Significantly (severely) degraded mission capability or unit readiness, permanent partial disability, temporary total disability exceeding 3 months time (accident risk), extensive (major) damage to equipment or systems, significant damage to property or the environment, security failure, significant collateral damage
Marginal (III)	Degraded mission capability or unit readiness; minor damage to equipment or systems, property, or the environment; lost day due to injury or illness, not exceeding 3 months (accident risk); minor damage to property or the environment
Negligible (IV)	Little or no adverse impact on mission capability, first aid or minor medical treatment (accident risk), slight equipment or system damage but fully functional and serviceable, little or no property or environmental damage

Figure 13-5. Hazard severities

The following are examples of hazard severities:

- Catastrophic—a spill of significant quantity in an unconfined area, such as a river or other water source, causing widespread pollution/health hazard to friendly forces and/or civilian personnel, as well as making cleanup extremely difficult, costly, and long-term. Will require notifying a higher HQ, public affairs, and outside agencies. Significant assistance from outside agencies is required. Widespread public concern is expected.

- Critical—a spill of more than 5 gallons or in an unconfined area such as drainage area, wetlands, rivers, or other water sources causing pollution and possible health hazards. Cleanup is difficult and costly and may require assistance and notification of outside agencies.

- Marginal–a small spill of less than 5 gallons in an area where the spill may not be as easily contained making spill cleanup efforts more difficult. No long-term, widespread environmental, or health effects are anticipated. Cleanup can be accomplished with available assets. Unit procedures may require reporting the spill to a higher HQ.

- Negligible–a small spill of less than 5 gallons in an area where the spill can be contained and immediately cleaned up using unit spill kits and available personnel.

From the information in Figure 13-2, page 13-4, a commander would make the following determinations:

- Based on experience and the information in Figure 13-4, page 13-6, he determines that a vehicle accident or breakdown causing a fuel and/or HM spill could be significant and cause major damage to the environment. The severity would be critical.

- Based on his judgment and the information in Figure 13-4, he determines that spills during refueling stops could cause minor damage to the environment. The severity would be marginal.

- Based on his working knowledge and the information in Figure 13-4, he determines that maneuver damage from off-road movement would cause little or no environmental damage. The severity would be negligible.

A commander uses the determinations from stage 1 with the severity caused by an occurrence in stage 2 to determine the overall risk of each hazard.

Stage 3

First a commander determines the risk level of each hazard. Then, using the defined degrees of probability and severity from above and the risk-assessment matrix (see Figure 13-6), he determines the overall environmental-related risk level.

For the hazards identified in Figure 13-3, page 13-5, a commander would make the following determinations and enter the assessments in block G of the risk-management work sheet (see Figure 13-2, page 13-4).

Risk-Assessment Matrix					
	Probability				
Severity	Frequent (A)	Likely (B)	Occasional (C)	Seldom (D)	Unlikely (E)
Catastrophic (I)	E	E	H	H	M
Critical (II)	E	H	H	M	L
Marginal (III)	H	M	M	L	L
Negligible (IV)	M	L	L	L	L

Risk Categories
Extremely High (E)
Mission failure if hazardous incidents occur during mission; a frequent or likely probability of catastrophic loss (IA or IB) or frequent probability of critical loss (IIA) occurs.
High (H)
Significantly degraded mission capabilities in terms of required mission standard or not accomplishing all parts of the mission, not completing the mission to standard (if hazards occur during mission); occasional to seldom probability of catastrophic loss (IC or ID); a likely to occasional probability of a critical loss occurring (IIB or IIC) with material and soldier system; frequent probability of marginal (IIIA) losses.
Moderate (M)
Expected degraded mission capabilities in terms of required mission standard; will have reduced mission capability (if hazards occur during mission); unlikely probability of catastrophic loss (IE). The probability of a critical loss occurring is seldom (IID). Marginal losses occur with a probability of no more often than likely (IIIB or IIIC). Negligible (IVA) losses are a frequent probability.
Low (L)
Expected losses have little or no impact on accomplishing the mission. The probability of critical loss is unlikely (IIE), while that of marginal loss is no more often than seldom (IIID through IIIE).

Figure 13-6. Risk-assessment matrix

- Vehicle accidents and breakdowns causing fuel and/or HM spills would seldom happen, but if they did, the severity could be critical. Based on this information and Figure 13-6 (severity row, critical, and probability column, seldom), he determines the overall assessment to be moderate.

- Spills during refueling stops will happen occasionally; when they do, the severity will marginal. Based on this information and Figure 13-6 (severity row, marginal, and probability column, occasional), he determines the overall assessment to be moderate.

- Maneuver damage from vehicle off-road movement will happen frequently. The damage caused by this movement will be negligible. Based on this information and Figure 13-6 (severity row, negligible, and probability column, frequent), he determines the overall assessment to be moderate.

Environmental-Risk Management 13-9

Step 3. Develop the Controls and Make a Decision

Controls eliminate or reduce the probability or severity of each hazard, thereby lowering the overall risk. Controls can consist of one of the categories listed in Figure 13-7, which also lists examples.

Control Type	Environmental-Related Examples
Educational	Conducting unit environmental-awareness training Conducting an environmental briefing before deployment Performing tasks to environmental standards Reviewing environmental considerations in AARs Reading unit's environmental SOPs and policies Conducting spill-prevention training Publishing an environmental annex/appendix to the OPORD/OPLAN
Physical	Providing spill-prevention equipment Establishing a field trash-collection point and procedures Establishing a field satellite-accumulation site and procedures Policing field locations Practicing good field sanitation Filling in fighting positions Posting signs and warnings for off-limit areas
Avoidance	Maneuvering around historical/cultural sites Establishing refueling and maintenance areas away from wetlands and drainage areas Crossing streams at approved sites Preventing pollution Limiting noise in endangered and threatened species habitats Avoiding refueling over water sources Curtailing live vegetation use for camouflage

Figure 13-7. Environmental-related controls

Many environmental-risk controls are simply extensions of good management, housekeeping, operations security (OPSEC), and leadership practices. Risk-reduction controls can include conducting rehearsals, changing locations, establishing procedures, and increasing supervision. Using the information from Figure 13-7, a commander fills in block H of the risk-management work sheet (see Figure 13-2, page 13-4).

Once all practicable risk-control measures are in place, some risk will always remain. Based on the controls that he develops, a commander reassesses the hazards using the procedures from step 2. Once he determines the residual risk for each hazard, he fills in block I in the risk-management work sheet (see Figure 13-2). The residual risk requires his attention. He decides whether or not to accept the risk. The commander may direct his staff to consider additional controls or a change in the COA. In the example below,

13-10 Environmental-Risk Management

where the risk is low, the commander accepts the risk and proceeds to implement the controls.

STEP 4. IMPLEMENT THE CONTROLS

Implementing the controls requires informing all subordinates of the risk-control measures. To do this, a commander defines the controls by filling in block J of the risk-management work sheet (see Figure 13-2, page 13-4). He states how each control will be implemented and assigns responsibility for implementing the controls. For example, if the control measures for a fuel-spill hazard are to ensure that operators are properly trained to dispense fuel and ensure that appropriate spill equipment is available, then he must ensure that these controls are in place before an operation.

A commander must anticipate environmental requirements and incorporate them as part of his long-, short-, and near-term planning. The key to success is identifying the who, what, where, when, and how aspects of each control and entering the information in the work sheet.

STEP 5. SUPERVISE AND EVALUATE

A commander and his staff continuously monitor controls throughout an operation to ensure their effectiveness and to modify them as required. The commander–

- Makes on-the-spot corrections and evaluates individual and collective performances.
- Holds those in charge accountable.
- Requires that all tasks be performed to applicable environmental standards.
- Ensures that the AAR process includes an evaluation of environmental-related hazards, controls, soldiers' performance, and leaders' supervision.
- Ensures that environmental lessons learned are developed for use in future operations.

SUMMARY

A commander uses risk assessment to estimate the impact of his unit's activities on the natural environment. Environmental-related risk is part of the risk-management process, as detailed in FM 100-14. Knowledge of environmental factors is key to planning and decision-making. Risk management does not convey authority to deliberately disobey local, state, national, or HN laws and regulations. A commanders uses the risk-management guidelines to help him comply with environmental regulatory and legal

FM 5-34

requirements and operate within the higher commander's intent. He should complete the risk assessments before conducting any training, operations, or logistical activities. Risk assessments help a commander and his staff identify potential environmental hazards, develop controls, make risk decisions, implement those controls, and ensure proper supervision and evaluation. Unit staffs consolidate environmental risks, as well as all other risks, into the overall unit risk-management plan for an operation.

Chapter 14

Miscellaneous Field Data

This chapter includes miscellaneous information, mainly figures and tables, that an engineer may need to do calculations in the field. The areas addressed include construction material, lumber data, trigonometric functions and geometric figures, weapons information, and vehicle classification.

WEIGHT AND GRAVITY

Table 14-1. Specific weights and gravities

Substance	Weight (lb per cu ft)	Specific Gravity
Aluminum, cast, hammered	165	2.55 to 2.75
Copper, cast, rolled	556	8.80 to 9.00
Iron, cast, pig	450	7.20
Lead	710	11.37
Magnesium alloys	112	1.74 to 1.83
Steel, rolled	490	7.85
Limestone, marble	165	2.50 to 2.80
Sandstone, bluestone	147	2.20 to 2.50
Riprap, limestone	80 to 85	
Riprap, sandstone	90	
Riprap, shale	105	
Glass, common	156	2.40 to 2.60
Hay and straw (bales)	20	
Paper	58	0.70 to 1.15
Stone, quarried, piles— •Basalt, granite, gneiss •Greenstone, hornblende •Limestone, marble, quartz •Sandstone •Shale	 96 107 90 82 92	
Excavations in water— •Clay •River mud •Sand or gravel •Sand or gravel and clay •Soil or gravel and clay •Stone riprap	 80 90 60 65 70 65	

Table 14-1. Specific weights and gravities (continued)

Substance	Weight (lb per cu ft)	Specific Gravity
Timber, US, seasoned (moisture content by weight: 15 to 50%)—		
•Soft wood	25	0.40
•Medium wood	40	0.63
•Hard wood	55	0.87
Asphaltum	81	1.10 to 1.50
Petroleum, gasoline, and diesel	42	0.66 to 0.69
Tar, bituminous	75	1.20
Cement, portland, loose	94	
Cement, portland, set	183	2.70 to 3.20
Clay, damp, plastic	110	
Clay, dry	63	
Earth, dry, loose	76	
Earth, dry, packed	96	
Earth, moist, loose	78	
Earth moist, packed	96	
Sand, gravel, dry, loose	90 to 105	
Sand, gravel, dry, packed	100 to 120	
Sand, gravel, wet	118 to 120	
Water, 4°C (max density)	62.428	1.00
Water, ice	56	0.88 to 0.92
Masonry, ashlar—		
•Granite, syenite, gneiss	165	2.30 to 3.00
•Limestone, marble	160	2.30 to 2.80
•Sandstone, bluestone	140	2.10 to 2.40
Masonry, brick—		
•Pressed brick	140	2.20 to 2.30
•Common brick	120	1.80 to 2.00
•Soft brick	100	1.50 to 1.70
Masonry, concrete—cement, stone, sand	144	2.20 to 2.40
Masonry, dry rubble—		
•Granite, syenite, gneiss	130	1.90 to 2.30
•Limestone, marble	125	1.90 to 2.10
•Sandstone, bluestone	110	1.80 to 1.90
Masonry, mortar, rubble—		
•Granite, syenite, gneiss	155	2.20 to 2.80
•Limestone, marble	150	2.20 to 2.60
•Sandstone, bluestone	130	2.00 to 2.20

14-2 Miscellaneous Field Data

CONSTRUCTION MATERIAL

ELECTRICAL WIRE

Convert load to amperes using the following formula:

$$amperes = \frac{total\ wattage\ required}{voltages} = \frac{voltage}{resistance\ (ohms)} = \frac{745.7 \times horsepower}{voltages}$$

Enter Table 14-2 or 14-3, page 14-4, using computed amperes and distance to load to obtain wire size. Use this procedure when you need to furnish power to a specific load such as a motor or a group of lights. See FM 5-424 for more details.

Table 14-2. Wire sizes for 110-volt single-phase circuits

Load (amp)	50	75	100	125	150	200	250	300	400	500
15	10/12	8/10	8/10	6/8	6/8	4/6	4/6	3/4	2/4	1/3
20	10/12	8/10	6/8	6/8	4/6	4/6	3/4	2/4	1/3	0/2
25	8/10	6/8	6/8	4/6	4/6	3/4	2/4	1/3	0/2	2/0 / 1
30	6/10	6/8	4/6	4/6	3/4	2/4	1/3	0/2	2/0 / 1	3/0 / 0
40	6/8	4/6	4/6	3/4	2/4	1/3	0/2	2/0 / 1	3/0 / 0	4/0 / 2/0
50	4/8	4/6	3/4	2/4	1/3	0/2	2/0 / 1	3/0 / 0	4/0 / 2/0	300/3/0
60	4/6	2/4	2/4	1/3	0/2	2/0 / 1	3/0 / 0	4/0 / 2/0	250/3/0	350/4/0
70	4/6	2/4	1/3	0/2	2/0 / 2	3/0 / 0	4/0 / 2/0	250/2/0	300/4/0	400/250
80	4/6	2/4	1/3	0/2	2/0 / 1	3/0 / 0	4/0 / 2/0	250/3/0	350/4/0	500/250
90	2/4	1/3	0/2	2/0 / 1	3/0 / 1	4/0 / 2/0	250/3/0	300/3/0	400/250	500/300
100	2/4	1/3	0/2	2/0 / 1	3/0 / 0	4/0 / 2/0	300/3/0	350/4/0	500/250	600/350

NOTE:
Top number = aluminum wire
Bottom number = copper wire

FM 5-34

Table 14-3. Wire sizes for 220-volt three-phase circuits

Load (amp)	100	200	300	400	500	600	700	800	900	1,000
15	12/12	8/10	6/8	4/6	4/6	3/4	2/4	2/4	1/3	1/3
20	10/12	6/8	4/6	4/6	3/4	2/4	1/3	1/3	0/2	0/2
25	8/10	6/8	4/6	3/4	2/4	1/3	0/2	0/2	2/0, 1	2/0, 1
30	6/10	4/6	3/4	2/4	1/3	0/2	2/0, 2	2/0, 1	3/0, 0	3/0, 0
40	4/8	4/6	2/4	1/3	0/2	2/0, 1	3/0, 0	3/0, 0	4/0, 2/0	4/0, 2/0
50	4/8	3/4	1/3	0/2	2/0, 1	3/0, 0	4/0, 2/0	4/0, 2/0	250/3/0	300/3/0
60	4/6	2/4	0/2	2/0, 1	3/0, 0	4/0, 2/0	250/2/0	250/3/0	300/4/0	350/4/0
70	4/6	1/3	2/0, 2	3/0, 0	4/0, 2/0	250/2/0	300/3/0	300/4/0	350/4/0	400/250
80	4/6	1/3	2/0, 1	3/0, 0	4/0, 2/0	250/3/0	300/4/0	350/4/0	400/250	500/250
90	2/4	0/2	3/0, 0	4/0, 2/0	250/3/0	300/4/0	350/4/0	400/250	500/300	500/300
100	2/4	0/2	3/0, 0	4/0, 2/0	300/3/0	350/4/0	400/250	500/250	500/300	600/350

NOTE:
10 = aluminum wire
12 = copper wire

14-4 Miscellaneous Field Data

LUMBER DATA

Table 14-4. Properties of southern pine

Nominal Size (in)	Actual Size Dressed (in)	Section Area (sq in)	Weight per Foot (lb)
2 x 4	1 5/8 x 3 5/8	5.89	1.63
4 x 4	3 5/8 x 3 5/8	13.14	3.64
2 x 6	1 5/8 x 5 5/8	9.14	2.53
6 x 6	5 5/8 x 5 5/8	31.64	8.76
2 x 8	1 5/8 x 7 1/2	12.19	3.38
4 x 8	3 5/8 x 7 1/2	27.19	7.55
6 x 8	5 5/8 x 7 1/2	42.19	11.72
8 x 8	7 1/2 x 7 1/2	56.25	15.58
2 x 10	1 5/8 x 9 1/2	15.44	4.28
6 x 10	5 5/8 x 9 1/2	53.44	14.84
10 x 10	9 1/2 x 9 1/2	90.25	25.00
2 x 12	1 5/8 x 11 1/2	18.69	5.18
3 x 12	2 5/8 x 11 1/2	30.19	8.39
6 x 12	5 5/8 x 11 1/2	64.69	17.96
8 x 12	7 1/2 x 11 1/2	86.25	23.89
10 x 12	9 1/2 x 13 1/2	109.25	30.26
2 x 14	1 5/8 x 13 1/2	21.94	6.09
3 x 14	2 5/8 x 13 1/2	35.44	9.84
6 x 14	5 5/8 x 13 1/2	75.94	21.09
10 x 14	9 1/2 x 13 1/2	128.25	35.53
14 x 14	13 1/2 x 13 1/2	182.25	50.48
2 x 16	1 5/8 x 15 1/2	25.19	7.00
3 x 16	2 5/8 x 15 1/2	40.69	11.30
8 x 16	7 1/2 x 15 1/2	116.25	32.20
12 x 16	11 1/2 x 15 1/2	178.25	49.37
14 x 16	13 1/2 x 15 1/2	209.25	57.96
16 x 16	15 1/2 x 15 1/2	240.25	66.55
4 x 18	3 5/8 x 17 1/2	63.44	17.62
8 x 18	7 1/2 x 17 1/2	131.25	36.36
12 x 18	11 1/2 x 17 1/2	201.25	55.75

NOTE: In some species, 5 1/2 inches is the dressed size for nominal 6 x 6 inches and larger.

FM 5-34

Table 14-5. Wood-screw diameters

Size (in)	Diameter—D (in)	$D^2 (in^2)$
1/2—No. 4	0.1105	0.0122
3/4—No. 8	0.1631	0.0266
1—No. 10	0.1894	0.0359
1 1/2—No. 12	0.2158	0.0466
2—No. 14	0.2421	0.0586
2 1/2—No. 16	0.2684	0.0720
3—No. 18	0.2947	0.0868

Table 14-6. Nail and spike sizes

Size	Length (in)	Gauge	Common			Finishing		Flooring	
			No. per lb	Diameter —D (in)	D3/2	Gauge	No. per lb	Gauge	No. per lb
3D	1 1/4	14	568	0.0800	0.0226	15 1/2	807		
4D	1 1/2	12 1/2	316	0.0985	0.0309	15	584		
6D	2	11 1/2	181	0.1130	0.0380	13	309	11	157
8D	2 1/2	10 1/4	106	0.1314	0.0476	12 1/2	189	10	99
10D	3	9	69	0.1483	0.0570	11 1/2	121	9	69
12D	3 1/4	9	63	0.1552	0.0611	11 1/2	113	8	54
16D	3 1/2	8	49	0.1620	0.0652	11	90	7	43
20D	4	6	31	0.1920	0.0841	10	61	6	31
30D	4 1/2	5	24	0.2070	0.0942				
40D	5	4	18	0.2253	0.1066				
60D	6	2	11	0.2625	0.1347				
Spikes									
7"	7	5/16"		5/16"	0.1750				
8"	8	3/8"		3/8"	0.2295				
9"	9	3/8"		3/8"	0.2295				
10"	10	3/8"		3/8"	0.2295				
12"	12	3/8"		3/8"	0.2295				

NOTE: To avoid splitting, nail diameters should not exceed 1/7 the thickness of lumber to be nailed.

14-6 Miscellaneous Field Data

To determine the approximate number of nails you need, use the following formulas:

$$\text{Number of pounds (12D to 60D, framing)} = \frac{D}{6} \times \frac{BF}{100}$$

or

$$\text{Number of pounds (2D to 12D, sheathing)} = \frac{D}{4} \times \frac{BF}{100}$$

where—

D = size of desired nail, in pennies

BF = total board feet to be nailed

TRIGONOMETRIC FUNCTIONS AND GEOMETRIC FIGURES

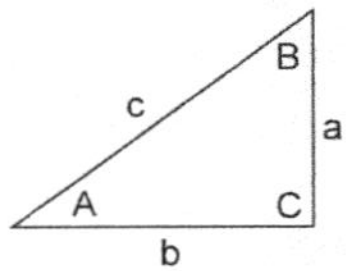

$a^2 = c^2 - b^2$ $\sin A = a/c$
$b^2 = c^2 - a^2$ $\cos A = b/c$
$c^2 = a^2 + b^2$ $\tan A = a/b$

Right Triangle

Given	A	B	C	To Find a	b	c	Area
a, b	$\tan A = \dfrac{a}{b}$	$\tan B = \dfrac{b}{a}$	90°			$\sqrt{a^2 + b^2}$	$\dfrac{ab}{2}$
a, c	$\sin A = \dfrac{a}{c}$	$\cos B = \dfrac{a}{c}$	90°		$\sqrt{c^2 - a^2}$		$\dfrac{a}{2}\left(\sqrt{c^2 - a^2}\right)$
A, a		90° - A	90°		$a \cot A$	$\dfrac{a}{\sin A}$	$\dfrac{a^2 \cot A}{2}$
A, b		90° - A	90°	$b \tan A$		$\dfrac{b}{\cos A}$	$\dfrac{b^2 \tan A}{2}$
A, c		90° - A	90°	$c \sin A$	$c \cos A$		$\dfrac{c^2 \sin 2A}{4}$

Figure 14-1. Trigonometric functions

14-8 Miscellaneous Field Data

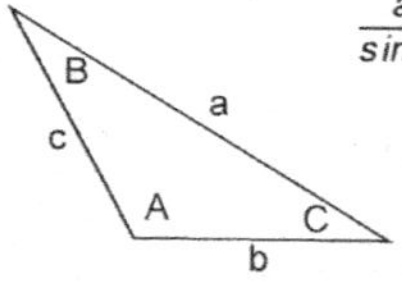

$$\frac{a}{\sin A} = \frac{b}{\sin B} = \frac{c}{\sin C}$$

$$a^2 = b^2 + c^2 - 2bc \cos A$$

$$b^2 = a^2 + c^2 - 2ac \cos B$$

$$c^2 = a^2 + b^2 - 2ab \cos C$$

$$S = \frac{a+b+c}{2}$$

Right Triangle

Given	To Find					
	A	B	C	a	b	c	Area
a, b, c	$\cos\frac{A}{2} = \sqrt{\frac{s(s-a)}{bc}}$	$\cos\frac{B}{2} = \sqrt{\frac{s(s-b)}{ac}}$	$\cos\frac{C}{2} = \sqrt{\frac{s(s-c)}{ab}}$				$\sqrt{s(s-a)(s-b)(s-c)}$
a, A, B			$180° - (A + B)$		$\frac{a \sin B}{\sin A}$	$\frac{a \sin C}{\sin A}$	$\frac{a^2 \sin B \sin C}{2 \sin A}$
a, b, A		$\sin B = \frac{b \sin A}{a}$				$\frac{b \sin C}{\sin B}$	
a, b, C		$\tan B = \frac{a \sin C}{b - a \cos C}$				$\sqrt{a^2 + b^2 - 2ab \cos c}$	$\frac{ab \sin C}{2}$

Figure 14-1. Trigonometric functions (continued)

Table 14-7. Trigonometric functions

Degree of Angle	Sine	Cosecant	Tangent	Cotangent	Secant	Cosine	Degree of Angle
0					1.000	1.000	90
1	0.017	57.30	0.017	57.29	1.000	1.000	89
2	0.035	28.65	0.035	28.64	1.001	0.999	88
3	0.052	19.11	0.052	19.08	1.001	0.999	87
4	0.070	14.34	0.070	14.30	1.002	0.998	86
5	0.087	11.47	0.087	11.43	1.004	0.996	85
6	0.105	9.567	0.105	9.514	1.006	0.995	84
7	0.122	8.206	0.123	8.144	1.008	0.993	83
8	0.139	7.185	0.141	7.115	1.010	0.990	82
9	0.156	6.392	0.158	6.314	1.012	0.988	81
10	0.174	5.759	0.176	5.671	1.015	0.985	80
11	0.191	5.241	0.194	5.145	1.019	0.982	79
12	0.208	4.810	0.213	4.705	1.022	0.978	78
13	0.225	4.445	0.231	4.331	1.026	0.974	77
14	0.242	4.134	0.249	4.011	1.031	0.970	76
15	0.259	3.864	0.268	3.732	1.035	0.966	75
16	0.276	3.628	0.287	3.487	1.040	0.961	74
17	0.292	3.420	0.306	3.271	1.046	0.956	73
18	0.309	3.236	0.325	3.078	1.051	0.951	72
19	0.326	3.072	0.344	2.904	1.058	0.946	71
20	0.342	2.924	0.364	2.747	1.064	0.940	70
21	0.358	2.790	0.384	2.605	1.071	0.934	69
22	0.375	2.669	0.404	2.475	1.079	0.927	68
23	0.391	2.559	0.424	2.356	1.086	0.921	67
24	0.407	2.459	0.445	2.246	1.095	0.914	66
25	0.423	2.366	0.466	2.145	1.103	0.906	65
26	0.438	2.281	0.488	2.050	1.113	0.899	64
27	0.454	2.203	0.510	1.963	1.122	0.891	63
28	0.469	2.130	0.532	1.881	1.133	0.883	62
29	0.485	2.063	0.554	1.804	1.143	0.875	61
30	0.500	2.000	0.577	1.732	1.155	0.866	60
31	0.515	1.942	0.601	1.664	1.167	0.857	59
32	0.530	1.887	0.625	1.600	1.179	0.848	58
33	0.545	1.836	0.649	1.540	1.192	0.839	57
34	0.559	1.788	0.675	1.483	1.206	0.829	56
35	0.574	1.743	0.700	1.428	1.221	0.819	55
36	0.588	1.701	0.727	1.376	1.236	0.809	54
37	0.602	1.662	0.754	1.327	1.252	0.799	53
38	0.616	1.624	0.781	1.280	1.269	0.788	52
39	0.629	1.589	0.810	1.235	1.287	0.777	51
40	0.643	1.556	0.839	1.192	1.305	0.766	50
41	0.656	1.542	0.869	1.150	1.325	0.755	49
42	0.669	1.494	0.900	1.111	1.346	0.743	48
43	0.682	1.466	0.933	1.072	1.367	0.731	47
44	0.695	1.440	0.966	1.036	1.390	0.719	46
45	0.707	1.414	1.000	1.100	1.414	0.707	45

14-10 Miscellaneous Field Data

(1) Any triangle:
$$A = 1/2 bh$$
$$\sin\gamma = \frac{c\sin\phi}{a}$$

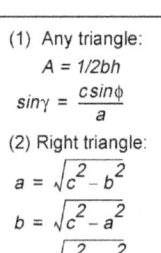

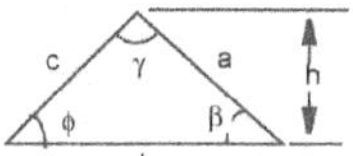

(2) Right triangle:
$$a = \sqrt{c^2 - b^2}$$
$$b = \sqrt{c^2 - a^2}$$
$$c = \sqrt{a^2 + b^2}$$

(3) Circle:
$$A = \pi r^2$$
$$A = 0.7854\, D^2$$
$$C = \pi D$$

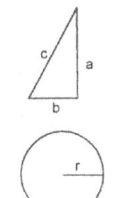

(4) Segment of circle:
$$A = \frac{\pi r^2 a}{360} - \frac{r^2 \sin a}{2}$$
$$L = \frac{2\pi r a}{360}$$

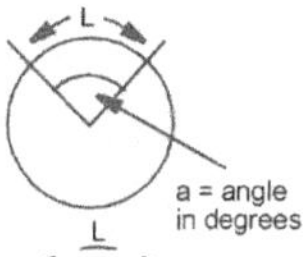

a = angle in degrees

(5) Sector of circle:
$$A = \frac{rL}{2} = \frac{\pi r^2 a}{360}$$

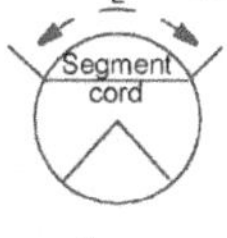

A = area
h = height
b = length of base
c = hypotenuse
C = circumference
V = volume

r = radius
D = diameter
π = 3.1416
L = length of arc

(6) Regular polygons. The area of any regular polygon (all sides equal, all angles equal) is equal to the product of the square of the lengths of one side and the factors. **Example problem:** Area of a regular octagon having 6-inch sides is 6 x 6 x 4.828 or 173.808 square inches. See factors in table.

No. of Sides	Factor	No. of Sides	Factor
3	0.433	8	4.828
4	1.000	9	6.182
5	1.720	10	7.694
6	2.598	11	9.366
7	3.634	12	11.196

Polygon Factors

(9) Cube:
$$V = b^3$$

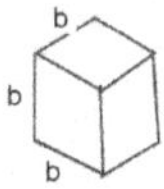

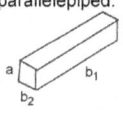

(10) Rectangular parallelepiped:
$$V = ab_1 b_2$$

(7) Rectangular parallelogram:
$$A = ab$$

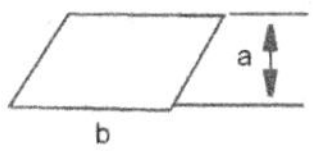

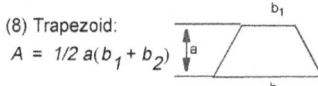

(8) Trapezoid:
$$A = 1/2\, a(b_1 + b_2)$$

(11) Prism or cylinder:
$$V = a \times area\ of\ base$$

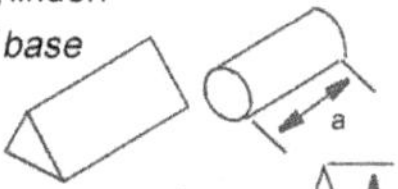

(12) Pyramid or cone:
$$V = (1/3)a \times area\ of\ base$$

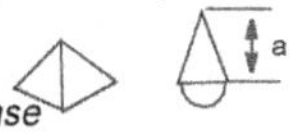

(13) Sphere:
$$V = (4 \S 3)\pi r^3 = \frac{\pi D^3}{6}$$
$$18(A = 4\pi r^3)$$

Figure 14-2. Geometric figures and formulas

Table 14-8. Time-distance conversion

Mph	Knots	FPS	Kmph	MPS
1	0.87	1.47	1.61	0.45
2	1.74	2.93	3.22	0.894
3	2.61	4.40	4.83	1.34
4	3.47	5.87	6.44	1.79
5	4.34	7.33	8.05	2.24
6	5.21	8.80	9.66	2.68
7	6.08	10.27	11.27	3.13
8	6.95	11.73	12.87	3.58
9	7.82	13.20	14.48	4.02
10	8.68	14.67	16.09	4.47
15	13.03	22.00	24.14	6.71
20	17.37	29.33	32.19	8.94
25	21.71	36.67	40.23	11.18
30	26.05	44.00	48.28	13.41
35	30.39	51.33	56.33	15.64
40	34.74	58.67	64.37	17.88
45	39.08	66.00	72.42	20.12
50	43.42	73.33	80.47	22.35
55	47.76	80.67	88.51	24.59
60	52.10	88.00	96.56	26.82
65	56.45	95.33	104.61	29.06
70	60.79	102.67	112.65	31.29
75	65.13	110.00	120.70	33.53
100	86.84	146.67	160.94	44.70

US EQUIPMENT AND WEAPONS CHARACTERISTICS

VEHICLE DIMENSIONS AND CLASSIFICATIONS

14-12 Miscellaneous Field Data

Table 14-9. Vehicle dimensions and classification

Nomenclature	Height (in)	Width (in)	Length (in)	MLC (C)	Max Speed (mph)
AVLB	200.0	158.0	439.0	59	30
Carrier, cargo 6-ton, M548	116.0	110.0	248	13	43
Carrier, command post, M577A1	106.0	106.0	226.5	13	8
Carrier, mortar: 81 mm, M125A1	86.5	106.0	191.5	13	40
Carrier, mortar: 107 mm, M106A1	86.5	113.0	194	14	40
Carrier, personnel, M113A2	86.5	106.0	191.5	13	40
Cavalry fighting vehicle, M3	118.0	126.0	258	24	45
Crane, boom, 20 ton RT	163.0	128.0	522	30	35
Crane, 25-ton hydraulic, MT-250	118.0	97.0	542	31	45
Dozer, D7 with blade	120.0	137.0	230	28	6.2
Howitzer, 155 mm (SP), M109A3	130.0	143.0	355	28	35
Howitzer, 8 in (SP), M110A2	135.0	140.0	392	29	32
Infantry fighting vehicle, M2	118.0	126.0	258	24	45
Loader, scoop, 2 1/2 C7, w/o roll cage	102.0	102.0	300	20	--
MLRS	108.0	115.0	274	27	36
M992 CATV (FAAS V)	127.0	125.0	269	28	35
Tank, combat 105 mm, M1	118.0	145.0	332	60	45
Tank, combat 105 mm, M48A5	129.5	143.0	325	54	30
Tank, combat, 105 mm, M60A1	129.5	143.0	325	54	30
Tank, combat, 105 mm, M60A2	130.5	143.0	300.5	57	30
Tank, combat, 105 mm, M60A3	130.0	143.0	325	55	30
Trailer, low-bed, 25 ton, M172	67.0	115.0	416	9	--
Trailer, water (400 gal), M149 w/o water	76.5	82.5	83	4	--
Truck, ambulance, M997	101.0	36.0	204.0	4	55
Truck, cargo (HEMTT), M977	108.0	97.0	403	16	55
Truck, cargo, 2 1/2 ton, M35A2	112.0	96.0	278.5	8	56
Truck, cargo, 5 ton, 6 x 6, M54A2	116.0	97.0	315	15	54
Truck, dump, 5 ton, 6 x 6, M930	111.0	98.0	282	17	30
Truck, fuel (2,500 gal), M559	134.0	108.0	391	23	30
Truck, tanker (HEMTT), M978	108.0	97.0	403	15	55
Truck, tractor, 20 ton, M920	144.0	132.0	320	15	--
Truck, wrecker, 5 ton, 6 x 6, M816	114.0	98.0	356	18	52
Truck, wrecker, 10 ton, 4 x 4, M553	134.0	108.0	401	23	30
Vehicle, M9 ACE	110.0	150.0	246	18	30
Vehicle, (light) recovery, M578	130.5	124.0	250	25	37
Vehicle, (med) recovery, M88A1	123.5	135.0	325.5	55	31

NOTE: MLC is for laden cross country or off highway (C).

Miscellaneous Field Data 14-13

EXPEDIENT VEHICLE CLASSIFICATION

> This section implements STANAG 2021.

In an emergency, you can do a temporary vehicle classification using expedient classification methods. However, you should reclassify the vehicle using the analytical method (see FM 5-446) or by referencing FM 5-170 as soon as possible to obtain a permanent classification number.

Wheeled

You can do an expedient classification for wheeled vehicles by doing either of the following:

- One: Compare the wheel and axle loadings and spacings of the unclassified vehicle with those of a classified vehicle of similar design and then assign a temporary class number.

- Two: Assign a temporary class number using the formulas below:

$$W_T = \frac{A_T P_T N_T}{2,000}$$

where—

W_T = gross weight of vehicle, in tons

A_T = average tire-contact area (hard surface), in square inches

P_T = tire pressure, in psi

N_T = number of tires

Estimated classification (wheeled vehicles) = $0.85\ W_T$

> NOTE: Assume the tire pressure to be 75 psi for 2 $^1/_2$-ton vehicles or larger if no tire gauge is available. For vehicles having unusual load characteristics or odd axle spacings, you will need a more deliberate vehicle-classification procedure as outlined in STANAG 2021.

Tracked

You can do an expedient classification for tracked vehicles using the following methods:

- Compare the ground-contact area of the unclassified tracked vehicle with that of a previously classified vehicle to obtain a temporary class number.

    or

14-14 Miscellaneous Field Data

- Assign a temporary class number using the formula below:

  temporary class (tracked vehicles) = W_T

where—

W_T = *gross weight, in tons; estimate the gross weight by measuring the total ground-contact area of the tracks (square feet) and equating this to the gross weight in tons.*

Example: An unclassified tracked vehicle has a ground-contact area of 5,500 square inches. Therefore, the area is about 38.2 square feet. The class of the vehicle is 38.2 or 39, since the ground-contact area in square feet equals the approximate weight of a tracked vehicle in tons, which is about equal to the class number.

Nonstandard Combinations

You can obtain the class number of nonstandard combinations of vehicles as follows:

Combination class = $0.9(A + B)$, *if* $A + B \leq 60$

or

Combination class = $A + B$, *if* $A + B > 60$

where—

A = *classification of the first vehicle*

B = *classification of the second vehicle*

Other-Than-Rated Load

You can assign an expedient classification to overloaded or underloaded vehicles by adding to or subtracting the difference in loading (in tons) from the normally assigned vehicle classification. Mark the expedient classification number with a standard vehicle classification sign, which indicates that it is a temporary classification (see Figure 14-3).

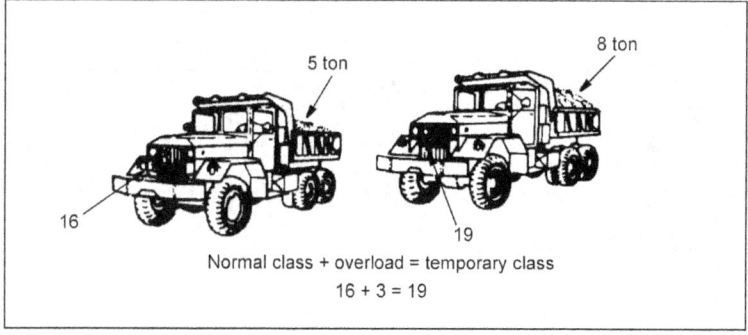

Figure 14-3. Single-vehicle expedient-class overload

FM 5-34

US WEAPONS

Table 14-10. Ranges of common weapons

Weapon		Maximum Effective Range	Planning Range*
FRIENDLY WEAPON SYSTEMS			
M16A2		580 m	400 m
M249 SAW		1,000 m	800 m
M60		1,100 m	1,100 m
M203	Area	350 m	350 m
	Point	160 m	160 m
M2, .50 cal	Area	1,830 m	1,830 m
	Point	1,200 m	1,200 m
MK19	Area	2,200 m	2,200 m
	Point	1,600 m	1,600 m
AT4		300 m	300 m
M47 Dragon		1,000 m	800 m
Javelin		2,000 m	2,000 m
M1 Abrams tank	105 mm	2,500 m	2,000 m
	120 mm	3,000 m	2,500 m
M2 Bradley ITV	25 mm (APDS)	3,000 m	1,700 m
	25 mm (HEI-T)	3,000 m	1,700 m
	TOW2	3,750 m	3,750 m
60-mm mortar	HE	3,400 m	50 m (min)
	WP	4,800 m	50 m (min)
	ILLUM	931 m	50 m (min)
81-mm mortar	HE	4,595 m	75 m (min)
	WP	4,595 m	75 m (min)
	ILLUM	3,150 m	75 m (min)
4.2-in mortar	HE	6,840 m	770 m (min)
	WP	5,650 m	920 m (min)
	ILLUM	5,490 m	400 m (min)
SOVIET-STYLE WEAPON SYSTEMS			
BMP, 73 mm		800 m	800 m
AT3 missile		3,000 m	3,000 m
AT5 missile		4,000 m	4,000 m
BMP-2		2,000 m	2,000 m
BTR, 14.5 mm		2,000 m	1,000 m
T-72 tank, 125 mm		2,100 m	2,000 m
T-80 tank, 125 mm		2,400 m	2,000 m
T-80 AT8		4,000 m	4,000 m

*The planning range is based on ideal weather conditions during daylight.

14-16 Miscellaneous Field Data

Table 14-11. US tanks

	M1	M1A1	M60	M60A1	M60A3	M551
Length (m)	7.92	7.92	6.95	6.95	6.95	6.30
Width (m)	3.65	3.66	3.63	3.63	3.63	2.82
Firing height (m)	1.89	1.89	2.10	2.10	2.10	
Max speed (kmph)	72.4	66.8	48.3	48.3	48.3	70
Fuel capacity (gal)	503.8	503.8	384.9	375.1	375.1	158
Max range (km)	498	465	500	500	480	600
Fording (m)	1.22	1.22	1.22	1.22	1.22	Amphibious
Gradient (percent)	60	60	60	60	60	60
Vertical obstacle (m)	1.24	1.07	0.91	0.91	0.91	0.84
Primary armament	105 mm	120 mm	105 mm	105 mm	105 mm	152 mm
Ammunition capacity- •Main gun •12.7 mm •7.62 mm	55 1,000 11,400	40 1,000 12,400	57 900 6,000	63 900 6,000	63 900 6,000	30 1,000 3,080

Table 14-12. US antiarmor missiles

Missile	Prime Mover	Weight (lb)	Guidance Linkage	Rounds Aboard	Range (m)
Shillelagh	M60A2 tank	61.3 (round only)	Infrared	13	3,000 max 800 min
TOW	M2 or M3 AH-1S atk hel	40 (round only)	Wire	10	3,000 max 65 min 3,750 max 65 min
Dragon	Individual soldier	32 (carry weight) 25.2 (round only)	Wire	6	1,000 max 65 min

Miscellaneous Field Data 14-17

FM 5-34

Table 14-13. US field artillery and air-defense weapons

Weapon	Rd on Veh	Rd on Carrier	Range (m)	Weight (lb)	Emplace-ment Time (min)	Max Rate of Fire Rd (3 min)	Sustained Fire Rd/hr	Ammunition		
								Types	Fuzes	
105-mm how, towed, M102	Sit dep	NA	11,500	3,170	2	30	180	WP, HE, HEAT, CML, illum, smk, ICM, APERS, HEP	Quick, delay, VT, time, con-crete, piercing	
155-mm how, towed, M114A1/A2	48	NA	14,600	12,700	3.5	12	60	SCATMINE HE WP	Quick, delay, VT	
155-mm how, SP, M109A1	28	96	18,100 24,000 (RAP)	53,940	0.5	12	60	CML, illum, smk	Time, concrete piercing	
155-mm how, SP, M109A2/A3	36	96	18,100 24,000 (RAP)	53,940	0.5 (RAP)	12	20	Nuc ICM RAP		
155-mm how, towed, M198	48	NA	24,000 30,000 (RAP)	15,500	5	12	Variable			
8-in how, SP, M102A2	2	36	22,900 30,000 (RAP)	62,500	2.5	4.5	30	HE, nuc, CML, ICM, spot	Quick, delay, VT, time, con-crete, piercing	
Vulcan, CM741		4,200	1,200 AD 4,500 Surface	26,000	NA	3,000	NA	HEI	PD	

NOTE: Refer to page 1-28, this manual, for additional fire-support munitions information.

14-18 Miscellaneous Field Data

FM 5-34

OPERATIONAL SYMBOLS

Size Indicator	Meaning
▬	Installation
⌀	Team/crew
●	Squad
● ●	Section
● ● ●	Platoon/detachment
I	Company/battery/troop
I I	Battalion/squadron
I I I	Regiment/group
X	Brigade
X X	Division
X X X	Corps
X X X X	Army
X X X X X	Army group/front
X X X X X X	Region

Figure 14-4. Unit size and installation indicator

Miscellaneous Field Data 14-19

FM 5-34

Unit	Symbol	Unit	Symbol
Airborne	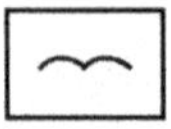	Infantry (basic)	
Air defense	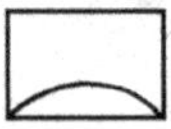	Light infantry	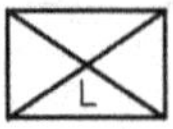
Air assault with organic lift	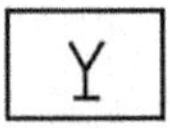	Maintenance	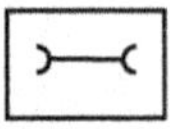
Mechanized or armor	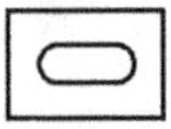	Mechanized infantry (in tracked APC)	
Armored (tracked IFV) reconnaissance or scouts	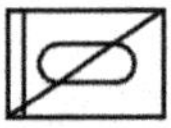	Medical (basic symbol)	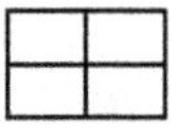
Aviation		Law enforcement - Army MP	
Attack helicopter		Mountain infantry	
Engineer bridge		Class IV (petroleum supply)	
Reconnaissance cavalry or scouts (basic or dismounted)	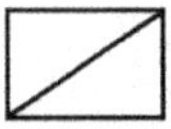	Class II (clothing, individual equipment, tentage, organizational tool sets)	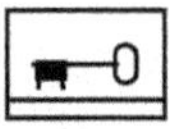
Nuclear, biological, chemical		Signal (basic symbol)	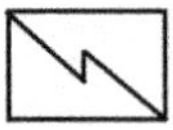
Engineer (basic)	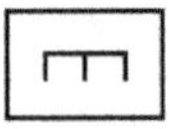	Transportation (basic symbol)	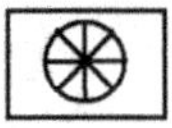
Field artillery (basic)(towed)	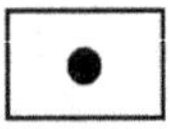		

Figure 14-5. Unit identification symbols

14-20 Miscellaneous Field Data

FM 5-34

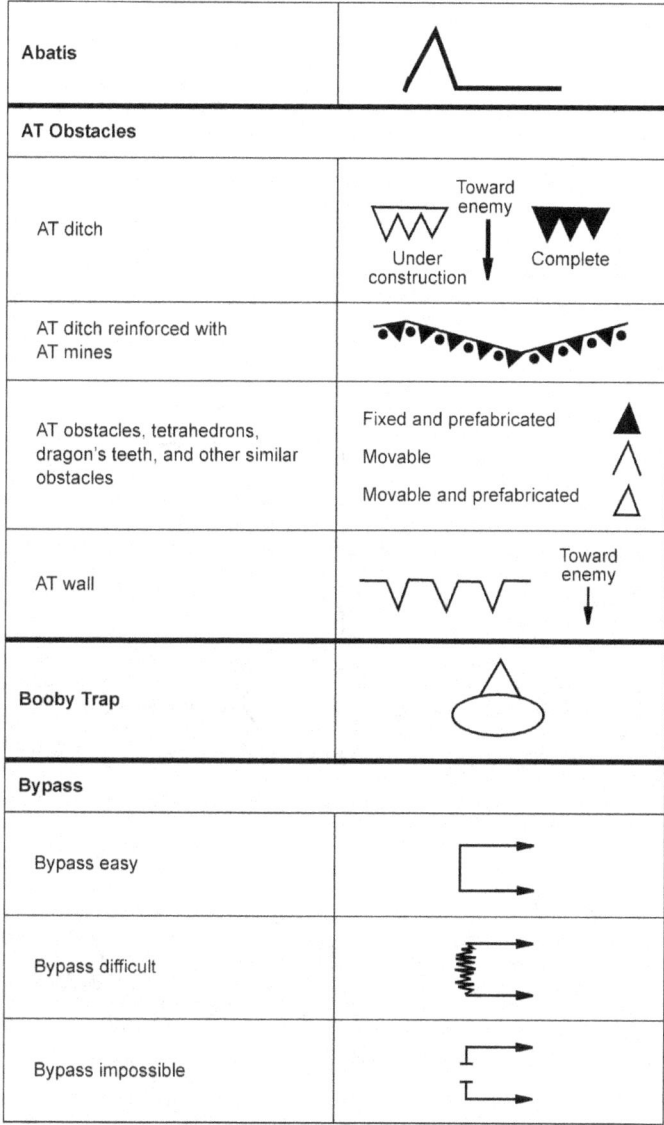

Figure 14-6. Obstacle symbols

Miscellaneous Field Data 14-21

FM 5-34

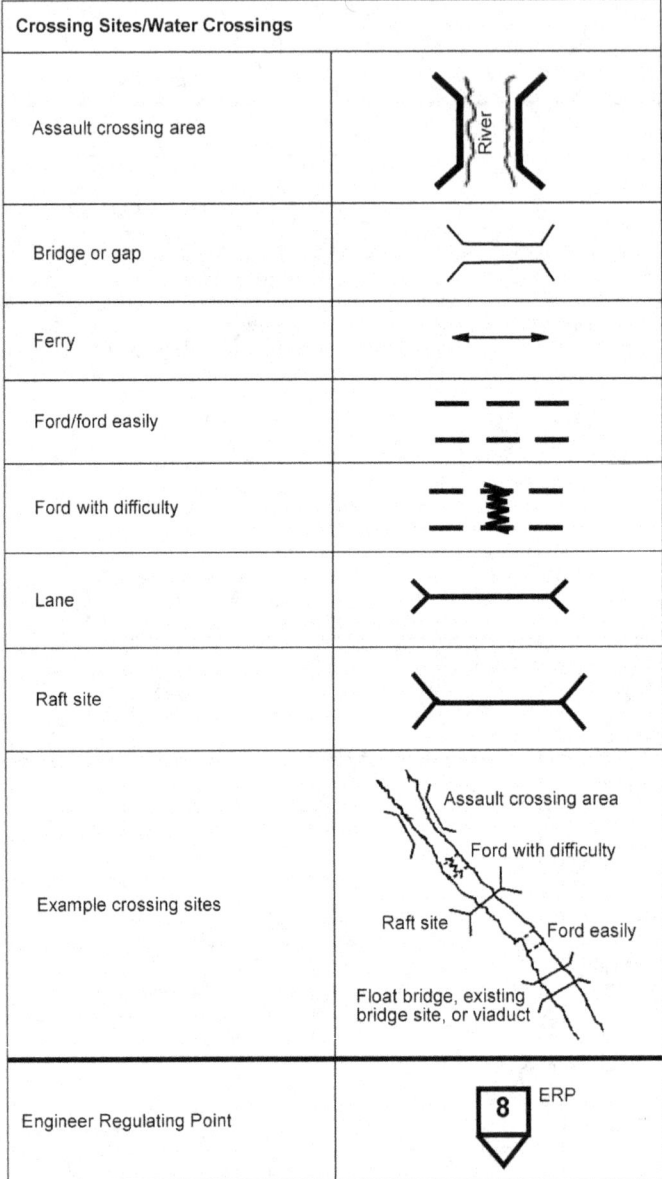

Figure 14-6. Obstacle symbols (continued)

14-22 Miscellaneous Field Data

FM 5-34

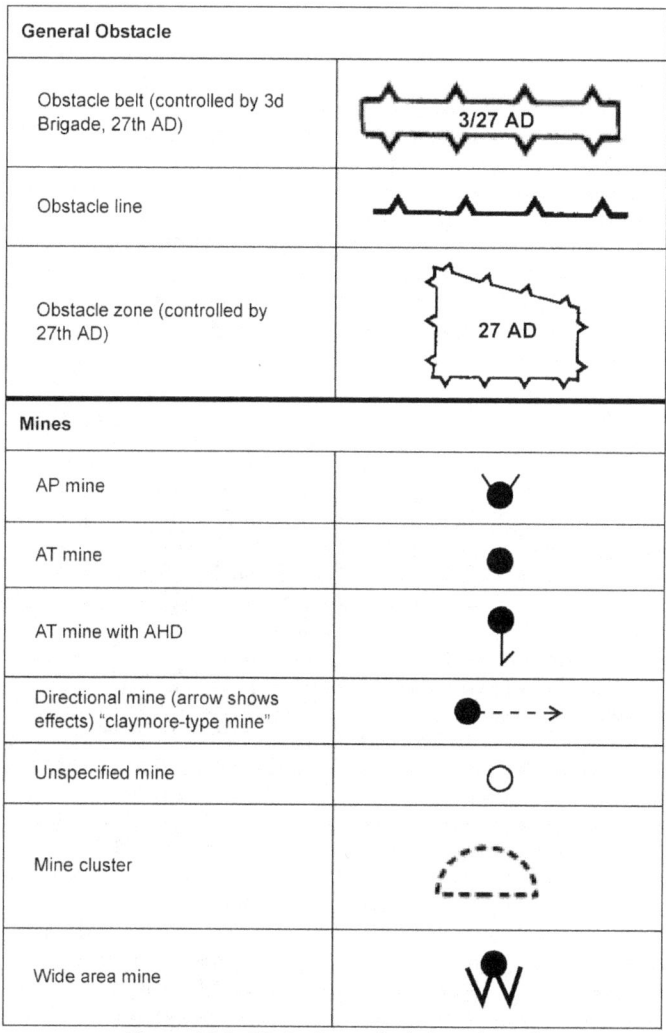

General Obstacle	
Obstacle belt (controlled by 3d Brigade, 27th AD)	
Obstacle line	
Obstacle zone (controlled by 27th AD)	
Mines	
AP mine	
AT mine	
AT mine with AHD	
Directional mine (arrow shows effects) "claymore-type mine"	
Unspecified mine	
Mine cluster	
Wide area mine	

Figure 14-6. Obstacle symbols (continued)

Miscellaneous Field Data 14-23

FM 5-34

Minefields	
Planned minefield (unspecified mines)	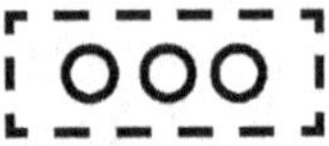 ⌐ ‾ ‾ ‾ ‾ ¬ │ ○ ○ ○ │ └ _ _ _ _ ┘
Completed minefield (unspecified mines)	 ○ ○ ○
AP minefield	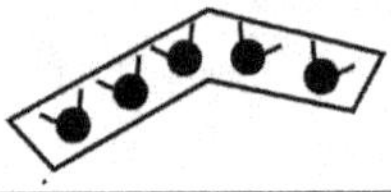 (symbol)
AT minefield with gap (show effective time and name of gap)	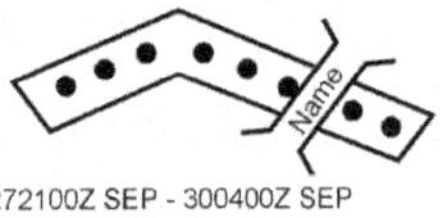 (symbol) 272100Z SEP - 300400Z SEP
AT minefield (line points to center of mass of minefield)	 ○○○
Scatterable minefield (unspecified mines with self-destruct DTG)	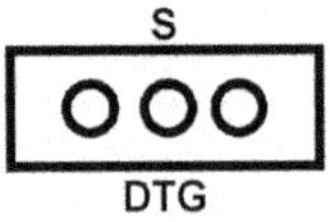 S ○ ○ ○ DTG
AP minefield reinforced with scatterable and self-destruct date-time-group	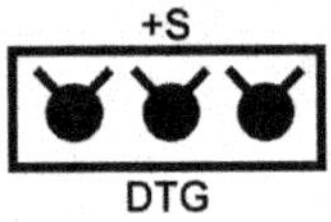 +S (symbol) DTG
Scatterable minefield (AT mines) with self-destruct date-time-group	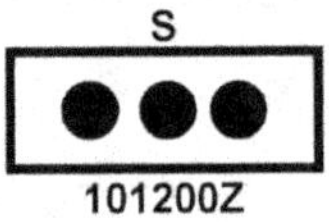 S ● ● ● 101200Z
Mined area	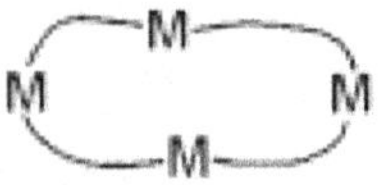 M M M M

Figure 14-6. Obstacle symbols (continued)

14-24 Miscellaneous Field Data

C1, FM 5-34

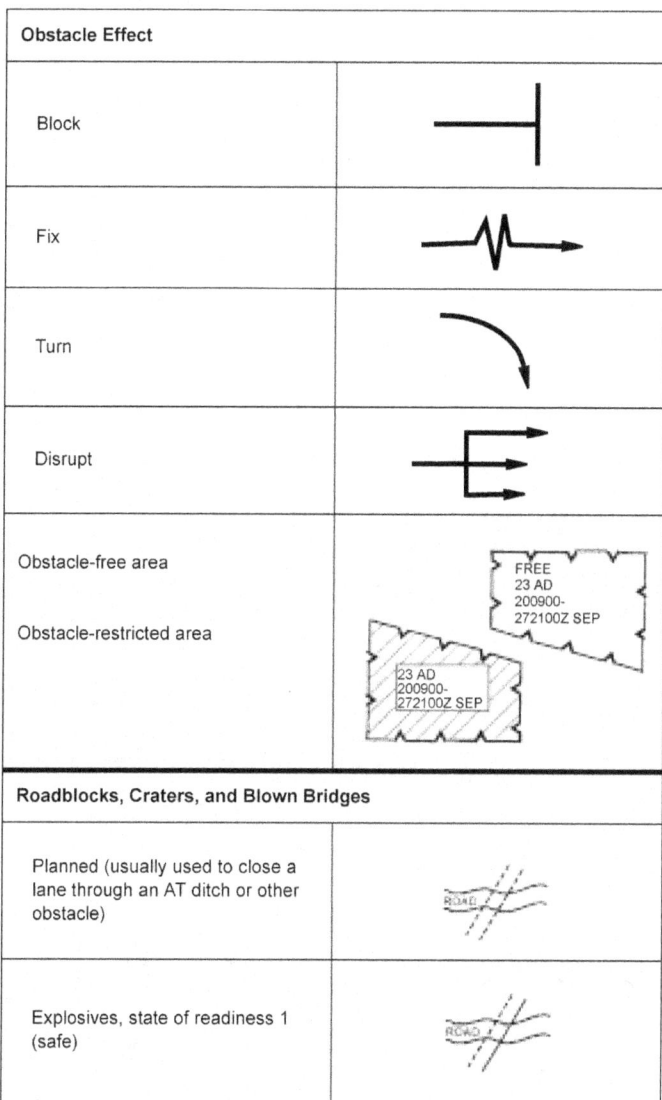

Figure 14-6. Obstacle symbols (continued)

Miscellaneous Field Data 14-25

FM 5-34

Roadblock (Continued)	
Explosives, state of readiness 2 (armed but passable)	
Roadblock complete (executed)	
Wire Obstacles	
Unspecified	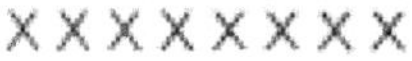 X X X X X X X X
Single fence	 X———X———X
Double fence	 XX——XX——XX
Double apron fence	 XXXXXXXX
Low wire fence	 X X X X X X X X
High wire fence	 XXXXXXXX
Single concertina	 OOOOOOOO
Double-strand concertina	 OOOOOOOO
Triple-strand concertina	 OOOOOOOO
Executed Volcano minefield	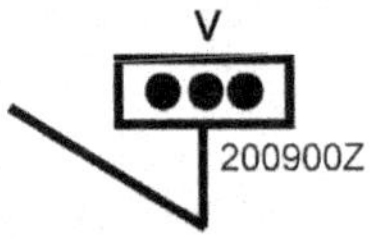 V / ●●● / 200900Z

Figure 14-6. Obstacle symbols (continued)

14-26 Miscellaneous Field Data

C1, FM 5-34

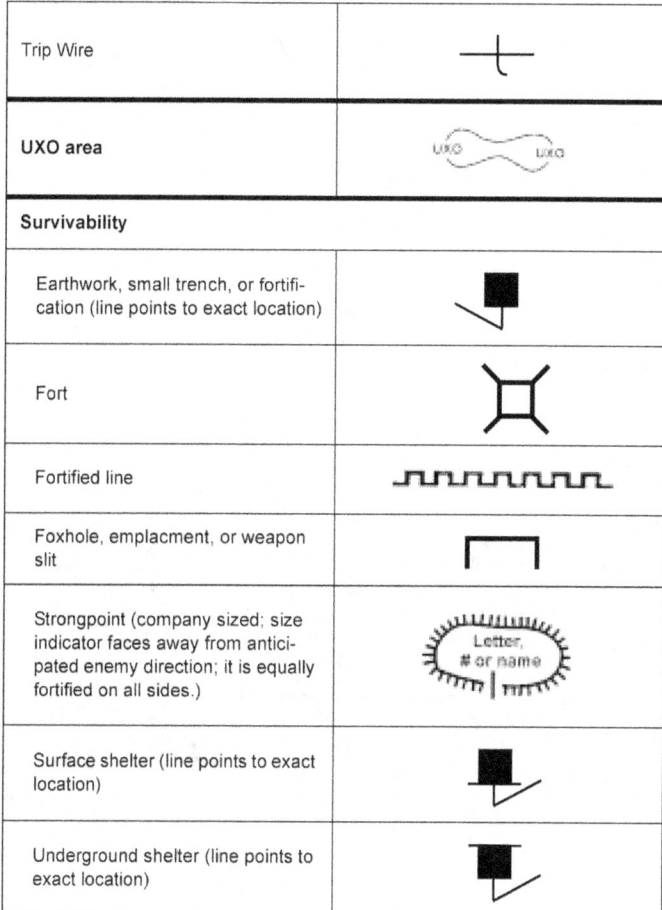

Trip Wire	
UXO area	
Survivability	
Earthwork, small trench, or fortification (line points to exact location)	
Fort	
Fortified line	
Foxhole, emplacment, or weapon slit	
Strongpoint (company sized; size indicator faces away from anticipated enemy direction; it is equally fortified on all sides.)	
Surface shelter (line points to exact location)	
Underground shelter (line points to exact location)	

Figure 14-6. Obstacle symbols (continued)

Miscellaneous Field Data 14-27

FM 5-34

IFV	 ⬡
APC	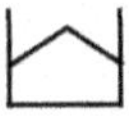 ⌂
Cargo or personnel carrier	 ⊔
Train engine/locomotive	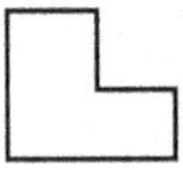 ⌐
ACE	
Grizzley (M1 breacher)	
AVLB	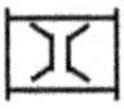
Hovercraft	
AVLM Trailer-mounted MICLIC	 
Tractor, full tracked, low speed (dozer)	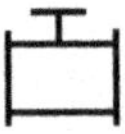
Armored carrier with Volcano Truck-mounted Volcano	 
Smoke generator	

Figure 14-7. Weapon symbols

14-28 Miscellaneous Field Data

FM 5-34

	Light	Medium	Heavy
Flamethrower	 ⌐	 ⌐ Vehicle	
Howitzer	 (light howitzer symbol)	 (medium howitzer symbol)	 (heavy howitzer symbol)
Mortar	 (light mortar symbol)	 (medium mortar symbol)	 (heavy mortar symbol)
Multibarrel rocket launcher	 (light symbol)	 (medium symbol)	 (heavy symbol)
Surface-to-surface missile	 (light symbol)	 (medium symbol)	 (heavy symbol)
Tank (friendly)	 (light tank symbol)	 (medium tank symbol)	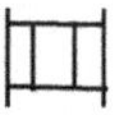 (heavy tank symbol)

Figure 14-7. Weapon symbols (continued)

Miscellaneous Field Data 14-29

CONVERSION FACTORS

Table 14-14. Conversion factors

Multiply	By	To Obtain	Multiply	By	To Obtain
Acres	4,046.9 or 4,047	Square meters	Kilograms per meter	9.302×10^{-3}	BTU
Atmospheres	14.70	Pounds per sq inch	Kilograms per sq meter	9.678×10^{-5}	Atmospheres
Centimeters	0.3937	Inches	Kilometers	3281	Feet
Centimeters of mercury	0.01316	Atmospheres	Meters	3.2808	Feet
Cemtimeters of mercury	0.1934	Pounds per sq inch	Miles	1.6093	Kilometers
Cubic feet	7.481	Gallons	Miles per hour	1.467	Feet per sec
Cubic meters	264.2	Gallons	Millimeters	0.03937	Inches
Degrees (angle)	0.01745	Radians	Nautical miles	1.152	Miles
Feet	0.3048	Meters	Ounces	28.35	Grams
Feet per min	0.5080	Centimeters per sec	Ounces	0.0625	Pounds
Feet per min	0.01136	Miles per hour	Pounds per sq inch	0.06804	Atmospheres
Feet per sec	1.097	Kilometers per hour	Radians	57.30	Degrees
Gallons	3.785×10^{-3}	Cubic meters	Square centimeters	0.1550	Square inches
Grams	0.03527	Ounces	Square feet	0.09290	Square meters
Grams	2.205×10^{-3}	Pounds	Square meters	10.764	Square feet
Grams-calories	3.968×10^{-3}	BTUs	Square miles	2.590	Square kilometers
Horsepower	42.44	BTUs per min	Temp (degs C) + 273	1	Abs temp (degs C)
Horsepower	745.7	Watts	Temp (degs C) + 17.8	1.8	Temp (degs F)
Inches	2.540	Centimeters	Temp (degs F) + 460	1	Abs temp (degs F)
Inches of water	0.002458	Atmospheres	Temp (degs F) - 32	5/9	Temp (degs C)
Joules	9.486×10^{-4}	BTUs	Tons (short)	907.2	Kilograms
Kilograms	2.2046	Pounds	Tons (short)	2000	Pounds
Kilograms	1.102×10^{3}	Tons (short)	Watts	0.05692	BTUs per min
Kilogram-calories	3.968	BTUs	Weeks	168	Hours
			Yards	0.9144	Meters

14-30 Miscellaneous Field Data

FM 5-34

LEVELS OF RISK MANAGEMENT

RISK-MANAGEMENT PROCESS

1. Identify the hazards.
2. Assess the risk of each hazard
3. Make a risk decision (see Figure 14-9).
   — Develop controls to reduce risks.
   — Reassess the risk with control measures (see Figure 4-10).
   — Make risk decision based on the residual risk (see Figure 14-11, page 14-32).
4. Implement controls to reduce level of risk.
5. Supervise and enforce risks, hazards, and control measures (see Figure 14-12, page 14-33).

RISK-MANAGEMENT RULES

Rule #1: Integrate protection into planning.
Rule #2: Accept no unnecessary risks.
Rule #3: Make risk decision at the proper level.
Rule #4: Accept risks only if the benefits outweight the potential costs.

Figure 14-8. Risk management

Risk	Division	Brigade	Battalion	Company	Platoon
Extremely high	Corps	Division	Division	Brigade	Brigade
High	Corp	Division	Brigade	Brigade	Battalion
Moderate	Division	Brigade	BAttalion	Battlaion	Company
Low	Division	Brigade	Battalion	Company	Platoon

Once the residual risks have been determined by applying the control measures, the risk decision must be presented to the proper level of command for final risk decision on mission execution.

References:
AR 385-10
DA Pamphlet 385-1

Figure 14-9. Levels of decision matrix

Miscellaneous Field Data 14-31

FM 5-34

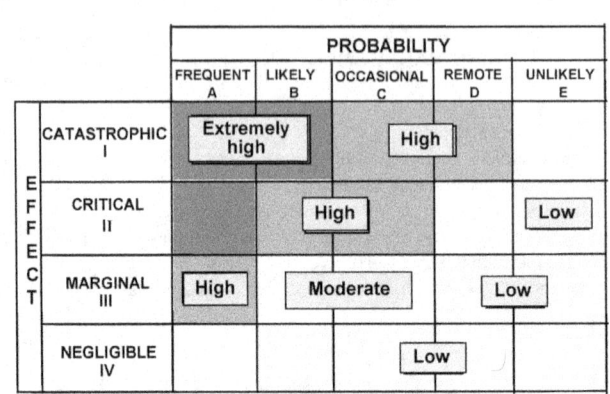

NOTE: Assess the hazards based on the probability of occurrence and the overall effect on the operation. Analyze all the effects of all variables on the operation, including such things as weather and equipment availability.

Figure 14-10. Risk-assessment matrix

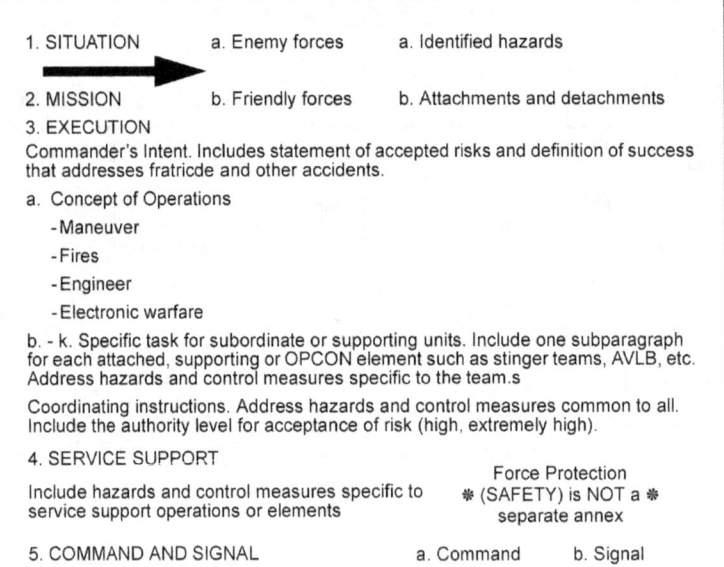

Figure 14-11. Steps in risk management

14-32 Miscellaneous Field Data

FM 5-34

Hazard	Risk Assessment	Controls Implemented	Residual Risk Level
Inexperienced military drivers	High	New or inexperienced drivers are: - Identified by the commander - Trained and licensed according to AR 600-55 - Assigned an experienced assistant driver or senior occupant (Officer/NCO)	Moderate
Excessive speed	High	All speed limits enforced	Low
Adverse environmental conditions	Moderated	If hazardous driving conditions are encountered or visibility drops below 50 meters convoy will stop at the nearest rest area	Low
Soldier fatigue	High	Rest periods every two hours in a planned rest area	Low
Highway congestion and construction zones	High	Convoy movements will scheduled to avoid peak traffic periods	Low
Inadequate protective clothing and equipment	High	All soldiers will wear protective clothing, KEVLAR helmets, and hearing protection	Low

Figure 14-12. Risk-management work sheet

Miscellaneous Field Data 14-33

Glossary

°C	degree(s) Celsius
°F	degree(s) Fahrenheit
1LT	first lieutenant
1SG	first sergeant
1st	first
2nd	second
A5	type of explosive (composition A5)
A&O	assault and obstacle platoon
AA	avenue of approach
AAR	after-action review
abs	absolute
AC	hydrogen cyanide
ACE	armored combat earthmover, M9
ACR	armored cavalry regiment
act	actual
AD	air defense
ADA	air-defense artillery
ADAM	area denial artillery munition
AFB	assault float bridge
AFV	armored fighting vehicle
AHD	antihandling device
alt	alternate
alum	aluminum
AM	ante meridian
amp	amphere(s)
ANT	antenna
AO	area of operations

FM 5-34

AP	antipersonnel
APC	armored personnel carrier
APDS	armor-piercing discarding sabot
APERS	antipersonnel
approx	approximately
ar	armor
arm	armored
ARTEP	Army Training and Evaluation Program
arty	artillery
ASAP	as soon as possible
ASL	assistant squad leader
asslt	assault
AT	antitank
ATACMS	Army Tactical Missile System
ATD	antitank ditch
atk	attack
atom	atomic
ATTN	attention
AUD	audio
AUG	August
avail	available
AVLB	armored vehicle-launched bridge
AVLM	armored vehicle-launched MICLIC
AXP	ambulance exchange point
B4	type of explosive (composition B4)
B/P	be prepared
bde	brigade
bdg	bridge
BDU	battle-dress uniform
BEB-SD	bridge-erection boat-shallow draft
BIFV	Bradley infantry fighting vehicle
bio	biological

2-Glossary

blk	blocks
BMP	amphibious infantry combat vehicle (Soviet threat vehicle)
bn	battalion
bo	blackout
BP	battle position
bps	bits per second
brg	bridge
BTR	amphibious armored personnel carrier (Soviet threat vehicle)
BTU	British thermal unit(s)
℄	centerline
C^2	command and control
c-to-c	center-to-center
C4	composition 4
cal	caliber
CAM	chemical agent monitor
CASEVAC	casualty evacuation
CATK	counterattack
CATV	community antenna television
cav	cavalry
CBR	California bearing ratio
cbt	combat
CBU	cluster bomb unit
CCIR	commander's critical information requirements
CCP	casualty collection point
CDF	cumulative distribution function
CDM	chemical downwind message
cdr	commander
CDS	container delivery system
CE	compaction effort

Glossary-3

FM 5-34

CEO	communication-electronics operation
CEV	combat engineer vehicle
cfc	chlorofluorocarbon
cfs	cubic foot (feet) per second
CG	phosgene
cgo	cargo
cGyph (cGy)	centigray (centigrade per hour)
CHAN	channel
chem	chemical
CINC	commander in chief
CK	cyanogen chloride
cl	class
CLR	clear
cm	centimeter(s)
CML	chemical
co	company
COA	course of action
COMSEC	communications security
cos	cosine
CDS	container delivery service
CP	command post
CPR	cadiopulmonary resuscitation
CPT	captain
CPU	chemical protective undergarment
CRB	change review board
CS	combat support
CSS	combat service support
CT	cipher text
cu	cubic
cu ft	cubic foot (feet)
CUE	setting on a radio
CX	phosgene oxime

4-Glossary

DA	Department of the Army
DC	District of Columbia
DD	double double (Bailey bridge)
DEC	December
dec	decontaminate
deg	degree
degs C	degrees Celcius
degs F	degrees Fahrenheit
demo	demolition
dep	dependent
det	detonator/detonating
DGN	degrees grid north
diam	diameter
dist	district
div	division
DMG	degrees magnetic north
DODIC	Department of Defense Identification Code
DPICM	dual-purpose, improved conventional munition
DS	double single (Bailey bridge)
DT	double triple (Bailey bridge)
DTG	date-time group
DTN	degrees true north
ea	each
EA	engagement area
ECCM	electronic counter-countermeasure
EEI	essential elements of information
el	elevation
elev	elevation
EM	enlisted member
EN	engineer
eng	engineer

engr	engineer
ENGR	engineer
EO	Executive Order
EOD	explosive ordnance detachment
EPW	enemy prisoner of war
eq	equipment
equip	equipment
ERF	electronic remote fill
ERI	engineer restructure initiative
ERP	engineer regulation point
est	estimated
FCTN	function
FB	far bank
FDC	fire-direction center
FEBA	forward edge of the battle area
FH	frequency hopping
FIST	fire-support team
FM	frequency modulated
FM	field manual
FO	forward observer
FPF	final protective fire
FPL	final protection line
FPOL	forward passage of lines
fps	foot (feet) per second
frag	fragmentation
FRAGO	fragmentary order
freq	frequency
FS	factor of safety
FS	far shore
FSE	fire-support element
FSO	fire-support officer
ft	foot (feet)

ft-lb	foot (feet) pound
Ft	fort
g	gram(s)
GA	tabun (chemical)
gal	gallon(s)
GB	sarin (chemical)
GD	soman
GEMMS	Ground-Emplaced Mine-Marking System
GI	government issue
gm	gram(s)
GN	grid north
GPBTO	general-purpose barbed-tape obstacle
GPS	Global Positioning System
GWL	groundwater level
HA	hazard area
HC	hydrogen chloride
HD	mustard
HDP	hull defilade postion
HE	high explosive
HEAT	high-explosive antitank
HEI	high-explosive incendiary
HEI-T	high-explosive incendiary tracer
hel	helicopter
HEMMS	hand-emplaced minefield marking set
HEMTT	heavy, expanded, mobility tactical truck
HEP	high-explosive plastic
HEP-T	high-explosive plastic tracer
HE-WAM	hand emplaced-widemarea munition
HHC	headquarters and headquarters company
HI	high

FM 5-34

HL	mustard lewisite
HM	hazardous material
HMMWV	high-mobility, multipurpose wheeled vehicle
HN	host nation
HN	nitrogen mustard
how	howitzer
HP	holding power
HQ	headquarters
hr	hour(s)
hvy	heavy
HW	hazardous waste
ICM	improved capability missile
ID	identification
ID	inside diameter
IDA	improved dog-bone assembly
IEDK	individual equipment decontamination kit
IFV	infantry fighting vehicle
illum	illumination
in	inche(s)
inf	infantry
IOE	irregular outer edge
IPB	intelligence preparation of the battlefield
IPS	improved plough steel
ISU	integrated sight unit
ITV	improved TOW vehicle
JUN	June
kbps	kilobits per second
kg	kilogram(s)
km	kilometer(s)
km/hr	kilometer(s) per hour

kmph	kilometer(s) per hour
kt	kiloton(s)
L	left
L	lewisite
LAW	light antitank weapon
LB	local battery
lb	pound(s)
LD	load
LET	light equipment transporter
LMTV	light mobile tactical vehicle
LO	low
LOC	lines of communication
LOGPAC	logistics package
LOS	line of sight
LP	listening post
LRA	local reproduction authorized
LRP	logistics release point
LRS	link-reinforcement set
LTR	light tactical raft
lube	lubrication
M	medium
m	meter(s)
M-S	Miznay-Scherdin
m/hr	meter(s) per hour
maint	maintenance
MAN	manual
MAR	March
max	maximum
mc	megacycle
MCRP	Marine Corps reference publication

FM 5-34

MDI	modernized demolition initiators
MDMP	military decision-making process
mech	mechanized
MEDEVAC	medical evacuation
MEL	maximum engagement line
MET	management engineer team
METT-T	mission, enemy, terrain, troops, and time available
MG	machine gun
MGB	medium girder bridge
MGN	mils grid north
MHz	megahertz
MICLIC	mine-clearing line charge
MID	mechanized infantry division
min	minute(s)
min	minimum
MKT	mobile kitchen trailer
MLG	millimeters grid north
MLM	millimeters magnetic north
MLRS	Multiple-Launch Rocket System
MLT	millimeters true north
mm	millimeter(s)
MMN	mils magnetic north
MN	magnetic north
MOPMS	Modular-Pack Mine System
MOPP	mission-oriented protective posture
MOS	minimum operating strip
MOUT	military operations on urbanized terrain
MP	military police
mph	mile(s) per hour
MPS	meter(s) per second
MPS	modified plough steel
MRD	motorized rifle division
MRR	motorized rifle regiment

MSR	main supply route
MT	maneuver target
mt	metric ton
mtd	mounted
MTN	mils true north
MTO	message to observer
MTOE	modified table of organization and equipment
MTP	mission training plan
MTV	medium tactical vehicle
mtzd	motorized
NA	not applicable
NATO	North Atlantic Treaty Organization
NB	near bank
NBC	nuclear, biological, chemical
NCO	noncommissioned officer
NCOIC	noncommissioned officer in charge
NCS	National Communications System
NCS	network control station
NLT	not later than
NM	nautical mile
No.	number
NP	nonpersistent
NS	near shore
nuc	nuclear
NVD	night-vision device
O/O	on orders
OB	obstruction
obj	objective
OBM	outboard motor
obs	obstacle

Glossary-11

FM 5-34

OCOKA	observation covering concealment, obstacles key terrain
OD	outside dose
OEG	operational exposure guidance
OIR	other intelligence requirements
OMC	optimum moisture content
OP	observation post
OPCON	operational control
OPLAN	operation plan
opns	operations
OPORD	operation order
ops	operations
OPSEC	operations security
org	organization
organ	organization
ORP	object release point
OT	observer target
oz	ounce(s)
P	persistent
PA	power amplification
para	paragraph
PD	pressure detonating
PDF	principle direction of fire
PDM	pursuit deterrent munition
PIP	product improvement plan
PIR	priority intelligence requirement
pkg	package
PL	phase line
plat	platoon.
pls	palatized load system
plt	platoon
POL	petroleum, oils, and lubricants

PS	plough steel
psi	pound(s) per square inch
PSYOP	psychological operations
PT	plain text
PTT	push-to-talk
PWR	power
R	right
RAAM	remote, antiarmor mine
RAP	rocket-assisted projectile
RCLR	recoilless rifle
rd	round
RDX	cyclonite
RE	relative effectiveness
recon	reconnaissance
REM	remote
RF	radio frequency
ROE	rules of engagement
RP	reference point
rt	route
RT	radio transmitter
RTO	radio/telephone operator
RV	receive variable
RXMT	retransmit
S&T	supply and transport
S	south
S2	Intelligence Officer (US Army)
S3	Operations and Training Officer (US Army)
S4	Supply Officer (US Army)
SALUTE	size, activity, location, unit, time, and equipment
SAW	squad automatic weapon

SC	single channel
SCATMINE	scatterable mine
SCATMINWARN	scatterable minefield warning
SD	self-destruct
SDK	skin decontamination kit
sec'y	security
sec	section
sec	second(s)
SEE	small emplacement excavator
sep	separate
SEP	September
SIG	signal
sin	sine
SINCGARS	Single-Channel, Ground-to-Air Radio System
sit	situation
SITEMP	situation template
SL	squad leader
SLAM	selectable lightweight attack munition
smk	smoke
SOEO	scheme of engineer operations
SOF	special operations forces
SOI	signal operations instructions
SOP	standing operating procedures
SOSR	suppress, obsucre, secure, reduce
sp	self-propelled
spot	round sent up to help aid in target (spot) acquisitions
spt	support
sq	square
SQ	squelch
sqd	squad
sqdn	squadron
SS	single single (Bailey bridge)
SSN	social security number

14-Glossary

STANAG	standardization agreement
STB	super tropical bleach
STBY	standby
STO	store
susp	suspicious
SWC	safe working capacity
SYNC	synchronize
TACSOP	tactical standing operating procedures
tan	tangent
TBD	to be determined
TC	tank commander
TC	training circular
TCP	tactical command post
TD	tank division
TD	triple double (Bailey bridge)
TDP	turret defilade position
temp	temperature
TF	transmission factor
TF	task force
tgt	target
TLP	troop-leading procedure
tm	team
TM	technical manual
TMAS	Thermal Mine Acquisition System
TNT	trinitrotoluene
TO	theater of operations
TOC	tactical operations center
TOE	table of organization and equipment
TOW	tube-launched, optically-tracked, wire-guided missile
TRADOC	United States Army Training and Doctrine Command
trk	truck

TRP	target reference point
trp	troop
TS	triple single (Bailey bridge)
TST	test
TT	triple triple (Bailey bridge)
TTP	tactics, techniques, and procedures
UMCP	unit maintenance collection point
unk	unknown
US	United States
USAES	United States Army Engineer School
USAF	United States Air Force
USCS	Unified Soil Classification System
UTM	universe traverse mercator
UW	upwind
UXO	unexploded ordnance
VDR2	radiac set
veh	vehicle
VHF	very high frequency
vic	vicinity
VOL	volume
VS	thickened G-agent
VT	virtual terminal
VX	thickened G-agent
W	west
w/	with
w/bo	with blackout
w/o	without
W/T	wheeled/tracked
WASPM	wide-area side-penetrator mine, M84
WHSP	whisper

16-Glossary

WO	warning order
WP	white phosphorus
WRP	weapon reference point
wt	weight
X	completed/executed obstacle
XO	executive officer
Z	zulu

References

SOURCES USED

These are the sources quoted or paraphrased in this publication.

International Standardization Agreements

STANAG 2002 NBC (Edition 7). *Warning Signs for the Marking of Contaminated or Dangerous Land Areas, Complete Equipments Supplies and Stores.* 26 November 1980

STANAG 2021 ENGR (Edition 5). *Computation of Bridge, Ferry, Raft, and Vehicle Classifications.* 18 September 1990.

STANAG 2036 ENGR (Edition 4). *Land Mine Laying, Marking, Recording, and Reporting Procedures.* 2 December 1987.

STANAG 2047 NBC (Edition 6). *Emergency Alarms of Hazard or Attack (NBC and Air Attack Only).* 27 March 1985.

Joint and Multiservice Publications

FM 20-400/MCRP 4-21C. *Military Environmental Protection.* To be published within six months.

FM 5-430-00-1/AFJPAM 32-8013, Vol I. *Planning and Design of Roads, Airfields, and heliports in the Theater of Operations—Road Design.* 26 August 1994.

Army Publications

AR 385-10. *The Army Safety Program.* 23 May 1988.

AR 385-63. *Policies and Procedures for Firing Ammunition for Training, Target Practice and Combat.* 15 October 1983.

AR 600-55. *The Army Driver and Operator Standardization Program (Selection, Training, Testing, and Licensing).* 31 December 1993.

ARTEP 5-145-32-MTP. *MTP for the Engineer Bridge Company.* 19 July 1991.

DA Pam 385-1. *Small Unit Safety Officer/NCO Guide.* 22 September 1993.

FM 5-36. *Route Reconnaissance and Classification.* 10 May 1989.

FM 5-125. *Rigging Techniques, Procedures, and Applications.* 3 October 1995.

FM 5-170. *Engineer Reconnaissance.* 5 May 1998.

FM 5-34

FM 5-277. *M2 Bailey Bridge.* 9 May 1986.

FM 5-250. *Explosives and Demolitions.* 30 July 1998

FM 3-7. *NBC Field Handbook.* 29 September 1994.

FM 5-424. *Theater of Operations Electrical Systems.* 25 June 1997.

FM 5-446. *Military Nonstandard Fixed Bridging.* 3 June 1991.

FM 6-30. *Tactics, Techniques, and Procedures for Observed Fire.* 16 July 1991.

FM 7-8. *Infantry Rifle Platoon and Squad.* 22 April 1992.

FM 10-71, *Petroleum Tank Vehicle Operations.* 12 May 1978.

FM 20-31. *Electric Power Generation in the Field.* 9 October 1987.

FM 20-32. *Mine/Countermine Operations.* 29 May 1998.

FM 21-10. *Field Hygiene and Sanitation.* 22 November 1988.

FM 90-7. *Combined Arms Obstacle Integration.* 29 September 1994.

FM 90-13-1. *Combined Arms Breaching Operations.* 28 February 1991.

FM 100-14. *Risk Management.* 23 April 1998.

FM 101-5. *Staff Organization and Operations.* 31 May 1997.

TC 5-210. *Military Float Bridging Equipment.* 27 December 1988.

TC 20-32-3. *Foreign Mine Handbook (Balkan States).* 15 August 1997.

TM 5-5420-209-12. *Operator's and Unit Maintenance Manual for Improved Float Bridge (Ribbon Bridge) Consisting of: Transporter CONDEC Model 2280 (NSN 5420-00-071-5321) CONDEC Model 2305 (5420-01-173-2020) PACAR Model 9999 (5420-01-175-6523) Southwest Model RBT (5420-01-175-6524) Interior Bay CONDEC Model 2282 (5420-00-071-5322) CONDEC Model 2307 (5420-01-173-2022) Space Model 66981 (5420-01-175-6526) Ramp Bay CONDEC Model 2281 (5420-497-5276) CONDEC Model 2306 (5420-01-174-8084) Space Model 6698R (5420-01-175-6525).* 15 September 1993.

TM 5-5420-212-12. *Maintenance Manual for Medium Girder Bridge (MGB) (NSN 5420-00-172-3520).* 18 April 1985.

TM 5-5420-212-12-1. *Operator's and Organizational Maintenance Manual Link Reinforcement Set for the Medium Girder Bridge (NSN 5420-01-139-1503).* 5 October 1984.

DOCUMENTS NEEDED

These documents must be available to the intended users of this publication.

DA Form 1248. *Road Reconnaissance Report.* July 1960.

DA Form 1249. *Bridge Reconnaissance Report.* 1 July 1960.

DA Form 1250. *Tunnel Reconnaissance Report.* 1 January 1955.

DA Form 1251. *Ford Reconnaissance Report.* 1 January 1955.

DA Form 1355. *Minefield Record.* March 1987.

DA Form 1355-1-R. *Hasty Protective Row Minefield Record (LRA).* October 1997.

DA Form 1711-R. *Engineer Reconnaissance Report.* May 1985.

DA Form 2028. *Recommended Changes to Publications and Blank Forms.* 1 February 1974.

DA Form 2203-R. *Demolition Reconnaissance Record.* June 1998.

DA Form 5517-R. *Standard Range Cards (LRA).* February 1986.

Index

A
abatis, 9-13
AP SCATMINEs. *See* scatterable mines, AP
AP. *See* minefields, antipersonnel
AT SCATMINEs. *See* scatterable mines, AT
AT. *See* minefields, antitank

B
battle tracking, 5-6
borehole method, 9-51
bridges
    medium girder, 10-20
    rope, 12-14

C
calculations
    ACE/ACE team HDP, 8-34
    ACE/ACE team TDP, 8-31
    charge, 9-8
    dozer team HDP, 8-28
    dozer team TDP, 8-25
camouflage, 8-48
cardiopulmonary resuscitation procedures, 1-47
charges
    ammonium nitrate satchel, 9-50
    bangalore torpedo, 9-51
    boulder-blasting, 9-19
    breaching, 9-14
    counterforce, 9-18
    cratering, 9-20, 9-47
    platter, 9-49
    shaped, 9-48
    steel-cutting, 9-8
    timber-cutting, 9-11
    triple-nickle forty, 9-39
checkpoints, 8-55
chemical agents, 1-30
chemical-agent detector kit, 1-40

Claymore, 7-14
communications, 1-54
    equipment, 1-57
    visual signals, 1-68
conversion factors, 14-30
CPR. *See* cardiopulmonary resuscitation procedures

D
demolitions, 9-1
    abutment, 9-39
    bridge, 9-25
    intermediate-support, 9-39
    reconnaissance, 9-41
detonating-cord wick. *See* borehole method

E
electrical wire, 14-3
equipment
    threat, 2-10
expedient surfaces
    over mud, 4-19
    over sand, 4-24
explosive characteristics, 9-4

F
fire team, 1-15
fire-support
    equipment, 1-28
    procedures and characteristics, 1-20
formulas
    abatis, 9-12
    adjustment for lateral shift, 1-23
    antenna length, 1-55
    any triangle, 14-11
    breaching, 9-14
        number of charges, 9-17
    bridge-abutment demolition, 9-40
    cables

cable clips, 10-11
clip spacing, 10-11
deadman length, 10-16
distance between towers, 10-10
distance from tower to waterline, 10-15
FB bearings, 10-29
H for jack launch, 10-26
H for push launch, 10-25
initial sag, 10-14
length of master cable, 10-10
mean depth of a deadman, 10-15
minimum thickness of deadman, 10-16
NB bearings, 10-29
tower height, 10-14
tower offset, 10-15
tower-to-deadman distance, 10-17
tower-to-deadman offset, 10-17
circle, 14-11
class number of nonstandard combinations of vehicles, 14-15
convert load to amperes, 14-3
counterforce charge, 9-18
cratering, number of holes, 9-20
cube, 14-11
culverts, 11-9
external charges, 9-12
internal charges, 9-12
minimum safe distance, detonating explosives, 9-2
number of mines and minefield rows, 7-2
number of nails needed, 14-7
percent of S, 11-2
percent of slope, 3-22
prism or cylinder, 14-11
protective wire, 6-3
pyramid or cone, 14-11
radius-of-curvature, 3-2
range deviation, 1-24
rectangular parallelepiped, 14-11
rectangular parallelogram, 14-11
regular polygons, 14-11
ribbon bridge, number of interior bays, 10-3
right triangle, 14-11
road gradient, 3-3
route-classification, 3-2
runoff estimate, 11-8
sector of circle, 14-11
segment of circle, 14-11
slope computation, 3-3
sphere, 14-11
steel-cutting, 9-9
stream velocity, 3-23
supplementary wire, 6-3
tactical wire, 6-3
temporary class number, 14-14
trapezoid, 14-11
trigonometric functions, 14-8
water source capacity, 3-24
fuses, 7-28

G

Gator. *See* minefields, scatterable

H

haul capacity, Class IV/V, 7-3
Hornet, 7-15

K

knots, 12-3
  Baker bowline, 12-12
  baker bowline, 12-3
  bowline, 12-3, 12-7
  bowline on a bight, 12-3, 12-8
  butterfly, 12-3, 12-11
  carrick bend, 12-3, 12-6
  cat's paw, 12-3, 12-10
  double bowline, 12-3, 12-7
  double sheet bend, 12-3, 12-6
  figure eight, 12-3, 12-4
  figure eight with an extra turn, 12-3, 12-11
  French bowline, 12-3, 12-9
  overhand, 12-3, 12-4
  running bowline, 12-3, 12-8
  single sheet bend, 12-3, 12-5
  Spanish bowline, 12-3, 12-9

Speir knot, 12-10
spier, 12-3
square, 12-3
wall, 12-3, 12-5

L

lane-marking
    levels, 4-14
    standard, 4-9
        full, 4-12
        initial, 4-10
        intermediate, 4-10
        North Atlantic Treaty Organization (NATO) Standard Marking, 4-17
latrines, 1-52
L_c values
    arch and portal bridge attacks, 9-27
    top attack, 9-26
lines of communication, 2-1
LOC. *See* lines of communication

M

M18A1. *See* Claymore
M86. *See* pursuit deterrent munition
M93. *See* Hornet
MDI, 9-53
    firing systems, 9-53
MDI firing system (stand-alone), 9-55
MDI. *See* modernized demolition initiators
MEDEVAC. *See* medical evacuation
medical evacuation, 1-48
MGB. *See* bridges, medium girder
minefield, 2-4
    detection and removal, 4-2
minefield markings, 7-26
minefields
    antipersonnel, 2-4, 2-5
    antitank, 2-5
    conventional, 7-1
        hasty protective row, 7-8
        row, 7-1
        standard-pattern, 7-7
    scatterable

ADAM/RAAM, 7-12
Gator, 7-13
Modular Pack Mine System, 7-9
Volcano, 7-11
scatterable (*See also* scatterable mines), 7-9
mines, 7-28
modernized demolition initiators, 9-1
MOPMS. *See* minefields, scatterable, Modular Pack Mine System

O

obstacle, 5-11
    C2, 5-4
    emplacement authority, 5-4
    classification, 5-4, 5-5
    control, 5-6
obstacle-control measures, 5-6
obstacle-effect graphics, 5-6
obstacles, 6-1
    antivehicular, 6-10
    antivehicular wire, 6-9
    barbed-wire, 6-3
    cattle fence, 6-7
    concertina, 6-4
    hedghogs, 6-12
    knife rest, 6-8
    log hurdles, 6-12
    post, 6-12
    tanglefoot, 6-8
    tetrahedrons, 6-12
    trestle-apron fence, 6-9
    wire, 6-1
obstacle-type abbreviations, 5-11
operational symbols, 14-19
operations
    dismounted, 1-15
        squad, 1-15
    mounted, 1-15
orders, 1-1
    combat, 1-1
    fragmentary, 1-1
    movement, 1-12
    operation, 1-1
    warning, 1-1

overwatch
  bounding, 1-15
  traveling, 1-15

P

patrol
  combat, 1-17
  reconnaissance, 1-17
PDM. *See* pursuit deterrent munition
positions, 8-1
  fighting
    deliberate, 8-40
    hasty, 8-37
    individual, 8-1
    modified, 8-37
    protective, 8-42
    weapons, 8-1
  vehicle, 8-23
procedures
  breaching, 9-22
pursuit deterrent munition (M86), 7-14

R

ranges of common weapons, 14-16
Raptor, 7-19
reconnaissance, 3-5
  bridge, 3-6
  demolition, 9-41
  engineer, 3-24
  ford, 3-24
  road, 3-5
  tunnel, 3-18
  water-crossing, 3-18
risk assessment, environmental, 13-5
risk management, 14-31
  environmental, 13-1
river crossing, 10-1
  anchorage systems, 10-8
  bridging, 10-2
  rafting, 10-2
rocks
  engineering properties, 11-5
rope, 12-1
  wire, 12-2
route classification, 3-1

S

safe bearing capacity, 10-53
SCATMINE. *See* scatterable mines
scatterable mines
  AP, 7-36
    characteristics, 7-36
  AT, 7-37
    characteristics, 7-37
  emplacement authority, 5-4, 5-5
selectable lightweight attack munition, 7-15
SINCGARS. *See* Single-Channel, Ground-to-Air Radio System
Single-Channel, Ground-to-Air Radio System, 1-60
SLAM. See selectable lightweight attack munition
soil characteristics, 11-1
specific weights and gravities, 14-1

T

task force, 5-2
TF. *See* task force
threat, 2-1
  mid- to high-intensity, 2-3
  offensive operations, 2-11
  organization, 2-5
    armor and mechanized based, 2-8
    infantry based, 2-6
time-distance conversion, 14-12

V

vehicle dimensions and classifications, 14-12

FM 5-34
30 August 1999

By Order of the Secretary of the Army:

DENNIS J. REIMER
General, United States Army
Chief of Staff

Official:

Joel B. Hudson
JOEL B. HUDSON
Administrative Assistant to the
Secretary of the Army

DISTRIBUTION:

Active Army, Army National Guard, and U.S. Army Reserve: To be distributed in accordance with DA Form 12-11a, Requirements for Engineer Field Data (Qty rqr block No. 110026).

- Combat Operations
- Threat
- Reconnaissance
- Mobility
- Defensive Operations and Obstacle Integration Framework
- Constructed and Preconstructed Obstacles
- Landmine and Special-Purpose Munition Obstacles
- Survivability
- Demolitions and Modernized Demolition Initiators (MDI)
- Bridging
- Roads and Airfields
- Rigging
- Environmental-Risk Management
- Miscellaneous Field Data

PIN: 021493-003

www.ingramcontent.com/pod-product-compliance
Lightning Source LLC
Chambersburg PA
CBHW050044230526
45470CB00004B/1405